Number Theory for the Millennium III

Number Theory for the Millennium III

Edited by

M. A. Bennett, B.C Berndt, N. Boston, H. G. Diamond, A. J. Hildebrand and W. Philipp

CRC Press
Taylor & Francis Group
Boca Raton London New York

CRC Press is an imprint of the
Taylor & Francis Group, an **informa** business

CRC Press
Taylor & Francis Group
6000 Broken Sound Parkway NW, Suite 300
Boca Raton, FL 33487-2742

© 2002 by Taylor & Francis Group, LLC
CRC Press is an imprint of Taylor & Francis Group, an Informa business

No claim to original U.S. Government works

ISBN 13: 978-1-56881-152-9 (hbk)

Visit the Taylor & Francis Web site at
http://www.taylorandfrancis.com

and the CRC Press Web site at
http://www.crcpress.com

Contents of Volume 1

Contents of Volume 2

Contents of Volume 3

Preface

The Millennial Conference on Number Theory was held May 21 - 26, 2000 on the campus of the University of Illinois at Urbana-Champaign. The meeting was organized by M. A. Bennett, B. C. Berndt, N. Boston, H. G. Diamond, and A. J. Hildebrand of the UIUC Mathematics Department, and W. Philipp, of the UIUC Statistics Department. A total of 276 mathematicians from 30 countries were present at the meeting.

The conference featured 157 talks, of which 19 were one-hour plenary talks, 73 were half-hour invited talks given in four parallel sessions, and 65 were contributed talks given in five parallel sessions. In addition, there were three 40 minute talks on history and reminiscences, and several after dinner speeches honoring Professor Emeritus Paul Bateman on his 80th birthday. Other evening activities included a reception, a banquet, and musical and theatrical performances.

The conference was underwritten by financial grants from the Number Theory Foundation, the National Science Foundation, the National Security Agency, the College of Liberal Arts and Sciences of the University of Illinois, and the Institute for Mathematics and Applications. Also, the University of Illinois Mathematics Department provided financial and logistical support for the conference along with its facilities in Altgeld Hall. We thank the staff and graduate students of the Mathematics Department for the many hours of work they contributed to the conference. Finally, we recognize the efforts of Betsy Gillies, Conference Coordinator of the Mathematics Department, who worked tirelessly to ensure that the conference would run smoothly. We are grateful to all these organizations and people for their assistance and support.

Urbana, Illinois
7 December 2001

M. A. Bennett
B. C. Berndt
N. Boston
H. G. Diamond
A. J. Hildebrand
W. Philipp

List of Contributors

R. C. Baker, Department of Mathematics, Brigham Young University, Provo, UT 84602, USA, `baker@math.byu.edu`

Michel Balazard, C.N.R.S., Laboratoire d'algorithmique arithmétique, Mathématiques, Université de Bordeaux 1, 351, Cours de la Libération, 33405 Talence Cedex, France, `balazard@math.u-bordeaux.fr`

Daniel J. Bernstein, Department of Mathematics, University of Illinois at Chicago, Chicago, IL 60607-7045, USA, `djb@math.uic.edu`

F. Beukers, Department of Mathematics, Rijksuniversiteit te Utrecht, 3508 TA Utrecht, The Netherlands, `beukers@math.uu.nl`

Jean Bourgain, School of Mathematics, Institute for Advanced Study, Princeton, NJ 08540, USA, `bourgain@ias.edu`

John Boxall, Département de Mathématiques et de Mécanique, CNRS FRE 2271, Université de Caen, Boulevard Maréchal Juin, B.P. 5186, 14032 Caen Cedex, France, `boxall@math.unicaen.fr`

David W. Boyd, Department of Mathematics, University of British Columbia, Vancouver, BC V6T 1Z2, Canada, `boyd@math.ubc.ca`

Yann Bugeaud, Université Louis Pasteur, U.F.R. de Mathématiques, 7, rue René Descartes, 67084 Strasbourg, France, `bugeaud@math.u-strasbg.fr`

David A. Cardon, Department of Mathematics, Brigham Young University, Provo, UT 84602, USA, `cardon@math.byu.edu`

Heng Huat Chan, Department of Mathematics, National University of Singapore, Kent Ridge, Singapore 119260, Republic of Singapore, `matchh@nus.edu.sg`

Robin Chapman, School of Mathematical Sciences, University of Exeter, Exeter EX4 4QE, United Kingdom, `rjc@maths.ex.ac.uk`

Ted Chinburg, Department of Mathematics, University of Pennsylvania, Philadelphia, PA 19104, USA, `ted@math.upenn.edu`

Fan Chung, Department of Mathematics, University of California at San Diego, San Diego, CA 92110, USA, `fan@math.ucsd.edu`

Todd Cochrane, Department of Mathematics, Kansas State University, Manhattan, KS 66506, USA, `cochranc@math.ksu.edu`

J. B. Conrey, American Institute of Mathematics, 360 Portage Ave., Palo Alto, CA 94306, USA, `conrey@aimath.org`

Morley Davidson, Department of Mathematics, Kent State University, Kent, OH 44240, USA, davidson@math.kent.edu

Harold G. Diamond, Department of Mathematics, University of Illinois, 1409 West Green Street, Urbana, IL 61801, USA, diamond@math.uiuc.edu

Karl Dilcher, Department of Mathematics and Statistics, Dalhousie University, Halifax, Nova Scotia B3H 3J5, Canada, dilcher@mathstat.dal.ca

Darrin Doud, Department of Mathematics, Brigham Young University, Provo, UT 84602, USA, doud@math.byu.edu

Minkin Eie, Department of Mathematics, National Chung Cheng University, Ming-Hsiung Chia-Yi 621, Taiwan, mkeie@math.ccu.edu.tw

P.D.T.A. Elliott, Department of Mathematics, University of Colorado, Boulder, CO 80309, USA, pdtae@euclid.colorado.edu

Christian Elsholtz, Institut für Mathematik, Technische Universität Clausthal, Erzstrasse 1, D-38678 Clausthal-Zellerfeld, Germany, elsholtz@math.tu-clausthal.de

Paul Erdős[†], Hungarian Academy of Sciences, Budapest, H-1053, Hungary

Ronald Evans, Department of Mathematics, University of California at San Diego, La Jolla, CA 92093-0112, USA, revans@ucsd.edu

Jan-Hendrik Evertse, Universiteit Leiden, Mathematisch Instituut, Postbus 9512, 2300 RA Leiden, The Netherlands, evertse@math.leidenuniv.nl

Michael Filaseta, Mathematics Department, University of South Carolina, Columbia, SC 29208, USA, filaseta@math.sc.edu

Kevin Ford, Department of Mathematics, University of Illinois, Urbana, IL 61801, USA, ford@math.uiuc.edu

D. Eric Freeman, Department of Mathematics, University of Colorado, Boulder, CO 80309, USA, freem@euclid.colorado.edu

Emiliano Gómez, Department of Mathematics, University of California at Berkeley, Berkeley, CA 94720, USA, emgomez@math.berkeley.edu

F. G. Garvan, Department of Mathematics, University of Florida, Gainesville, FL 32611, USA, frank@math.ufl.edu

Ronald Graham, Department of Computer Science and Engineering, University of California at San Diego, La Jolla, CA 92093-0114, graham@ucsd.edu

---

[†]Paul Erdős passed away on September 23, 1996.

David Grant, Department of Mathematics, University of Colorado, Boulder, CO 80309, USA, `grant@euclid.colorado.edu`

George Greaves, School of Mathematics, Cardiff University, 23 Senghenydd Rd., Cardiff CF2 4YH, United Kingdom, `greaves@cf.ac.uk`

James Lee Hafner, IBM Research, Almaden Research Center, K53-B2, 650 Harry Road, San Jose, CA 95120, USA, `hafner@almaden.ibm.com`

H. Halberstam, Department of Mathematics, University of Illinois, 1409 West Green Street, Urbana, IL 61801, USA, `heini@math.uiuc.edu`

Glyn Harman, Department of Mathematics, Royal Holloway, University of London, Egham Hill, Egham, Surrey TW20 0EX, United Kingdom, `G.Harman@rhul.ac.uk`

Charles Helou, Penn State Univ., 25 Yearsley Mill Rd, Media, PA 19063, USA, `cxh22@psu.edu`

Doug Hensley, Department of Mathematics, Texas A&M University, College Station, TX 77843, USA, `doug.hensley@math.tamu.edu`

Noriko Hirata-Kohno, Nihon University, College of Science and Technology, Department of Mathematics, Surugudai, Tokyo 101-8308, Japan, `hirata@math.cst.nihon-u.ac.jp`

C. Hooley, School of Mathematics, Cardiff University, 23 Senghenydd Rd., Cardiff CF2 4YH, United Kingdom

James G. Huard, Department of Mathematics and Statistics, Canisius College, Buffalo, NY 14208, USA, `huard@canisius.edu`

M. N. Huxley, School of Mathematics, University of Cardiff, 23, Senghenydd Road, Cardiff CF24 4YH Wales, United Kingdom, `huxley@cf.ac.uk`

Georg Illies, Fachbereich Mathematik, Universität-Gesamthochschule Siegen, 57068 Siegen, Germany, `illies@math.uni-siegen.de`

J. P. Keating, School of Mathematics, University of Bristol, Bristol BS8 1TW, England, `j.p.keating@bristol.ac.uk`

A. Kumchev, Department of Mathematics, University of Toronto, Toronto, ON M5S 3G3, Canada, `kumchev@math.toronto.edu`

M. B. S. Laporta, Dipartimento di Matematica e Appl. "R. Caccioppoli", Complesso Universitario di Monte S. Angelo, Via Cinthia, 80126 Napoli, Italy, `laporta@matna2.dma.unina.it`

Xian-Jin Li, Department of Mathematics, Brigham Young University, Provo, UT 84602, USA, `xianjin@math.byu.edu`

Wen-Ching Winnie Li, Department of Mathematics, Pennsylvania State University, University Park, PA 16802, USA, `wli@math.psu.edu`

Lisa Lorentzen, Department of Mathematical Sciences, Norwegian University of Science and Technology, N-7491 Trondheim, Norway, `lisa@math.ntnu.no`

Lutz G. Lucht, Institut für Mathematik, Technische Universität Clausthal, Erzstraße 1, 38678 Clausthal-Zellerfeld, Germany, `lucht@math.tu-clausthal.de`

Eugenijus Manstavičius, Vilnius University, Department of Mathematics and Informatics, Naugarduko Str. 24, LT 2600 Vilnius, Lithuania, `eugenijus.manstavicius@maf.vu.lt`

Greg Martin, Department of Mathematics, University of Toronto, Toronto, ON M5S 3G3, Canada, `gerg@math.toronto.edu`

Kohji Matsumoto, Graduate School of Mathematics, Nagoya University, Chikusa-ku, Nagoya 464-8602, Japan, `kohjimat@math.nagoya-u.ac.jp`

Jeffrey L. Meyer, Department of Mathematics, Syracuse University, 215 Carnegie Hall, Syracuse, NY 13244-1150, USA, `jlmeye01@mailbox.syr.edu`

Maurice Mignotte, Université Louis Pasteur, U.F.R. de Mathématiques, 7, rue René Descartes, 67084 Strasbourg, France, `mignotte@math.u-strasbg.fr`

Siguna Müller, University of Klagenfurt, Department of Mathematics, A-9020 Klagenfurt, Austria, `siguna.mueller@uni-klu.ac.at`

V. Kumar Murty, Department of Mathematics, University of Toronto, Toronto, ON M5S 3G3, Canada, `murty@math.toronto.edu`

Ram Murty, Department of Mathematics, Queen's University, Kingston, ON K7L 3N6, Canada, `murty@mast.queensu.ca`

J.-L. Nicolas, Institut Girard Desargues, UMR 5028, Université Claude Bernard (Lyon 1), F-69622 Villeurbanne Cedex, France, `jlnicola@in2p3.fr`

Pace P. Nielsen, Department of Mathematics, Brigham Young University, Provo, UT 84602, USA

Cormac O'Sullivan, Department of Mathematics and Computer Science, Bronx Community College, City University of New York, Bronx, NY 10453, `cormac@math.umd.edu`

Yao Lin Ong, Department of Mathematics, National Chung Cheng University, Ming-Hsiung Chia-Yi 621, Taiwan

Ken Ono, Department of Mathematics, University of Wisconsin, Madison, WI 53706, USA, `ono@math.wisc.edu`

Zhiming M. Ou, Department of Basic Science, Beijing University of Posts and Telecommunications, Beijing 100876, People's Republic of China

Matthew A. Papanikolas, Department of Mathematics, Brown University, Providence, RI 02912-1917, USA, `map@math.brown.edu`

Georgios Pappas, Department of Mathematics, Michigan State University, East Lansing, MI 48824, USA, `pappas@math.msu.edu`

Scott T. Parsell, Department of Mathematics, Penn State University, University Park, PA 16802, USA, `parsell@alum.mit.edu`

Bernadette Perrin-Riou, Mathématiques, Bat. 425, Université Paris-Sud, F-91 405 Orsay, France, `bpr@math.u-psud.fr`

Bjorn Poonen, Department of Mathematics, University of California, Berkeley, CA 94720-3840, USA, `poonen@math.berkeley.edu`

Alfred J. van der Poorten, Centre for Number Theory Research, Macquarie University, Sydney, Australia 2109, `alf@mpce.mq.edu.au`

Igor E. Pritsker, Department of Mathematics, 401 Mathematical Sciences, Oklahoma State University, Stillwater, OK 74078-1058, USA, `igor@math.okstate.edu`

Robert A. Rankin[†], Department of Mathematics, University of Glasgow, Glasgow G12 8QT, Scotland, UK

David P. Roberts, Division of Science and Mathematics, University of Minnesota-Morris, Morris, MN 56267, USA, `roberts@mrs.umn.edu`

Fernando Rodriguez-Villegas, Department of Mathematics, University of Texas at Austin, Austin, TX 78712, USA, `villegas@math.utexas.edu`

Michael Rubinstein, Department of Mathematics, University of Texas at Austin, Austin, TX 78712, USA, `miker@fireant.ma.utexas.edu`

A. Sárközy, Department of Algebra and Number Theory, Eötvös Loránd University, Pázmány Péter sétány 1/C, H-1117 Budapest, Hungary, `sarkozy@cs.elte.hu`

C. J. Smyth, Department of Mathematics and Statistics, University of Edinburgh, JCMB, King's Buildings, Mayfield Road, Edinburgh EH9 3JZ, United Kingdom, `chris@maths.ed.ac.uk`

[†]Robert Rankin passed away on January 27, 2001.

N. C. Snaith, School of Mathematics, University of Bristol, Bristol BS8 1TW, England, `n.c.snaith@bristol.ac.uk`

Blair K. Spearman, Department of Mathematics and Statistics, Okanagan University, College, Kelowna, BC V1V 1V7, Canada, `bspearman@okanagan.bc.ca`

Anupam Srivastav, Department of Mathematics and Statistics, State University of New York, Albany, NY 12222, USA, `anupam@math.albany.edu`

Kenneth B. Stolarsky, Department of Mathematics, University of Illinois, 1409 West Green Street, Urbana, IL 61801, USA, `stolarsk@math.uiuc.edu`

Martin J. Taylor, UMIST, Manchester, M60 1QD, United Kingdom, `Martin.Taylor@umist.ac.uk`

Jeffrey Lin Thunder, Department of Mathematics, Northern Illinois University, DeKalb, IL 60115, USA, jthunder@math.niu.edu

R. Tijdeman, Mathematical Institute, Leiden University, Postbus 9512, 2300 RA Leiden, The Netherlands, `tijdeman@math.leidenuniv.nl`

Alain Togbé, Department of Mathematics and Comp. Sci., Greenville College, 315 E. College Ave., Greenville, IL 62246, USA, `atogbe@greenville.edu`

R. C. Vaughan, Department of Mathematics, McAllister Building, Pennsylvania State University, University Park, PA 16802, USA, `rvaughan@math.psu.edu`

V. H. Vu, Department of Mathematics, University of California at San Diego, La Jolla, CA 92093-0112, USA, `vanvu@ucsd.edu`

Samuel S. Wagstaff, Jr., Center for Education and Research in Information Assurance and Security, and Department of Computer Sciences, Purdue University, West Lafayette, IN 47907-1398, USA, `ssw@cerias.purdue.edu`

Lynne H. Walling, Department of Mathematics, University of Colorado, Boulder, CO 80309, USA, `walling@euclid.colorado.edu`

P. G. Walsh, Department of Mathematics, University of Ottawa, 585 King Edward, Ottawa, ON K1S 5B6, Canada, `gwalsh@mathstat.uottawa.ca`

Kenneth S. Williams, Centre for Research in Algebra and Number Theory, School of Mathematics and Statistics, Carleton University, Ottawa, ON K1S 5B6, Canada, `williams@math.carleton.ca`

H. C. Williams, Department of Mathematics and Statistics, University of Calgary, Calgary, AB T2N 1N4, Canada, `williams@math.ucalgary.ca`

Siman Wong, Department of Mathematics, University of Massachusetts, Amherst, MA 01003, USA, `siman@math.umass.edu`

T. D. Wooley, Department of Mathematics, University of Michigan, East Hall, 525 E. University Avenue, Ann Arbor, MI 48109-1109, USA, `wooley@math.lsa.umich.edu`

Wen-Bin Zhang, Department of Mathematics, University of the West Indies, Mona, Kingston 7, Jamaica, `wbzhang@uwimona.edu.jm`

Zhiyong Zheng, Department of Mathematics, Tsinghua University, Beijing 100084, People's Republic of China, `zzheng@math.tsinghua.edu.cn`

Some Remarks on Primality Testing Based on Lucas Functions

Siguna Müller[1]

1 Motivation

A number of primality tests exist that determine, with certainty, whether a large integer n is prime. However, the theory and implementation of these tests is rather involved. By contrast, pseudoprimality tests, which determine whether, with high probability (but not certainty), an integer is prime, are a lot faster and can much more easily be implemented. Methods for modular exponentiation and evaluation of recurrence sequences are widely employed in this context and are very efficient. The catch of tests based on these methods is the existence of "pseudoprimes," i.e., composite numbers that are identified as primes by the test.

In this paper we investigate several new and efficient methods for probable prime testing. Our methods are based on the very strong test developed by J. Grantham [3], but they will detect most of the pseudoprimes that pass Grantham's test. In fact, we have not found a composite integer for which the proposed test of Section 4.6 fails.

The main steps in Grantham's test can be summarized as follows: As with the Miller-Rabin test, his *fundamental testing condition* relies on the fact that, when n is prime, both roots $y \in \mathbb{F}_{n^2}$ of $x^2 - Px + Q$ with $\left(\frac{P^2 - 4Q}{n}\right) = -1$, satisfy $y^s \equiv 1 \bmod n$, or $y^{2^t s} \equiv -1 \bmod n$ for some $0 \le t \le r - 2$, where $n^2 - 1 = 2^r s$ with s odd. However, the exponent $2^t s = \frac{n^2 - 1}{2^j}$ is still too large to obtain strong testing conditions. More restrictive properties of the roots y constitute the *second step*. Consider the exponent $n' = n - \epsilon(n)$, where $D = P^2 - 4Q$ and $\left(\frac{D}{n}\right) = \epsilon(n)$. Then $y^{n'}$ is congruent to 1 or Q modulo n according as $\epsilon(n)$ in n' is 1 or -1. As the former case constitutes an ordinary Fermat condition, Grantham only tests for the latter case. Moreover, if $\left(\frac{Q}{n}\right) = 1$ and if α and $\overline{\alpha}$ denote the two roots y, then the two roots need to evaluate to the same value, even with the smaller exponent $n'/2$ in place of n', i.e., we must have $\alpha^{n'/2} \equiv \overline{\alpha}^{n'/2} \bmod n$. This

[1]Research supported by the Austrian Science Fund (FWF), FWF-Projects no. P 13088-MAT and P 14472-MAT

1

stronger condition is being tested in the *third step*; it also justifies the bound $t \leq r - 2$ above.

However, Grantham does not consider the nature of this common value. Mo and Jones (cf. [3]) observed that, if $i = \nu_2(n')$ and $\alpha^{n'/2^i}$ and $\overline{\alpha}^{n'/2^i}$ are congruent modulo n, then both expressions must be congruent to ± 1 when n is a prime. Unfortunately their result only holds for $Q = 1$.

In [7] the author gave an explicit formula for the value of $\alpha^{n'/2} \equiv \overline{\alpha}^{n'/2}$ modulo a prime n, for any Q with $\left(\frac{Q}{n} \right) = 1$, and showed how this leads to a better error-probability estimate than the Grantham test, in the case when $n \equiv 1 \bmod 4$. This case is easier to deal with, since then, for $\epsilon(n) = -1$, the number $n' = n + 1$ is divisible by 2 but not by any higher power of 2.

The case $n \equiv 3 \bmod 4$ cannot be treated in the same way, since for $\epsilon(n) = -1$, $n' = n + 1$ is at least divisible by 4. This leads to a number of complications, which we will deal with in this paper.

Underlying Grantham's test is the fact that when $2^i || n'$, then the strong condition $\alpha^{n'/2^i} \equiv \overline{\alpha}^{n'/2^i} \bmod n$ might hold. However, depending on the values of P and Q that determine the roots $\alpha, \overline{\alpha}$, it is also possible that for some $j < \nu_2(n')$ we have

$$\alpha^{n'/2^j} \equiv \overline{\alpha}^{n'/2^j} \bmod n, \tag{1.1}$$

$$\alpha^{n'/2^{j+1}} \equiv -\overline{\alpha}^{n'/2^{j+1}} \bmod n. \tag{1.2}$$

This condition can be phrased in terms of the strong Lucas test. The Lucas functions $U_k(P, Q) = (\alpha^k - \overline{\alpha}^k)/(\alpha - \overline{\alpha})$ and $V_k(P, Q) = \alpha^k + \overline{\alpha}^k$ of degree k have integer values which can also be described via second-order recurrences [12][16]. If n is a prime and $n' = n - \epsilon(n) = 2^i u$, $2 \nmid u$, then either $U_u(P, Q) \equiv 0$, or $V_{2^t u}(P, Q) \equiv 0 \bmod n$ for some t, $0 \leq t < i$. However, if the U–function vanishes (modulo n), it is not known what the value of the corresponding V–function is. Moreover, for given P and Q, there is no efficient way to determine a priori which of the two functions will vanish, let alone the degree of vanishing $2^t u = n'/2^j$. Finding P and Q for which $U_u(P, Q) \equiv 0 \bmod n$ requires more effort as $i = \nu_2(n')$ gets larger.

We now describe the main results of this paper. In the first part, we consider the case when n is an odd prime. We develop a method to determine the exponent j for which (1.1) holds, and the value of $\alpha^{n'/2^j}$ modulo n. We also present a method that allows us to efficiently find parameters P and Q for which the strongest possible case, namely $j = i = \nu_2(n')$ in (1.1), holds.

The results will be simple functions of P and Q. In the second part of the paper, we will use these results to obtain highly efficient pseudoprimality testing conditions for the case $n \equiv 3$ mod 4, which complement the methods developed in [7] for the case $n \equiv 1$ mod 4. Although our methods are rooted in Lucas–based primality tests, the conditions we obtain turn out to be a combination of both the Fermat and the Lucas primality tests, and are more efficient than iterations of either one of these tests.

While the motivation behind this work was theoretical, our test performs well in practice. Any pseudoprimes for the proposed test of Section 4.6 must have properties very similar to those of the celebrated Baillie-PSW test [1][11]. However, for the latter test no pseudoprimes are known.

Our test will turn out to be more reliable than the one introduced by Grantham [3], while the running time is similar to that of Grantham's test. The main advantage of the latter test is that an estimate for the error probability is known for random input parameters. On the other hand, this random selection might result in the choice of weak parameters. In fact, we will list several Grantham pseudoprimes and parameters for which the test in [3] fails quite often.

We propose a method for avoiding such weak parameters. Instead of using random parameters (which might be weak), we propose to use certain special parameters which turn out to be very strong. The method will utilize the results developed for obtaining the strong case $j = i = \nu_2(n')$ in (1.1).

2   A Refinement of Euler's Criterion in $\mathbb{F}_{p^2}$

Let $D = P^2 - 4Q$, let $\alpha, \bar{\alpha}$ be the roots of $x^2 - Px + Q$, $\epsilon(p) = \left(\frac{D}{p}\right)$, where p is an odd prime, and define $\epsilon(n) = \left(\frac{D}{n}\right)$ for any integer $n \equiv 3$ mod 4 with $\gcd(n, 2D) = 1$. For simplicity of notation, we write throughout $p' = p - \epsilon(p)$, and $n' = n - \epsilon(n)$.

2.1   The Euler Criterion in $\mathbb{F}_{p^2}$

Grantham's second step, $\alpha^{p'} \equiv \bar{\alpha}^{p'} \equiv Q$ mod p for (P, Q) with $\epsilon(p) = -1$, can be viewed as the $\mathbb{F}_{p^2}$ based analog of the Fermat condition $a^{p-1} \equiv 1$ mod p. Similarly, for the case when $(Q/p) = 1$ one can consider the $\mathbb{F}_{p^2}$ based analog of the Euler condition $a^{\frac{p-1}{2}} \equiv \left(\frac{a}{p}\right)$ mod p. (If $(Q/p) = -1$ then $\alpha^{p'/2} \equiv -\bar{\alpha}^{p'/2}$ mod p.) The following result was proved in [7].

Proposition 2.1. *If $\alpha, \bar{\alpha} \in \mathbb{F}_{p^2}$ are the roots of $x^2 - Px + Q$, where $a^2 \equiv Q$ mod p, then $\alpha^{p'/2} \equiv \bar{\alpha}^{p'/2}$ is congruent to $\left(\frac{P+2a}{p}\right)$ mod p if $\epsilon(p) = 1$, and to $\left(\frac{P+2a}{p}\right) a$ mod p if $\epsilon(p) = -1$.*

2.2 The Transformation of the Parameters

Proposition 2.1 was proved in [7] by means of the following result, which enables one to represent powers of α in terms of powers of another root related to α in such a way that a factor of 2 is preserved in the exponent.

Proposition 2.2.   1. Let $\alpha \in \mathbb{F}_{p^2}$ be any root of $x^2 - Px + Q$ and suppose that $a^2 \equiv Q \bmod p$. Then $(P + 2a)\alpha \equiv (\alpha + a)^2 \bmod p$.

2. If $\alpha, \overline{\alpha}$ are the roots of $x^2 - Px + Q$ and a is a square root of Q modulo p then $\alpha + a$ and $\overline{\alpha} + a$ are the roots of $x^2 - (P + 2a)x + (P + 2a)a$.

In what follows, we consider the case when $\nu_2(p')$ is 'large'. Proposition 2.2 can be generalized to obtain explicit formulae for $\alpha^{p'/2^j}$ for $j > 1$. We note that, while for $j = 1$ the value of $\alpha^{p'/2^j} \equiv \overline{\alpha}^{p'/2^j} \bmod p$ in Proposition 2.1 depends only on the quadratic character of Q modulo p, this is not the case for larger values of j. We first consider the case when $j = 2$.

Suppose that $a^2 \equiv Q \bmod p$. For $j = 2$ we require that $U_{p'/4}(P, Q) \equiv 0 \bmod p$. This condition involves both arguments P and Q. We now transform this condition to one that depends only on the second argument.

By Proposition 2.2, we have $(\alpha + a)^2 \equiv \alpha(P + 2a) \bmod p$, which implies $U_m(P, Q)(P + 2a)^m \equiv U_{2m}(P', Q') \bmod p$, where $P' \equiv P + 2a$, $Q' \equiv a(P + 2a) \bmod p$. By hypothesis, $P^2 \not\equiv 4Q \bmod p$, or equivalently, $P \not\equiv \pm 2a \bmod p$, so that $U_{p'/4}(P, Q) \equiv 0 \bmod p$ if and only if $U_{p'/2}(P', Q') \equiv 0 \bmod p$. The latter condition is satisfied if and only if $\left(\frac{Q'}{p}\right) = \left(\frac{a(P+2a)}{p}\right) = 1$; otherwise we have $V_{p'/4}(P, Q) \equiv 0 \bmod p$.

If we suppose that $b^2 \equiv a(P + 2a) \bmod p$, then, by Proposition 2.1, we have $(\alpha + a)^{p'/2} \equiv \left(\frac{(P+2a)+2b}{p}\right) b^{\frac{1-\epsilon(p)}{2}} \bmod p$. On the other hand, Proposition 2.2 yields $\left(\alpha(P + 2a)\right)^{p'/4} \equiv (\alpha + a)^{p'/2} \bmod p$. This gives an explicit formula for $\alpha^{p'/4}$ (see Theorem 2.1 below).

The same reasoning can be applied to obtain a formula for $\alpha^{p'/2^j}$, for any j such that $2^j | p'$: we keep transforming the parameters P and Q until we obtain a Q-parameter that is not a quadratic residue modulo p. This process is formalized in the following definition.

Definition 2.1. Let $P_1 = P$, $Q_1 = Q \equiv a_1^2 \bmod p$ and let $\alpha = \alpha_1(P_1, Q_1)$ and $\overline{\alpha} = \overline{\alpha}_1(P_1, Q_1)$ be the roots of $x^2 - P_1 x + Q_1$. Set

$$P_j \equiv P_{j-1} + 2a_{j-1} \bmod p, \qquad Q_j \equiv P_j a_{j-1} \bmod p, \qquad (2.1)$$

$$a_j^2 \equiv Q_j \bmod p \qquad \text{if } \left(\frac{Q_j}{p}\right) = 1, \qquad (2.2)$$

for $1 < j \le t$, where a_j is the smaller of the two roots of $Q_j \bmod p$ (without loss of generality – see Lemma 2.3 below) and t is the smallest subscript with $\left(\frac{Q_{t-1}}{p}\right) = 1$ and $\left(\frac{Q_t}{p}\right) = -1$. We call t the Legendre Q–index with respect to P. Also, we let $\alpha_j = \alpha_j(P_j, Q_j)$ and $\overline{\alpha}_j = \overline{\alpha}_j(P_j, Q_j)$ be the roots corresponding to the parameters P_j and Q_j.

Using the above argument and induction, we then obtain the following results.

Lemma 2.1. *If $D_j = P_j^2 - 4Q_j$ then $D_j \equiv D \equiv P^2 - 4Q \bmod p$.*

Lemma 2.2. *Let Q_{j-1} be a quadratic residue modulo p for all $j \le \nu_2(p')$ (so that Q_j exists). Then $U_{p'/2^j}(P, Q) \equiv 0 \bmod p$ iff $U_{p'/2}(P_j, Q_j) \equiv 0 \bmod p$, and $V_{p'/2^j}(P, Q) \equiv 0 \bmod p$ iff $V_{p'/2}(P_j, Q_j) \equiv 0 \bmod p$.*

Remark 2.1. By [7] it suffices to deal with the case $p \equiv 3 \bmod 4$, and we can further assume that $\epsilon(p) = -1$, since for $\epsilon(p) = 1$ we are in the Fermat case where $\alpha, \overline{\alpha} \in \mathbb{F}_p$. If $p \equiv 3 \bmod 4$, the two roots $\pm a \bmod p$ of Q in the case $\left(\frac{Q}{p}\right) = 1$ can be computed efficiently via the formula $a \equiv Q^{\frac{p+1}{4}} \bmod p$. We will use this observation below to establish a link between Lucas-based and Fermat-based tests.

Due to the ambiguity in the selection of the square roots modulo p, the above process is not unique. However, as the following lemma shows, different Q_j's resulting from selecting different roots of Q_{j-k} all have the same Legendre symbols.

Lemma 2.3. *Let $p \equiv 3 \bmod 4$, $\epsilon(p) = -1$, $r \le \nu_2(p')$, and suppose that Q_r exists. Let $j \le r$, and let $\hat{P}_j$, $\hat{Q}_j$ be the values obtained in Definition 2.1, if instead of a_{j-1}, $\hat{a}_{j-1} \equiv -a_{j-1} \bmod p$ is taken as the root of Q_{j-1}. Then $\hat{Q}_t$ exists for all $j \le t \le r$, and $\left(\frac{\hat{Q}_t}{p}\right) = \left(\frac{Q_t}{p}\right)$ for all t.*

Proof. For $j = r$, we have $\left(\frac{\hat{Q}_j}{p}\right)\left(\frac{Q_j}{p}\right) = \left(\frac{P_{j-1}^2 - 4Q_{j-1}}{p}\right)\left(\frac{-1}{p}\right)\left(\frac{Q_{j-1}}{p}\right) = 1$, and hence $\left(\frac{\hat{Q}_j}{p}\right) = \left(\frac{Q_j}{p}\right)$. We can therefore assume that $j < r$.

The existence of Q_r implies $\left(\frac{Q_{r-1}}{p}\right) = 1$ and $U_{p'/2}(P_{r-1}, Q_{r-1}) \equiv 0 \bmod p$, so that $U_{p'/2^{r-1}}(P, Q) \equiv U_{p'/2^{r-s}}(P, Q) \equiv 0 \bmod p$ for $1 \le s \le r$. To get a contradiction, assume that, for some $j < r$, $\left(\frac{\hat{Q}_j}{p}\right) = -1$. Then $U_{p'/2}(\hat{P}_j, \hat{Q}_j) \not\equiv 0 \bmod p$. Reversing the inductive steps in Definition 2.1, we get back the original values of P and Q, regardless of whether we start with P_j and Q_j, or with $\hat{P}_j$ and $\hat{Q}_j$ (see also Lemma 4.1 (a) below). But then our assumption implies that $U_{p'/2^j}(P, Q) \not\equiv 0 \bmod p$, which contradicts the congruence obtained above. □

By the lemma, there is no need to distinguish between the different roots of the Q_j, and we will therefore let a_j denote the smaller root, provided that $\left(\frac{Q_j}{p}\right) = 1$.

2.3 2^jth Characters Modulo a Prime

Obviously, a formula for (1.1) becomes more interesting for large values of j. To obtain such a formula, we proceed as in the argument prior to Definition 2.1, to express powers of the α_j in terms of the numbers α_{j-k}. Proposition 2.2 leads to the relation

$$\alpha_j^2 \equiv \alpha_{j-1} P_j \bmod p, \tag{2.3}$$

which can be iterated to express α_j in terms of $\alpha = \alpha_1$. For example,

$$\alpha_3^{p'/2} \equiv \alpha_2^{p'/4} P_3^{p'/4} \equiv (\alpha_2^2)^{p'/8} P_3^{p'/4}$$
$$\equiv \alpha_1^{p'/8} P_2^{p'/8} P_3^{p'/4} \bmod p.$$

If $\left(\frac{Q_3}{p}\right) = 1$, then $a_3 \bmod p$ exists, and Proposition 2.1 then yields a relation between powers of α and a_3. In this manner, we obtain the following result.

Theorem 2.1. *Suppose that $2^l \| p'$. Then for all j with $2 \le j \le l$ for which $\left(\frac{Q_j}{p}\right) = 1$ we have $U_{p'/2^j}(P,Q) \equiv 0 \bmod p$. Moreover,*

$$\alpha^{p'/2^j} \equiv \overline{\alpha}^{p'/2^j} \equiv \frac{\left(\frac{P_j + 2a_j}{p}\right)}{P_2^{p'/2^j} \cdot \ldots \cdot P_j^{p'/2^2}} \cdot a_j^{\frac{1-\epsilon(p)}{2}} =: T_j \bmod p. \tag{2.4}$$

Lemma 2.4. *Let p be an odd prime. Then the roots α and $\overline{\alpha}$ corresponding to a pair (P,Q) satisfy the congruence (2.4) for some j, if and only if Q_j with $\left(\frac{Q_j}{p}\right) = 1$ exists and $\alpha_r^{p'/2} \equiv \left(\frac{P_r + 2a_r}{p}\right) a_r^{\frac{1-\epsilon(p)}{2}} \bmod p$ holds for all $r \le j$.*

Proof. The necessity of the condition follows as above, by recursively applying Proposition 2.1 and (2.3).

To prove the sufficiency, we focus on the case $j = 3$; the general case can be proved similarly. We also assume that $\epsilon(p) = -1$ (so that $p' = p - \epsilon(p) = p + 1$) as the case $\epsilon(p) = 1$ is analogous. By hypothesis, and by double squaring, we obtain in this case $\alpha^{p'/2} = \alpha_1^{p'/2} \equiv Q_3^2 \left(\frac{P_2}{p}\right) / P_3^2 P_2 \bmod p,$

since $P_3^{p-1} \equiv 1 \bmod p$ and $P_2^{(p-1)/2} \equiv \left(\frac{P_2}{p}\right) \bmod p$. This yields $(\alpha_1)^{p'/2} \equiv \left(\frac{P_1+2a_1}{p}\right) a_1 \bmod p$, which proves the result for $r = 1$. For $r = 2$ we obtain $(\alpha_1 P_2)^{p'/2^2} \equiv \frac{Q_3\left(\frac{P_3}{p}\right)}{P_3} \bmod p$. Since $P_3^{\frac{p-1}{2}} \equiv \left(\frac{P_3}{p}\right) \bmod p$, this implies $(\alpha_2)^{p'/2} \equiv \left(\frac{P_2+2a_2}{p}\right) a_2 \bmod p$, as desired. The result for $r = 3$ follows immediately from (2.3). □

2.4 The Strongest Condition

Definition 2.2. Let t be the Legendre Q–index with respect to P. If $t \leq \nu_2(p')$, where $p' = p - \epsilon(p)$, we call t the V–index of (P, Q), and denote its value by k; otherwise we say that the V–index (or k) does not exist.

The most stringent condition in (1.1) is intimately related to this value k. If k exists, then $\alpha^{p'/2^k} \equiv -\overline{\alpha}^{p'/2^k} \bmod p$ and thus $V_{p'/2^k}(P, Q) \equiv 0 \bmod p$, and $U_{p'/2^j}(P, Q) \equiv 0 \bmod p$ for all $j < k$.

The reason for restricting k to the range $k \leq \nu_2(p')$ is that otherwise the relation in Lemma 2.2 would not be satisfied.

We have been using the fact that $\alpha^d \equiv \overline{\alpha}^d \bmod p$ for some divisor $d|p'/2$, when $\left(\frac{Q}{p}\right) = 1$; it is known that such a value d always exists (cf. [12][16]).

Definition 2.3. Let m be any integer with $\gcd(Q, m) = 1$. The rank of appearance, $\rho(m, P, Q)$ modulo m, is the smallest integer l such that $U_l(P, Q) \equiv 0 \bmod m$.

In particular, we have $\nu_2(\rho(p, P, Q)) = \nu_2(p'/2^{k-1})$, or equivalently, $k = \nu_2(p'/\rho(p, P, Q)) + 1$. This specifies the degree to which the U– and V–functions vanish.

Clearly, k is uniquely determined by P and Q. Note that if $\left(\frac{Q}{p}\right) = 1$, then $\nu_2(\rho(p, P, Q)) < \nu_2(p')$, so that $k \geq 2$. The largest possible value of k occurs when P and Q are such that $\nu_2(\rho(p, P, Q)) = 1$; in this case, $k = \nu_2(p')$. Furthermore, k does not exist if and only if $\nu_2(\rho(p, P, Q)) = 0$, because then $U_u(P, Q) \equiv 0 \bmod p$, where u is the odd part of p'.

If we take i as the largest value of j in Theorem 2.1 (i.e., $i = k - 1$ or $\nu_2(p')$, depending on whether or not k exists), we obtain the following corollary.

Corollary 2.1. Let $i = \nu_2(p')$, if the V–index k of (P, Q) does not exist, and $i = k - 1$ otherwise. Then the following conditions are equivalent.

(a) $\begin{cases} \alpha_j^{p'/2} \equiv \overline{\alpha}_j^{p'/2} \equiv \left(\frac{P_j+2a_j}{p}\right) a_j^{\frac{1-\epsilon(p)}{2}} \bmod p, & \text{for all } j \leq i, \text{ and} \\ \alpha_k^{p'/2} \equiv -\overline{\alpha}_k^{p'/2} \bmod p & \text{(if } k \text{ exists)}, \end{cases}$

(b) $\begin{cases} \alpha^{p'/2^i} \equiv \overline{\alpha}^{p'/2^i} \equiv T_i \bmod p & and \\ \alpha^{p'/2^k} \equiv -\overline{\alpha}^{p'/2^k} \bmod p & (if\ k\ exists). \end{cases}$

3 Applications to Probable Prime Testing

In practice, $\alpha^j \bmod n$ can be evaluated very fast [3][15][16]. Thus, the above conditions, with p replaced by n, yield efficient tests for pseudoprimality. For simplicity we use the abbreviations of [12], $psp(a)$, $epsp(a)$, $spsp(a)$, to denote, respectively, a pseudoprime, an Euler pseudoprime, and a strong pseudoprime, to base a. Also, as in [12], we let $lpsp(P,Q)$, $elpsp(P,Q)$, $slpsp(P,Q)$, denote, respectively, a Lucas pseudoprime, an Euler Lucas pseudoprime, and a strong Lucas pseudoprime.

As noted above, Grantham's second step can be viewed as the $\mathbb{F}_{p^2}$ analog of the Fermat condition. This criterion gives rise to another class of pseudoprimes. Composite integers with the property $\alpha^{n+1} \equiv \overline{\alpha}^{n+1} \equiv Q \bmod n$ have been studied extensively. Following [2], we call such integers *quadratic field (QF)–based pseudoprimes* with respect to the parameters P and Q, relative to the roots α and $\overline{\alpha}$, and use the notation $QFpsp(P,Q)$ to denote such a pseudoprime.

3.1 The Euler-QF-Test

The above argument relies on finding a root a of $Q \bmod p$. When applied as a probable prime test modulo n, we need to ensure that the value of a, obtained by the above square root finding mechanism for the prime p, is indeed a square root modulo n. If this is not the case, then n is certainly not prime. It follows immediately that if $a \equiv Q^{(n+1)/4} \bmod n$, then a is a correct square root of Q modulo n exactly when n is $epsp(Q)$.

On the other hand, Proposition 2.1 immediately implies the well-known Cipolla–Lehmer root finding algorithm: in the case when $\epsilon(p) = -1$, $\alpha^{p'/2}$ mod p is a square root of Q modulo a prime p. This property can also be used as a pseudoprimality testing condition (cf. [7][10]).

Proposition 3.1. *Suppose n is a composite integer, $\left(\frac{Q}{n}\right) = 1$, and $\epsilon(n) = -1$ in $n' = n - \epsilon(n)$. If $a \equiv \alpha^{n'/2} \equiv \overline{\alpha}^{n'/2} \bmod n$ is such that $a^2 \equiv Q \bmod n$, then n is $QFpsp(P,Q)$, $epsp(Q)$, $psp(a)$, and $elpsp(P,Q)$.*

Proof. The hypothesis implies that $U_{n'/2}(P,Q) \equiv 0 \bmod n$, and hence that n is $elpsp(P,Q)$. Obviously, n is also $QFpsp(P,Q)$. Furthermore, we have $(\alpha\overline{\alpha})^{n'/2} \equiv a^2 \equiv Q \bmod n$. Hence, $Q^{\frac{n+1}{2}} \equiv Q \bmod n$ and $Q^{\frac{n-1}{2}} \equiv 1 \equiv \left(\frac{Q}{n}\right) \bmod n$. $\square$

Definition 3.1. Let $\left(\frac{Q}{n}\right) = 1$ and $a^2 \equiv Q \bmod n$. We call a composite number n an Euler $QFpsp(P,Q)$ if n is $epsp(Q)$ and if

$$\alpha^{n'/2} \equiv \overline{\alpha}^{n'/2} \equiv \left(\frac{P+2a}{n}\right) a^{\frac{1-\epsilon(n)}{2}} \bmod n.$$

If n is an Euler $QFpsp(P,Q)$, then n is $elpsp(P,Q)$. In the following we investigate conditions under which n is also $slpsp(P,Q)$.

3.2 The Strong-QF-Test

The strong Lucas test can be considerably strengthened by testing for the value of the congruence $\alpha^{n'/2^i} \equiv \overline{\alpha}^{n'/2^i} \bmod n$ with the largest possible exponent i. In its original form, the test only checks whether the congruence holds for $i = k - 1$, or $i = \nu_2(n')$, respectively, but it does not exploit the nature of the common value if the congruence holds. To test for the desired value modulo n (e.g., via Corollary 2.1), we need to ensure that the values a_j obtained are correct roots of Q_j modulo n. Then the Q_{j+1} will be correct, and we need to check whether or not a_{j+1} is a root of Q_{j+1}, and so forth. If, at some point, this condition is violated, n is revealed as a composite number. When applied as a pseudoprimality test, we check, in each step of Definition 2.1, the correctness of the root a_j. To this end we run a 'Q_j-routine' by evaluating $a_j \equiv Q_j^{\frac{n+1}{4}} \bmod n$ and testing whether the square of this expression is congruent to Q_j modulo n. Alternatively, n has to be $epsp(Q_j)$ for all Q_j.

Definition 3.2. Let $i = k - 1$, or $i = \nu_2(n')$, respectively, where k is the V–index of (P,Q). We call a composite number n a strong $QFpsp(P,Q)$ if n is $epsp(Q_j)$ for all $j \leq i$, and satisfies $\alpha^{n'/2^i} \equiv \overline{\alpha}^{n'/2^i} \equiv T_i \bmod n$.

The equivalence in Corollary 2.1 is no longer valid if p is replaced by a composite number n. While (a) implies (b), the converse need not be true, as can be seen from the proof of Lemma 2.4. The additional condition that $P_r^{\frac{n-1}{2}} \equiv \left(\frac{P_r}{n}\right) \bmod n$ for all r is required. The following result can easily be verified (cf. [7]).

Proposition 3.2. *Let* $n \equiv 3 \bmod 4$ *and* $j \geq 2$. *Suppose* $Q_{j-1}^{\frac{n+1}{4}} \equiv a_{j-1} \bmod n$ *is such that* $a_{j-1}^2 \equiv Q_{j-1} \bmod n$. *Then* $(Q_{j-1}/p) = 1$ *for all* $p|n$, *n is* $spsp(Q_{j-1})$ *and* $spsp(a_{j-1})$, *and we have* $Q_j \equiv (P_{j-1} + 2a_{j-1})a_{j-1} \bmod n$, *and* $P_j \equiv P_{j-1} + 2a_{j-1} \bmod n$. *Moreover, if n is* $epsp(Q_j)$, *then n is also* $epsp(P_j)$.

Corollary 3.1. *A composite number n is a strong $QFpsp(P,Q)$ if it is $epsp(Q_j)$ for all $j \leq i$ and satisfies any of the two conditions of Corollary 2.1.*

In Theorem 4.1 below we will give a closed and simple expression for T_i in the case when p is a prime. This yields a practical and efficient probable prime test. Unless otherwise stated, we assume throughout that $\epsilon(n) = -1$.

To apply the strong QF–test directly, one approach would be to successively test whether the individual Q_j are correct until either this is not satisfied, in which case n is composite, or the index j reaches $i = \nu_2(n') = \nu_2(n+1)$, resp. $i = k-1$, in which case the Q_j–routine terminates. In a second step, it is then tested whether Theorem 2.1, resp. Theorem 4.1, holds. This method turns out to be very efficient when $\nu_2(n+1)$ is not too large. The drawback of this approach is that when n is $epsp(Q_r)$ for many bases Q_r, the Q_j–routine might have to be carried out to the maximal index. When i is large, this could involve too many modular exponentiations.

As a combined Fermat–Lucas test is much more reliable than any Euler iteration, we propose the following more efficient ordering of the conditions for the **strong** QF–**test**. Note that the value $a^{n'/2^i} = \alpha_1^{n'/2}$ in Step 4 already has been computed in Step 2. Also, if Step 3 is passed, then all the $a = a_1, ..., a_i$ are correct roots, so that the value Q_{i+1} is also correct.

The strong QF-test (sqf-test):

1. Select P_1, Q_1 such that $\left(\frac{Q_1}{n}\right) = 1$, $\left(\frac{P_1^2 - 4Q_1}{n}\right) = -1$.

2. Test if n is $slpsp(P_1, Q_1)$. If not, declare n composite, and stop. Let $i = k - 1$, if the V-index k of (P_1, Q_1) exists, and $i = \nu_2(n') = \nu_2(n+1)$, otherwise.

3. For $1 \leq j \leq i$, evaluate $a_j \equiv Q_j^{(n+1)/4} \bmod n$, P_{j+1}, and Q_{j+1}. If for some j, $a_j^2 \not\equiv Q_j \bmod n$, declare n composite and stop.

4. With p replaced by n, test for the corresponding value T_i of Theorem 4.1 below.

Any pseudoprime that passes this test must have several very restrictive properties. From property (4.1) and the fact that, in the case $n \equiv 3 \bmod 4$, an $epsp(Q_i)$ pseudoprime is also a $spsp(Q_i)$ pseudoprime, it follows from Proposition 3.2 that the above test is stronger than that of Grantham.

Theorem 3.1. *Suppose $n \equiv 3 \bmod 4$ is a strong $QFpsp(P_1, Q_1)$. Then n is also $slpsp(P_1, Q_1)$, and for all $1 \leq j \leq i$, where $i = \nu_2(n+1)$ or*

$k - 1$, respectively, n is $elpsp(P_j, Q_j)$, $spsp(Q_j)$, and $spsp(a_j)$, and for all $2 \leq j \leq i$, n is $spsp(P_j)$. Moreover, for all $1 \leq j \leq i$ we have $(Q_j/p) = 1$ for all $p|n$, and n passes the Grantham test for (P_1, Q_1).

Remark 3.1. The Fermat-Lucas combination for $\epsilon(n) = -1$ is very strong. A composite integer n would have to pass the tests of Theorem 3.1 for all $j \leq i$ pairs of parameters (P_j, Q_j). In contrast to the Lucas tests for $\epsilon(n) = -1$, which themselves are independent of the Fermat test, this combination arises here naturally from the above condition on the correctness of the roots a_j.

4   Some Properties of the Strong-QF-Test

4.1   A Simpler Formula

We now show that the right-hand side of (2.4) can be replaced by a much simpler expression.

Theorem 4.1. *Suppose that* $2^j | p'$, $a^2 \equiv Q \bmod p$ *and that* Q_{j+1} *exists. Then*

$$\alpha^{p'/2^j} \equiv \overline{\alpha}^{p'/2^j} \equiv T_j \equiv \begin{cases} a^{p'/2^j} & \bmod\ p, \ \text{if} \ \left(\frac{Q_{j+1}}{p}\right) = 1, \\ -a^{p'/2^j} & \bmod\ p, \ \text{if} \ \left(\frac{Q_{j+1}}{p}\right) = -1. \end{cases} \tag{4.1}$$

Proof. The hypothesis that Q_{j+1} exists implies $\left(\frac{Q_j}{p}\right) = 1$. Consider first the case when $2^{j+1} | p'$. Then, by the identity (cf. [13])

$$U_m(P, Q) = \left(\alpha^{2m} - Q^m\right) / \left(\alpha^m(\alpha - \overline{\alpha})\right), \tag{4.2}$$

we have $\alpha^{p'/2^j} \equiv Q^{p'/2^{j+1}} \equiv a^{p'/2^j} \bmod p$, as desired, provided that we have $U_{p'/2^{j+1}}(P, Q) \equiv 0 \bmod p$, or equivalently, $U_{p'/2}(P_{j+1}, Q_{j+1}) \equiv 0 \bmod p$. This condition is satisfied if and only if $\left(\frac{Q_{j+1}}{p}\right) = 1$. On the other hand, if $\left(\frac{Q_{j+1}}{p}\right) = -1$, then $U_{p'/2^m}(P, Q)$ is congruent to $0 \bmod p$ for $m = j$, but not for $m = j + 1$. In particular, $\alpha^{p'/2^j} \equiv \overline{\alpha}^{p'/2^j} \bmod p$. By (4.2) we have $\alpha^{p'/2^j} \not\equiv Q^{p'/2^{j+1}} \bmod p$, since otherwise $U_{p'/2^{j+1}}(P, Q)$ would be congruent to $0 \bmod p$. Furthermore, since $U_{p'/2^j}(P, Q) \equiv 0 \bmod p$, (4.2) yields

$$\left(\alpha^{p'/2^j}\right)^2 \equiv \left(\overline{\alpha}^{p'/2^j}\right)^2 \equiv Q^{p'/2^j} \equiv \left(\pm a^{p'/2^j}\right)^2 \bmod p,$$

so that $\alpha^{p'/2^j} \equiv -a^{p'/2^j} \bmod p$.

It remains to deal with the case when $j = \nu_2(p')$. By Proposition 2.2 we have

$$(\alpha + a)^{p'/2^{j-1}} = \left((\alpha + a)^2\right)^{p'/2^j} \equiv (P + 2a)^{p'/2^j} \alpha^{p'/2^j} \bmod p.$$

Moreover, from the same proposition and the first part of the above argument we see that $(\alpha + a)^{p'/2^{j-1}} \equiv \left((P + 2a)a\right)^{p'/2^j} \bmod p$ holds whenever $U_{p'/2^j}(P', Q') \equiv 0 \bmod p$, where $P' \equiv P + 2a$ and $Q' \equiv (P + 2a)a \bmod p$. As before, this is equivalent to $U_{p'/2}(P'_j, Q'_j) \equiv 0 \bmod p$. Otherwise we have $V_{p'/2}(P'_j, Q'_j) \equiv 0 \bmod p$ and the result follows since $(Q'_j/p) = (Q_{j+1}/p)$.   □

4.2   The Role of the Parameters

For pseudoprimality testing, the most interesting values of P and Q are those for which (4.1) (with p replaced by n) can be tested with the maximal value of j, i.e., $j = i = \nu_2(n')$. This requires that Q_{i+1} exists. Alternatively, we need to have $U_u(P, Q) \equiv 0 \bmod n$ (we will call this the U-case'), where u is the odd part of n'. On the other hand, if the V-index k corresponding to the pair (P, Q) exists (the V-case'), then we have $V_{n'/2^k}(P, Q) \equiv 0 \bmod n$, which yields the weaker condition $\alpha^{n'/2^{k-1}} \equiv \overline{\alpha}^{n'/2^{k-1}} \bmod n$. Indeed, in this case, n satisfies (4.1) only for indices j that are less than k. Hence, in (4.1) the maximal value $j = i = \nu_2(n')$ can only be tested in the U-case, but not in the V-case.

Example 4.1. Let $n = 671 = 11 \cdot 61$, $\epsilon(n) = -1$. If $P = 27$, $Q = 58$, then n is $slpsp(P, Q)$ in the U-case and $spsp(Q)$. Thus, for $a \equiv Q^{(n+1)/4} \bmod n$, we have $a^2 \equiv Q \bmod n$. Here, n satisfies $\alpha^{n'/2^j} \equiv \overline{\alpha}^{n'/2^j} \equiv a^{n'/2^j} \bmod n$ for $j = 1, 2, 3, 4$, but $\alpha^{n'/2^j} \equiv \overline{\alpha}^{n'/2^j} \not\equiv \pm a^{n'/2^j} \bmod n$ for $j = \nu_2(n + 1) = 5$, and n is therefore composite. In particular, $\gcd(\alpha^{n'/2^5} - a^{n'/2^5}, n) = 61$.

Alternatively, for $P = 633$, $Q = 34$, n is $slpsp(P, Q)$ in the V-case, and n cannot be tested for the maximal value of j in (4.1). Also, n is $spsp(Q)$, so that for $a \equiv Q^{(n+1)/4} \bmod n$, $a^2 \equiv Q \bmod n$. However, n satisfies (4.1): we have $\alpha^{n'/2^j} \equiv \overline{\alpha}^{n'/2^j} \bmod n$, which is congruent to $a^{n'/2^j} \bmod n$ for $j = 1, 2, 3$, and to $-a^{n'/2^j} \bmod n$ for $j = 4$. Here, (4.1) cannot be tested for $j = 5$, as $\alpha^{n'/2^5} \equiv -\overline{\alpha}^{n'/2^5} \bmod n$.

4.3   The Euler Test in the U-Case

The following example illustrates a situation with $U_u(P, Q) \equiv 0 \bmod n$ (defined as above), where n passes Grantham's test, but is revealed to be composite by the strong QF-test. Note that both prime factors of n are greater than $B = 50000$, the trial division bound in Grantham's test.

Example 4.2. Let $n = p \cdot q = 50129 \cdot 50131 = 2513016899$. It can be shown [8] that in this case for $Q = 1$ the number of liars P to the Grantham test becomes $\frac{1}{2}\big(\gcd((n+1)/4, p+1) - 1\big) \cdot \big(\gcd((n+1)/4, q-1) - 1\big) = \frac{1}{2}\frac{p-1}{2}\frac{p-1}{4} = 314102048$. However, n is easily disclosed as composite via the 'Q_j–routine' in the strong QF–test. Consider the smallest liars $P = 1, 9, 10, 23, 37, 42, 60, 65, 73, 79, 85, 90, 97, 98, \ldots$. Although n satisfies (4.1) for $j = \nu_2(n+1) = 2$, we have $a_2^2 \not\equiv Q_2 \bmod n$ in each of these cases. The same result holds for all of the 220458 Grantham liars $P < 10^6$ that were tested.

4.4 Some Numerical Data

In this section we analyze all composites $n < 1000$ that pass the shortened Grantham test, without the trivial division up to 50000 required in [3]. In this case, an exhaustive search over all parameters with $\left(\frac{Q}{n}\right) = 1$ and $\left(\frac{D}{n}\right) = -1$ can be performed. It turns out that the results are quite similar to those for the 'real' pseudoprimes in Section 5 below (apart from their size).

The first row gives the number of such pairs that pass the test in [3] (excluding those composites that pass only for the special parameters $(P, Q) = (1, 1)$ and $(-1, 1)$, which according to Definition 2.1 are a bad choice). The second row shows how many of these pairs are strong QF–liars, and the final two rows describe how the latter pairs are distributed among U– and V–cases. A star ($\star$) means that there are no corresponding pseudoprimes. The symbol $1^\dagger$ indicates that the only sqf–pseudoprime in the U–case is the one corresponding to the pair $(-1, 1)$.

Grantham Pseudoprimes n (without trial division) < 1000																	
n	119	143	275	287	299	319	323	343	391	399	407	455	511	527	539	559	575
no. of G-liars	6	10	4	6	10	8	32	6	4	4	10	4	36	22	4	162	4
no. of sqf-liars	4	$\star$	3	$\star$	9	$\star$	$1^\dagger$	6	4	$\star$	$\star$	$\star$	$\star$	16	3	$\star$	$\star$
V–case	4	$\star$	2	$\star$	8	$\star$	$\star$	6	4	$\star$	$\star$	$\star$	$\star$	16	2	$\star$	$\star$
U–case	$\star$	$\star$	$1^\dagger$	$\star$	$1^\dagger$	$\star$	$1^\dagger$	$\star$	$\star$	$\star$	$\star$	$\star$	$\star$	$\star$	$1^\dagger$	$\star$	$\star$

n	611	623	671	679	767	779	783	791	799	803	851	899	923	935	943	959	979
no. of G-liars	10	6	250	36	10	34	6	6	20	10	10	98	28	10	4	6	8
no. of sqf-liars	9	⋆	⋆	36	⋆	25	⋆	4	⋆	9	9	$1^\dagger$	9	⋆	⋆	⋆	⋆
V–case	8	⋆	⋆	36	⋆	24	⋆	4	⋆	8	8	⋆	8	⋆	⋆	⋆	⋆
U–case	$1^\dagger$	⋆	⋆	⋆	⋆	$1^\dagger$	⋆	⋆	⋆	$1^\dagger$	$1^\dagger$	$1^\dagger$	$1^\dagger$	⋆	⋆	⋆	⋆

The table shows that the sqf–test for the U–case is very reliable. Apart from the few examples marked by $1^\dagger$, there are no sqf–pseudoprimes. For the U–case, as well as for most of the V–cases, Q_j–testing is not required for all indices $j \leq i$. Instead, almost always the conditions will fail in Step 1 or 2 already, so that the algorithm terminates and n is known to be composite. We have found no examples in the U–case that require more than three steps in the Q_j–routine.

4.5 Eliminating the V-Case

Since, apart from the few cases indicated by $1^\dagger$, all of the above pseudo-primes arise from the V–case, it makes sense to try to avoid parameters P and Q giving rise to the V–case. This would also enable testing for the maximal value $j = i$ in (4.1). When $i = \nu_2(p')$ is large, finding such parameters P and Q by trial and error alone appears to be difficult. However, we present below an efficient algorithm that achieves this, for any given value of i.

Lemma 4.1. *Select P, Q and let $Q^{(0)} := Q$, $P^{(0)} := P$, and for $j \geq 1$, define $a^{(j)} \equiv \frac{Q^{(j-1)}}{P^{(j-1)}} \bmod p$, $Q^{(j)} \equiv \left(a^{(j)}\right)^2 \bmod p$, $P^{(j)} \equiv P^{(j-1)} - 2a^{(j)} \bmod p$. (Assume here that the multiplicative inverses exist; otherwise discard P and Q and repeat the process.)*

(a) The sequences defined in this way are inverses of a_j, P_j, Q_j, respectively, in the sense that if $\widetilde{P} \equiv P_j, \widetilde{Q} \equiv Q_j$, then $\widetilde{P}^{(j)} \equiv P, \widetilde{Q}^{(j)} \equiv Q \bmod p$.

(b) For $j \leq \nu_2(p')$, let $\widehat{P}_1 \equiv P^{(j)}$ and $\widehat{Q}_1 \equiv Q^{(j)}$. Then $\widehat{Q}_j$ exists and
$$\left(\frac{\widehat{Q}_j}{p}\right) = \left(\frac{Q}{p}\right).$$

Proof. Part (a) follows from the definition. To prove (b), recall that, by Lemma 2.3, the Legendre symbols are independent of the choice of the square roots of the Q_j. Therefore, in order to evaluate these symbols, we can take $\hat{a}_1 = a^{(j)}$, $\hat{P}_2 = P^{(j-1)}$, $\hat{Q}_2 = Q^{(j-1)}$, $\hat{a}_2 = a^{(j-1)}$, etc., and by induction we obtain the desired assertion. □

If n is composite and if the corresponding Q_j are 'correct' mod n, then the assertions of Lemma 4.1 also hold modulo n.

Unfortunately, the above method requires the evaluation of a multiplicative inverse in each step. We now describe a more efficient method.

It is convenient to define the function $W_n(P,Q) \equiv V_{2n}(P,Q)Q^{-n} \bmod n$ (cf. [16]). From the relation $V_{2n}(P,Q) = V_n(P^2 - 2Q, Q^2) = Q^n V_n(P^2/Q - 2, 1)$ it then follows that $W_n(P,Q) \equiv V_n(P^2/Q - 2, 1) \bmod n$, $W_0 = 2$, $W_1 \equiv P^2/Q - 2$, $W_{n+1} \equiv W_1 W_n - W_{n-1} \bmod n$, and, most importantly,

$$W_{2n} \equiv W_n^2 - 2 \bmod n, \tag{4.3}$$

which allows a very fast evaluation. We also write

$$X_j(P,Q) = W_{2^j}(P,Q). \tag{4.4}$$

Then the above quantities $a^{(j)}, Q^{(j)}, P^{(j)}$ can be evaluated much more efficiently.

Lemma 4.2. *We have* $a^{(1)} = \frac{Q}{P}$, $P^{(1)} = a^{(1)} X_0$, $Q^{(1)} = (a^{(1)})^2$, *and for* $i \geq 2$, *we have* $a^{(i)} \equiv \frac{Q}{P} \frac{1}{X_0 \cdots X_{i-2}} \equiv a^{(i-1)} \frac{1}{X_{i-2}} \bmod n$, $P^{(i)} \equiv a^{(i)} X_{i-1} \bmod n$, $Q^{(i)} \equiv (a^{(i)})^2 \bmod n$. *(Again assume that the multiplicative inverses exist.)*

Proof. This follows easily by induction using (4.3). □

Remark 4.1. The requirement that the inverses exist is not a real restriction in primality testing. It follows from [9] that for each Q there are at most 2^{j+1} parameters P such that $X_j \equiv 0 \bmod n$, when n is prime. Modulo a composite n, $\gcd(X_j, n)$ often yields a factor of n. If for some Q, $X_j \equiv 0 \bmod n$ for more than 2^{j+1} different P, then n is disclosed as composite.

In the lemma, the evaluation of X_j can be carried out using only the operations of squaring and multiplying. To obtain $a^{(i)}, P^{(i)}, Q^{(i)}$ only *one* multiplicative inverse, i.e., that of X_{i-2}, has to be computed.

For any predetermined value of i, and in particular for $i = \nu_2(n')$, this now enables us to very rapidly compute parameters (P,Q) for which Q_{i+1} exists and thus to test for the condition $U_u(P,Q) \equiv 0 \bmod n$. This provides an efficient way to eliminate the pseudoprimes arising from the V-case above.

4.6   A More Practical Implementation

Since the parameters in the U–case yield the optimal testing condition, we propose the following alternative test. The test is presented here in terms of Lucas functions, which allow very fast evaluation and can easily be implemented. Alternatively, the test could be described within the setting of finite fields. As in [3], we assume that n is not a perfect square and not divisible by primes up to 50000.

Note that if n is prime then by Lemma 4.1 we have $\left(\frac{Q_i}{n}\right) = \left(\frac{Q_0}{n}\right) = 1$, so that Q_{i+1} exists and can be obtained as above.

1. Select P_0, Q_0 such that $\left(\frac{Q_0}{n}\right) = 1$, $\epsilon(n) = \left(\frac{P_0^2 - 4Q_0}{n}\right) = -1$.

2. For $i = \nu_2(n') = \nu_2(n+1)$ compute $Q \leftarrow (Q_0)^{(i)}$, $P \leftarrow (P_0)^{(i)}$.
   (In the rare event that $(P, Q) = (-1, 1)$ select another pair P_0, Q_0.)

3. Let $a \equiv Q^{(n+1)/4} \bmod n$. If $a^2 \not\equiv Q \bmod n$, declare n composite and stop.

4. If $U_{n'/2^i}(P/a, 1) \not\equiv 0 \bmod n$ or $V_{n'/2^i}(P/a, 1) \not\equiv 2\sigma \bmod n$, for $\sigma \in \{1, -1\}$, declare n composite and stop.

5. Let $a_1 \leftarrow a$, $Q_1 \leftarrow Q$, $P_1 \leftarrow P \bmod n$. For $2 \le j \le i$ evaluate P_j, Q_j, $a_j \equiv Q_j^{(n+1)/4} \bmod n$, testing in each step whether $a_j^2 \equiv Q_j \bmod n$. If not, declare n composite and stop.

6. If $\left(\frac{Q_{i+1}}{n}\right) \neq \sigma$ declare n composite; otherwise declare n a probable prime.

If n passes Step 4, then with $\alpha = \alpha(P, Q)$, $\bar{\alpha} = \bar{\alpha}(P, Q)$, we have $\alpha^{n'/2^i} \equiv \bar{\alpha}^{n'/2^i} \equiv \pm a^{n'/2^i} \bmod n$ and $a^2 \equiv Q \bmod n$. This follows from the identities $V_m(ca, a^2) = a^m V_m(c, 1)$ and $aU_m(ca, a^2) = a^m U_m(c, 1)$. The correctness of the signs is being tested in Step 6.

Running time analysis: The evaluation of the above Lucas functions can be performed very fast [15]. In particular, we only need the values of the U– and V–functions to the same degree. As the U–sequence needs to vanish, for Step 4 only the evaluation of two consecutive V–sequences is required [7]. In the case $Q = 1$, the combined evaluation of $V_m(P, Q)$ and $V_{m+1}(P, Q)$ requires on average 1.33 times as many multiplications than exponentiation [5], and in the worst case less than $2 \log_2(m)$ multiplications. In addition, we need one exponentiation in Step 3 and several squarings modulo n, depending on the value of $i = \nu_2(n+1)$. For most composite numbers, Steps 2–4 disclose n as composite and Step 5 is not required.

Remark 4.2. We suggest aborting the Q_j–routine in Step 5 after one or two steps, as composites requiring a value $j > 3$ are likely to be very scarce (if there are any). For this shortened test, Step 6 should be skipped as well.

Lemma 4.3. *Suppose* $n \equiv 3 \bmod 4$ *passes the shortened test above, where the* Q_j*–routine is only performed for one step, namely* $j = 2$. *Then,* n *is* $spsp(P_2)$ *and for* $l = 1, 2$, n *is simultaneously* $spsp(Q_l)$, $spsp(a_l)$, *and for all* $p|n$, $\left(\frac{Q_l}{p}\right) = 1$. *Also,* n *is* $slpsp(P,Q)$ *for the special case* $U_u(P,Q) \equiv 0 \bmod n$, *where* u *is the odd part of* $n + 1$. *In addition,* n *passes the Grantham test for* (P,Q).

Proof. This follows from Proposition 3.2 and the fact that for integers $n \equiv 3 \bmod 4$ $epsp(Q)$ implies $spsp(Q)$. □

5 Comparison of the Pseudoprimes

5.1 Grantham Pseudoprimes and sqf-Pseudoprimes

Although the Grantham test is already very reliable, in many cases pseudoprimes occur frequently.

For $Q = 1$ it suffices to have $\alpha^{n'/2^i} \equiv \overline{\alpha}^{n'/2^i} \equiv \pm 1 \bmod n$ for $i = \nu_2(n') = \nu_2(n+1)$ or alternatively, $\alpha^{n'/2^j} \equiv \overline{\alpha}^{n'/2^j} \equiv -1 \bmod n$ for some $1 \leq j < i$. Then, when $n \equiv 3 \bmod 4$, exponentiation with $(n-1)/2$ will preserve the sign and yield the desired value. This condition will usually follow if Grantham's second step, $\alpha^{n'} \equiv \overline{\alpha}^{n'} \equiv 1 \bmod n$, is fulfilled. (If this is not the case, one would be able to factorize n.) Hence, the question reduces to finding P with $\rho(p, P, 1)|n'$, that is, $\rho(p, P, 1)|n+1$. The number of the liars will depend on $\prod_{p|n} \gcd(p'/2, n+1)$, since $\rho(p, P, 1)|p'/2$. In particular, a product of twin primes allows a large number of liars (Example 4.2), as do Carmichael Lucas numbers [14].

For general Q a necessary condition for the existence of Grantham liars for (P,Q) is that n is $epsp(Q)$. This follows, since any Grantham pseudoprime is $QFpsp(P,Q)$ and therefore $U_n(P,Q) \equiv \left(\frac{D}{n}\right) \bmod n$. Moreover, $U_{n'/2}(P,Q) \equiv U_{(n+1)/2}(P,Q) \equiv 0 \bmod n$ and thus, since $U_n(P,Q) - Q^{(n-1)/2}\left(\frac{D}{n}\right) = U_{(n+1)/2}(P,Q)V_{(n+1)/2}(P,Q)$, one has $Q^{(n-1)/2} \equiv 1 \equiv \left(\frac{Q}{n}\right) \bmod n$.

Another necessary condition is that $\left(\frac{Q}{p}\right) = 1$ for all $p|n$ ([3, p. 39]).

The next lemma follows from the formula $V_{2m}(P,Q) = V_m(P,Q)^2 - 2Q^m$ and the fact that $(n-1)/2$ is odd when $n \equiv 3 \bmod 4$.

Lemma 5.1. *Let $n \equiv 3 \bmod 4$, and suppose $V_{n'/2^k}(P, Q) \equiv 0 \bmod n$ in the strong Lucas test. Then $\alpha^{n'/2^{k-1}} \equiv \overline{\alpha}^{n'/2^{k-1}} \equiv -Q^{n'/2^k} \bmod n$, where $\alpha = \alpha(P, Q)$, $\overline{\alpha} = \overline{\alpha}(P, Q)$. If n is $epsp(Q)$ for $\left(\frac{Q}{p}\right) = 1$ for all $p|n$ then n is also a Grantham pseudoprime corresponding to the pair (P, Q).*

Similarly, strong Lucas pseudoprimes in the U–case for integers n that are $epsp(Q)$ will often (but not always, as in Lemma 5.1) yield Grantham pseudoprimes.

There are several methods for constructing strong Lucas pseudoprimes for different P and a given Q (see, e.g., [6]). The problem again reduces to finding primes p with $\epsilon(n) = \prod_{p|n} \epsilon(p) = -1$ and $\gcd(p'/2, n+1) = g_1$ large for all $p|n$. In order for n to be $epsp(Q)$ for $Q \neq \pm 1$, the product $\prod_{p|n}(\gcd(p-1, n-1)) = g_2$ (resp. the odd part of it) cannot be too small.

The desired values of Q and P can be most efficiently found by looping over the different values and all possible multiplicative orders (which divide g_2) resp. ranks of appearance (which divide g_1). This can be done modulo each prime p, and by an application of the Chinese Remainder Theorem (see Example 5.1). However, if the corresponding gcd's, g_1 and g_2, get too large (as is the case of Example 5.2), a trial search for small liars is usually more economical, in order to avoid large memory requirements.

To find small Grantham pseudoprimes we simply looped over the smallest primes larger than 50000 until some of the required gcd's turned out to be large. In practice, this 'naive' approach turned out to be very efficient. More sophisticated methods (such as those in [4]) could certainly be applied and would yield larger pseudoprimes with more prime factors. Our motivation here was to show that many composites, which are a product of primes > 50000, have a large number of Grantham liars.

Example 5.1. For $n = p \cdot q = 53299 \cdot 81901 = 4365241399$ one has $\gcd(n + 1, (p + 1)/2) = \gcd(n + 1, (q - 1)/2) = 650$ and $\gcd(n - 1, p - 1) = \gcd(n - 1, q - 1) = 126$. This gives 3844 parameters Q, each of which allows Grantham liars P. For example, for $Q = 122000$ there are 156493 parameters P that pass the Grantham test. All of these are detected by the sqf–test (on input $P_1 = P$, $Q_1 = Q$), 52164 in the U–case and 104329 in the V–case, since for the corresponding values of a_2, $a_2^2 \not\equiv Q_2 \bmod n$. For the other values of Q there are analogous results.

Example 5.2. Let $n = p \cdot q = 52631 \cdot 87721 = 4616843951$. For $Q = 3689311$ there are 5666 liars $P < 10^6$ to the Grantham test. All of these are detected by the sqf–test (on input $P_1 = P$, $Q_1 = Q$), 488 in the U–case and 5178 in the V–case, since for all of these values the corresponding roots a_2 are not correct. For $Q = 559687173$ there are 5719 Grantham liars $P < 10^6$, but by the a_2–testing, no sqf–liars. For

$Q = 1081658870, 1123711039, 1386983924, \ldots$ the results are analogous.

For $n = p \cdot q = 51899 \cdot 72661 = 3771033239$ and $Q = 1639269$ there are 2618 Grantham liars P up to 10^6. In all of these cases the corresponding a_2 satisfy $a_2^2 \not\equiv Q_2 \bmod n$ and thus all cases are detected by the sqf–test. Of these, 859 fall into the U–case and 1759 into the V–case. Analogously, for $Q = 143624263$ there are 2705 Grantham liars in this range which are all detected via the individual a_2–testing. Similar results hold for the other bases $Q = 220253863, 307106113, 364378136, 547610131$ that were tested.

5.2 V-Case Pseudoprimes and the Reversion Method

Although the sqf–test is stronger than Grantham's, sqf–pseudoprimes occasionally arise from the V–case. In particular, a composite number as in Lemma 5.1 passes (4.1), so that it is only through the Q_j–routine that such numbers can be revealed as composite. The following lemma characterizes the integers for which the sqf–test in the V–case is not very strong.

Lemma 5.2. *Suppose that $n \equiv 3 \bmod 4$ is slpsp(P, Q) in the V–case, and let p be a prime factor of n. If $\nu_2(\rho(p, P, Q)) + 1 = \nu_2(n + 1)$, where $\left(\frac{P^2 - 4Q}{n}\right) = -1$, then $\nu_2(\rho(q, P, Q)) + 1 = \nu_2(\rho(n, P, Q)) + 1 = \nu_2(n + 1)$ for all prime divisors q of n. Further, the V–index k of (P, Q) equals 2 and the Q_j–routine reduces to a single step. In particular, Step 3 is only tested for $j = 1$. If n passes this step, then n passes Step 4 for $i = 1$.*

Proof. By Theorem 1 of [6], n is slpsp(P, Q) if and only if $\rho(p, P, Q)|n'$ and $\nu_2(\rho(p, P, Q))$ is a constant value for all primes p dividing n. The first assertion follows from the fact that the rank modulo n equals the least common multiple of the rank modulo each prime factor. The remaining assertions follow from Lemma 5.1. □

For such numbers the sqf–test will often not be stronger than the test by Grantham. Otherwise, the larger k in the V–case becomes, the more efficiently Grantham liars will be detected by the Q_j–routine.

Since $\rho(p, P, Q)|p'/2$ for $\left(\frac{Q}{p}\right) = 1$, it follows that any integer with $\nu_2(n+1) = \nu_2(p')$ for some $p|n$ potentially falls into the class described by Lemma 5.2. Then, the larger both $\prod_{p|n} \gcd(n + 1, p')$ and $\prod_{p|n} \gcd(n - 1, p - 1)$ become, the more likely it will be for n to pass the sqf–test.

The main advantage of the test of Section 4.6 is that any possibly V–cases can a priori be avoided. This can be achieved with little effort by several squarings modulo n.

Example 5.3. Let $n = p \cdot q = 65599 \cdot 78721 = 5164018879$. Note that $\gcd(n + 1, (p + 1)/2) = 6560$, $\gcd(n + 1, (q - 1)/2) = 13120$, so that there exist many parameters P with rank modulo p dividing $(p+1)/2$, and rank modulo q dividing $(q-1)/2$. Since $\nu_2(n+1) = \nu_2(p+1) = 6$, some of the V–case parameters will not be detected. For example, for $Q = 763094184$ there are 2678 liars P less than 10^6 that pass the Grantham test. Among those, 682 parameters P would have been detected by the sqf–test, as the corresponding a_2 satisfy $a_2^2 \not\equiv Q_2 \bmod n$. The remaining 1996 liars P in this range all fall into the V–case and pass the sqf–test. By applying the test of Section 4.6 (with $(P_0, Q_0) = (P, Q)$), all of these liars P are correctly identified. Similarly, the reversion method is always successful for the liars P corresponding to the other bases $Q = 1164625663, 3999393215, 4400924694$ that were tested.

For all of these values of Q and all $P < 10^6$ with $\epsilon(n) = -1$, the test of Section 4.6 with $(P_0, Q_0) = (P, Q)$ already terminates at Step 3 and discloses n as composite.

Example 5.4. For the composites n of Example 5.2, and the above values of Q, all $P < 10^6$ with $\epsilon(n) = -1$, the test of Section 4.6 with $(P_0, Q_0) = (P, Q)$ also terminates at Step 3 already and proves these n to be composite.

Any pseudoprimes for the test of Section 4.6, and even for the shortened version of this test, would have to be pseudoprimes on both the Fermat and the Lucas side (see Lemma 4.3). Note that, at the beginning of the process, for any given (P_0, Q_0), it is not known what the result (P, Q) of the reversion will be. Thus, the values P, Q, P_j, Q_j, and a_j, are not a priori known. It does not seem possible to generate a set of liars that would match these parameters. This is in contrast to the Grantham test, for which the liars were found via *any* parameters with suitable rank of appearance.

It is unlikely that the reversion procedure accidentally generates (P, Q) which are liars. This follows from the result in [9] which presents an analog to the number of bases with the same multiplicative order. More precisely, for any given Q and any $d|p'/2$ there are $\phi(d)$ parameters P modulo p with $\rho(p, P, Q) = d$. Since there are $\phi(d)$ bases Q modulo p with $\mathrm{ord}_p(Q) = d|p-1$, it follows that for arbitrary parameters, resp. bases, the order as well as the rank typically will be large.

Since any pseudoprime would need to satisfy both $\mathrm{ord}_p(Q)|n-1$ and $\rho(q, P, Q)|n+1$, we necessarily have $\gcd(\mathrm{ord}_p(Q), \rho(q, P, Q)) < 2$ for any primes p, q dividing n. This typically implies $\epsilon(p) = -1$ for all divisors p of n. Thus, such pseudoprimes seem to be closely related to numbers n satisfying $p-1|n-1$ and $p+1|n+1$ for all $p|n$. However, no composite integers n with this property are known. This suggests that pseudoprimes of the above type are very scarce, if they exist at all. We have not found any pseudoprime for the test of Section 4.6, nor for the shortened version (Remark 4.2).

Acknowledgements. I am deeply grateful to Prof. H. C. Williams for many very interesting conversations. His valuable comments have helped tremendously in writing this paper. I am also thankful to the anonymous referee for his insightful remarks, and to the editors for a number of suggestions that helped improve the presentation of the paper. Finally, I thank Prof. W. B. Müller for his support.

References

[1] R. Baillie and S.S. Wagstaff, Jr., *Lucas pseudoprimes*, Math. Comp. **35** (1980), 1391–1417.

[2] D. Bleichenbacher, *Efficiency and security of cryptosystems based on number theory*, Ph.D. thesis, ETH Zürich, 1996.

[3] J. Grantham, *A probable prime test with high confidence*, J. Number Theory **72** (1998), 32–47.

[4] D. Guillaume and F. Morain, *Building pseudoprimes with a large number of prime factors*, Appl. Algebra Engrg. Comm. Comput. **7** (1996), 263–277.

[5] W. More, *Der QNR-Primzahltest*, Ph.D. thesis, University of Klagenfurt, Department of Mathematics, 1994.

[6] S. Müller, *On strong Lucas pseudoprimes*, Contributions to general algebra, 10 (Klagenfurt, 1997), Heyn, Klagenfurt, 1998, pp. 237–249.

[7] ———, *On probable prime testing and the computation of square roots mod n*, Algorithmic Number Theory, 4th International Symposium, ANTS-IV, Leiden, The Netherlands, July 2000 (Wieb Bosma, ed.), Springer-Verlag, 2000, pp. 423–437.

[8] ———, *On the extra strong Lucas probable prime test*, manuscript, University of Klagenfurt, 2000.

[9] ———, *On the rank of appearance and the number of zeros of the Lucas sequences over $\mathbb{F}_q$*, Finite Fields and Applications (D. Jungnickel and Niederreiter H., eds.), Springer-Verlag, 2001, pp. 390–408.

[10] ———, *A probable prime test with very high confidence for $n \equiv 1 \bmod 4$*, Advances in Cryptology – ASIACRYPT '01 (Colin Boyd, ed.), Springer-Verlag, Berlin, 2001, pp. 87–106.

[11] C. Pomerance, J. L. Selfridge, and S.S. Wagstaff, Jr., *The pseudoprimes to $25 \cdot 10^9$*, Math. Comp. **35** (1980), 1003–1026.

[12] P. Ribenboim, *The new book of prime number records*, Springer-Verlag, New York, 1996.

[13] L. Somer, *The divisibility properties of primary Lucas recurrences with respect to primes*, Fibonacci Quart. **18** (1980), 316–334.

[14] H. Williams, *On numbers analogous to the Carmichael numbers*, Canad. Math. Bull. **20** (1977), 133–143.

[15] ———, *A p+1 method of factoring*, Math. Comp. **39** (1982), 225–234.

[16] ———, *Édouard Lucas and primality testing*, John Wiley & Sons Inc., New York, 1998, A Wiley-Interscience Publication.

Splitting of Primes in Infinite Extensions

V. Kumar Murty[1]

1 Introduction

In 1840, Dirichlet proved his famous theorem on the infinitude of primes in arithmetic progressions. This states that given integers $1 \leq a$, $q \in \mathbb{Z}$ with $(a, q) = 1$, there are infinitely many primes p with

$$p \equiv a \pmod{q}.$$

Davenport ([2, p. 1]) says this result is the beginning of analytic number theory. Dirichlet's theorem may be viewed as a special case of a result on the splitting of primes in finite extensions of the rational number field. The aim of this article is to consider possible generalizations to infinite extensions. After a brief discussion of the Chebotarev density theorem in number fields, we examine the case of infinite extensions, especially those of GL_2-type. We then consider a more general case by introducing certain zeta functions, the conjectural analytic properties of which should give information about the splitting of primes in certain infinite extensions. The conjectures we state are really to be viewed as working hypotheses for further investigations. The hope is to open up a line of investigation from which a more refined picture can emerge.

2 Chebotarev Density Theorem

Let K be a number field and F a finite Galois extension of K. To each prime ideal $\mathfrak{p}$ of the ring of integers $\mathcal{O}_K$ which does not ramify in F, there is a conjugacy class of automorphisms $\text{Frob}_{\mathfrak{p}}$. If we want to make explicit the field extension, we also write this as the Artin symbol $(\mathfrak{p}, F/K)$. If F/K is Abelian, then $\text{Frob}_{\mathfrak{p}}$ consists of a single element. If $F = K(\zeta_q)$, then $\text{Gal}(F/K)$ is a subgroup of $(\mathbb{Z}/q)^{\times}$ and $\text{Frob}_{\mathfrak{p}}$ is the image of the automorphism

$$\zeta_q \mapsto \zeta_q^a$$

[1]Research partially supported by a grant from NSERC.

where $p \equiv a \pmod{q}$ and p is the rational prime below $\mathfrak{p}$. Thus, distribution of Frobenius elements in the Abelian case is closely related to the problem of primes in arithmetic progressions.

A simple non-Abelian example where Frobenius classes can be described explicitly is given by $K = \mathbb{Q}$ and $F = \mathbb{Q}(2^{1/3}, \zeta_3)$. The Galois group is S_3 and there are three conjugacy classes:

$$C_1 = \{(1)\}$$
$$C_2 = \{(12), (23), (13)\}$$
$$C_3 = \{(123), (132)\}.$$

We have

$$\mathrm{Frob}_p = C_1 \Leftrightarrow p \equiv 1 \pmod{3} \text{ and } p = A^2 + 27B^2.$$

Also,

$$\mathrm{Frob}_p = C_2 \Leftrightarrow p \equiv 2 \pmod{3}$$

and

$$\mathrm{Frob}_p = C_3 \Leftrightarrow p \equiv 1 \pmod{3} \text{ and } p \neq A^2 + 27B^2.$$

Thus, in this case, distribution of Frobenius classes gives information about whether p can be expressed by a certain quadratic form.

The Chebotarev density theorem gives the asymptotic distribution of the primes $\mathfrak{p}$ with $\mathrm{Frob}_{\mathfrak{p}}$ equal to a given conjugacy class. If we set $\pi_C(x, F/K)$ to be the number of primes $\mathfrak{p}$ of K with $N_{K/\mathbb{Q}}(\mathfrak{p}) \leq x$, $\mathfrak{p}$ unramified in F and $\mathrm{Frob}_{\mathfrak{p}} = C$, then in its ineffective form, the theorem states that

$$\pi_C(x, F/K) \sim \frac{|C|}{|G|} \mathrm{Li}(x)$$

where $G = \mathrm{Gal}(F/K)$. Thus, if the finite group G is given the discrete measure, then Frobenius classes are equidistributed with respect to this measure. If C is a union of conjugacy classes C_i, set

$$\pi_C(x, F/K) = \sum \pi_{C_i}(x, F/K).$$

Then, we see that the above asymptotic holds for conjugacy sets (that is, unions of conjugacy classes) as well.

This has been made effective by Lagarias and Odlyzko [10]. If we assume the GRH, they showed that for any conjugacy set $C \subseteq G$,

$$\left| \pi_C(x, F/K) - \frac{|C|}{|G|} \mathrm{Li}(x) \right| \ll |C| n_K x^{1/2} (\log \mathcal{M}_{F/K} x)$$

where $n_K = [K : \mathbb{Q}]$ and $\mathcal{M}_{F/K}$ is a term involving the degree of F/K, the discriminant of K and the rational primes which lie below primes of K which are ramified in F. If we assume the GRH and Artin's conjecture (AC), Ram Murty, Saradha and the author [15] showed that

$$\left| \pi_C(x, F/K) - \frac{|C|}{|G|} \mathrm{Li}(x) \right| \ll |C|^{1/2} n_K x^{1/2} (\log \mathcal{M}_{F/K} x).$$

In fact, Artin's conjecture can be replaced by a weaker hypothesis involving the growth of Artin L-functions in $Re(s) > 1/2$.

Remark. Dirichlet's theorem raised a whole array of problems on the fine distribution of primes in arithmetic progressions. The Chebotarev density theorem leads to similar questions in the general case. Many of these are largely unexplored. A problem in which there has been some progress is the analogue of Linnik's theorem on the least prime in an arithmetic progression.

Estimates for the least prime with given Frobenius class[2] were given by Lagarias and Odlyzko [10] assuming the GRH. They showed that given any conjugacy class C, we can find a prime whose Frobenius class is C and whose norm is

$$\ll (\log d_F)^2.$$

Unconditionally, Lagarias, Odlyzko and Montgomery showed [9] that there is an absolute constant $c > 0$ such that the bound

$$\ll d_F^c$$

holds. Let us compare this with Linnik's famous theorem on the least prime in an arithmetic progression. This result states that if $F = \mathbb{Q}(\zeta_q)$, there is an absolute constant $c > 0$ such that in every class, there is a prime satisfying the bound

$$\ll (\log d_F)^c.$$

In the non-Abelian case, one expects that the bounds should depend on the size of the conjugacy class. As shown in [18, Theorem 3.1], the assumption of AC in addition to the GRH implies the bound

$$\ll \frac{n_K^2}{|C|} (n \log n + \log d_F)^2$$

---

[2]By abuse of terminology, we shall refer to this as the least prime in a conjugacy class.

where $n = [F : K]$. In the same article, it is conjectured [18, Conjecture 2.1], that there are constants $a, b > 0$ so that unconditionally the bound

$$\ll d_F^{a/n} (\log d_F)^b$$

should hold.

3   The Case of Infinite Extensions

Suppose now that F/K is of infinite degree. In this case, one has to introduce some topological and measure-theoretic notions. We know that $\mathrm{Gal}(F/K)$ is a compact, totally disconnected, Hausdorff topological group in which the basic open neighbourhoods of the identity are the closed subgroups of finite index. We have

$$\mathrm{Gal}(F/K) = \mathrm{inv}\lim \mathrm{Gal}(M/K)$$

where the inverse limit is over finite normal subextensions M/K. The topology described previously is the profinite topology. We can define a measure on G by the limit of the discrete measures on each finite subextension. Thus, if μ denotes the Haar measure on G which gives the whole group measure 1, and for M/K normal and finite, if μ_M denotes the discrete measure on $G_M = \mathrm{Gal}(M/K)$, then

$$\mu = \lim \mu_M$$

where the limit is taken in the sense of weak convergence.

A prime of F is now a system $\{\mathfrak{p}_M : M/F \text{ normal and finite}\}$, with $\mathfrak{p}_M$ a prime ideal of M, satisfying the compatibility that if there are two extensions $K \subset M_1 \subset M_2 \subset F$ with M_1/K and M_2/K both normal and of finite degree, then

$$\mathfrak{p}_{M_2}|_{M_1} = \mathfrak{p}_{M_1}.$$

(The notation on the left hand side indicates the intersection of $\mathfrak{p}_{M_2}$ with the ring of integers of M_1.) We can define a Frobenius element of $\mathrm{Gal}(F/K)$ as the sequence

$$\{\mathrm{Frob}_{\mathfrak{p}_M} : M/K \text{ normal and finite}\}.$$

We write $\mathrm{Frob}_{\mathfrak{q}}$ for the conjugacy class of all sequences with $\mathfrak{q} = \mathfrak{p}_K$. Frobenius elements are dense in the Galois group. However, note that as G is uncountable if it is infinite, and the set of Frobenius classes are countable, we cannot expect every element of G to be a Frobenius element.

Given a conjugacy class C of G, denote by $\pi_C(x, F/K)$ the number of primes $\mathfrak{p}$ of K with $N_{K/\mathbb{Q}}\mathfrak{p} \le x$ and satisfying $\mathrm{Frob}_{\mathfrak{p}} = C$. More generally, we consider conjugacy sets C. Serre [20] showed that if C is a conjugacy class whose boundary has measure zero, then

$$\pi_C(x, F/K) \sim \mu(C)\mathrm{Li}(x).$$

If C is a "small" conjugacy class, in the sense that $\mu(C) = 0$, then the main term is zero, so Serre's theorem does not give asymptotics about such primes. For example, if $C = \{1\}$, it only says that the set of primes of K which split completely in F has density zero. It does not tell us if there are infinitely many. If we consider $F = \mathbb{Q}(\mu_{p^\infty})$, then no rational prime splits completely in F. On the other hand, there are examples of Ihara [6] which show that one can have an arbitrarily large finite set of primes splitting completely.

In the study of Artin's primitive root conjecture, one considers a lattice of Kummer extensions (see the papers of Ram Murty [13], [14]). The conjugacy set in question is the one corresponding to primes which do not split completely at any field in the lattice.

A curious example involving a similar class in a tower of Abelian extensions occurs in the thesis of Ram Murty (see [13]). Let $p_1 = 3$ and for each j, let p_j be the smallest prime which is not congruent to 1 (mod p_i) for each $i < j$. For each i, let $M_i = \mathbb{Q}(\zeta_{p_i})$ and set $F = \cup M_i$. If the prime q doesn't split completely in any M_i then $q \not\equiv 1$ (mod p_i) for any i. Say $p_j < q \le p_{j+1}$. By definition of p_{j+1}, we must have $q = p_{j+1}$. Thus, the set of primes that do not split completely in any of the M_i is precisely the set $\{p_i, i = 1, 2, \cdots\}$. By a result of Erdös, the number of $p_i \le x$ is

$$\sim \frac{x}{(\log x)(\log \log x)}.$$

Erdös apparently studied these primes in response to a problem posed by Golomb. It is interesting to note that in this case, all the primes of interest ramify in the tower.

The question of ramification in an infinite tower is very interesting. It is known that in a $\mathbb{Z}_p$ extension, apart from a finite number of exceptions, only primes above p can ramify. Serre [20] has given an example of a Galois extension whose group is a reducible subgroup of $\mathrm{GL}_2(\mathbb{Z}_p)$ and which is everywhere ramified. The situation is quite different if the subgroup is irreducible. Indeed, Khare and Rajan [8] have recently shown that if one has a semisimple representation, then the set of primes which ramify has density zero. It would be very interesting to get an explicit upper bound on this set. On the other hand, Ravi Ramakrishna [19] has constructed examples of GL_2 extensions which are ramified at infinitely many primes.

In our discussion, we will only work with extensions F/K that are unramified outside a finite set of primes. Thus, we are excluding extensions such as

$$\mathbb{Q}\left(\sqrt{2}, \sqrt{3}, \cdots, \sqrt{p}, \cdots\right)$$

which are obtained from $\mathbb{Q}$ by adjoining the square root of each prime. By comparison, in the work of Ihara [6], he considers extensions which are unramified or at most tamely ramified at a finite set of primes.

In the remainder of this paper, we shall consider conjugacy classes which have measure zero.

4 GL_2 Extensions

It is possible to conjecture what happens for some GL_2 extensions. Let f be a normalized eigenform of weight $k > 1$, level N and character ϵ. (See Lang [11] for background information on modular forms.) The Fourier coefficients of f lie in some number field E_f (say). For each prime λ of E_f, there is associated a λ-adic representation

$$\rho_{f,\lambda}\colon \mathrm{Gal}(\overline{\mathbb{Q}}/\mathbb{Q}) \longrightarrow \mathrm{GL}_2(\mathcal{O}_\lambda)$$

with the property that it is unramified outside ℓN and for p unramified, the characteristic polynomial of Frob_p is

$$X^2 - a_f(p)X + \epsilon(p)p^{k-1}.$$

Let us set

$$\pi_f(x, a) = \#\{p \le x\colon a_f(p) = a\}$$

If F is the fixed field of the kernel of $\rho_{f,\lambda}$, it is an infinite Galois extension of $\mathbb{Q}$ and $\pi_f(x, a)$ is the same as $\pi_C(x, F/\mathbb{Q})$ where C is the conjugacy set consisting of elements whose image under $\rho_{f,\lambda}$ has trace a.

If the "stable trace field" of f is $\mathbb{Q}$ and the weight $k = 2$,then the author [17] (and independently Bayer and Gonzalez i Rovira [1]) conjectured that

$$\pi_f(x, a) \sim (c_f(a) + \mathbf{o}(1))\sqrt{x}/\log x.$$

If f has integral Fourier coefficients, then this was conjectured by Lang and Trotter in 1976 [12].

What is known? In terms of theoretical results, one has not even been able to show that $\pi_f(x, a) \ll x^{1/2}$. The sharpest result to date is due to Ram Murty, Saradha, and the author who showed [15] that the GRH implies

$$\pi_f(x, a) \ll x^{4/5}.$$

In [16], it is shown that unconditionally, one has the estimate

$$\pi_f(x, a) \ll \frac{x(\log\log x)^2}{(\log x)^2}.$$

There does seem to be some hope for proving unconditionally a bound of the form

$$\pi_f(x, a) \ll x^{1-\alpha}$$

for some $\alpha > 0$. If f corresponds to an elliptic curve, then Elkies, Kaneko, and Ram Murty [4] showed that

$$\pi_f(x, 0) \ll x^{3/4}.$$

Fouvry and Ram Murty [5] showed that unconditionally

$$\pi_f(x, 0) \gg (\log\log\log x)^{1-\epsilon}$$

holds. They also showed that

$$\pi_f(x, 0) = \Omega(\log\log x).$$

Surprisingly, not much numerical work seems to have been done on the growth of $\pi_f(x, a)$. Recently, Kevin James, Cyrus Mehta and the author [7] looked at two cases. The first is the form

$$f = (\eta(z)\eta(11z))^2.$$

This form corresponds to the elliptic curve $X_0(11)$ which has equation

$$y^2 + y = x^3 - x^2 - 10x - 20$$

with discriminant $\Delta = -11^5$ and invariant $j = -2^{12}(31)^3/11^5$. The second is a form g which is of level 512, weight 2 and with Fourier coefficients in the field $\mathbb{Q}(\sqrt{2})$. It is a normalized eigenform with an inner twist in the sense that

$$g^\sigma = g \otimes \chi_4$$

where $1 \neq \sigma \in \text{Gal}(\mathbb{Q}(\sqrt{2})/\mathbb{Q})$ and χ_4 is the non-trivial character modulo 4. Thus, the stable trace field is $\mathbb{Q}$. In both cases, the form can be expressed explicitly in terms of theta functions and this makes them particularly amenable to calculation. In each case, we computed the first 67 million Fourier coefficients. The data, which will be presented in detail in [7] seems to support the conjectures. Looking at the numbers, one can speculate about the error terms as well. Indeed, there is some very small numerical data that suggests that the estimate

$$\pi_f(x, a) - c_f(a)\mathrm{Li}(\sqrt{x}) = \mathbf{o}(x^{1/4}(\log\log x)^{1/2}/(\log x)^{1/2})$$

might hold.

For general modular forms, it is possible to make a conjecture of a new kind. Let M be a subfield of $E_{f,st}$ (the stable field of Fourier coefficients). Suppose that the form is of weight ≥ 2 and not of CM-type. Then [17, Conjecture 3.4] states the following:

Conjecture 1. *There is a constant $c_{f,M}$ such that*

$$\#\{p \leq x \colon a_f(p) \in M\}$$

is asymptotic to

$$(1+o(1))c_{f,M}\begin{cases} x/\log x & \text{if } M = E_{f,st} \\ \sqrt{x}/\log x & \text{if } k = 2, M = \mathbb{Q} \text{ and } [E_{f,st} \colon \mathbb{Q}] = 2 \\ \log\log x & \text{if } k = 2 \text{ and } [M \colon \mathbb{Q}] = [E_{f,st} \colon M] = 2, \\ & \quad \text{or } k = 2 \text{ and } M = \mathbb{Q}, [E_{f,st} \colon M] = 3, \\ & \quad \text{or } k = 3 \text{ and } M = \mathbb{Q}, [E_{f,st} \colon \mathbb{Q}] = 2 \\ 1 & \text{otherwise.} \end{cases}$$

By the same method as in the proof of [17, Theorem 3.6], we have the following result.

Theorem 1. *Assume the GRH. Then*

$$\#\{p \leq x \colon \mathbb{Q}(a_f(p)) \subset E_{f,st}\} \ll x^{7/8}.$$

5 An Approach Using L-Functions

The constant $c_f(a)$ is prescribed using local information about the Galois representation attached to f, in much the similar way as the constant in the Hardy-Littlewood "Partitio Numerorum" conjectures. Does there exist an L-function formalism to "explain" the conjecture? This is one of the motivating questions of this article.

In the spirit of the works of Taniyama [22] and Ihara [6], we attempt to introduce an L-function formalism to "explain" the conjectures of Lang-Trotter type. In doing so, we propose an alternate expression for the constant $c_f(a)$ which depends on zeros of certain Artin L-functions. In some sense, it is dual to the expression in terms of local densities. It seems to support an idea expressed by Ram Murty that there should exist a new kind of explicit formula in this context.

Consider an infinite Galois extension F/K which is unramified outside a finite set of primes S.

Let $G = \mathrm{Gal}(F/K)$ and let C be a conjugacy set. Set

$$Z(F, C, s) = \sum_{(\mathfrak{p}^m, F/K) = C} \frac{\log N\mathfrak{p}}{(N\mathfrak{p})^{ms}}$$

where in the sum, $\mathfrak{p}$ is a prime ideal of K.

By analogy with Serre [21], we say that the Minkowski dimension of C (written $\dim_M C$) is $\leq \alpha$ if for each intermediate extension

$$K \subseteq M \subseteq F$$

with M/K normal and finite, the projection C_M of C under the quotient map

$$G = \mathrm{Gal}(F/K) \longrightarrow \mathrm{Gal}(M/K) = G_M$$

satisfies

$$|C_M| \ll |G_M|^{\alpha} .$$

We have the following easy result. (In all that follows, by the GRH, we shall mean the Riemann Hypothesis for all Dedekind zeta functions.)

Proposition 1. *Assume the GRH and suppose that* $\dim_M C \leq \alpha$. *Then* $Z(F, C, s)$ *has an analytic continuation to the half plane* $\mathrm{Re}(s) > \frac{1+\alpha}{2}$. *If we assume in addition that AC holds, then this can be extended to* $\mathrm{Re}(s) > \frac{1}{2} + \frac{\alpha}{2(2-\alpha)}$.

Proof. Set

$$\psi(F, C, x) = \sum_{\substack{N\mathfrak{p}^m \leq x \\ (\mathfrak{p}^m, F/K) = C}} \log N\mathfrak{p}.$$

Then, in an obvious notation, for all finite M (that is, $K \subseteq M \subseteq F$ with M/K normal and of finite degree), we have the inequality

$$\psi(F, C, s) \leq \psi(M, C_M, x).$$

By the effective Chebotarev density theorem,

$$\psi(M, C_M, x) = \frac{|C_M|}{|G_M|} x + \mathbf{O}\left(|C_M| n_K x^{\frac{1}{2}} (\log \mathfrak{M}_{M/K} x)(\log x)\right)$$

where
$$\log \mathcal{M}_{M/K} = \log[M:K] + \frac{1}{n_K}\log d_K + \sum_{p\in\mathcal{P}(M/K)}\log p$$

and $\mathcal{P}(M/K)$ is the set of rational primes which lie below primes of K ramified in M. As we are assuming that F/K is unramified outside the set S, the last term is bounded in terms of S. Choose M so that

$$|G_M| \asymp \frac{x^{\frac{1}{2}}}{(\log x)^2}.$$

The symbol means that

$$\frac{x^{\frac{1}{2}}}{(\log x)^2} \ll_{K,S} |G_M| \ll_{K,S} \frac{x^{\frac{1}{2}}}{(\log x)^2}$$

with the implied constants depending on K and S. Then

$$\log \mathcal{M}_{M/K} \ll_{K,S} \log x$$

and
$$\psi(F,C,x) \ll \frac{1}{|G_M|^{1-\alpha}}x + \mathbf{O}_{K,S}\left(|G_M|^\alpha\, x^{\frac{1}{2}}(\log x)^2\right).$$

We see that this is
$$\ll_{K,S} x^{\frac{1+\alpha}{2}}(\log x)^{2-2\alpha}.$$

Hence, the integral
$$\int_1^\infty \frac{\psi(F,C,x)}{x^{s+1}}dx$$

converges absolutely for $Re(s) > (1+\alpha)/2$. Since

$$Z(F,C,s) = s\int_1^\infty \frac{\psi(F,C,x)}{x^{s+1}}dx,$$

the result is proved. If we assume AC, the result follows on using the estimate

$$\psi(M,C_M,x) = \frac{|C_M|}{|G_M|}x + \mathbf{O}_{K,S}\left(|C_M|^{\frac{1}{2}}\,x^{\frac{1}{2}}(\log x)^2\right)$$

of [15]. $\qquad\qquad\qquad\qquad\qquad\qquad\qquad\qquad\qquad\qquad\qquad\square$

It is natural to ask whether $Z(F,C,s)$ has a continuation to a wider region. Since $Z(F,C,s)$ is a Dirichlet series with non-negative coefficients,

Landau's theorem implies that it has a singularity at the real point on the abscissa of convergence. If this abscissa is $\sigma = \sigma_0$, then

$$\psi(F, C, x) = \Omega(x^{\sigma_0}).$$

We consider the following hypothesis.

Hypothesis 1. $Z(F, C, s)$ *has a continuation as a meromorphic function in the half plane* $\mathrm{Re}(s) \geq 1/2$. *It is analytic there apart from a simple pole at* $s = 1/2$.

Assuming this hypothesis, denote by $\rho(F, C)$ the residue of $Z(F, C, s)$ at $s = \frac{1}{2}$. Then we see that

$$\psi(F, C, x) \sim \rho(F, C)x^{\frac{1}{2}}.$$

By partial summation, it also follows that

$$\pi(F, C, x) \sim \rho(F, C)\frac{x^{\frac{1}{2}}}{\log x}.$$

In particular, if $K = \mathbb{Q}$ and F is a GL_2 extension arising from a modular form f as in the previous section, and C is the conjugacy set consisting of elements of a fixed trace a, then we must have $\rho(F, C) = c_f(a)$.

6 An Expression for $\rho(\mathbf{F}, \mathbf{C})$ in Terms of Zeros

Now we shall derive a formula for the residue in terms of zeros of Artin L-functions. For each finite M/K, set

$$Z(M, C_M, s) = \sum_{\substack{(\mathfrak{p}^m, M/K) \subseteq C_M \\ p(\mathfrak{p}) \notin S}} \frac{\log N\mathfrak{p}}{(N\mathfrak{p})^{ms}}$$

where the sum ranges over primes $\mathfrak{p}$ of K which do not lie over primes in S and which have the specified Frobenius class. In terms of Artin L-functions, we have the expression

$$Y(M, C_M, s) = -\sum_{D \subseteq C_M} \frac{|D|}{|G_M|} \sum_{\chi \in \mathrm{Irr}(G_M)} \bar{\chi}(D)\frac{L'}{L}(s, \chi)$$

where the outer sum is over conjugacy classes of G_M contained in C_M. In fact, $Z(M, C_M, s)$ and $Y(M, C_M, s)$ agree apart from ramified terms, and we have

$$Z(M, C_M, s) = Y(M, C_M, s) + \sum_{D \subseteq C_M} \frac{|D|}{|G_M|} \sum_{\chi \in \mathrm{Irr}(G_M)} \bar{\chi}(D) \sum_{\substack{\mathfrak{p}, m \\ p(\mathfrak{p}) \in S}} \frac{\chi(\sigma_\mathfrak{p}^m)}{(N\mathfrak{p})^{ms}}.$$

Set also
$$Z^0(M, C_M, s) = Z(M, C_M, s) - \frac{|C_M|}{|G_M|} \frac{1}{s-1}$$

and
$$\psi^0(M, C_M, x) = \psi(M, C_M, x) - \frac{|C_M|}{|G_M|} x.$$

Let us suppose that
$$\dim_M C \leq \alpha < 1.$$

Then
$$\frac{|C_M|}{|G_M|} \to 0$$

as $M \to F$. Thus, for $Re(s) > 1$, we certainly have
$$Z(F, C, s) = \lim_{M \to F} Z^0(M, C_M, s).$$

Hypothesis 2. *For $Re(s) > 1/2$, we have*
$$Z(F, C, s) = \lim_{M \to F} Z^0(M, C_M, s).$$

Assuming a stability hypothesis on the extension F/K, we can prove that the above holds for $Re(s) > \frac{1}{2}(1 + \alpha)$. The stability hypothesis is as follows: given x, there exists a finite normal extension M/K with $K \subseteq M \subseteq F$ and so that for all primes $\mathfrak{p}$ of K with $N_{K/\mathbb{Q}}(\mathfrak{p}) \leq x$, if $(\mathfrak{p}, M/K) \subseteq C_M$, then $(\mathfrak{p}, F/K) = C$.

Consider a finite extension M/K. For a character χ of G_M, denote by $\mathfrak{f}_\chi$ the Artin conductor χ. It is an ideal of K. Denote by V_χ the underlying space of χ. For each infinite real place v of K, let $\chi_v^{\pm}(1)$ be the dimension of the subspace of V_χ on which σ_v acts by ± 1. Set
$$a_\chi = \sum \chi_v^+(1)$$

and
$$b_\chi = \sum \chi_v^-(1).$$

To count the contribution of complex places, set
$$c_\chi = \chi(1) r_2(K)$$

where $r_2(K)$ denotes the number of complex places of K. Set
$$A_\chi = d_K^{\chi(1)} N_{K/\mathbb{Q}}(\mathfrak{f}_\chi)$$

and

$$\Gamma(s,\chi) = (\pi^{-s/2}\Gamma(s/2))^{a_\chi}(\pi^{-(s+1)/2}\Gamma((s+1)/2))^{b_\chi}((2\pi)^{-s}\Gamma(s))^{c_\chi}.$$

Set as usual

$$\Lambda(s,\chi) = A_\chi^{s/2}\Gamma(s,\chi)L(s,\chi).$$

Then we have the functional equation

$$\Lambda(s,\chi) = w_\chi\Lambda(1-s,\overline{\chi})$$

and also the Hadamard factorization into zeros

$$\Lambda(s,\chi) = (s(s-1))^{-\delta_\chi}\exp\{\alpha_\chi + \beta_\chi s\}\prod_\rho\left(1-\frac{s}{\rho}\right)e^{s/\rho}$$

where the product is over non-trivial zeros of $L(s,\chi)$ and

$$\delta_\chi = \begin{cases} 1 & \text{if } \chi \text{ is the trivial character} \\ 0 & \text{otherwise} \end{cases}.$$

Logarithmically differentiating, we get

$$\frac{L'}{L}(s,\chi) = \sum_\rho\frac{1}{s-\rho} - \delta_\chi\left(\frac{1}{s} + \frac{1}{s-1}\right) - \frac{1}{2}\log A_\chi - \frac{\Gamma'}{\Gamma}(s,\chi).$$

Let D be a conjugacy class in $G_M = \text{Gal}(M/K)$. For simplicity of notation, we shall denote G_M by G until further notice. Then, summing over irreducible characters χ of G, we have

$$-\frac{|D|}{|G|}\sum\overline{\chi}(D)\frac{L'}{L}(s,\chi)$$

is equal to

$$-\frac{|D|}{|G|}\sum\overline{\chi}(D)\sum_{\rho_\chi}\frac{1}{s-\rho_\chi} + \frac{|D|}{|G|}\left(\frac{1}{s} + \frac{1}{s-1}\right)$$

$$+\frac{1}{2}\frac{|D|}{|G|}\sum\overline{\chi}(D)\log A_\chi + \frac{|D|}{|G|}\sum\overline{\chi}(D)\frac{\Gamma'}{\Gamma}(s,\chi).$$

To estimate the last two terms in the above, we have the following result.

Proposition 2. *We have*

$$\frac{|D|}{|G|}\sum_{\chi\in\text{Irr}(G)}\overline{\chi}(D)\log A_\chi \ll \log d_K + \sum_{v\in S(M/K)}\log Nv$$

where $S(M/K)$ is the set of primes of K ramified in M.

Proof. By definition

$$\log A_\chi = \chi(1) \log d_K + \log N\mathfrak{f}_\chi.$$

We have

$$\sum_\chi \overline{\chi}(D)\chi(1) = \begin{cases} 0 & \text{if } D \neq \{1\} \\ |G| & \text{if } D = \{1\}. \end{cases}$$

Hence

$$\frac{|D|}{|G|} \sum_\chi \overline{\chi}(D)\chi(1) \log d_K = \begin{cases} 0 & \text{if } D \neq \{1\} \\ \log d_K & \text{if } D = \{1\}. \end{cases}$$

Again, by definition

$$\log N\mathfrak{f}_\chi = \sum_{v \in S(M/K)} n(\chi, v) \log Nv.$$

The integer $n(\chi, v)$ is given by the formula

$$n(\chi, v) = \sum_{i=0}^\infty \frac{\left|G_v^{(i)}\right|}{\left|G_v^{(0)}\right|} \left\{ \chi(1) - \frac{1}{\left|G_v^{(i)}\right|} \sum_{a \in G_v^{(i)}} \chi(a) \right\}.$$

Here, $G_v^{(0)}$ denotes the inertia group at v and $G_v^{(i)}$ is the $i-th$ ramification group. Then, for a fixed $v \in S$, we have

$$\sum_\chi \overline{\chi}(D)n(\chi, v) = \sum_{i=0}^\infty \frac{\left|G_v^{(i)}\right|}{\left|G_v^{(0)}\right|} \left\{ \sum_\chi \overline{\chi}(D)\chi(1) - \frac{1}{\left|G_v^{(i)}\right|} \sum_{a \in G_v^{(i)}} \sum_\chi \overline{\chi}(D)\chi(a) \right\}.$$

Let us suppose that $D \neq \{1\}$. Then the first sum over χ on the right is zero. Thus,

$$\sum_\chi \overline{\chi}(D)n(\chi, v) = -\sum_{i=0}^\infty \frac{1}{\left|G_v^{(0)}\right|} \sum_{a \in G_v^{(i)} \cap D} \frac{|G|}{|D|} = -\frac{|G|}{|D|} \sum_{i=0}^\infty \frac{\left|G_v^{(i)} \cap D\right|}{\left|G_v^{(0)}\right|}.$$

Putting all of this together, we get for $D \neq \{1\}$,

$$\frac{|D|}{|G|} \sum_\chi \overline{\chi}(D) \log N\mathfrak{f}_\chi = -\sum_{v \in S(M/K)} (\log Nv) \sum_{i=0}^\infty \frac{\left|G_v^{(i)} \cap D\right|}{\left|G_v^{(0)}\right|}.$$

Now, we know that

$$\sum_{i=0}^{\infty} \left(\left| G_v^{(i)} \right| - 1 \right) = \mathrm{ord}_w \mathfrak{D}$$

where $\mathfrak{D}$ is the different of M/K at w. Thus, we have the crude bound

$$\sum_{i=0}^{\infty} \frac{\left| G_v^{(i)} \cap D \right|}{\left| G_v^{(0)} \right|} \leq \frac{\mathrm{ord}_w \mathfrak{D}}{\left| G_v^{(0)} \right|} \leq 2.$$

Thus, if $D \neq \{1\}$,

$$\frac{|D|}{|G|} \sum_{\chi} \overline{\chi}(D) \log A_\chi \leq 2 \sum_{v \in S(M/K)} \log Nv.$$

If $D = \{1\}$, then

$$\frac{|D|}{|G|} \sum_{\chi} \overline{\chi}(D) \log A_\chi = \frac{1}{|G|} \log d_M.$$

By Hensel's bound,

$$\log d_M \leq |G| \left\{ \log d_K + \sum_{v \in S(M/K)} \log Nv \right\}.$$

Hence, in all cases, the bound holds and this proves the result. □

By a similar method, we can show that the following holds.

Proposition 3. *We have*

$$\frac{|D|}{|G|} \sum_{\chi \in \mathrm{Irr}(G)} \overline{\chi}(D) \frac{\Gamma'}{\Gamma}(s, \chi) \ll n_K \left| \log \pi + \frac{\Gamma'}{\Gamma}(s) \right|.$$

Proof. Summing over real primes v of K, we see that

$$\sum_{\chi} \overline{\chi}(D) \sum_{v} \chi_v^{\pm}(1) = \frac{1}{2} \sum_{v} \sum_{\chi} \overline{\chi}(D) \left(\chi(1) + \chi(\sigma_v) \right)$$

where σ_v denotes Frobenius at v. The above sum is clearly

$$\leq \frac{1}{2} \frac{|G|}{|D|} r_1(K)$$

where $r_1(K)$ denotes the number of real places of K. Thus,

$$\frac{|D|}{|G|} \sum_\chi \overline{\chi}(D) \left\{ a_\chi \left(-\frac{1}{2} \log \pi + \frac{1}{2} \frac{\Gamma'}{\Gamma} \left(\frac{s}{2} \right) \right) \right.$$

$$\left. + b_\chi \left(-\frac{1}{2} \log \pi + \frac{1}{2} \frac{\Gamma'}{\Gamma} \left(\frac{s+1}{2} \right) \right) \right\}$$

$$\ll r_1(K) \left| \log \pi + \frac{\Gamma'}{\Gamma}(s) \right|.$$

The calculation is similar for the complex places. (In fact, unless $D = \{1\}$, the average over the complex places will be zero.) $\square$

The calculations above, together with the GRH and AC imply that for $\sigma > \frac{1}{2}$, we have

$$Z^0(M, C_M, s) = -\sum_{D \subseteq C_M} \frac{|D|}{|G_M|} \sum_\chi \overline{\chi}(D) \sum_{\rho_\chi} \frac{1}{s - \rho_\chi} + \frac{|C_M|}{|G_M|} \frac{1}{s}$$

$$+ \sum_{D \subseteq C_M} \frac{|D|}{|G_M|} \sum_\chi \overline{\chi}(D) \sum_{\substack{\mathfrak{p}, m \\ p(\mathfrak{p}) \in S}} \frac{\chi(\sigma_\mathfrak{p}^m)}{(N\mathfrak{p})^{ms}}$$

$$+ \mathbf{O}_{K, S}(\log(|s| + 2)).$$

Note that for $Re(s) > 0$, we have

$$\left| \sum_{D \subseteq C_M} \frac{|D|}{|G_M|} \sum_\chi \overline{\chi}(D) \sum_{\substack{\mathfrak{p}, m \\ p(\mathfrak{p}) \in S}} \frac{\chi(\sigma_\mathfrak{p}^m)}{(N\mathfrak{p})^{ms}} \right| \leq \sum_{\substack{\mathfrak{p}, m \\ p(\mathfrak{p}) \in S}} \frac{1}{(N\mathfrak{p})^{m\sigma}}.$$

Notice that the sum is over primes of K. Thus, the right hand side above is bounded as $M \to F$. Thus,

$$Z^0(M, C_M, s) + \sum_{D \subseteq C_M} \frac{|D|}{|G_M|} \sum_\chi \overline{\chi}(D) \sum_{\rho_\chi} \frac{1}{s - \rho_\chi}$$

is holomorphic for $Re(s) > 0$ and if

$$\lim_{M \to F} \left\{ Z^0(M, C_M, s) + \sum_{D \subseteq C_M} \frac{|D|}{|G_M|} \sum_\chi \overline{\chi}(D) \sum_{\rho_\chi} \frac{1}{s - \rho_\chi} \right\}$$

exists, then it defines a function which is holomorphic for $\sigma > 0$. Now, *assuming* that we are allowed to remove the parentheses, we would have

$$Z(F, C, s) + \lim_{M \to F} \sum_{D \subseteq C_M} \frac{|D|}{|G_M|} \sum_{\chi} \overline{\chi}(D) \sum_{\rho_\chi} \frac{1}{s - \rho_\chi}$$

is holomorphic for $\sigma > 0$. Thus, we are led to state the following.

Conjecture 2. *We have*

$$\rho(F, C) = -\lim_{\sigma \to \frac{1}{2}} \left(\sigma - \frac{1}{2} \right) \lim_{M \to F} \sum_{D \subseteq C_M} \frac{|D|}{|G_M|} \sum_{\chi} \overline{\chi}(D) \sum_{\rho_\chi} \frac{1}{\sigma - \rho_\chi}.$$

We expect that the contribution of zeros ρ_χ whose imaginary part is bounded away from zero is negligible. Thus, we expect that in order for $Z(F, C, s)$ to have a pole at $s = \frac{1}{2}$, the zeta functions $\zeta_M(s)$ at finite layers must acquire zeros near $s = \frac{1}{2}$ as $M \to F$.

For each finite layer M, consider the set $Z_M(\chi, C)$ of L-functions $L(s, \chi)$ with $\chi \in \mathrm{Irr}(G_M)$ and $\chi(C) \neq 0$. Define

$$\delta_M = \min |\gamma|$$

where γ is the imaginary part of some L-function in $Z_M(\chi, C)$. Define

$$\delta_F = \liminf \delta_M.$$

The above suggests that if $\delta_F > 0$, then

$$\pi(F, C, x) = \mathrm{o}(\sqrt{x}/\log x).$$

This then raises the following question. In the case of a GL_2 extension F which arises from an elliptic curve, consistency with the Lang-Trotter conjecture would require that if $\delta_F > 0$ then $c_f(a) = 0$. We close with the question of whether this can be verified even in special cases.

References

[1] P. Bayer and G. Rovira, *On the Hasse-Witt invariants of modular curves*, Exp. Math. **6** (1997), 57–76.

[2] H. Davenport, *Multiplicative number theory*, 2nd ed., Springer Verlag.

[3] N. Elkies, *The existence of infinitely many supersingular primes for every elliptic curve over* $\mathbb{Q}$, Invent. Math. **89** (1987), 561–568.

[4] _____, *Distribution of supersingular primes*, Journées Arithmetiques, (Luminy 1989), Asterisque, 1992, pp. 127–132.

[5] E. Fouvry and M. Ram Murty, *On the distribution of supersingular primes*, Canadian J. Math. **48** (1996), 81–104.

[6] Y. Ihara, *How many primes decompose completely in an infinite unramified Galois extension of a global field*, J. Math. Soc. Japan **35** (1983), 693–709.

[7] K. James, C. Mehta, and V. Kumar Murty, *Frobenius distributions and Galois representations II: Numerical evidence*, in preparation.

[8] C. Khare and C. S. Rajan, *The density of ramified primes in semisimple p-adic Galois representations*, Intl. Math. Res. Notices **12** (2001), 601–607.

[9] J. Lagarias, H. Montgomery, and A. Odlyzko, *A bound for the least prime ideal in the Chebotarev density theorem*, Invent. Math. **54** (1979), 271–296.

[10] J. Lagarias and A. Odlyzko, *Effective versions of the Chebotarev density theorem*, Algebraic Number Fields (A. Fröhlich, ed.), Academic Press, New York, 1977, pp. 409–464.

[11] S. Lang, *Introduction to modular forms*, Springer, New York, 1976.

[12] S. Lang and H. Trotter, *Frobenius distributions in* GL_2 *extensions*, Lecture Notes in Mathematics, vol. 504, Springer, Heidelberg, 1976.

[13] M. Ram Murty, *On Artin's conjecture*, J. Number Theory **16** (1983), 147–167.

[14] _____, *Artin's conjecture and elliptic analogues*, Sieve Methods, Exponential Sums and their Applications (G. R. H. Greaves, G. Harmon, and M. N. Huxley, eds.), Cambridge University Press, Cambridge, 1996, pp. 325–344.

[15] M. Ram Murty, V. Kumar Murty, and N. Saradha, *Modular forms and the Chebotarev density theorem*, Amer. J. Math. **110** (1988), 253–281.

[16] V. Kumar Murty, *Modular forms and the Chebotarev density theorem II*, Analytic number theory (Y. Motohashi, ed.), London Mathematical Society Lecture Notes 247, Cambridge University Press, Cambridge, 1997, pp. 287–308.

[17] ———, *Frobenius distributions and Galois representations*, Automorphic forms, Automorphic representations, and Arithmetic (R. Doran, Z.-L. Dou, and G. Gilbert, eds.), Proc. Symp. Pure Math., vol. 66, part 1, American Math. Soc., Providence, 1999, pp. 193–211.

[18] ———, *The least prime in a conjugacy class*, C. R. Math. Rep. Acad. Sci. Canada **22** (2000), 129–146.

[19] R. Ramakrishna, *Infinitely ramified Galois representations*, Annals Math. **151** (2000), 793–815.

[20] J.-P. Serre, *Abelian ℓ-adic representations and elliptic curves*, Benjamin, New York, 1968.

[21] ———, *Quelques applications du théorème de densité de Chebotarev*, Publ. Math. IHES **54** (1981), 123–201.

[22] Y. Taniyama, *L-functions of number fields and zeta functions of abelian varieties*, J. Math. Soc. Japan **9** (1957), 330–366.

The ABC Conjecture and Prime Divisors of the Lucas and Lehmer Sequences

Ram Murty[1] and Siman Wong

1 Introduction

A straightforward consequence of Thue's pioneering work on Diophantine approximation [26] is: Let m be a non-zero integer and let $f \in \mathbf{Z}[x,y]$ be a binary form. Then the equation $f(x,y) = m$ has finitely many integer solutions, unless f is the multiple of either a power of a linear form, or a power of a binary quadratic form with positive nonsquare discriminant. Thue's techniques were refined by Pólya [17] and Siegel [20] to show that if $f \in \mathbf{Z}[x]$ has at least two distinct roots, then

$$P(f(x)) \to \infty \tag{1.1}$$

as $|x| \to \infty$, where $P(a)$ denotes the largest prime divisor of a non-zero integer a. Using the Gelfond-Baker method, Shorey, van der Poorten, Tijdeman and Schinzel [19] showed that if $f \in \mathbf{Z}[x,y]$ is a binary form with at least three distinct linear factors (over $\mathbf{C}$), then for any positive integer d and all pairs of integers x, y with $(x,y) = d$ and $\max(|x|,|y|) > e$,

$$P(f(x,y)) \gg_{f,d} \log\log\max(|x|,|y|). \tag{1.2}$$

with an effective constant. Using the so-called sharpening of Baker ([2]–[4]), in the case of binomials, Stewart [22] improved this to

$$P(ax^n - by^n) \gg_{a,b} \sqrt{n/\log n}.$$

Furthermore, if we let n run through the set S_κ of integers n with at most $\kappa \log\log n$ distinct prime divisors ($0 < \kappa < 1/\log 2$, fixed), Stewart ([23], [24]) showed that there exists an effective constant $C(\kappa, a, b) > 0$ such that

[1]Ram Murty's research was supported in part by a Killam Research Fellowship and Bankers Trust Company Foundation by a grant to the Institute for Advanced Study. The first author also thanks Brown University where this work was initiated and the Institute for Advanced Study where it was completed.

$$\frac{P(a^n - b^n)}{n} > C(\kappa, a, b)\frac{\log^{1-\kappa \log 2} n}{\log \log \log n} \tag{1.3}$$

(note that the set S_κ has density 1 if $\kappa > 1$). There are also similar results when n is restricted to the set of primes. In 1962, Erdös [5, p. 218] conjectured that

$$\lim_{n \to \infty} \frac{P(2^n - 1)}{n} = \infty. \tag{1.4}$$

While heuristic arguments (see section 4 below) suggest the stronger result

$$P(a^n - 1) \overset{?}{\gg} a^{n^{1-\epsilon}}, \tag{1.5}$$

currently we do not even know that $P(a^n - b^n) > C(a, b)n^\theta$ for some $\theta > 1$. For more information on the history of these problems, see [5] and the introduction in Shorey et. al. [19].

Let $Q(n)$ denote the largest prime power divisor of n. Clearly, $Q(n) \geq P(n)$ and so a consequence of the conjecture of Erdös is that $Q(2^n - 1)/n \to \infty$ as $n \to \infty$. In this paper, we prove this. In fact, we can prove a more general result:

Theorem 1. *For any $\epsilon > 0$ and any integers $a > b > 0$, we have $Q(a^n - b^n) \gg n^{2-\epsilon}$, where the implied constant depends on a, b and ϵ.*

One can sharpen the theorem so that

$$Q(a^n - b^n) \gg n^{2-c/\log \log n}$$

for a suitable constant $c > 0$. Most likely, $Q(2^n - 1) = P(2^n - 1)$ but we are unable to prove this. If we write $2^n - 1 = u_n v_n$ with u_n squarefree, v_n squarefull and $(u_n, v_n) = 1$, then a result of Silverman [21] states that u_n is "large" under the ABC conjecture. We therefore apply the ABC conjecture (see section 2) to resolve Erdös conjecture (1.4), in this way. More precisely, we prove assuming ABC that for any $\epsilon > 0$,

$$P(2^n - 1) \gg n^{2-\epsilon},$$

for n sufficiently large. Under the same hypothesis, we deduce sharper forms of (1.1) and (1.2) above. A key role is played by the Brun-Titchmarsh theorem on primes in arithmetic progressions which is a familiar theorem in sieve theory. More generally, we prove:

Theorem 2. *Assume the ABC conjecture. For any $\epsilon > 0$ and any integers $a > b > 0$, we have*

$$P(a^n - b^n) \gg n^{2-\epsilon},$$

where the implied constant depends on a, b and ϵ.

Recall that if α, β are algebraic integers such that $\alpha + \beta$ and $\alpha\beta$ are coprime, non-zero rational integers and that α/β is not a root of unity, then the n-th Lucas number with respect to α, β is defined to be

$$t_n = (\alpha^n - \beta^n)/(\alpha - \beta)$$

Now, if α, β are algebraic integers subjected to the weaker conditions that $(\alpha + \beta)^2$ and $\alpha\beta$ are coprime, non-zero rational integers and that α/β is not a root of unity, then the n-th Lehmer number with respect to α, β is defined to be

$$u_n = (\alpha^n - \beta^n)/(\alpha^{\delta_n} - \beta^{\delta_n}),$$

where $\delta_n = 1$ if n is odd, and $\delta_n = 2$ if n is even. These sequences arise naturally in the study of primality testing [27]. Using the same techniques as in ([23], [24]), Stewart proved that for almost all integers n and κ as in (1.3),

$$P(u_n), P(t_n) \gg_{\alpha,\beta,\kappa} C \frac{\log^{1-\kappa \log 2} n}{\log \log \log n} n \tag{1.6}$$

Using Frey's refined ABC conjecture for number fields (cf. section 2) and the Brun-Titchmarsh theorem for number fields [11], the same argument for Theorem 2 yields a corresponding improvement of (1.6) (we will omit the proof).

Theorem 3. *Let α, β be non-zero algebraic integers in a number field K such that α/β is not a root of unity. Fix an integer $\delta > 0$. Under Frey's refined ABC conjecture, for any $\epsilon > 0$, if n is sufficiently large (depending on α, β and δ), then*

$$P\left(\mathrm{Norm}_{K/\mathbf{Q}}\left(\frac{\alpha^n - \beta^n}{\alpha^\delta - \beta^\delta}\right)\right) > n^{2-\epsilon}. \qquad \square$$

Remark 1. As the referee points out, Theorem 1 can be readily generalized to more general sequences such as those given in Theorem 3. Cf. also the recent results of Ribenboim and Walsh [18] along these lines.

For prime divisors of polynomial values we have the following strengthening of (1.1).

Theorem 4. *Assume the ABC conjecture. Let $f(x) \in \mathbf{Z}[x]$ be a non-constant polynomial with no repeated roots. Then for any $\epsilon > 0$ and $n \gg_{\epsilon,f} 1$, we have the lower bound*

$$P(f(n)) > (\deg(f) - 1 - \epsilon) \log n.$$

Erdös [5, p. 218] also raised the question of studying $P(n! + 1)$. The best results to date [6] are that $P(n! + 1) > n + (1 - o(1))\frac{\log n}{\log \log n}$, and that $\limsup_{n\to\infty} P(n! + 1)/n > 2 + \delta$ for some $\delta > 0$. The ABC conjecture yields the following improvement.

Theorem 5. *Under the ABC conjecture, for every $\epsilon > 0$ there exists a constant $c(\epsilon) > 0$ such that*

$$P(n! + 1) > \left(n + \frac{1}{2}\right)\log n - (2 + \epsilon)n + c(\epsilon).$$

In the same paper, Erdös asked if numbers of the form $2^n \pm 1$ represent infinitely many k-th power-free integers. In the case of $2^n - 1$ there is a curious relation with Artin's conjecture on primitive roots. More generally, for any integer d, the Artin conjecture with index d states that for any positive square-free integer $a \neq 1$, there exists infinitely many primes p such that $a \pmod{p}$ generates a subgroup of index d in $(\mathbf{Z}/p\mathbf{Z})^\times$. The argument in [13] readily shows that this generalized Artin conjecture follows from the generalized Riemann hypothesis. We have the following curious theorem which is suggested by the previous discussion and is of independent interest. It can be viewed as a variation on the work of Murty and Srinivasan [14].

Theorem 6. *At least one of the following conditions hold:*

(a) *Artin's conjecture on primitive roots with index 2 holds for $a = 2$.*

(b) *There exist infinitely many primes p such that $2^p - 1$ is composite.*

Remark 2. Most likely, both possibilities of the Theorem are true. However, at present, neither assertion has been established unconditionally so the Theorem is of some interest.

2   Statements of the ABC Conjectures

We begin by stating the usual ABC conjecture [16]. See [15] for a survey of its many conjectural applications.

Conjecture 1 (The ABC conjecture). *For every $\epsilon > 0$ there exists a constant $c(\epsilon) > 0$, such that for any triple of nonzero, pairwise coprime integers A, B, C with $A + B + C = 0$, we have*

$$\max(|A|, |B|, |C|) < c(\epsilon) \prod_{p|ABC} p^{1+\epsilon},$$

where the product is taken over the distinct prime divisors of ABC.

Now, let $K/\mathbf{Q}$ be a number field with discriminant Δ_K. Denote by $\mathcal{O}_K$ the ring of integers of K. For any $x \in K^{\times}$, denote by $h_K(x)$ the exponential height of x over K. Frey [7] proposed the following refined version of the ABC conjecture:

Conjecture 2 (Frey). *For any $\epsilon > 0$ and any ideal $\mathfrak{a} \subset \mathcal{O}_K$ there exists a constant $c(\epsilon, \Delta_K, \mathfrak{a})$ which depends linearly on $\log|\Delta_K|$ and $\log \operatorname{Norm} \mathfrak{a}$ but does not otherwise depend on K, such that for any pair of elements $A, B \in \mathcal{O}_K$ which generate the ideal $\mathfrak{a}$, we have the upper bound*

$$\max\left(h_K(A), h_K(B), h_K(A - B)\right)$$

$$\leq (1 + \epsilon)\left(\prod_{\mathfrak{p}|AB(A-B)} \operatorname{Norm}(\mathfrak{p})\right) + c(\epsilon, \Delta_K, \mathfrak{a}).$$

Remark 3. It is a standard fact from commutative algebra that every non-zero ideal in a Dedekind domain can be generated by two elements (cf. [1, Exercise 7]).

3   Proof of Theorem 1

Denote by $\Phi_d(x, y)$ the homogenized d-th cyclotomic polynomial, so

$$a^n - b^n = \prod_{d|n} \Phi_d(a, b).$$

Fix $\epsilon > 0$. Recall the fact from elementary number theory that the number of divisors of n, denoted $d(n)$, satisfies the estimate $d(n) = O(n^{\epsilon})$. Thus, for any $z \geq 1$ and $a > b > 0$,

$$\begin{aligned}
a^n - b^n &= \prod_{\substack{d|n \\ d \leq z}} \Phi_d(a, b) \cdot \prod_{\substack{d|n \\ d > z}} \Phi_d(a, b) \\
&\leq a^{zn^{\epsilon}} \cdot \prod_{\substack{d|n \\ d > z}} \Phi_d(a, b).
\end{aligned}$$

Let p be a prime not dividing ab. It is an easy exercise to show that $p|\Phi_d(a, b)$ implies $p \equiv 1 \pmod{d}$ or $p = P(d)$, and this occurs to at most the first power. Let $Q = Q(a^n - b^n)$. As usual, write

$$\psi(x, d, 1) = \sum_{\substack{n \leq x \\ n \equiv 1 \pmod{d}}} \Lambda(n),$$

where $\Lambda(n)$ is the von Mangoldt function. Then, for $z = n^{1-2\epsilon}$, we obtain

$$
\begin{aligned}
\log(a^n - b^n) \quad &\ll \quad n^{1-\epsilon} + \sum_{\substack{d|n \\ d>z}} \log \Phi_d(a,b) \\
&\ll \quad n^{1-\epsilon} + O(n^\epsilon) + \sum_{\substack{d|n \\ d>z}} \psi(Q,d,1),
\end{aligned}
$$

where the implied constants depend only on a, b and ϵ. The last sum is

$$
\leq \sum_{\substack{d|n \\ d>z}} \log Q \frac{Q}{d} \ll \frac{Q}{z} d(n) \log Q \ll Q(\log Q)n^{-1+3\epsilon}
$$

so that $Q \gg n^{2-3\epsilon}$.   □

4   Prime Divisors of Binomials

For any integer $n > 1$, define the powerful part of n to be the product $\kappa(n) := \prod_{p:p^2|n} p^{ord_p(n)}$. The quotient $n/\kappa(n)$ is called the powerfree part of n.

Denote by $\Phi_d(x,y)$ the homogenized d-th cyclotomic polynomial. For any integers $a > b > 0$, write $\Phi_d(a,b) = U_d(a,b)V_d(a,b)$, where $U_d(a,b)$ (resp. $V_d(a,b)$) is the power-free part of $\Phi_d(a,b)$ (resp. powerful part), so $(U_d(a,b), V_d(a,b)) = 1$.

Lemma 1. *Under the ABC conjecture, for every $\epsilon > 0$ there exists an absolute constant $c_1(\epsilon) > 0$ such that, for any integers $a > b > 0$,*

$$
\prod_{d|n} V_d(a,b) < c_1(\epsilon)(ab)^{1+\epsilon} a^{\epsilon n}.
$$

Proof. This is essentially Lemma 7 in [21]. For the sake of completeness, we shall briefly review the proof.

Applying the ABC conjecture, we get that for every $\epsilon > 0$ there exists an absolute constant $c_0(\epsilon) > 0$ such that

$$
a^n = \max(a^n, b^n, a^n - b^n) < c_0(\epsilon) \left(\prod_{p|ab(a^n-b^n)} p \right)^{1+\epsilon}.
$$

Since the square-free part of $V_d(a,b)$ is $\leq \sqrt{V_d(a,b)}$, this becomes

$$
\begin{aligned}
a^n \;&<\; c_0(\epsilon)(ab)^{1+\epsilon}\left(\prod_{d|n} U_d(a,b)\sqrt{V_d(a,b)}\right)^{1+\epsilon} \\
&<\; c_0(\epsilon)(ab)^{1+\epsilon}a^{n(1+\epsilon)}\left(\prod_{d|n} V_d(a,b)\right)^{-(1+\epsilon)/2}.
\end{aligned}
$$

Rearrange the terms and we are done. □

Proof of Theorem 2. Lemma 1 shows that, under the ABC conjecture,

$$
\prod_{d|n} U_d(a,b) = \frac{a^n - b^n}{\prod_{d|n} V_d(a,b)} \gg_\epsilon (a^n - b^n)a^{-n\epsilon} > a^{n-1-n\epsilon} = a^{n(1-\epsilon)-1}.
$$

For any d, we have the trivial estimate $U_d(a,b) < \Phi_d(a,b) < a^d$. For any $z \geq 1$, we have

$$
a^{n(1-\epsilon)} \ll_{\epsilon,a} \prod_{\substack{d|n \\ d\leq z}} U_d(a,b) \prod_{\substack{d|n \\ d>z}} U_d(a,b) \ll_{\epsilon,a} \prod_{\substack{d|n \\ d\leq z}} a^d \prod_{\substack{d|n \\ d>z}} U_d(a,b).
$$

Since $\sum_{d|n} 1 \ll_\epsilon n^\epsilon$, upon letting $z = n^{1-2\epsilon}$,

$$
a^{n(1-\epsilon)} \ll_{\epsilon,a} n^\epsilon a^{n^{1-\epsilon}} \prod_{\substack{d|n \\ d>z}} U_d(a,b),
$$

whence

$$
a^{n(1-2\epsilon)} \ll_{\epsilon,a} \prod_{\substack{d|n \\ d>z}} U_d(a,b). \tag{4.1}
$$

Now, U_d is the square-free part of the value of the homogeneous d-th cyclotomic polynomial, so that if $p \nmid ab$ divides $U_d(a,b)$ then $p \equiv 1 \pmod{d}$. Now, suppose that $P(a^n - b^n) \leq N := n^{2-5\epsilon}$. Then for $d > z$, the Brun-Titchmarsh theorem [25, p. 73] gives

$$
U_d(a,b) \ll_{a,b} \prod_{\substack{p\leq N \\ p\equiv 1(d)}} p \ll_{a,b,\epsilon} \exp(2N/\varphi(d)) \ll \exp(2N \log\log d/d)
$$

$$
\leq \exp(n^{1-3\epsilon} \log\log n),
$$

upon using $\varphi(d) \gg d/\log\log d$. Combining this with (4.1), we obtain

$$
a^{n(1-\epsilon)} \ll_{\epsilon,a,b} \prod_{\substack{d|n \\ d>z}} \exp(n^{1-3\epsilon} \log\log n).
$$

Taking logs on both sides, we get

$$n(1 - \epsilon) \ll_{\epsilon,a,b} \sum_{\substack{d|n \\ d>z}} n^{1-3\epsilon} \log\log n \ll_{\epsilon,a,b} n^{1-2\epsilon},$$

a contradiction if $n \gg_{\epsilon,a,b} 1$.   □

Proof of Theorem 5. The ABC conjecture gives

$$n! \ll_{\epsilon} \left(\prod_{p|(n!)(n!+1)} p \right)^{1+\epsilon} \ll_{\epsilon} \left(e^n \prod_{\substack{p|(n!+1) \\ p>n}} p \right)^{1+\epsilon}.$$

Taking logs and applying the Stirling approximation, we get

$$(n + 1/2) \log n - n \leq (1 + \epsilon)\left(n + \sum_{\substack{p|(n!+1) \\ p>n}} \log p\right),$$

and the theorem follows.   □

5 Heuristics for Erdös' Conjecture

For any prime p and any integer a not divisible by p, denote by $f_a(p)$ the order of $a \pmod p$. Define

$$P_x = \max_{n \leq x} P(a^n - 1).$$

Under the ABC conjecture, the argument for Lemma 1 shows that $a^n - 1 = u_n v_n$ with u_n square-free and $v_n \ll_{a,\epsilon} a^{\epsilon n}$. Thus

$$\sum_{\substack{p \leq P_x \\ f_a(p) \leq x}} [x/f_a(p)] \log p \sim \frac{1}{2} x^2 \log a,$$

whence

$$(\log P_x) \sum_{p: f_p(a) < x} [x/f_p(a)] \geq \frac{1}{2} x^2 \log a.$$

On the other hand,

$$\sum_{\substack{p \leq P_x \\ f_a(p) \leq x}} [x/f_p(a)] = \sum_{n \leq x} v(a^n - 1), \tag{5.1}$$

where $v(m)$ denotes the number of distinct prime divisors of m. A classical result of Hardy and Ramanujan [10] says that the average order of $v(m)$ is $\log \log m$. If we assume that $v(a^n - 1)$ behaves like this, then it follows that $a^n - 1$ usually has $O(\log n)$ prime divisors, in which case the sum on the right side of (5.1) behaves like $x \log x$. This gives a more precise form of (1.5):

$$P_x \geq a^{x/2 \log x}.$$

In fact, we can do even better. The Hardy-Ramanujan result is based on the heuristic that the probability a prime divides n is $1/p$. Thus the number of prime divisors of n should be roughly

$$\sum_{p \leq n} \frac{1}{p} \sim \log \log n.$$

In our case, $a^n - 1 = \prod_{d|n} \phi_d(a)$, so we should really be looking at prime divisors of $\phi_d(a)$. But then all prime divisors are $\equiv 1 \pmod{d}$, so the average number of prime divisors of $\phi_d(a)$ should be

$$\sum_{\substack{p \leq a^d \\ p \equiv 1 \pmod{d}}} \frac{1}{p} \sim \frac{\log d}{\phi(d)}.$$

This improves the heuristic above to

$$P_x \geq a^{cx}$$

for some constant $c > 0$.

6   Prime Divisors of Polynomials

The following result was noted independently by Langevin [12] and Granville [8].

Theorem 7. *Assume the ABC conjecture. Suppose that $g \in \mathbf{Z}[x]$ has distinct roots. Then for any $\epsilon > 0$ there exists a constant $c(\epsilon, g) > 0$ such that for any integer m,*

$$\prod_{p|g(m)} p \leq c(\epsilon, g)|m|^{\deg g - 1 - \epsilon}. \qquad \square$$

Proof of Theorem 4. Suppose that $P(f(n)) \ll_{\epsilon,f} (\deg(f) - 1 - \epsilon) \log n$ for $n \gg_{\epsilon,f} 1$. Then under the ABC conjecture, Theorem 7 gives

$$\left(\deg(f) - 1 - \frac{\epsilon}{2}\right) \log n \ll_{\epsilon,f} \sum_{p|f(n)} \log p \ll \sum_{p < (\deg f - 1 - \epsilon) \log n} \log p$$

$$\ll (\deg f - 1 - \epsilon) \log n$$

for n sufficiently large, a contradiction. $\qquad\qquad\qquad\qquad\square$

7   Compositeness of $2^q - 1$

Denote, by $f_a(p)$ the order of $a \pmod p$, as before.

Proof of Theorem 6. Define a set of primes (with $\eta > 0$)

$$S = \{p \le x : p - 1 = 2l_1 \text{ or } 2l_1l_2, l_2 > x^{1/2-\eta} > l_1 > x^{1/4-\eta}\},$$

where 2 is a quadratic residue mod p. If the set of primes $p \in S$ with $p - 1 = 2l_1$ is infinite, then Artin's conjecture holds for index 2. So now suppose that all primes $p \in S$ are of the form $2l_1l_2$ with l_1, l_2 as indicated in S. Then, the order of 2 mod p for $p \in S$ is either l_1 or l_2. In either case, $2^{l_1} - 1$ or $2^{l_2} - 1$ is divisible by p and hence composite since each of these numbers is larger than p. This completes the proof. $\qquad\qquad\square$

A similar argument can be applied to show that if it is **not** the case that 2 is a primitive root for infinitely many primes p then $(2^p+1)/3$ is composite for infinitely many primes p. We leave the details to the interested reader.

References

[1] M. F. Atiyah and I. G. MacDonald, *Introduction to commutative algebra*, Addison-Wesley, 1969.

[2] A. Baker, *A sharpening of the bounds for linear forms in logarithms I*, Acta Arith. **21** (1972), 117–129.

[3] ———, *A sharpening of the bounds for linear forms in logarithms II*, Acta Arith. **24** (1973), 33–36.

[4] ———, *A sharpening of the bounds for linear forms in logarithms III*, Acta Arith. **27** (1975), 247–252.

[5] P. Erdős, *Some recent advances and current problems in number theory*, Lectures on modern mathematics, III (T. L. Saaty, ed.), John Wiley & Sons, 1965, pp. 196–244.

[6] P. Erdős and C. L. Stewart, *On the greatest and least prime factors of* $n! + 1$, J. London Math. Soc. (2) **13** (1976), 513–515.

[7] G. Frey, *On ternary equations of Fermat type and relations with elliptic curves*, Modular forms and Fermat's last theorem, Springer-Verlag, 1997, pp. 527–548.

[8] A. Granville, *ABC allows us to count squarefrees*, Inter. Math. Research Notices (1998), no. 19, 991–1009.

[9] R. Gupta and R. Murty, *A remark on Artin's conjecture*, Invent. Math. **78** (1984), 127–130.

[10] G. H. Hardy and S. Ramanujan, *The normal number of prime factors of a number n*, Quarterly J. Math. **48** (1920), 76–92.

[11] Jürgen Hinz and M. Lodemann, *On Siegel zeros of Hecke-Landau zeta-functions*, Monatsh. Math. **118** (1994), 231–248.

[12] M. Langevin, *Cas d'egalité pour le théorème de Mason et applications de la conjecture (abc)*, C. R. Acad. Sci. Paris Ser. I Math. **317** (1993), 441–444.

[13] R. Murty, *On Artin's conjecture*, J. Number Theory **16** (1983), 147–168.

[14] R. Murty and S. Srinivasan, *Some remarks on Artin's conjecture*, Canad. Math. Bulletin **30** (1987), 80–85.

[15] A. Nitaj, *La conjecture abc*, Enseign. Math. (2) **42** (1996), 1–24.

[16] J. Oesterlé, *Nouvelles approches du théorème de Fermat*, Astérisque (1988), Exp. 694, 165–186 (1989), Séminaire Bourbaki, Vol. 1987/88.

[17] G. Pólya, *Zur arithmetischen Untersuchung der Polynome*, Math. Z. **1** (1918), 143–148.

[18] P. Ribenboim and G. Walsh, *The ABC conjecture and the powerful part of terms in binary recurring sequences*, J. Number Theory **74** (1999), 134–147.

[19] T. N. Shorey, A. J. van der Poorten, R. Tijdeman, and A. Schinzel, *Applications of the Gelfond-Baker method to Diophantine equations*, Transcendence theory: advances and applications (A. Baker and D. W. Masser, eds.), Academic Press, 1977, pp. 59–77.

[20] C. L. Siegel, *Approximation algebraischer Zahlen*, Math. Z. **10** (1921), 173–213.

[21] J. Silverman, *Wieferich's criterion and the abc-conjecture*, J. Number Theory **30** (1988), 226–237.

[22] C. L. Stewart, *Divisor properties of arithmetic sequences*, Ph.D. thesis, University of Cambridge, 1976.

[23] _____, *On divisors of Fermat, Fibonacci, Lucas, and Lehmer numbers*, Proc. London Math. Soc. (3) **35** (1977), 425–477.

[24] _____, *Primitive divisors of Lucas and Lehmer numbers*, Transcendence theory: advances and applications (A. Baker and D. W. Masser, eds.), Academic Press, 1977, pp. 79–92.

[25] G. Tenenbaum, *Introduction to analytic and probabilistic number theory*, Cambridge Univ. Press, 1995.

[26] A. Thue, *Über Annäherungswerte algebraischer Zahlen*, J. reine angew. Math. **135** (1909), 284–305.

[27] H. C. Williams, *Edouard Lucas and primality testing*, Wiley, 1998.

On the Parity of Generalized Partition Functions

J.-L. Nicolas[1] and A. Sárközy[1]

1   Introduction

$\mathbb{N}_0$ and $\mathbb{N}$ denote the sets of the non-negative resp. positive integers; $\mathcal{A}, \mathcal{B} \ldots$ denote sets of positive integers, and their counting functions are denoted by $A(x), B(x), \ldots$, so that, e.g.,

$$A(x) = |\{a : a \leq x, a \in \mathcal{A}\}|.$$

If $\mathcal{A} = \{a_1, a_2, \ldots\} \subset \mathbb{N}$ (where $a_1 < a_2 < \ldots$), then $p(\mathcal{A}, n)$ denotes the number of partitions of n with parts in $\mathcal{A}$, that is, the number of solutions of the equation

$$a_1 x_1 + a_2 x_2 + \cdots = n$$

in non-negative integers $x_1, x_2, \ldots$. As usual, we set $p(\mathcal{A}, 0) = 1$.

For $i = 0$ or 1, if $\mathcal{A} \subset \mathbb{N}$ and there is a number N such that

$$p(\mathcal{A}, n) \equiv i \,(\mathrm{mod}\, 2) \quad \text{for all} \quad n \in \mathbb{N}, n > N,$$

then $\mathcal{A}$ is said to possess property P_i. If $i = 0$ or 1, $\mathcal{B} = \{b_1, \ldots, b_k\} \neq \emptyset$ (where $b_1 < \cdots < b_k$) is a finite set of positive integers, $N \in \mathbb{N}$ and $N \geq b_k$, then there is a unique set $\mathcal{A} \subset \mathbb{N}$ such that

$$\mathcal{A} \cap \{1, 2, \ldots, N\} = \mathcal{B}$$

and

$$p(\mathcal{A}, n) \equiv i \,(\mathrm{mod}\, 2) \quad \text{for} \quad n \in \mathbb{N}, n > N.$$

We will denote this set $\mathcal{A}$ by $\mathcal{A}_i(\mathcal{B}, N)$ and, in particular, we will write $\mathcal{A}_i(\mathcal{B}, b_k) = \mathcal{A}_i(\mathcal{B})$. The construction of the set $\mathcal{A}_i(\mathcal{B}, N)$ is described in [3]; let us recall it when, for instance, $i = 0$. The set $\mathcal{A} = \mathcal{A}_0(\mathcal{B}, N)$ will be defined by recursion. We write $\mathcal{A}_n = \mathcal{A} \cap \{1, 2, \ldots, n\}$, so that

$$\mathcal{A}_N = \mathcal{A} \cap \{1, 2, \ldots, N\} = \mathcal{B}.$$

[1]Research partially supported by Hungarian National Foundation for Scientific Research, Grant No. T 029759, MKM fund FKFP-0139/1997, French-Hungarian APAPE-OMFB exchange program F-5/1997 and CNRS, Institut Girard Desargues, UMR 5028.

Assume that $n \geq N+1$ and $\mathcal{A}_{n-1}$ has been defined so that $p(\mathcal{A}, m)$ is even for $N+1 \leq m \leq n-1$. Then set

$$n \in \mathcal{A} \quad \text{if and only if} \quad p(\mathcal{A}_{n-1}, n) \quad \text{is odd.}$$

It follows from the construction that for $n \geq N+1$ we have $p(\mathcal{A}, n) = 1 + p(\mathcal{A}_{n-1}, n)$ if $n \in \mathcal{A}$, and $p(\mathcal{A}, n) = p(\mathcal{A}_{n-1}, n)$ if $n \notin \mathcal{A}$. This shows that $p(\mathcal{A}, n)$ is even for $n \geq N+1$.

Note that, in the same way, any finite set $\mathcal{B} = \{b_1, b_2, \ldots, b_k\}$ can be extended to an infinite set $\mathcal{A}$ so that $\mathcal{A}_{b_k} = \mathcal{B}$ and the parity of $p(\mathcal{A}, n)$ is given for $n \geq N+1$ (where N is any integer such that $N \geq b_k$). The problem we will consider here is the estimation of $A(x)$.

In [4] we initiated the study of sets $\mathcal{A}$ possessing property P_0 or P_1. In [3] we asked the following question: *But what can one say on such a set $\mathcal{A}\ldots$? In particular, how thin, or how dense can a set of this type be?* All we could prove in this direction was that there is an infinite set $\mathcal{A}$ which possesses property P_0 and for which $A(x) \gg x / \log x$; more precisely, $p(\mathcal{A}, n)$ is even for $n \geq 4$ and

$$\liminf_{x \to \infty} \frac{A(x) \log x}{x} \geq \frac{1}{2}. \tag{1.1}$$

Indeed, we showed that the set

$$\mathcal{A} = \mathcal{A}_0(\mathcal{B}), \quad \text{where} \quad \mathcal{B} = \{1, 2, 3\} \tag{1.2}$$

has these properties. In [3] we wrote regarding this set $\mathcal{A}$: *First we thought that perhaps even*

$$A(x) = \left(\frac{1}{2} + o(1)\right) x$$

holds. However, computing the elements of $\mathcal{A}$ up to 10000, it turned out that $A(10000) = 2204$ so that, probably,

$$\liminf_{x \to \infty} \frac{A(x)}{x} < \frac{1}{2}.$$

In this paper we will first continue the study of the sequence $\mathcal{A}$ in (1.2). Then, in Section 3, we will show that there are numerous sequences $\mathcal{A}$ which possess property P_0 or P_1 and whose counting function grows very slowly: namely, we have

$$A(x) \ll \log x.$$

(Computer experiments lead us to the construction of sets $\mathcal{A}$ with those properties; it surprised us very much that such sets $\mathcal{A}$ exist.) In Section 4 we will prove a criterion which can be used to show that for fixed i, $\mathcal{B}$, N the set $\mathcal{A} = \mathcal{A}_i(\mathcal{B}, N)$ satisfies an inequality like (1.1), i.e., we have

$$A(x) \gg \frac{x}{\log x}. \tag{1.3}$$

This criterion will suggest that for the most $\mathcal{B}$, N the set $\mathcal{A} = \mathcal{A}_i(\mathcal{B}, N)$ (for both $i = 0$ and 1) satisfies (1.3). In Section 5 we will improve on (1.3) by constructing a set $\mathcal{A} = \mathcal{A}_0(\mathcal{B}, N)$ with

$$A(x) \gg \frac{x}{(\log x)^{1-c}}$$

for some $c > 0$. Finally, in Section 6 we will formulate several problems and conjectures based on computer experiments.

By using modular forms, K. Ono has obtained in [5] and [6] nice results about the distribution of the values of the classical partition function $p(n) = p(\mathcal{N}, n)$ in the different residues classes modulo m. By the above algorithm, it is possible to construct sets $\mathcal{A}$ such that, for $n \equiv a \pmod{m}$ and $n > N$, the parity of $p(\mathcal{A}, n)$ is fixed.

2   Further Study of the Set $\mathcal{A}$ in (1.2)

We will use the following notation: If $\mathcal{A} \subset \mathbb{N}$, then $\chi(\mathcal{A}, n)$ denotes the characteristic function of $\mathcal{A}$, i.e.,

$$\chi(\mathcal{A}, n) = \begin{cases} 1 & \text{if } n \in \mathcal{A} \\ 0 & \text{if } n \notin \mathcal{A}. \end{cases}$$

Moreover, we write

$$\sigma(\mathcal{A}, n) = \sum_{d|n} \chi(\mathcal{A}, d)d = \sum_{d|n,\ d\in\mathcal{A}} d. \tag{2.1}$$

By (4.5) in [3] we have

$$np(\mathcal{A}, n) = \sum_{k=0}^{n-1} p(\mathcal{A}, k)\sigma(\mathcal{A}, n - k). \tag{2.2}$$

Let μ denote the Möbius function. We shall need the following lemma which allows us to determine $\chi(\mathcal{A}, n)$ for n odd if the σ function is known:

Lemma 1. *If n is odd, then*

$$\chi(\mathcal{A}, n) \equiv \sum_{d|n} \mu(d)\sigma(\mathcal{A}, n/d) \pmod{2}, \tag{2.3}$$

while if $n = 2^\alpha m$, $\alpha \geq 1$, and m is odd, then

$$n\chi(\mathcal{A}, n) = -\sum_{\beta=0}^{\alpha-1} 2^\beta m\chi(\mathcal{A}, 2^\beta m) + \sum_{d|m} \mu(d)\sigma(\mathcal{A}, n/d). \qquad (2.4)$$

Proof. Applying the Möbius inversion formula, it follows from (2.1) that

$$n\chi(\mathcal{A}, n) = \sum_{d|n} \mu(d)\sigma(\mathcal{A}, n/d), \qquad (2.5)$$

which gives (2.3) for n odd. When n is even, we write the divisors d of n in the form $d = 2^\beta \delta$, where $\beta \leq \alpha$ and $\delta|m$, so that (2.5) can be written as

$$\begin{aligned}
n\chi(\mathcal{A}, n) &= \sum_{\delta|m} \sum_{\beta=0}^{\alpha} \mu(2^\beta \delta)\sigma(\mathcal{A}, n/2^\beta \delta) \\
&= \sum_{\delta|m} \mu(\delta)\sigma(\mathcal{A}, n/\delta) - \sum_{\delta|m} \mu(\delta)\sigma(\mathcal{A}, n/2\delta).
\end{aligned}$$

Here the last sum is

$$\begin{aligned}
\sum_{\delta|m} \mu(\delta) \sum_{a|(n/2\delta)} \chi(\mathcal{A}, a)a &= \sum_{a|(n/2)} a\chi(\mathcal{A}, a) \sum_{\delta|(n/(2a),m)} \mu(\delta) \\
&= \sum_{\beta=0}^{\alpha-1} 2^\beta m\chi(\mathcal{A}, 2^\beta m).
\end{aligned}$$

This completes the proof of Lemma 1.   □

From now on $\mathcal{A}$ denotes the set (1.2). In [3] we showed that $\sigma(\mathcal{A}, n)$ modulo 2 is periodic with period 7. More precisely, as

$$n \equiv 0, 1, 2, 3, 4, 5 \text{ and } 6 \pmod 7,$$

we have

$$\sigma(\mathcal{A}, n) \equiv 1, 1, 1, 0, 1, 0 \text{ and } 0 \pmod 2.$$

This can be expressed in the following form:

$$\sigma(\mathcal{A}, n) \equiv 1 + \frac{1}{2}\left(\left(\frac{n}{7}\right) - \left(\frac{n}{7}\right)^2\right) \pmod 2, \qquad (2.6)$$

where $\left(\frac{n}{7}\right)$ is the Legendre symbol for $(n, 7) = 1$, and $\left(\frac{n}{7}\right) = 0$ for $7|n$.

In [3] we proved that a prime p belongs to $\mathcal{A}$ if and only if $p \equiv 3, 5$ or $6 \pmod 7$ (i.e., if $\left(\frac{p}{7}\right) = -1$). We will prove:

Theorem 1. *The odd elements of A are of the following form: $n = 1$, or $n = p^\alpha$ or $n = 7p^\alpha$, where p is a prime $\equiv 3, 5$ or 6 (mod 7) and $\alpha \geq 1$.*

Proof. By Lemma 1 and (2.6) we have, for n odd, $n > 1$,

$$\chi(A, n) \equiv \sum_{d|n} \mu(d) \left(1 + \frac{1}{2} \left(\left(\frac{n/d}{7}\right) - \left(\frac{n/d}{7}\right)^2 \right) \right)$$

$$\equiv \frac{1}{2}(f_1(n) - f_2(n)) \pmod 2 \tag{2.7}$$

with

$$f_i(n) = \sum_{d|n} \mu(d) \left(\frac{n/d}{7}\right)^i.$$

But $f_1(n)$ and $f_2(n)$ are multiplicative functions, and $f_2(n) = 0$ for all n except for $n = 1$ and $n = 7$ when $f_2(1) = +1$ and $f_2(7) = -1$. Further, $f_1(p^\alpha) = 0$ for $\left(\frac{p}{7}\right) = +1$ and $f_1(p^\alpha) = (-1)^\alpha \cdot 2$ for $\left(\frac{p}{7}\right) = -1$, and $f_1(7) = -1$ and $f_1(7^\alpha) = 0$ for $\alpha \geq 2$. Thus it follows from (2.7) that $7 \notin A$, and for n odd, $n \neq 1, 7$,

$$\chi(A, n) \equiv \frac{1}{2} f_1(n) \pmod 2,$$

so that by using the multiplicativity of $f_1(n)$ and the values of $f_1(p^\alpha)$, $f_1(n) \equiv 2 \pmod 4$ holds only if $n = p^\alpha$ or $7p^\alpha$ with $\left(\frac{p}{7}\right) = -1$. This completes the proof of Theorem 1. $\qquad \square$

The even elements of A could be determined if the following conjecture holds:

Conjecture. *If n is even then*

$$\sigma(A, n) \equiv 2, 3, 1 \pmod 4 \quad for \quad \left(\frac{n}{7}\right) = -1, +1, 0, \quad respectively. \tag{2.8}$$

More generally, if $k \geq 1$, $u_k = \sigma(A, 3 \cdot 2^k)$, $v_k = \sigma(A, 2^k)$, and n is a multiple of 2^k, then

$$\sigma(A, n) \equiv u_k, v_k, -3 \pmod{2^{k+1}} \quad for \quad \left(\frac{n}{7}\right) = -1, +1, 0, \quad respectively. \tag{2.9}$$

This conjecture has been checked up to $n = 10000$ by computer. By an argument similar to the proof of Theorem 1, one may deduce from the validity of (2.8) for $n \leq n_0$ that the elements n of A with $n \equiv 2 \pmod 4$ and $n \leq n_0$ are $n = 2$; $n = 2p^\alpha 7^\gamma$, $\left(\frac{p}{7}\right) = -1$, $p \equiv 1 \pmod 4$, α odd, $\gamma = 0$ or 1; or $n = 2p^\alpha q^\beta 7^\gamma$, $\left(\frac{p}{7}\right) = \left(\frac{q}{7}\right) = -1$, $p \neq q$, $\alpha \geq 1$, $\beta \geq 1$, $\gamma = 0$ or 1.

3   Thin Sets with Properties P_0, P_1

We will show that there are sets $\mathcal{B}$, $\mathcal{C}$ such that $\mathcal{A}_0(\mathcal{B})$ and $\mathcal{A}_1(\mathcal{C})$ are geometric progressions (apart from a single exceptional element):

Theorem 2. *(i) For all $a, b \in \mathbb{N}$ such that $a | b$, we have*

$$\mathcal{A}_0(\{a, b\}) = \{a, b, 2b, \ldots, 2^k b, \ldots\}. \tag{3.1}$$

(ii) We have

$$\mathcal{A}_1(\{1\}) = \{1\} \tag{3.2}$$

and, for all $k \in \mathbb{N}$,

$$\mathcal{A}_1(\{2, 2k+1\}) = \{2, 2k+1, 2(2k+1), \ldots, 2^\ell(2k+1), \ldots\}. \tag{3.3}$$

Proof. (i) By the uniqueness of $\mathcal{A}_0(\{a, b\})$, it suffices to show that, writing $\mathcal{D} = \{a, b, 2b, \ldots, 2^k b, \ldots\}$, we have

$$\mathcal{D} \cap \{1, 2, \ldots, b\} = \{a, b\} \tag{3.4}$$

and

$$p(\mathcal{D}, n) \equiv 0 \pmod{2} \quad \text{for} \quad n > b. \tag{3.5}$$

(3.4) is trivial, so it remains to prove (3.5). Clearly we have

$$
\begin{aligned}
\sum_{n=0}^{+\infty} p(\mathcal{D}, n) x^n &= \prod_{d \in \mathcal{D}} \frac{1}{1 - x^d} = \frac{1}{1 - x^a} \prod_{k=0}^{+\infty} \frac{1}{1 - x^{2^k b}} \\
&\equiv \frac{1}{1 - x^a} \prod_{k=0}^{+\infty} \frac{1}{1 + x^{2^k b}} \\
&= \frac{1 - x^b}{1 - x^a} = 1 + x^a + x^{2a} + \cdots + x^{b-a} \pmod{2},
\end{aligned}
$$

which proves (3.5). (Here the notation $\equiv \pmod{2}$ means that the corresponding coefficients are congruent modulo 2.)

(ii) (3.2) is trivial, while (3.3) can be proved in the same way as (3.1). $\square$

The sets constructed in Theorem 2, possessing properties P_0, resp. P_1, consist of a single geometric progression, apart from their smallest elements. We can show that a set possessing property P_0 or P_1 may consist of arbitrarily many geometric progressions. Here we will consider only the even case (P_0), since the other case is similar but slightly more complicated.

Theorem 3. *Let $k \in \mathbb{N}$ and let $q_1 < q_2 < \cdots < q_k$ be arbitrary positive odd integers. Then defining $\mathcal{D}$, M and $\mathcal{B}$ by $\mathcal{D} = \bigcup_{i=1}^{k} \{q_i, 2q_i, \ldots, 2^\ell q_i, \ldots\}$, $M = \sum_{i=1}^{k} q_i$ and $\mathcal{B} = \mathcal{D} \cap \{1, 2, \ldots, M\}$, respectively, we have*

$$A_0(\mathcal{B}, M) = \mathcal{D}. \tag{3.6}$$

Note that the function $\sigma(\mathcal{D}, n)$ defined by (2.1) satisfies

$$\sigma(\mathcal{D}, n) \equiv \sum_{i=1, q_i \mid n}^{k} 1 \pmod{2}$$

and is periodic in n with period $\mathrm{lcm}(q_1, q_2, \ldots, q_k)$.

Proof. To prove (3.6) we have to show that

$$\mathcal{D} \cap \{1, 2, \ldots, M\} = \mathcal{B} \tag{3.7}$$

and

$$p(\mathcal{D}, n) \equiv 0 \pmod{2} \text{ for } n > M. \tag{3.8}$$

(3.7) holds by the definition of $\mathcal{B}$. Thus it remains to show that (3.8) also holds.

Clearly we have

$$
\begin{aligned}
\sum_{n=0}^{+\infty} p(\mathcal{D}, n) x^n &= \prod_{d \in \mathcal{D}} \frac{1}{1 - x^d} \\
&= \prod_{i=1}^{k} \prod_{\ell=0}^{+\infty} \frac{1}{1 - x^{2^\ell q_i}} \equiv \prod_{i=1}^{k} \prod_{\ell=0}^{+\infty} \frac{1}{1 + x^{2^\ell q_i}} \\
&= \prod_{i=1}^{k} (1 - x^{q_i}) = a_0 + a_1 x + \cdots + a_M x^M \pmod{2},
\end{aligned}
$$

where $a_0, a_1, \ldots, a_M$ are integers, and this proves (3.8). $\square$

4 Dense Sets with Properties P_0, P_1

We believe that the sets $\mathcal{A}_0$, $\mathcal{A}_1$ of "geometric progression type", described in Section 3, are exceptional, and that typically, the sets $\mathcal{A} = \mathcal{A}_0(\mathcal{B}, N)$, $\mathcal{A} = \mathcal{A}_1(\mathcal{B}, N)$ are "dense" in the sense that they satisfy (1.3). We will prove a criterion which provides a simple algorithm to show that, for fixed $\mathcal{B}$, N, the sets $\mathcal{A}_0(\mathcal{B}, N)$ and $\mathcal{A}_1(\mathcal{B}, N)$ are indeed of this type:

Theorem 4. *For every finite set* $\mathcal{B} = \{b_1, \ldots, b_k\}$ *(where* $b_1 < \cdots < b_k$*), every* $N \in \mathbb{N}$*,* $N \geq b_k$*, and for both* $\mathcal{A} = \mathcal{A}_0(\mathcal{B}, N)$ *and* $\mathcal{A} = \mathcal{A}_1(\mathcal{B}, N)$*, there is a* $q = q(\mathcal{A}) \in \mathbb{N}$ *such that* q *is odd,*

$$q(\mathcal{A}_0(\mathcal{B}, N)) \leq 2^N, \qquad q(\mathcal{A}_1(\mathcal{B}, N)) \leq 2^{N+1} \qquad (4.1)$$

and

$$\sigma(\mathcal{A}, n) \equiv \sigma(\mathcal{A}, n + q) \pmod 2 \qquad for \ n \geq 1; \qquad (4.2)$$

i.e., $\sigma(\mathcal{A}, n)$ *is periodic modulo* 2 *for* $n \geq 1$ *with period* q *satisfying* (4.1).

The proof will be based on the following lemma:

Lemma 2. *For every finite set* $\mathcal{B} = \{b_1, \ldots, b_k\}$ *(where* $b_1 < \cdots < b_k$*) and every* $N \in \mathbb{N}$*,* $N \geq b_k$*, both* $\mathcal{A} = \mathcal{A}_0(\mathcal{B}, N)$ *and* $\mathcal{A} = \mathcal{A}_1(\mathcal{B}, N)$ *satisfy a congruence of form*

$$\sigma(\mathcal{A}, n) \equiv \varepsilon_0 + \sum_{j=1}^{J} \varepsilon_j \sigma(\mathcal{A}, n - j) \pmod 2 \ for \ n = J+1, J+2, .., \quad (4.3)$$

where each of $\varepsilon_0, \varepsilon_1, \ldots, \varepsilon_{J-1}$ *is equal to* 0 *or* 1*,* $\varepsilon_J = 1$*, and* J *is a positive integer satisfying* $J \leq N$ *if* $\mathcal{A} = \mathcal{A}_0(\mathcal{B}, N)$ *and* $J \leq N+1$ *if* $\mathcal{A} = \mathcal{A}_1(\mathcal{B}, N)$*.*

Proof. (i) Consider first the case $\mathcal{A} = \mathcal{A}_0(\mathcal{B}, N)$ ("even case") where $p(\mathcal{A}, n) \equiv 0 \pmod 2$ for $n \geq N + 1$. Let us define J as the smallest integer such that $p(\mathcal{A}, J) \equiv 1 \pmod 2$ and $p(\mathcal{A}, j) \equiv 0 \pmod 2$ for $j \geq J + 1$. (Note that $J \geq b_1 = \min \mathcal{B}$, since $p(\mathcal{A}, b_1) = 1$.) From the definition of J it follows that $J \leq N$ and

$$p(\mathcal{A}, n) \equiv 0 \pmod 2 \ for \ n = J+1, J+2, \ldots. \qquad (4.4)$$

The proof will be based on identity (2.2), which can be rewritten as

$$np(\mathcal{A}, n) = \sigma(\mathcal{A}, n) + \sum_{k=1}^{n-1} p(\mathcal{A}, k)\sigma(\mathcal{A}, n - k), \quad n \geq 1. \qquad (4.5)$$

By (4.4), it follows that for $n \geq J + 1$ we have

$$0 \equiv \sigma(\mathcal{A}, n) + \sum_{k=1}^{J} p(\mathcal{A}, k)\sigma(\mathcal{A}, n - k) \pmod 2. \qquad (4.6)$$

Writing $\varepsilon_k = p(\mathcal{A}, k) \pmod 2$, that is

$$\varepsilon_k = \begin{cases} 1 & \text{if } p(\mathcal{A}, k) \text{ is odd,} \\ 0 & \text{if } p(\mathcal{A}, k) \text{ is even,} \end{cases}$$

for $k = 1, 2, \ldots, J$, it follows from (4.6) that

$$\sigma(\mathcal{A}, n) \equiv \sum_{k=1}^{J} \varepsilon_k \sigma(\mathcal{A}, n - k) \pmod{2},$$

which is a congruence of form (4.3) with $\varepsilon_0 = 0$.

(ii) Consider now the odd case, i.e., a set $\mathcal{A} = \mathcal{A}_1(\mathcal{B}, N)$ so that

$$p(\mathcal{A}, n) \equiv 1 \pmod{2} \text{ for } n = N + 1, N + 2, \ldots. \tag{4.7}$$

Replacing n by $n - 1$ in (2.2) we obtain for $n \geq 1$

$$(n - 1)p(\mathcal{A}, n - 1) = \sum_{k=0}^{n-2} p(\mathcal{A}, k)\sigma(\mathcal{A}, n - 1 - k) = \sum_{j=1}^{n-1} p(\mathcal{A}, j - 1)\sigma(\mathcal{A}, n - j)$$
$$\tag{4.8}$$

Subtracting (4.8) from (2.2) yields for $n \geq 1$

$$np(\mathcal{A}, n) - (n - 1)p(\mathcal{A}, n - 1) = \sigma(\mathcal{A}, n) + \sum_{j=1}^{n-1} t_j \sigma(\mathcal{A}, n - j) \tag{4.9}$$

with $t_j = p(\mathcal{A}, j) - p(\mathcal{A}, j - 1)$. Here we define J as the smallest integer such that $p(\mathcal{A}, J - 1) \equiv 0 \pmod{2}$ and $p(\mathcal{A}, j) \equiv 1 \pmod{2}$ for $j \geq J$. Except for the case $\mathcal{B} = \{1\}$ (which leads to $\mathcal{A}_1(\mathcal{B}, N) = \{1\}$ for all $N \geq 1$), such a J always exists: if $1 \notin \mathcal{B}$, $p(\mathcal{A}, b_1 - 1) = 0$ so that $J \geq b_1$, while, if $1 = b_1 \in \mathcal{B}$, $p(\mathcal{A}, b_2) = 2$ and $J \geq b_2 + 1$. From (4.7), $J \leq N + 1$ holds, and for $j \geq J + 1$, we have $t_j \equiv 0 \pmod{2}$. Defining ε_j by

$$\varepsilon_j = \begin{cases} 0, & \text{if } t_j = p(\mathcal{A}, j) - p(\mathcal{A}, j - 1) \text{ is even} \\ 1, & \text{if } t_j = p(\mathcal{A}, j) - p(\mathcal{A}, j - 1) \text{ is odd} \end{cases} \quad (\text{for } j = 1, \ldots, J),$$

(4.9) implies

$$\sigma(\mathcal{A}, n) \equiv 1 + \sum_{j=1}^{J} \varepsilon_j \sigma(\mathcal{A}, n - j) \pmod{2} \quad (\text{for } n \geq J + 1) \tag{4.10}$$

which is again of form (4.3). This completes the proof of Lemma 2.   □

Proof of Theorem 4. We start out from the characteristic polynomial of the linear recurrence relation (4.3):

$$P(X) = X^J + \sum_{k=1}^{J} \varepsilon_k X^{J-k}. \tag{4.11}$$

Let us consider this polynomial on the finite field $\mathbb{F}_2$, and let $K = \mathbb{F}_{2^u}$ be a finite extension of $\mathbb{F}_2$ on which P splits into linear factors. Let $\xi_1, \ldots, \xi_J$ be the (not necessarily distinct) roots of P on K and

$$S_n = \xi_1^n + \xi_2^n + \cdots + \xi_J^n, \quad n = 1, 2, 3, \ldots$$

the associated Newton sums. These sums belong to $\mathbb{F}_2$ and, by a classical result in elementary algebra, they satisfy the following identities:

$$S_1 + \varepsilon_1 = 0,$$

$$S_2 + \varepsilon_1 S_1 + 2\varepsilon_2 = 0,$$

$$\cdot \quad \cdot \quad \cdot \quad \cdot \quad \cdot \quad \cdot \quad \cdot \quad \cdot \quad \cdot \quad \cdot \quad \cdot \quad \cdot$$

$$S_n + \varepsilon_1 S_{n-1} + \cdots + \varepsilon_{n-1} S_1 + n\varepsilon_n = 0, \quad \text{for } 1 \leq n \leq J,$$

$$S_n + \varepsilon_1 S_{n-1} + \cdots + \varepsilon_{J-1} S_{n-J+1} + \varepsilon_J S_{n-J} = 0, \quad \text{for } n \geq J + 1.$$

In the even case, since $\varepsilon_k \equiv p(\mathcal{A}, k) \pmod{2}$, it follows by induction on n from (4.5) that

$$\sigma(\mathcal{A}, n) \equiv S_n \pmod{2}. \tag{4.12}$$

But each non-zero root ξ_j has an order in K which divides $2^u - 1$, and so S_n is periodic in $n \geq 1$ with a period dividing $2^u - 1$. Then it follows from (4.12) that the period q of $\sigma(\mathcal{A}, n) \bmod 2$ is a divisor of $2^u - 1$ and so is odd, and that (4.2) holds.

The odd case is similar, with (4.12) replaced by

$$\sigma(\mathcal{A}, n) \equiv 1 + S_n \pmod{2}.$$

If the polynomial P is irreducible over $\mathbb{F}_2$, we can choose $K = \mathbb{F}_{2^J}$, since J is the degree of P. If P is reducible, let us write its factorisation as

$$P = P_1 P_2 \ldots P_r,$$

where $P_1, P_2, \ldots, P_r$ are the (not necessary distinct) irreducible factors of P over $\mathbb{F}_2$. If we denote by $S_n^{(i)}$ the Newton sum of index n associated to the polynomial P_i, $S_n^{(i)}$ is periodic in n with period q_i dividing $2^{d_i} - 1$, where d_i is the degree of P_i. Clearly,

$$S_n = S_n^{(1)} + S_n^{(2)} + \cdots + S_n^{(r)},$$

and the period of S_n is a divisor of $\mathrm{lcm}\,(q_1, q_2, \ldots, q_r)$ so that

$$\begin{aligned} q \leq q_1 q_2 \ldots q_r &\leq \left(2^{d_1} - 1\right)\left(2^{d_2} - 1\right)\ldots\left(2^{d_r} - 1\right) \\ &< 2^{d_1 + d_2 + \cdots + d_r} = 2^J, \end{aligned}$$

which, from the definition of J, implies (4.1). This completes the proof of Theorem 4. $\qquad\qquad\qquad\qquad\qquad\qquad\qquad\qquad\qquad\qquad\qquad\quad\Box$

Theorem 5. *Let $\mathcal{A} = \mathcal{A}_0(\mathcal{B}, N)$ or $\mathcal{A} = \mathcal{A}_1(\mathcal{B}, N)$, let $q = q(\mathcal{A})$ be the period of $\sigma(\mathcal{A}, n)$ (mod 2) as described in Theorem 4, and c the number of m with $1 \le m \le q$, such that*

$$(m, q) = 1 \quad and \quad \sigma(\mathcal{A}, m) \equiv 1 - \chi(\mathcal{A}, 1) \pmod{2}, \tag{4.13}$$

(where $\chi(\mathcal{A}, n)$ is the characteristic function of the set $\mathcal{A}$ as in Section 2). Then any prime $p \equiv m \pmod{q}$, where m is any integer satisfying (4.13), belongs to $\mathcal{A}$, and thus

$$\liminf_{x \to \infty} \frac{A(x) \log x}{x} \ge \frac{c}{\varphi(q)}, \tag{4.14}$$

where φ is Euler's function.

Note that Theorems 4 and 5 also provide a simple algorithm to show that for fixed $\mathcal{B}$, N, (4.14) holds for both sets $\mathcal{A} = \mathcal{A}_0(\mathcal{B}, N)$ and $\mathcal{A} = \mathcal{A}_1(\mathcal{B}, N)$ (and, indeed, for the most $\mathcal{B}$ and N this is expected to happen). Namely, we first look for a period q satisfying (4.1) and (4.2), and then we count the m's satisfying $1 \le m \le q$ and (4.13) to get c; if $c \ne 0$, then (4.14) is proved.

Proof of Theorem 5. If p is a prime congruent to m modulo q, then by (4.2) and (4.13) we have

$$\sigma(\mathcal{A}, p) \equiv \sigma(\mathcal{A}, m) \equiv 1 - \chi(\mathcal{A}, 1) \pmod{2},$$

so that by Lemma 1

$$
\begin{aligned}
\chi(\mathcal{A}, p) &\equiv \sum_{d|p} \mu(d) \sigma(\mathcal{A}, p/d) \\
&\equiv \sigma(\mathcal{A}, p) - \sigma(\mathcal{A}, 1) \equiv 1 - \chi(\mathcal{A}, 1) - \sigma(\mathcal{A}, 1) \equiv 1 \pmod{2},
\end{aligned}
$$

whence $p \in \mathcal{A}$. By the prime number theorem for arithmetic progressions, it follows that for each m coprime to q,

$$|\{p : p \le x, \ p \equiv m \pmod{q}\}| = (1 + o(1)) \frac{x}{\varphi(q) \log x},$$

whence the result follows. $\qquad\square$

5 Improving on (1.3)

We will prove:

Theorem 6. *There is an absolute constant $c > 0$ such that for $\mathcal{A} = \mathcal{A}_0(\{1,2,3,4,5\})$ we have*

$$A(x) > \frac{x}{(\log x)^{1-c}} \quad \text{for } x > x_0 \tag{5.1}$$

Proof. $\mathcal{A}$ simple computation (cf. Example 2 in Section 7) shows that for this set $\mathcal{A}$ the period q defined in Theorem 4 is $q = 31$, and $\sigma(\mathcal{A}, n) \equiv 0 \pmod 2$ if and only if n is congruent to 3, 5, 6, 7, 9, 10, 12, 14, 17, 18, 19, 20, 24, 25, 28 modulo 31. Thus for

$$n = q_1 q_2 \ldots q_k, \tag{5.2}$$

where $q_1, q_2, \ldots, q_k$ are distinct primes $\equiv 5 \pmod{31}$, we have

$$\sigma(\mathcal{A}, n) \equiv 1 \pmod 2 \quad \text{if} \quad \text{and} \quad \text{only} \quad \text{if } 3|k. \tag{5.3}$$

By (2.3) in Lemma 1, for the n in (5.2) we have

$$\chi(\mathcal{A}, n) \equiv \sum_{d|q_1 q_2 \ldots q_k} \sigma(\mathcal{A}, q_1 q_2 \ldots q_k/d) \pmod 2$$

so that, by (5.3),

$$\chi(\mathcal{A}, n) \equiv \sum_{\substack{0 \le r \le k \\ r \equiv 0 \,(\mathrm{mod}\,3)}} \binom{k}{r} \pmod 2. \tag{5.4}$$

Now we need the following lemma:

Lemma 3. *Write*

$$S(a, k) = \sum_{\substack{0 \le r \le k \\ r \equiv a \,(mod\,3)}} \binom{k}{r}.$$

Then for $k \in \mathbb{N}$ we have

$$S(a, k) \equiv \begin{cases} 0 \pmod 2 & \text{if } a + k \equiv 0 \pmod 3 \\ 1 \pmod 2 & \text{if } a + k \equiv 1 \text{ or } 2 \pmod 3. \end{cases} \tag{5.5}$$

Proof. By the identity

$$\binom{t}{i} = \binom{t-1}{i} + \binom{t-1}{i-1},$$

we have for $k \ge 4$

$$S(a,k) = S(a,k-1) + S(a-1,k-1)$$
$$= S(a,k-2) + 2S(a-1,k-2) + S(a-2,k-2)$$
$$= S(a,k-3) + 3S(a-1,k-3) + 3S(a-2,k-3) + S(a-3,k-3)$$
$$= 2S(a,k-3) + 3S(a-1,k-3) + 3S(a-2,k-3)$$
$$= 3 \sum_{0 \le r \le k-3} \binom{k-3}{r} - S(a,k-3) = 3 \cdot 2^{k-3} - S(a,k-3),$$

and finally

$$S(a,k) \equiv S(a,k-3) \pmod 2. \tag{5.6}$$

(5.5) follows by induction from (5.6) and

$$S(0,1) = 1, \quad S(1,1) = 1, \quad S(2,1) = 0$$
$$S(0,2) = 1, \quad S(1,2) = 2, \quad S(2,2) = 1$$
$$S(0,3) = 2, \quad S(1,3) = 3, \quad S(2,3) = 3. \qquad \square$$

By Lemma 3, it follows from (5.4) that if n is of the form (5.2), where

$$k \equiv 1 \text{ or } 2 \pmod 3, \tag{5.7}$$

then we have $n \in \mathcal{A}$. The following lemma then completes the proof of Theorem 6. $\qquad \square$

Lemma 4. *For $x > x_0$ the number of the integers n of form (5.2), where $n \le x$, $q_1 < q_2 < \cdots < q_k$ are primes $\equiv 5 \pmod{31}$ and $k \equiv 1,2 \pmod 3$, is $\gg \frac{x}{(\log x)^{1-c}}$ for a positive constant c.*

Lemma 4 will follow from the following theorem:

Theorem 7. *Let ℓ and m be two positive coprime integers. Let ρ be the multiplicative function defined by*

$$\begin{cases} \rho(p) = 1 & \text{if } p \equiv \ell \pmod m \\ \rho(p) = 0 & \text{if } p \not\equiv \ell \pmod m \end{cases}$$

and $\rho(p^\alpha) = 0$ for all primes p and all exponents $\alpha \ge 2$. Let $\omega(n)$ denote the number of prime factors of n, let φ be Euler's function, z any complex number and

$$U(x,z) = \sum_{n \le x} \rho(n) z^{\omega(n)}.$$

Then, for x going to infinity, we have

$$U(x,z) \sim \frac{C^z}{\Gamma(z/\varphi(m))} \prod_{p \equiv \ell \pmod m} \left(1 + \frac{z}{p}\right)\left(1 - \frac{1}{p}\right)^z \frac{x}{(\log x)^{1-z/\varphi(m)}},$$
$$\tag{5.8}$$

where

$$C := \left\{ \frac{\varphi(m)}{m} g(1) \prod_{\chi \neq \chi_0} (L(1,\chi))^{\overline{\chi}(l)} \right\}^{1/\varphi(m)}, \tag{5.9}$$

$L(s,\chi)$ *is the Dirichlet function associated to a character χ modulo m, and g is the function (holomorphic in $\Re s > 1/2$)*

$$g(s) = \exp \left(\sum_{\chi} \overline{\chi}(l) \sum_{p} \sum_{j=2}^{\infty} \frac{\chi(p) - \chi(p^j)}{jp^{js}} \right). \tag{5.10}$$

Proof. Theorem 7 is an extension of the so-called Selberg-Delange formula (cf. [7], II.5) by considering only the squarefree integers composed of primes congruent to ℓ modulo m. A sketch of the proof is given (for the case $z = 1$) in [8], as the solution to Exercise II.8.6, p. 124-125. A detailed proof will appear in [1]. □

Proof of Lemma 4. Let us set $\xi = e^{2i\pi/3}$. In Theorem 7, let us fix $\ell = 5$ and $m = 31$. The number $V(x,a)$ of integers $n \leq x$ satisfying (5.2) with $k = \omega(n) \equiv a \pmod{3}$ is, by (5.8), equal to

$$V(x,a) = \sum_{\substack{n \leq x \\ \omega(n) \equiv a \pmod{3}}} \rho(n) = \frac{1}{3} \sum_{n \leq x} \sum_{r=0}^{2} \xi^{r(\omega(n)-a)} = \frac{1}{3} \sum_{r=0}^{2} \xi^{-ra} U(x,\xi^r). \tag{5.11}$$

But, from (5.8) it follows that, for $r = 1$ or 2,

$$U(x,\xi^r) = O\left(x(\log x)^{\Re \xi/\varphi(31)-1}\right) = O\left(x(\log x)^{-\frac{61}{60}}\right)$$

while $U(x,1) \asymp x(\log x)^{-\frac{29}{30}}$. Therefore, (5.11) yields for $a = 0, 1$ or 2

$$V(x,a) \sim \frac{1}{3} U(x,1) \gg x(\log x)^{-\frac{29}{30}}.$$

This completes the proof of Lemma 4. □

An improvement of Theorem 6 is given in [2].

6　Problems

In this section we list several unsolved problems and conjectures based on the computer experiments carried out by us (see the examples below).

- *Is it true that for all $\mathcal{B}$ and $\mathbb{N}$, and for both $\mathcal{A} = \mathcal{A}_0(\mathcal{B}, N)$ and $\mathcal{A} = \mathcal{A}_1(\mathcal{B}, N)$, we have $A(x) = o(x)$? We believe that $A(x) \ll x/(\log x)^c$* with some $c > 0$. However, we cannot even show that there is an $\mathcal{A}$ with $A(x) \neq O(\log x)$, and

$$\liminf_{x \to \infty} \frac{A(x)}{x} < \frac{1}{2}.$$

To show this, it would suffice to show that for the set $\mathcal{A} = \mathcal{A}_0(\{1, 2, 3\})$ studied in Section 2, the number of the even elements of $\mathcal{A}$ not exceeding x is $\leq (\frac{1}{2} - \varepsilon)x$ for infinitely many $x \in \mathbb{N}$.

- *Is it true that if $A(x) \neq O(\log x)$ so that $\mathcal{A}$ is not of the "geometric progression type" (see Section 3), then we have $\frac{A(x)}{\log x} \to \infty$? Perhaps, in this case even*

$$\lim_{x \to \infty} \frac{A(x) \log x}{x} = \infty$$

must hold.

- *Is it true that for all $\mathcal{B}$ and N, and for both $\mathcal{A} = \mathcal{A}_0(\mathcal{B}, N)$ and $\mathcal{A} = \mathcal{A}_1(\mathcal{B}, N)$, denoting the smallest period of $\sigma(\mathcal{A}, n)$ by q we have*

$$\sigma(\mathcal{A}, 2(n + q)) \equiv \sigma(\mathcal{A}, 2n) \pmod 4$$

and more generally,

$$\sigma(\mathcal{A}, 2^{h-1}(n + q)) \equiv \sigma(\mathcal{A}, 2^{h-1}n) \pmod{2^h} ?$$

7　Examples

By computer, we have studied all sets $\mathcal{A}_i(\mathcal{B}, N)$ for $\mathcal{B} \subset \{1, 2, 3, 4, 5\}$, $i = 0$ or 1 and $\max_{b \in \mathcal{B}} b \leq N \leq 10$. For all of these sets, we have computed the period q of $\sigma(\mathcal{A}, n)$ mod 2, the constants c and $c/\varphi(q)$ introduced in Theorem 5, the characteristic polynomial P defined by (4.11) and its factorisation into irreducible factors over $\mathbb{F}_2$, the values of the first elements of $\mathcal{A}$ (up to 1000), and the values of $p(\mathcal{A}, n)$ for small n.

We give below the description of some of these sets which seem to us particularly interesting: in Examples 1 and 7, the elements greater than 5 of $\mathcal{A}$ coincide; in Examples 3 and 8, we have $c/\varphi(q) \neq 0, 1/2$; the sets $\mathcal{A}$ in Examples 5 and 6 coincide apart from the first element; in Example 5, the elements are twice the elements of $\mathcal{A}$ of Example 4.

Example 1: $\mathcal{B} = \{1, 2, 3\}$; $N = 3$; $i = 0$.

$$q = 7, \ c = 3, \ c/\varphi(q) = 1/2,$$

$$P = X^3 + X^2 + 1 : \text{ irreducible },$$

$$A = \{1, 2, 3, 5, 8, 9, 10, 13, 14, 16, 17, 19, 20, 21, 24, 25, 26, 27, 28, 30, 31,$$
$$34, 35, 36, 40, 41, 47, 48 \ldots\}; \quad A(1000) = 293.$$

Example 2: $\mathcal{B} = \{1, 2, 3, 4, 5\}$; $N = 5$; $i = 0$.

$$q = 31, \ c = 15, \ c/\varphi(q) = 1/2,$$

$$P = X^5 + X^4 + X^2 + X + 1 : \text{ irreducible },$$

$$A = \{1, 2, 3, 4, 5, 7, 8, 10, 12, 14, 16, 17, 19, 20, 22, 26, 27, 28, 33, 34, 36,$$
$$37, 38, 39, 41, 42, 43, 45, 46, 48, 50 \ldots\}; \quad A(1000) = 480.$$

Example 3: $\mathcal{B} = \{1, 2, 4\}$; $N = 8$; $i = 0$.

$$q = 63, \ c = 24, \ c/\varphi(q) = 2/3,$$

$$P = X^8 + X^7 + 1 = (X^2 + X + 1)(X^6 + X^4 + X^3 + X + 1),$$

$$A = \{1, 2, 4, 9, 10, 11, 12, 13, 14, 15, 18, 19, 22, 23, 25, 26, 28, 29, 31, 32$$
$$33, 34, 36, 37, 41, 43, 44, 45, 46, 47, 48, 50, \ldots\}; \quad A(1000) = 496.$$

Example 4: $\mathcal{B} = \{1, 2\}$; $N = 4$; $i = 0$.

$$q = 15, \ c = 4, \ c/\varphi(q) = 1/2,$$

$$P = X^4 + X^3 + 1 : \text{ irreducible },$$

$$A = \{1, 2, 5, 6, 7, 10, 11, 13, 14, 16, 21, 22, 24, 28, 29, 33, 35, 37, 39,$$
$$41, 42, 43, 48, 49, \ldots\}; \quad A(1000) = 307.$$

Example 5: $\mathcal{B} = \{2, 4\}$; $N = 8$; $i = 0$.

$$q = 1, \ c = 0, \ c/\varphi(q) = 0,$$

$$P = X^8 + X^6 + 1 = (X^4 + X^3 + 1)^2,$$

$$A = \{2, 4, 10, 12, 14, 20, 22, 26, 28, 32, 42, 44, 48, \ldots\}; \ A(1000) = 171.$$

Example 6: $\mathcal{B} = \{1, 4\}$; $N = 9$; $i = 0$.

$$q = 1, \ c = 0, \ c/\varphi(q) = 0,$$

$$P = X^9 + X^8 + X^7 + X^6 + X + 1 = (X + 1)(X^4 + X^3 + 1)^2,$$

$$A = \{1, 4, 10, 12, 14, 20, 22, 26, 28, 32, 42, 44, 48, \ldots\}; \ A(1000) = 171.$$

Example 7: $\mathcal{B} = \{3, 4\}$; $N = 4$; $i = 1$.

$$q = 7, \ c = 3, \ c/\varphi(q) = 1/2,$$

$$P = X^3 + X^2 + 1 : \text{ irreducible },$$

$$A = \{3, 4, 5, 8, 9, 10, 13, 14, 16, 17, 19, 20, 21, 24, 25, 26, 27, 28, 30, 31,$$
$$34, 35, 36, 40, 41, 47, 48 \ldots\}; \quad A(1000) = 292.$$

Example 8: $\mathcal{B} = \{6\}$; $N = 9$; $i = 1$,

$$q = 31, \ c = 10, \ c/\varphi(q) = 1/3,$$

$$P = X^{10} + X^9 + X^4 + X^3 + 1 = (X^5 + X^3 + X^2 + X + 1)(X^5 + X^4 + X^3 + X + 1).$$

$$A = \{6, 10, 11, 13, 14, 15, 20, 21, 22, 23, 27, 29, 30, 31, 32, 33,$$
$$34, 38, 39, 40, 45, 46, 48, \ldots\}; \quad A(1000) = 479.$$

References

[1] F. Ben Saïd and J.-L. Nicolas, *Sur une extension de la formule de Selberg-Delange*, to appear.

[2] J.-L. Nicolas, *On the parity of generalized partition functions. II*, Period. Math. Hungar. **43** (2001), 177–189.

[3] J.-L. Nicolas, I. Z. Ruzsa, and A. Sárközy, *On the parity of additive representation functions*, J. Number Theory **73** (1998), 292–317, With an appendix in French by J.-P. Serre.

[4] J.-L. Nicolas and A. Sárközy, *On the parity of partition functions*, Illinois J. Math. **39** (1995), 586–597.

[5] K. Ono, *Parity of the partition function in arithmetic progressions*, J. Reine Angew. Math. **472** (1996), 1–15.

[6] _____, *Distribution of the partition function modulo m*, Ann. of Math. (2) **151** (2000), 293–307.

[7] G. Tenenbaum, *Introduction to analytic and probabilistic number theory*, Cambridge University Press, Cambridge, 1995, Translated from the second French edition (1995) by C. B. Thomas.

[8] G. Tenenbaum and J. Wu, *Exercices corrigés de théorie analytique et probabiliste des nombres*, Société Mathématique de France, Paris, 1996.

Quadratic Twists of Modular Forms and Elliptic Curves

Ken Ono[1] and Matthew A. Papanikolas

1 Introduction

Here we summarize the results presented in the first author's lecture at the Millennial Conference on Number Theory. These results appear in [16] in full detail. In addition, we present a new result regarding the growth of Tate-Shafarevich groups of certain elliptic curves over elementary abelian simple 2-extensions.

We begin by fixing notation. Suppose that

$$F(z) = \sum_{n=1}^{\infty} a(n)q^n \in S_{2k}(\Gamma_0(M)) \ (q := e^{2\pi i z} \text{ throughout})$$

is an even weight newform on $\Gamma_0(M)$ with trivial Nebentypus character. As usual, let $L(F, s)$ denote its L-function which is defined by analytically continuing

$$L(F, s) = \sum_{n=1}^{\infty} \frac{a(n)}{n^s}.$$

If D is a fundamental discriminant of the quadratic field $\mathbb{Q}(\sqrt{D})$, then let $\chi_D = \left(\frac{D}{\bullet}\right)$ denote its usual Kronecker character. Let $F \otimes \chi_D$ denote the newform that is the D-quadratic twist of F, and let $L(F \otimes \chi_D, s)$ denote the associated L-function. In particular, if $\gcd(D, M) = 1$, then

$$(F \otimes \chi_D)(z) = \sum_{n=1}^{\infty} \chi_D(n)a(n)q^n.$$

Given a fixed F, we consider the behavior of the central critical values $L(F \otimes \chi_D, k)$ as D varies. In an important paper, Goldfeld [7] conjectured that

$$\sum_{\substack{|D|<X \\ \gcd(D,M)=1}} \mathrm{ord}_{s=k}\left(L(F \otimes \chi_D, s)\right) \sim \frac{1}{2} \sum_{\substack{|D|\leq X \\ \gcd(D,M)=1}} 1. \qquad (1.1)$$

---

[1]The first author thanks the National Science Foundation, the Alfred P. Sloan Foundation and the David and Lucile Packard Foundation for their generous research support.

(This conjecture was originally formulated for weight 2 newforms associated to modular elliptic curves.) Obviously, this conjecture implies the weaker statement

$$\#\{|D| \le X : L(F \otimes \chi_D, k) \ne 0\} \gg_F X. \tag{1.2}$$

Using a variety of methods, early works by Bump, Friedberg, Hoffstein, Iwaniec, Murty, Murty and Waldspurger (see [2], [9], [15], [23]) produced a number of important nonvanishing theorems in the direction of (1.2). More recently, Katz and Sarnak [11] provided (among many other results) conditional proofs of (1.2). However, this claim has only been proven for certain special newforms by the works of James [10], Kohnen [12] and Vatsal [22]. These cases require that the modular forms possess exceptional mod 3 Galois representations. The best unconditional general result in the direction of (1.2) is due to the first author and Skinner [17]. They proved that

$$\#\{|D| \le X : L(F \otimes \chi_D, k) \ne 0\} \gg_F X/\log X. \tag{1.3}$$

We obtain, for almost every F, the following minor improvement of (1.3).

Theorem 1.1. *Let $F(z) = \sum_{n=1}^{\infty} a(n)q^n \in S_{2k}(\Gamma_0(M))$ be an even weight newform and let K be a number field containing the coefficients $a(n)$. If v is a place of K over 2 and there is a prime $p \nmid 2M$ for which*

$$\mathrm{ord}_v(a(p)) = 0, \tag{1.4}$$

then there is a rational number $0 < \alpha < 1$ for which

$$\#\{|D| \le X : L(F \otimes \chi_D, k) \ne 0\} \gg_F \frac{X}{\log^{1-\alpha} X}.$$

This result has immediate implications for elliptic curves. We begin by fixing notation. Suppose that $E/\mathbb{Q}$ is an elliptic curve

$$E : y^2 = x^3 + ax + b,$$

and let $L(E, s) = \sum_{n=1}^{\infty} a_E(n)n^{-s}$ be its Hasse-Weil L-function. For integers d which are not perfect squares, let $E(d)$ denote the d-quadratic twist of E,

$$E(d) : dy^2 = x^3 + ax + b.$$

Moreover, if E is an elliptic curve defined over a number field K, then let $\mathrm{rk}(E, K)$ denote the rank of the Mordell-Weil group $E(K)$. Similarly, let $\mathrm{III}(E, K)$ denote the Tate-Shafarevich group of E/K.

By a celebrated theorem of Kolyvagin [13] and the modularity of E, (1.2) implies the widely held speculation that

$$\#\{|D| \le X : \mathrm{rk}(E(D), \mathbb{Q}) = 0\} \gg_E X. \tag{1.5}$$

Heath-Brown [8] confirmed (1.5) for the congruent number elliptic curves, and subsequent works by James [12], Vatsal [22] and Wong [24] confirm this assertion for a variety of families of quadratic twists which contain an elliptic curve with a rational torsion point of order 3. However, (1.5) remains open for most elliptic curves. In this direction, Theorem 1.1. implies the following result.

Corollary 1.2. *If $E/\mathbb{Q}$ is an elliptic curve without a $\mathbb{Q}$-rational torsion point of order 2, then there is a number $0 < \alpha(E) < 1$ for which*

$$\#\{|D| \leq X \colon \operatorname{rk}(E(D), \mathbb{Q}) = 0\} \gg_E \frac{X}{\log^{1-\alpha(E)} X}.$$

Theorem 1.1 and Corollary 1.2 depend on a nonvanishing theorem (see Theorem 2.1) which guarantees the existence of a fat set of discriminants D, which is closed under multiplication, for which $L(F \otimes \chi_D, k) \neq 0$. The most interesting consequence of Theorem 2.1 may be the following result concerning the triviality of the rank of the Mordell-Weil group of most elliptic curves E over prescribed elementary abelian 2-extensions of $\mathbb{Q}$ of arbitrarily large degree.

Theorem 1.3. *Let $E/\mathbb{Q}$ be an elliptic curve without a $\mathbb{Q}$-rational torsion point of order 2. Then there is a fundamental discriminant D_E and a set of primes S_E of positive density with the property that for every positive integer j we have*

$$\operatorname{rk}(E(D_E), \mathbb{Q}(\sqrt{m_1}, \sqrt{m_2}, \ldots, \sqrt{m_j})) = \operatorname{rk}(E(D_E), \mathbb{Q}) = 0$$

whenever the integers $m_1, m_2, \ldots, m_j > 1$ satisfy the following conditions:

(1) *Each m_i is square-free with an even number of prime factors.*

(2) *All of the prime factors of each m_i are in S_E.*

Theorem 2.1 may also be used to prove the existence of non-trivial elements of Tate-Shafarevich groups of elliptic curves. Regarding Tate-Shafarevich groups, works by Bölling [1], Cassels [3], Kramer [14], and Rohrlich [18] yield a variety of results concerning the non-triviality of the 2 and 3-parts of Tate-Shafarevich groups for families of elliptic curves. Less is known about the non-triviality of p-parts of $\text{III}(E)$ for primes $p \geq 5$.

Under a natural hypothesis, Theorem 2.1 and a theorem of Frey yield a general result which holds for many (if not all) curves E whose Mordell-Weil group over $\mathbb{Q}$ has torsion subgroup $\mathbb{Z}/3\mathbb{Z}$, $\mathbb{Z}/5\mathbb{Z}$ or $\mathbb{Z}/7\mathbb{Z}$. For simplicity we present the following special case of this result.

Theorem 1.4. *Suppose that $E/\mathbb{Q}$ is an elliptic curve whose torsion sub-group over $\mathbb{Q}$ is $\mathbb{Z}/\ell\mathbb{Z}$ with $\ell \in \{3, 5, 7\}$. If E is good at ℓ (see §3), has good reduction at ℓ and has the property that there is an odd prime $p_0 \equiv -1$ (mod ℓ) of bad reduction with*

$$\mathrm{ord}_{p_0}(\Delta(E)) \not\equiv 0 \quad (\mathrm{mod}\ \ell),$$

where $\Delta(E)$ is the discriminant of E, then there are infinitely many negative square-free integers d for which

$$\mathrm{rk}(E(d), \mathbb{Q}) = 0 \quad and \quad \#\text{Ш}(E(d), \mathbb{Q}) \equiv 0 \quad (\mathrm{mod}\ \ell).$$

Remark. Earlier work of Wong [24] implied that every elliptic curve without a $\mathbb{Q}$-rational point of order 2 is good at ℓ. We employed his result in a crucial way to obtain [16, Th. 5] and [16, Cor. 6]. Unfortunately, we have been informed that there is a mistake in Wong's argument. Therefore, readers should be aware that [16, Th. 5] and [16, Cor. 6] are true for elliptic curves that are good at ℓ (see §3).

Example. Let E be the elliptic curve of conductor 26 given by

$$E\colon y^2 + xy + y = x^3 - x^2 - 3x + 3.$$

Its torsion subgroup is $\mathbb{Z}/7\mathbb{Z}$, and we have $\Delta(E) = -2^7 \cdot 13$. By Theorem 1.4, if E is good at 7, there are infinitely many negative fundamental discriminants D for which

$$\mathrm{rk}(E(D), \mathbb{Q}) = 0 \quad and \quad \#\text{Ш}(E(D), \mathbb{Q}) \equiv 0 \quad (\mathrm{mod}\ 7).$$

Using Theorems 1.3 and 1.4, we obtain the next theorem which shows, for certain elliptic curves $E/\mathbb{Q}$, that there are infinitely many number fields K for which both

$$\mathrm{rk}(E, K) \gg_E \log([K : \mathbb{Q}]),$$
$$\mathrm{rk}_p(\text{Ш}(E, K)) \gg_E \log([K : \mathbb{Q}]).$$

Theorem 1.5. *Let $E/\mathbb{Q}$ be an elliptic curve whose torsion subgroup over $\mathbb{Q}$ is $\mathbb{Z}/p\mathbb{Z}$ with $p \in \{3, 5, 7\}$. If E is good at p (see §3), has good reduction at p, and has the property that there is an odd prime $p_0 \equiv -1$ (mod p) of bad reduction with*

$$\mathrm{ord}_{p_0}(\Delta(E)) \not\equiv 0 \quad (\mathrm{mod}\ p),$$

where $\Delta(E)$ is the discriminant of E, then for every pair of non-negative integers r_m and r_s there are $r_m + r_s$ square-free integers $d_1, d_2, \ldots, d_{r_m+r_s}$ with

$$rk(E, \mathbb{Q}(\sqrt{d_1}, \sqrt{d_2}, \ldots, \sqrt{d_{r_m+r_s}})) \geq 2r_m,$$
$$\mathrm{rk}_p(\text{Ш}(E, \mathbb{Q}(\sqrt{d_1}, \sqrt{d_2}, \ldots, \sqrt{d_{r_m+r_s}}))) \geq 2r_s.$$

In §2 we describe Theorem 2.1 and give a brief sketch of its proof, and in §3 we sketch the proofs of Theorems 1.1, 1.3, 1.4 and 1.5.

2   The Crucial Nonvanishing Theorem

The next theorem is the main result which is vital for all of the results described in §1.

Theorem 2.1. *Let $F(z) = \sum_{n=1}^{\infty} a(n)q^n \in S_{2k}(\Gamma_0(M))$ be an even weight newform and let K be a number field containing the coefficients $a(n)$. If v is a place of K over 2 and there is a prime $p \nmid 2M$ for which*

$$\mathrm{ord}_v(a(p)) = 0, \qquad (2.1)$$

then there is a fundamental discriminant D_F and a set of primes S_F with positive density such that for every positive integer j we have

$$L(F \otimes \chi_{p_1 p_2 \cdots p_{2j} D_F}, k) \neq 0$$

whenever $p_1, p_2, \ldots, p_{2j} \in S_F$ are distinct primes not dividing D_F.

Sketch of the proof. We begin by recalling a theorem due to Waldspurger [23, Th. 1] on the Shimura correspondence [20]. This result expresses many of the central critical values $L(F \otimes \chi_D, k)$ in terms of the Fourier coefficients of certain half integral weight cusp forms.

For every fundamental discriminant D, define D_0 by

$$D_0 := \begin{cases} |D| & \text{if } D \text{ is odd,} \\ |D|/4 & \text{if } D \text{ if even.} \end{cases} \qquad (2.2)$$

If $\delta \in \{\pm 1\}$ is the sign of the functional equation of $L(F, s)$, then there is a positive integer N with $M | N$, a Dirichlet character χ modulo $4N$, a non-zero complex number Ω_F and a non-zero half integral weight eigenform

$$g_F(z) = \sum_{n=1}^{\infty} b_F(n)q^n \in S_{k+\frac{1}{2}}(\Gamma_0(4N), \chi)$$

with the property that if $\delta D > 0$, then

$$b_F(D_0)^2 = \begin{cases} \epsilon_D \cdot \dfrac{L(F \otimes \chi_D, k) D_0^{k-\frac{1}{2}}}{\Omega_F} & \text{if } \gcd(D_0, 4N) = 1, \\ 0 & \text{otherwise,} \end{cases} \qquad (2.3)$$

where ϵ_D is algebraic. Moreover, the coefficients $a(n), b_F(n)$ and the values of χ are in O_K, the ring of integers of some fixed number field K. In addition, if $p \nmid 4N$ is prime, then

$$\lambda(p) = \chi^2(p)a(p), \tag{2.4}$$

where $\lambda(p)$ is the eigenvalue of $g_F(z)$ for the half integer weight Hecke operator $T_k^\chi(p^2)$ on $S_{k+\frac{1}{2}}(\Gamma_0(4N), \chi)$.

Define the integer s_0 by

$$s_0 := \min\{\operatorname{ord}_v(b_F(n))\}. \tag{2.5}$$

In addition, let $G_F(z)$ be the integer weight cusp form defined by

$$G_F(z) = \sum_{n=1}^\infty b_g(n)q^n := g_F(z) \cdot \left(1 + 2\sum_{n=1}^\infty q^{n^2}\right). \tag{2.6}$$

It is easy to see that if n is a positive integer, then

$$b_g(n) = b_F(n) + 2\sum_{t=1}^\infty b_F(n - t^2) \equiv b_F(n) \pmod 2. \tag{2.7}$$

It is our goal to determine conditions under which $b_g(n)$ is non-zero modulo 2. To achieve this, we employ classical results regarding modular Galois representations due to Deligne and Serre [4], [5].

Suppose that $f(z) = \sum_{n=1}^\infty a_f(n)q^n \in S_k(\Gamma_0(M), \psi)$ is an integer weight newform and suppose that K is a number field whose ring of integers O_K contains the Fourier coefficients $a(n)$ and the values of ψ. If O_v is the completion of O_K at any finite place v of K, say with residue characteristic ℓ, then there is a continuous representation

$$\rho_{f,v} \colon \operatorname{Gal}(\overline{\mathbb{Q}}/\mathbb{Q}) \to GL_2(O_v)$$

with the property that if $p \nmid \ell M$ is prime, then

$$\operatorname{Tr}(\rho_{f,v}(\operatorname{Frob}(p))) = a_f(p).$$

Using these representations, we are able to study the arithmetic of the Fourier expansions of a collection of integer weight cusp forms with coefficients that are algebraic integers. Suppose that $f_1(z), f_2(z), \ldots, f_y(z)$ are integer weight cusp forms with

$$f_i(z) = \sum_{n=1}^\infty a_i(n)q^n \in S_{k_i}(\Gamma_0(M_i), \chi_i).$$

Suppose that the coefficients of all the $f_i(z)$ and the values of all the χ_i are in O_K (for some sufficiently large number field K), and let v be a finite place of K with residue characteristic ℓ. If $p_0 \nmid \ell M_1 M_2 \ldots M_y$ is prime and j is a positive integer, then the Chebotarev density theorem implies that there is a set of primes p, of positive density, such that for every $1 \le i \le y$ we have

$$\mathrm{ord}_v \left(f_i(z)|T_{p_0}^{k_i,\chi_i} - f_i(z)|T_p^{k_i,\chi_i} \right) > j. \tag{2.8}$$

In view of (2.4), there is a prime $p_0 \nmid 4N$ for which $\mathrm{ord}_v(\lambda(p_0)) = 0$. Applying (2.8) to $G_F(z)$ and $F(z)$, there is a set of primes S_{p_0} with positive density with the property that every prime $p \in S_{p_0}$ satisfies

$$\mathrm{ord}_v(\lambda(p)) = \mathrm{ord}_v(\lambda(p_0)) = 0, \tag{2.9}$$

and

$$\mathrm{ord}_v \left(G_F(z)|T_p^{k+1,\chi\chi_{-1}^{k+1}} - G_F(z)|T_{p_0}^{k+1,\chi\chi_{-1}^{k+1}} \right) > s_0. \tag{2.10}$$

Suppose that m is a positive integer for which $\mathrm{ord}_v(b_F(m)) = s_0$, and suppose that $q_1, q_2 \in S_{p_0}$ are distinct odd primes which are coprime to m. Using the definition of the integer and half integral weight Hecke operators, one easily checks that the coefficient of q^{mq_1} in $G_F(z)|T_{q_1}^{k+1,\chi\chi_{-1}^{k+1}}$ is

$$\lambda(q_1)b_g(m) + b_g(m)\chi^\star(q_1)q_1^{k-1}\left(\chi_{-1}(q_1)q_1 - \left(\frac{m}{q_1}\right)\right),$$

where $\chi^\star(p) := \chi(p)\left(\frac{-1}{p}\right)^k$. Since $\chi_{-1}(q_1)q_1 - \left(\frac{m}{q_1}\right) \equiv 0 \pmod 2$, by (2.9), we find that the coefficient of q^{mq_1} in $G_F(z)|T_{q_1}^{k+1,\chi\chi_{-1}^{k+1}}$ has ord_v equal to s_0. By (2.10), the coefficient of q^{mq_1} in $G_F(z)|T_{q_2}^{k,\chi\chi_{-1}^{k+1}}$ also has $\mathrm{ord}_v = s_0$, and this equals

$$b_g(mq_1q_2) + \chi(q_2)\chi_{-1}^{k+1}(q_2)q_2^k b_g(mq_1/q_2) = b_g(mq_1q_2).$$

This shows that if $\mathrm{ord}_v(b_F(m)) = s_0$ and $q_1, q_2 \in S_{p_0}$ are distinct odd primes which do not divide m, then $\mathrm{ord}_v(b_F(mq_1q_2)) = s_0$. In view of (2.3), the theorem follows by iterating this observation. $\square$

3 The Loose Ends

In this section we sketch the proofs of Theorems 1.1, 1.3 and 1.4.

Sketch of the proof of Theorem 1.1. Let T be a set of primes with density $0 < \alpha < 1$, and let $\mathbb{N}_T$ denote the set

$$\mathbb{N}_T := \{n \in \mathbb{N} \colon n = \prod_i p_i, \ p_i \in T, \text{ and } \mu(n) = 1\}. \qquad (3.1)$$

Here μ denotes the usual Möbius function.

Generalizing an argument of Landau, Serre [19, Th. 2.8] proved that

$$\#\{n \leq X \colon n = \prod_i p_i \text{ with } p_i \in T\} = c_T \cdot \frac{X}{\log^{1-\alpha} X} + O\left(\frac{X}{\log^{2-\alpha} X}\right)$$

for some positive constant c_T. A simple sieve argument yields

$$\#\{n \in \mathbb{N}_T \colon n \leq X\} \gg \frac{X}{\log^{1-\alpha} X}.$$

Theorem 1.1 follows immediately from this estimate and Theorem 2.1. □

Sketch of the proof of Theorem 1.3. Since $a_E(p)$ is even for all but finitely many primes p if and only if E has a rational point of order 2, we may freely apply Theorem 2.1. By the modularity of E and Kolyvagin's theorem, if $L(E(D), 1) \neq 0$, then $\mathrm{rk}(E(D), \mathbb{Q}) = 0$. Suppose that $S := \{m_1, m_2, \ldots, m_t\}$ is a set of square-free pairwise coprime integers > 1 where all the prime factors of each m_i are in the prescribed set of primes. If each m_i has an even number of prime factors, then for each $1 \leq s \leq t$ and any distinct $d_1, d_2, \ldots, d_s \in S$ we have

$$\mathrm{rk}(E(d_1 d_2 \ldots d_s), \mathbb{Q}) = 0.$$

Theorem 1.3 now follows from the fact that

$$rk(E, K(\sqrt{d})) = \mathrm{rk}(E, K) + \mathrm{rk}(E(D), K)$$

whenever $[K(\sqrt{d}) \colon K] = 2$. □

Let $E/\mathbb{Q}$ be an elliptic curve whose torsion subgroup over $\mathbb{Q}$ is $\mathbb{Z}/\ell\mathbb{Z}$ where $\ell \in \{3, 5, 7\}$. By Theorem 2.1, there is a discriminant d_E and a subset of primes of primes S_E such that for every set of distinct odd primes $p_1, \ldots, p_{2j} \in S_E$ coprime to d_E we have

$$L(E(d_E p_1 \ldots p_{2j}), 1) \neq 0.$$

We say that E is *good at* ℓ if there are infinitely many such negative fundamental discriminants $D = d_E p_1 \ldots p_{2j}$ for which the following hold:

(i) We have $\ell \mid \# \operatorname{Cl}(D)$, the ideal class group of the quadratic field $\mathbb{Q}(\sqrt{D})$.

(ii) We have $\gcd(D, \ell N(E)) = 1$, where $N(E)$ is the conductor of E.

(iii) We have $D \equiv 0 \pmod 4$ and $D/4 \equiv 3 \pmod 4$.

(iv) If $\operatorname{ord}_\ell(j(E)) < 0$, then $\left(\frac{D}{\ell}\right) = -1$.

(v) Every prime $p \mid N(E)$ with $p \notin \{2, \ell\} \cup S_E$ and with the additional property that $p \not\equiv -1 \pmod \ell$ or $\operatorname{ord}_p(\Delta(E)) \not\equiv 0 \pmod \ell$ satisfies:

$$\left(\frac{D}{p}\right) = \begin{cases} -1 & \text{if } \operatorname{ord}_p(j(E)) \geq 0, \\ -1 & \text{if } \operatorname{ord}_p(j(E)) < 0 \text{ and } E/\mathbb{Q}_p \text{ is a Tate curve,} \\ 1 & \text{otherwise.} \end{cases}$$

Sketch of the proof of Theorem 1.4. Let ℓ be an odd prime such that $E/\mathbb{Q}$ is an elliptic curve with good reduction at ℓ with a $\mathbb{Q}$-rational point of order ℓ. If D is a negative fundamental discriminant satisfying the conditions appearing in the definition above, then Frey [6] proved that

$$\# \operatorname{Cl}(D)_\ell \mid \# S_\ell(E(D), \mathbb{Q}).$$

Here $\operatorname{Cl}(D)_\ell$ denotes the ℓ-part of the ideal class group of $\mathbb{Q}(\sqrt{D})$, and $S_\ell(E(D), \mathbb{Q})$ is the ℓ-Selmer group of $E(D)$ over $\mathbb{Q}$.

Let S_E be the set of primes with positive density given in Theorem 2.1. Then there is a discriminant d_E, and a subset $\tilde{S}_E$ of S_E, such that for every set of distinct odd primes $p_1, p_2, \ldots, p_{2j} \in \tilde{S}_E$ which are coprime to d_E we have

$$L(E(d_E p_1 p_2 \ldots p_{2j}), 1) \neq 0, \tag{3.2}$$
$$\# \operatorname{Cl}(d_E p_1 p_2 \ldots p_{2j})_\ell \mid \# S_\ell(E(d_E p_1 p_2 \ldots p_{2j}), \mathbb{Q}). \tag{3.3}$$

By Kolyvagin's theorem, (3.2) implies that $E(d_E p_1 p_2 \ldots p_{2j})$ has rank zero. Therefore, we find that if $p_1, p_2, \ldots, p_{2j} \in \tilde{S}_E$ are distinct primes which do not divide d_E, then

$$\# \operatorname{Cl}(d_E p_1 p_2 \ldots p_{2j})_\ell \equiv 0 \pmod \ell$$
$$\Rightarrow \mathbb{Z}/\ell\mathbb{Z} \times \mathbb{Z}/\ell\mathbb{Z} \subseteq \text{Ш}(E(d_E p_1 p_2 \ldots p_{2j})).$$

Therefore, it suffices to prove that there are infinitely many suitable discriminants satisfying (3.2) and (3.3) with the additional property that $\ell \mid \# \operatorname{Cl}(d_E p_1 p_2 \ldots p_{2j})_\ell$. This is guaranteed since E is good at ℓ. $\square$

Now we prove Theorem 1.5. We begin by fixing notation. For an abelian group A and a positive integer m, we let $A[m]$ denote the m-torsion of A, and for a prime p, we let $\mathrm{rk}_p(A)$ denote the p-rank of $A[p]$. If A is finitely generated, we let $\mathrm{rk}(A)$ denote its rank. For a Galois extension of fields L/K, we let $G_{L/K}$ denote the Galois group. We let $G_K = G_{\overline{K}/K}$. For $d \in K$, we let $K_d = K(\sqrt{d})$ and $G_d = G_{K_d/K}$.

Our results depend on the following relations between p-ranks of Tate-Shafarevich and Mordell-Weil groups of elliptic curves upon a quadratic extension. Here $S_p(E, K)$ denotes the p-Selmer group of an elliptic curve E/K.

Lemma 3.1. *Let E be an elliptic curve defined over a number field K. Let p be an odd prime, and let d be a non-square in K. Let $r(E, K)$ denote either $\mathrm{rk}(E, K)$, $\mathrm{rk}_p(S_p(E, K))$, or $\mathrm{rk}_p(\text{Ш}(E, K))$. Then*

$$r(E, K_d) = r(E, K) + r(E(d), K).$$

Proof. The result for the rank of $E(K_d)$ is well-known. Let G_d be represented by $\{1, \rho\} \subset G_K$. Fix an isomorphism $\phi \colon E \to E(d)$ defined over K_d so that $\rho\phi(Q) = -\phi(\rho Q)$ for all $Q \in E(\overline{K})$. As p is odd, we note that

$$E(K_d)[p] \cong E(K)[p] \oplus E(d)(K)[p]$$

via the map $Q \mapsto \frac{1}{2}(\rho Q + Q, \phi(\rho Q - Q))$. Therefore the result also holds for $r(E, K_d) = \mathrm{rk}_p(E(K_d)/pE(K_d))$.

By the exact sequence

$$0 \to E(K_d)/pE(K_d) \to S_p(E, K_d) \to \text{Ш}(E, K_d)[p] \to 0,$$

it now suffices to prove the result for Selmer groups. The Selmer group decomposes as

$$S_p(E, K_d) = S_p(E, K_d)^{G_d,+} \oplus S_p(E, K_d)^{G_d,-},$$

where the first group is the G_d-invariants of $S_p(E, K_d)$ and the second comprises those elements on which ρ acts by -1. Since p is odd, the Galois cohomology groups $H^i(G_d, E(K_d)[p])$ are trivial for $i \geq 1$; therefore, by the Hochschild-Serre spectral sequence, the restriction map $H^1(G_K, E[p]) \to H^1(G_{K_d}, E[p])^{G_d}$ is an isomorphism. Thus it follows that

$$\mathrm{res} \colon S_p(E, K) \to S_p(E, K_d)^{G_d}$$

is an isomorphism. Likewise, $S_p(E(d), K) \cong S_p(E(d), K_d)^{G_d}$. If $\tilde{\phi}$ is the map induced by ϕ on cohomology, then one can check that

$$\tilde{\phi} \colon H^1(G_{K_d}, E[p])^{G_d,-} \to H^1(G_{K_d}, E(d)[p])^{G_d}$$

is well-defined and an isomorphism. For each place v of K_d the map $\tilde{\phi}: H^1(G_{K_{d,v}}, E(\overline{K}_{d,v})) \to H^1(G_{K_{d,v}}, E(d)(\overline{K}_{d,v}))$ is necessarily an isomorphism, so it follows from the definition of the Selmer group that

$$\tilde{\phi}: S_p(E, K_d)^{G_d,-} \to S_p(E(d), K_d)^{G_d}$$

is also an isomorphism. Therefore $S_p(E, K_d)^{G_d,-} \cong S_p(E(d), K)$, and we are done. $\qquad\square$

The following theorem is a weak version of one of the main results in [21].

Theorem 3.2. *If $E/\mathbb{Q}$ is an elliptic curve, then for every positive integer r_m there are distinct square-free integers $D_1, D_2, \ldots, D_{r_m}$ for which*

$$\mathrm{rk}(E(D_i), \mathbb{Q}) \geq 2.$$

Proof of Theorem 1.5. Using Theorem 3.2, let $D_1, D_2, \ldots D_{r_m}$ be distinct square-free integers for which $\mathrm{rk}(E(D_i), \mathbb{Q}) \geq 2$. Hence, Lemma 3.1 implies that

$$\mathrm{rk}(E, \mathbb{Q}(\sqrt{D_1}, \sqrt{D_2}, \ldots, \sqrt{D_{r_m}})) \geq \sum_{j=1}^{r_m} \mathrm{rk}(E(D_j), \mathbb{Q}) = 2r_m. \qquad (3.4)$$

Similarly, by Theorem 1.4 there are r_s many distinct square-free integers $d_1, d_2, \ldots, d_{r_s}$ for which

$$\mathrm{rk}(E(d_i), \mathbb{Q}) = 0 \text{ and } \#\mathrm{III}(E(d_i), \mathbb{Q}) \equiv 0 \pmod{p}.$$

By the proof of Theorem 1.4 and Kolyvagin's theorem on the Birch and Swinnerton-Dyer Conjecture (i.e. finiteness of $\mathrm{III}(E, \mathbb{Q})$ and the existence of the Cassels-Tate pairing implies that $\mathrm{rk}_p(\mathrm{III}(E, \mathbb{Q}))$ is even), it follows that for each such d_i that

$$\mathrm{rk}_p(\mathrm{III}(E(d_i), \mathbb{Q})) \geq 2.$$

Therefore, by Lemma 3.1 we get

$$\mathrm{rk}_p(\mathrm{III}(E, \mathbb{Q}(\sqrt{d_1}, \sqrt{d_2}, \ldots, \sqrt{d_{r_s}}))) \geq \sum_{j=1}^{r_s} \mathrm{rk}_p(\mathrm{III}(E(d_i), \mathbb{Q})) \geq 2r_s.$$

$$(3.5)$$

Consequently, (3.4), (3.5) and Lemma 3.1 imply that

$$\mathrm{rk}(E, \mathbb{Q}(\sqrt{D_1}, \sqrt{D_2}, \ldots, \sqrt{D_{r_m}}, \sqrt{d_1}, \sqrt{d_2}, \ldots, \sqrt{d_{r_s}}) \geq 2r_m,$$
$$\mathrm{rk}_p(E, \mathbb{Q}(\sqrt{D_1}, \sqrt{D_2}, \ldots, \sqrt{D_{r_m}}, \sqrt{d_1}, \sqrt{d_2}, \ldots, \sqrt{d_{r_s}}) \geq 2r_s.$$

This completes the proof. $\qquad\square$

References

[1] R. Bölling, *Die Ordung der Schafarewitsch-Tate Gruppe kann beliebig gross werden*, Math. Nachr. **67** (1975), 157–179.

[2] D. Bump, S. Friedberg, and J. Hoffstein, *Nonvanishing theorems for L-functions of modular forms and their derivatives*, Invent. Math. **102** (1990), 543–618.

[3] J. W. S. Cassels, *Arithmetic on curves of genus 1 (VI). The Tate-Shafarevich group can be arbitrarily large*, J. Reine. Angew. Math. **214/215** (1964), 65–70.

[4] P. Deligne, *Formes modulaires et représentations ℓ-adiques*, Sem. Bourbaki 1968/1969, Exposé 355, Lect. Notes, vol. 179, Springer, 1971, pp. 139–172.

[5] P. Deligne and J.-P. Serre, *Formes modulaires de poids 1*, Ann. Scient. Ec. Norm. Sup. Sér. 4 **7** (1974), 507–530.

[6] G. Frey, *On the Selmer group of twists of elliptic curves with $\mathbb{Q}$-rational torsion points*, Canad. J. Math. **40** (1988), 649–665.

[7] D. Goldfeld, *Conjectures on elliptic curves over quadratic fields*, Number Theory (Carbondale), Springer Lect. Notes, vol. 751, 1979, pp. 108–118.

[8] D. R. Heath-Brown, *The size of the Selmer groups for the congruent number problem, II*, Invent. Math. **118** (1994), 331–370.

[9] H. Iwaniec, *On the order of vanishing of modular L-functions at the critical point*, Sém. Théorie Nombres Bordeaux **2** (1990), 365–376.

[10] K. James, *L-series with nonzero central critical value*, J. Amer. Math. Soc. **11** (1998), 635–641.

[11] N. Katz and P. Sarnak, *Random matrices, Frobenius eigenvalues, and monodromy*, Amer. Math. Soc. Colloq. Publ., vol. 45, Amer. Math. Soc., Providence, 1999.

[12] W. Kohnen, *On the proportion of quadratic twists of L-functions attached to cusp forms not vanishing at the central point*, J. Reine. Angew. Math. **508** (1999), 179–187.

[13] V. Kolyvagin, *Finiteness of $E(\mathbb{Q})$ and $\text{III}_{E/\mathbb{Q}}$ for a subclass of Weil curves (Russian)*, Izv. Akad. Nauk., USSR, ser. Matem. **52** (1988), 522–540.

[14] K. Kramer, *A family of semistable elliptic curves with large Tate-Shafarevich groups*, Proc. Amer. Math. Soc. **89** (1983), 473–499.

[15] M. R. Murty and V. K. Murty, *Mean values of derivatives of modular L-series*, Annals of Math. **133** (1991), 447–475.

[16] K. Ono, *Nonvanishing of quadratic twists of modular L-functions and applications to elliptic curves*, J. Reine Angew. Math. **533** (2001), 81–97.

[17] K. Ono and C. Skinner, *Non-vanishing of quadratic twists of modular L-functions*, Invent. Math. **134** (1998), 651–660.

[18] D. Rohrlich, *Unboundedness of the Tate-Shafarevich group in families of quadratic twists*, Appendix to J. Hoffstein and W. Luo, *Nonvanishing of L-series and the combinatorial sieve*, Math. Res. Lett. **4** (1997), 435–444.

[19] J.-P. Serre, *Divisibilité de certaines fonctions arithmétiques*, L'Enseign. Math. **22** (1976), 227–260.

[20] G. Shimura, *On modular forms of half-integral weight*, Annals of Math. **97** (1973), 440–481.

[21] C. Stewart and J. Top, *On ranks of twists of elliptic curves and power-free values of binary forms*, J. Amer. Math. Soc. **8** (1995), 947–974.

[22] V. Vatsal, *Canonical periods and congruence formulae*, Duke Math. J. **98** (1999), 397–419.

[23] J.-L. Waldspurger, *Sur les coefficients de Fourier des formes modulaires de poids demi-entier*, J. Math. Pures et Appl. **60** (1981), 375–484.

[24] S. Wong, *Elliptic curves and class number divisibility*, Int. Math. Res. Notes **12** (1999), 661–672.

Identities from the Holomorphic Projection of Modular Forms

Cormac O'Sullivan[1]

1 Introduction

Set $\mathfrak{H} = \{z = x + iy \in \mathbb{C} : y > 0\}$ and let Γ be the modular group

$$\mathrm{PSL}_2(\mathbb{Z}) = \left\{ \begin{pmatrix} a & b \\ c & d \end{pmatrix} : a, b, c, d \in \mathbb{Z}, ad - bc = 1 \right\} / \pm 1.$$

The complex vector space of modular forms for Γ of weight $2k$, denoted by $\mathcal{M}_{2k}(\Gamma) = \mathcal{M}_{2k}$, consists of functions $f : \mathfrak{H} \to \mathbb{C}$ that satisfy

$$f\left(\begin{pmatrix} a & b \\ c & d \end{pmatrix} z \right) = f\left(\frac{az + b}{cz + d} \right) = (cz + d)^{2k} f(z) \tag{1.1}$$

for each $\begin{pmatrix} a & b \\ c & d \end{pmatrix} \in \Gamma$ and are holomorphic on $\mathfrak{H}$. We also require them to be holomorphic at infinity. In other words, f has the Fourier expansion $f(z) = \sum_{n=0}^{\infty} a_n e^{2\pi i n z}$. If $a_0 = 0$ then $f(z)$ has exponential decay as $y \to \infty$. We term the space of such forms $\mathcal{S}_{2k}$, the cusp forms. The spaces $\mathcal{M}_{2k}$ and $\mathcal{S}_{2k}$ with $k \in \mathbb{N}$ are finite dimensional with

$$\dim \mathcal{M}_{2k} = \left[\frac{2k}{12} \right] + 1 \ (\text{or} \ \left[\frac{2k}{12} \right] \ \text{if} \ 2k \equiv 2 \mod 12),$$

$$\dim \mathcal{S}_{2k} = \dim \mathcal{M}_{2k} - 1 \ (\text{or} \ 0 \ \text{if} \ 2k \leq 10).$$

For example, an element of $\mathcal{M}_{2k}$ is the Eisenstein series

$$E_{2k}(z) = 1 - \frac{4k}{B_{2k}} \sum_{m=1}^{\infty} \sigma_{2k-1}(m) q^m \tag{1.2}$$

where $k \geq 2$, B_k is the kth Bernoulli number ($B_2 = \frac{1}{6}$, $B_4 = \frac{-1}{30}$, $B_6 = \frac{1}{42}$, $B_8 = \frac{-1}{30}$, ...), $q = e^{2\pi i z}$ and σ_k the divisor sum defined by $\sigma_k(n) = \sum_{d|n} d^k$.

[1] I would like to express my thanks to Professor K. S. Williams for sending me his preprint and to the Millennial Conference on Number Theory organizers for a very enjoyable week.

If $f \in \mathcal{M}_k$ and $g \in \mathcal{M}_l$ then $fg \in \mathcal{M}_{k+l}$ so we get the well known result $E_4(z)E_4(z) = cE_8(z)$ since $\dim \mathcal{M}_8 = 1$. Comparing Fourier expansions we obtain the identity

$$\sigma_3(n) + 120 \sum_{i=1}^{n-1} \sigma_3(i)\sigma_3(n-i) = \sigma_7(n) \tag{1.3}$$

for all $n \geq 1$. What kinds of identities of this type are possible in more general settings?

2 The Space $\mathbf{C^\infty(\Gamma \backslash \mathfrak{H}, 2k)}$

This is the space of smooth functions Φ that transform as follows:

$$\Phi\left(\begin{pmatrix} a & b \\ c & d \end{pmatrix} z\right) = \left(\frac{cz+d}{|cz+d|}\right)^{2k} \Phi(z). \tag{2.1}$$

For example, if $f \in \mathcal{M}_{2k}$ then $y^k f(z) \in C^\infty(\Gamma \backslash \mathfrak{H}, 2k)$. For elements $\Phi_1, \Phi_2 \in C^\infty(\Gamma \backslash \mathfrak{H}, 2k)$ that do not grow large too quickly we have the inner product

$$\langle \Phi_1, \Phi_2 \rangle = \int_{\Gamma \backslash \mathfrak{H}} \Phi_1(z)\overline{\Phi_2(z)} \, d\mu z, \tag{2.2}$$

where $d\mu z = \frac{dx\,dy}{y^2}$. Also available are the Maass raising and lowering operators $R_{2k} = 2iy\frac{d}{dz} + k$ and $L_{2k} = -2iy\frac{d}{d\bar{z}} - k$, where

$$R_{2k} : C^\infty(\Gamma \backslash \mathfrak{H}, 2k) \to C^\infty(\Gamma \backslash \mathfrak{H}, 2k+2),$$
$$L_{2k} : C^\infty(\Gamma \backslash \mathfrak{H}, 2k) \to C^\infty(\Gamma \backslash \mathfrak{H}, 2k-2).$$

Finally we may also define the weight $2k$ hyperbolic Laplacian $\Delta_{2k} = -y^2(\frac{\partial^2}{\partial x^2} + \frac{\partial^2}{\partial y^2}) + 2iky\frac{\partial}{\partial x}$. It is related to the raising and lowering operators by

$$\Delta_{2k} = -L_{2k+2}R_{2k} - k(1+k) \tag{2.3}$$
$$= -R_{2k-2}L_{2k} + k(1-k). \tag{2.4}$$

See [2, Chapter 2] for more details.

3 Holomorphic Projection

To construct identities we need to project our results into the finite dimensional space $\mathcal{M}_{2k}$. Define the Poincare series

$$P_m(z, 2k) = \sum_{\left(\begin{smallmatrix} a & b \\ c & d \end{smallmatrix}\right) \in \Gamma_\infty \backslash \Gamma} (cz+d)^{-2k} e^{2\pi i m \frac{az+b}{cz+d}} \tag{3.1}$$

for $k \geq 2$ and $\Gamma_\infty = \{\gamma \in \Gamma : \gamma\infty = \infty\}$. It is a cusp form in S_{2k} for $m \geq 1$, and for any other $f \in S_{2k}$ we have

$$\langle y^k f(z), y^k P_m(z, 2k)\rangle = a_m \frac{(2k-2)!}{(4\pi m)^{2k-1}}$$

with a_m the mth Fourier coefficient of $f(z)$. We may use this feature of the Poincare series to define a projection map $\pi_{\text{hol}} : C^\infty(\Gamma\backslash\mathfrak{H}, 2k) \to M_{2k}$.

Lemma 3.1. *For $\Phi \in C^\infty(\Gamma\backslash\mathfrak{H}, 2k)$ satisfying $\frac{1}{y^k}\Phi(z) = c_0 + O(y^{-\varepsilon})$ as $y \to \infty$ with $k \geq 2$ and $\varepsilon > 0$ set*

$$\pi_{hol}(\Phi(z)) = c_0 + \sum_{m=1}^{\infty} \langle\Phi(z), y^k P_m(z, 2k)\rangle \frac{(4\pi m)^{2k-1}}{(2k-2)!} q^m.$$

Then $\pi_{hol}(\Phi(z)) \in M_{2k}$ and $\langle y^k f(z), \Phi\rangle = \langle y^k f(z), y^k \pi_{hol}\Phi\rangle$ for every $f \in S_{2k}$.

Note that if $g(z)$ is already an element of M_{2k} then $\pi_{\text{hol}}(y^k g(z)) = g(z)$, and in that sense it is a projection. This idea originated in [14]. See [15] for a proof of the above lemma.

In the rest of this paper we explore some of the combinatorial and number theoretic identities arising from the following operations:

(i) Multiplication: $M_{2k}(\Gamma) \times M_{2l}(\Gamma) \to M_{2k+2l}(\Gamma)$,

(ii) $R_{2k} : C^\infty(\Gamma\backslash\mathfrak{H}, 2k) \to C^\infty(\Gamma\backslash\mathfrak{H}, 2k+2)$,

(iii) $L_{2k} : C^\infty(\Gamma\backslash\mathfrak{H}, 2k) \to C^\infty(\Gamma\backslash\mathfrak{H}, 2k-2)$,

(iv) $\pi_{\text{hol}} : C^\infty(\Gamma\backslash\mathfrak{H}, 2k) \to M_{2k}(\Gamma)$.

4 Repeatedly Raising and Lowering Holomorphic Modular Forms

To see what happens when we repeatedly apply the Maass raising operator it is useful to re-express things in terms of the simpler operator $D_R = 2iy^2\frac{d}{dz}$. We obtain

$$R_{2(k+n-1)}R_{2(k+n-2)} \cdots R_{2k} = \sum_{j=0}^{n} \binom{n}{j}\frac{(j+k-1)!}{(k-1)!}y^{j-n}D_R^{n-j} \qquad (4.1)$$

for $k > 0$, and $R_{2(n-1)} \cdots R_0 = y^{-n}D_R^n$.

Similarly, for the lowering operator we have

$$L_{2(k-n+1)}L_{2(k-n+2)}\ldots L_{2k} = (-1)^n \sum_{j=0}^{n} \binom{n}{j} \frac{k!}{(k-j)!} y^{j-n} D_L^{n-j} \qquad (4.2)$$

for $k \geq n+1$ and $D_L = 2iy^2\frac{d}{dz}$. Formulas (4.1) and (4.2) may be verified by induction. For convenience, when it is clear that we are dealing with an element of $C^\infty(\Gamma\backslash\mathfrak{H}, 2k)$, we shall write R^n instead of $R_{2(k+n-1)}R_{2(k+n-2)}$ $\ldots R_{2k}$ and L^n for $L_{2(k-n+1)}L_{2(k-n+2)}\ldots L_{2k}$.

4.1   Lowering Modular Forms

For $f(z) = \sum_{m=0}^{\infty} a_m q^m \in M_{2k}$ we compute $L^n y^k f(z)$. Actually the answer is rather easy because

$$L_{2k} y^k f(z) = (-2iy\frac{d}{dz} - k)y^k f(z)$$

$$= -2iy(\frac{d}{dz}y^k)f(z) - 2iy^{k+1}\frac{d}{dz}f(z) - ky^k f(z) = 0.$$

Therefore $L^n y^k f(z) = 0$ for any $n > 0$. On the level of the Fourier coefficients we have

$$L^n y^k f(z) = \sum_{m=0}^{\infty} a_m \left((-1)^n \sum_{j=0}^{n} \binom{n}{j} \frac{k!}{(k-j)!} y^{j-n} D_L^{n-j}(y^k q^m) \right)$$

$$= \sum_{m=0}^{\infty} a_m q^m \left((-1)^n \sum_{j=0}^{n} \binom{n}{j} \frac{k!}{(k-j)!} y^{j-n} y^{k+n-j} \right.$$

$$\left. \times (-1)^{n-j} \frac{(k+n-j-1)!}{(k-1)!} \right)$$

since $D_L^n(y^k q^m) = y^{k+n}(-1)^n \frac{(k+n-1)!}{(k-1)!} q^m$. On simplifying we see that we must have the identity

$$\sum_{j=0}^{n} (-1)^j \binom{n}{j}(k-j+1)_{n-1} = 0 \qquad (4.3)$$

for $n \geq 1$, where we define

$$(a)_m = \frac{\Gamma(a+m)}{\Gamma(a)} = \begin{cases} 1 & \text{if } m = 0, \\ (a)(a+1)\ldots(a+m-1) & \text{if } m > 0. \end{cases} \qquad (4.4)$$

We can prove this more directly. One method is to consider $\frac{d^m}{dx^m}(1+x)^n x^r$ at $x = -1$. This may be evaluated in two ways. First use the binomial expansion of $(1+x)^n$ to get

$$\sum_{i=0}^{n} \binom{n}{i}(r-m+i+1)_m(-1)^{i+r-m}.$$

Secondly use Leibnitz' formula to show that

$$\frac{d^m}{dx^m}(1+x)^n x^r = \sum_{i=0}^{m} \binom{m}{i}(n-i+1)_i(1+x)^{n-i}(r-m+i+1)_{m-i}x^{r-m+i}.$$

At $x = -1$ the only non-zero term is when $i = n$ (provided $m \geq n$) and so we obtain

$$\sum_{i=0}^{n}(-1)^i \binom{n}{i}(r-m+i+1)_m = \begin{cases} (-1)^n \frac{m!}{(m-n)!}(r-m+n+1)_{m-n} & m \geq n, \\ 0 & m < n. \end{cases} \tag{4.5}$$

Replace i by $n-i$ and set $m = n-1, k = r+1$ to see (4.3).

4.2 Raising Modular Forms

As in the previous example apply (4.1) and the formula

$$D_R^n(y^k q^m) = y^{k+n} \sum_{l=0}^{n}(-1)^l \binom{n}{l}\frac{(k+n-1)!}{(k+l-1)!}y^l(4\pi m)^l q^m, \tag{4.6}$$

to get

$$R^n y^k f(z) = \sum_{m=1}^{\infty} a_m q^m \sum_{l=0}^{n} M_l(n,k)(4\pi m)^l y^{k+l},$$

for $f(z) = \sum_{m=1}^{\infty} a_m q^m$ with

$$M_l(n,k) = (-1)^l \frac{n!}{l!} \sum_{j=0}^{n-l} \binom{j+k-1}{j}\binom{k+n-j-1}{k+l-1}.$$

Now for $p, l \geq 0$ we have the identity

$$\sum_{j=0}^{m} \binom{p+j}{p}\binom{p+l+m-j}{p+l} = \binom{2p+l+m+1}{2p+l+1}. \tag{4.7}$$

To see this note that

$$\sum_{j=0}^{\infty} \binom{p+j}{p} x^j = \frac{1}{(1-x)^{p+1}}$$

for $|x| < 1$. Consequently,

$$\sum_{i=0}^{\infty} \binom{p+i}{p} x^i \sum_{j=0}^{\infty} \binom{p+l+j}{p+l} x^j = \frac{1}{(1-x)^{p+1+p+l+1}}$$

$$= \sum_{j=0}^{\infty} \binom{2p+l+1+j}{2p+l+1} x^j,$$

and comparing the coefficients of x^m on each side yields (4.7). Thus

$$M_l(n,k) = (-1)^l \frac{n!}{l!} \binom{2k-1+n}{2k-1+l}$$

and

$$R^n y^k f(z) = \sum_{m=1}^{\infty} a_m q^m \sum_{l=0}^{n} \frac{n!}{l!} \binom{2k-1+n}{2k-1+l} (-4\pi m)^l y^{k+l}. \qquad (4.8)$$

If $f(z) = \sum_{m=0}^{\infty} a_m q^m \in \mathcal{M}_{2k}$ then the above formula remains true with a small alteration. Define

$$E(m,l) = \begin{cases} m^l & \text{if } m > 0 \text{ or } l > 0, \\ 1 & \text{if } m = 0 \text{ and } l = 0. \end{cases} \qquad (4.9)$$

Then

$$R^n y^k f(z) = \sum_{m=0}^{\infty} a_m q^m \sum_{l=0}^{n} (-4\pi)^l \frac{n!}{l!} \binom{2k-1+n}{2k-1+l} E(m,l) y^{k+l}. \qquad (4.10)$$

To work out the holomorphic projection of $R^n y^k f(z)$ and similar functions we use the next result.

Lemma 4.1. *For $\Phi \in C^{\infty}(\Gamma\backslash\mathfrak{H}, 2k)$ with*

$$\Phi(z) = \sum_{m=0}^{\infty} u_m(y) q^m, \quad u_m(y) = \sum_{i=0}^{r_m} v_m(i) y^{l_m(i)}$$

and $l_0(i) \leq k$ then $\pi_{hol}\Phi(z) = \sum_{m=0}^{\infty} c_m q^m$, where c_0 is the constant part of $\frac{1}{y^k} u_0(y)$ and for $m \geq 1$

$$c_m = \sum_{i=0}^{r_m} v_m(i) \frac{\Gamma(l_m(i) + k - 1)}{(2k-2)!} (4\pi m)^{k - l_m(i)}.$$

Proof. To compute the coefficients c_m we evaluate $\langle \Phi, y^k P_m \rangle$ by unfolding it.

$$\langle \Phi(z), y^k P_m(z, 2k) \rangle = \int_{\Gamma \backslash \mathfrak{H}} \Phi(z) y^k \overline{P_m(z, 2k)} \, d\mu z$$

$$= \int_{\Gamma \backslash \mathfrak{H}} \Phi(z) y^k \sum_{\left(\begin{smallmatrix} a & b \\ c & d \end{smallmatrix} \right) \in \Gamma_\infty \backslash \Gamma} \overline{(cz + d)^{-2k} e^{2\pi i m \gamma z}} \, d\mu z$$

$$= \int_{\Gamma \backslash \mathfrak{H}} \sum_{\left(\begin{smallmatrix} a & b \\ c & d \end{smallmatrix} \right) \in \Gamma_\infty \backslash \Gamma} \Phi(\gamma z) \mathrm{Im}(\gamma z)^k \overline{(cz + d)^{-2k} e^{2\pi i m \gamma z}} \, d\mu z$$

$$= \int_0^\infty \int_0^1 \Phi(z) y^k e^{-2\pi i m \bar{z}} \frac{dx \, dy}{y^2}$$

$$= \int_0^\infty u_m(y) y^{k-2} e^{-4\pi m y} \, dy$$

$$= \sum_{i=0}^{r_m} v_m(i) \Gamma(l_m(i) + k - 1)(-4\pi m)^{-l_m(i) - k + 1}.$$

The result follows by applying Lemma 3.1. $\qquad \square$

Set

$$\Phi(z) = R^n y^k f(z) \text{ for } f(z) = \sum_{m=1}^\infty a_m q^m \in S_{2k}.$$

Then $\Phi \in C^\infty(\Gamma, 2(n+k))$ and we have by (4.8) that $\Phi(z) = \sum_{m=1}^\infty u_m(y) q^m$ with $u_m(y) = \sum_{i=0}^{r_m} v_m(i) y^{l_m(i)}$, where $r_m = n$,

$$v_m(i) = a_m \frac{n!}{i!} \binom{2k - 1 + n}{2k - 1 + i} (-4\pi m)^i$$

and $l_m(i) = k + i$. Thus

$$\pi_{\mathrm{hol}}(R^n y^k f(z)) = \pi_{\mathrm{hol}}(\Phi(z)) = \sum_{m=1}^\infty c_m q^m$$

with

$$c_m = a_m (4\pi m)^n \sum_{i=0}^n (-1)^i \frac{n!}{i!} \binom{2k - 1 + n}{2k - 1 + i} \frac{(2k + n + i - 2)!}{(2k + 2n - 2)!}. \qquad (4.11)$$

As with the previous example we must have $\pi_{\mathrm{hol}}(R^n y^k f(z)) = 0$. This time the reason is that

$$\langle R_{2k} \Phi(z), y^k P_m(z, 2k + 2) \rangle = \langle \Phi(z), -L_{2k+2} y^k P_m(z, 2k + 2) \rangle$$
$$= \langle \Phi(z), 0 \rangle = 0. \qquad (4.12)$$

This shows that π_{hol} composed with the raising operator is identically zero. (See [2, Prop. 2.1.3] for the above relation between R_{2k} and $-L_{2k+2}$.) Therefore each $c_m = 0$ and we obtain the identity

$$\sum_{i=0}^{n}(-1)^i \binom{n}{i}(2k+i)_{n-1} = 0,$$

which may be verified for $n \geq 1$ with (4.5).

5 Raising Maass Forms

A Maass form is an element $\eta(z,s)$ of $C^\infty(\Gamma\backslash\mathfrak{H}, 2k)$, that is an eigenfunction of the Laplacian Δ_{2k} so that

$$\Delta_{2k}\eta = \lambda\eta = s(1-s)\eta. \tag{5.1}$$

Maass cusp forms have zero constant terms in their Fourier expansions and are non holomorphic analogs of the elements of S_{2k}. From (5.1) it may be shown that a Maass cusp form of weight $2k = 0$ has Fourier expansion

$$\eta(z,s) = \sum_{m\neq 0} b_m \sqrt{|m|y} K_{s-1/2}(2\pi|m|y)e^{2\pi imx}, \tag{5.2}$$

where K is the K-Bessel function and we are summing over all non-zero integers; see, for example, [2], [7]. To find $R^n\eta(z,s)$ we first need to compute $D_R^n(y^{1/2}K_v(2\pi|m|y)e(mx))$ for $v = s - 1/2$. Use the fact that $\frac{d}{dy}K_v(y) = -(1/2)(K_{v-1}(y) + K_{v+1}(y))$ to show that

$$D_R^n(y^{1/2}K_v(2\pi|m|y)e(mx)) = \sum_{i=0}^{n}\sum_{j=-i}^{i} \alpha_j^n(i)y^{1/2+n+i}K_{v+j}(2\pi|m|y)e(mx).$$

The numbers $\alpha_j^n(i)$ depend on m and may be defined recursively. (Note that the superscript n is an index not an exponent.) For $n = 0$ we have $\alpha_j^0(i) = 0$ unless $i = j = 0$ in which case $\alpha_0^0(0) = 1$. For $n \geq 0$ and $\delta = |m|m^{-1}$ we have

$$\alpha_j^{n+1}(i) = -\pi m\left(\delta\alpha_{j-1}^n(i-1) + 2\alpha_j^n(i-1) + \delta\alpha_{j+1}^n(i-1)\right)$$
$$+ (i+n+1/2)\alpha_j^n(i).$$

Setting $\beta_j^n(i) = \frac{(-2)^n\delta^j}{(2\pi m)^i}\alpha_j^n(i)$ removes the m dependence and

$$\beta_j^{n+1}(i) = \beta_{j-1}^n(i-1) + 2\beta_j^n(i-1) + \beta_{j+1}^n(i-1) - (2i+2n+1)\beta_j^n(i).$$

To isolate the dependence on j set $\beta_j^n(i) = (-1)^{n+i}\gamma_i^n\binom{2i}{j+i}$ to obtain the relation

$$\gamma_i^{n+1} = \gamma_{i-1}^n + (2i + 2n + 1)\gamma_i^n. \tag{5.3}$$

We may solve the recurrence (5.3) with the initial conditions $\gamma_0^0 = 1$ and $\gamma_i^0 = 0$ for $i \neq 0$ to get

$$\gamma_i^n = \frac{(2n)!}{(n+i)!(2i)!2^{n-i}} \quad \text{for } n \geq 1,\ 0 \leq i \leq n.$$

Therefore

$$\alpha_j^n(i) = (-4\pi m)^i \frac{(2n)!\delta^j}{(i+j)!(i-j)!(n-i)!} \tag{5.4}$$

and

$$R^n\eta(z,s) = y^{-n}D_R^n\eta(z,s)$$

$$= \sum_{m\neq 0} b_m\sqrt{|m|}y^{-n}D_R^n(K_v(2\pi|m|y)e(mx))$$

$$= \sum_{m\neq 0} b_m\sqrt{|m|}\sum_{i=0}^{n} \frac{(-4\pi m)^i}{(n-i)!}y^{i+\frac{1}{2}}$$

$$\times \sum_{j=-i}^{i} \frac{(2n)!(|m|m^{-1})^j}{(i+j)!(i-j)!}K_{v+j}(2\pi|m|y)e(mx). \tag{5.5}$$

Incidentally, with (2.3) and (2.4) we can show that $R^n\eta$ is an eigenfunction of Δ_{2n} if and only if η is an eigenfunction of Δ_0. This means that every Maass cusp form of weight $2n$ and eigenvalue $1/4 - v^2$ has the Fourier expansion (5.5).

6 Holomorphic Projection of Raised Maass Forms

As we have seen, (4.12) forces $\pi_{\text{hol}}(R^n\eta(z,s))$ to be zero. To see how the Fourier coefficients vanish we calculate $\langle R^n\eta(z,s), y^n P_m(z,2n)\rangle$. Use the fact that

$$\int_0^\infty x^{r-1}K_v(x)e^{-x}\,dx = 2^{-r}\pi^{1/2}\Gamma(r+1/2)^{-1}\Gamma(r+v)\Gamma(r-v) \tag{6.1}$$

for $\text{Re}(r) > |\text{Re}(v)|$ to get $\langle R^n\eta(z,s), y^n P_m(z,2n)\rangle$ equaling

$$\frac{b_m}{2}(4\pi m)^{1-n}\sum_{i=0}^{n} \frac{(-1)^i(2n)!}{(n-i)!}\sum_{j=-i}^{i} \frac{\Gamma(n+v+i+j-\frac{1}{2})\Gamma(n-v+i-j-\frac{1}{2})}{(i+j)!(i-j)!\Gamma(n+i)}.$$

Consequently, $\pi_{\mathrm{hol}}(R^n\eta(z,s)) = \sum_{m=1}^{\infty} c_m q^m$ with c_m given by

$$\frac{b_m}{2}(4\pi m)^n \sum_{i=0}^{n} \frac{(-1)^i 2n(2n-1)}{(n+i-1)(n-i)!}$$

$$\times \sum_{j=-i}^{i} \frac{\Gamma(n+v+i+j-\frac{1}{2})\Gamma(n-v+i-j-\frac{1}{2})}{(i+j)!(i-j)!}.$$

To check that this is indeed zero we will prove it by hand.

Lemma 6.1. *For $n \geq 1, v \in \mathbb{C}$*

$$\sum_{i=0}^{n} \frac{(-1)^i}{(n+i-1)!(n-i)!} \sum_{j=-i}^{i} \frac{\Gamma(n+v+i+j-\frac{1}{2})\Gamma(n-v+i-j-\frac{1}{2})}{(i+j)!(i-j)!} = 0.$$

Proof. Note that

$$(1+x)^s = \sum_{l=0}^{\infty} \binom{s}{l} x^l$$

is valid for all $-1 < x < 1$ and $s \in \mathbb{C}$ if we set

$$\binom{s}{l} = \frac{\Gamma(s+1)}{\Gamma(s-l+1)\Gamma(l+1)}$$

and use the usual conventions for defining x^s. With the formula

$$\frac{\Gamma(l-s)}{l!} = \Gamma(-s)(-1)^l \binom{s}{l}$$

for $l \in \mathbb{N}$ we see that

$$\sum_{l=0}^{2i} \frac{\Gamma(n+s+l-1)\Gamma(n+2i-l-s)}{l!(2i-l)!}$$

$$= \Gamma(n+s-1)\Gamma(n-s) \sum_{l=0}^{2i} \binom{-n-s+1}{l}\binom{-n+s}{2i-l}.$$

Also, since

$$\sum_{l=0}^{\infty} \binom{-n-s+1}{l} x^l = (1+x)^{-n-s+1}$$

and

$$\sum_{l=0}^{\infty} \binom{-n+s}{l} x^l = (1+x)^{-n+s},$$

their product is $(1+x)^{1-2n}$, implying that the coefficient of x^{2i} in the above is

$$\sum_{l=0}^{2i} \binom{-n-s+1}{l}\binom{-n+s}{2i-l} = \binom{1-2n}{2i} = \binom{2n+2i-2}{2i}.$$

Therefore, to finish the proof, it suffices to show that

$$S = \sum_{i=0}^{n}(-1)^i\binom{2n-1}{n-i}\binom{2n+2i-2}{2i} = 0. \qquad (6.2)$$

By adapting the identity following (4.7) we have that

$$\sum_{l=0}^{\infty}\binom{2n-2+2l}{2l}x^l = \frac{1}{2}\left((1-\sqrt{x})^{-2n+1} + (1+\sqrt{x})^{-2n+1}\right).$$

Also,

$$\sum_{l=0}^{\infty}\binom{2n-1}{l}(-x)^l = (1-x)^{2n-1},$$

so that S is the coefficient of x^n in the product

$$\frac{1}{2}\left((1-\sqrt{x})^{-2n+1} + (1+\sqrt{x})^{-2n+1}\right)(1-x)^{2n-1}$$

$$= \frac{1}{2}\left((1+\sqrt{x})^{2n-1} + (1-\sqrt{x})^{2n-1}\right) = \sum_{i=0}^{n-1}\binom{2n-1}{2i}x^i.$$

Thus $S = 0$ completing the proof. □

The identity (6.2) also appears as a special case of the identity on p. 37 of [13].

7 Projecting Products of Raised Modular Forms

In order to get a non zero projection we try the following. For $f = \sum a_m q^m \in \mathcal{M}_{2k_1}$, $g = \sum b_m q^m \in \mathcal{M}_{2k_2}$ we examine

$$\pi_{\text{hol}}(R^{n_1}y^{k_1}f(z) \cdot R^{n_2}y^{k_2}g(z)).$$

Clearly it is an element of $\mathcal{M}_{2(K+N)}$ for $K = \sum_i k_i$, $N = \sum_i n_i$. From (4.10) we derive

$$R^{n_1}y^{k_1}f(z) \cdot R^{n_2}y^{k_2}g(z) = \sum_{m=0}^{\infty}u_m(y)q^m$$

with

$$u_m(y) = \sum_{l_1=0}^{n_1} \sum_{l_2=0}^{n_2} (4\pi)^L (-1)^L \frac{n_1! n_2!}{l_1! l_2!} \binom{2k_1 - 1 + n_1}{n_1 - l_1} \binom{2k_2 - 1 + n_2}{n_2 - l_2}$$

$$\times y^{K+L} \sum_{j=0}^{m} a_j b_{m-j} E(j, l_1) E(m - j, l_2) \tag{7.1}$$

where $L = \sum_i l_i$. (This convention for N, K and L will be in place from now on.) If we label the inner sum $T_m^*(l_1, l_2; f, g)$ then

$$T_m^*(l_1, l_2; f, g) = T_m(l_1, l_2; f, g)$$
$$+ a_0 b_m E(0, l_1) E(m, l_2) + a_m b_0 E(m, l_1) E(0, l_2) \tag{7.2}$$

where

$$T_m(l_1, l_2; f, g) = \sum_{j=1}^{m-1} a_j b_{m-j} j^{l_1} (m - j)^{l_2}. \tag{7.3}$$

Finally, applying Lemma 4.1, we have that

$$\pi_{\mathrm{hol}} \left(R^{n_1} y^{k_1} f(z) \cdot R^{n_2} y^{k_2} g(z) \right) = \sum_{m=0}^{\infty} c_m q^m$$

with

$$c_m = (4\pi)^N \sum_{l_1=0}^{n_1} \sum_{l_2=0}^{n_2} (-1)^L \frac{n_1! n_2!}{l_1! l_2!} \binom{2k_1 - 1 + n_1}{n_1 - l_1} \binom{2k_2 - 1 + n_2}{n_2 - l_2}$$

$$\times \frac{(2K + N + L - 2)!}{(2K + 2N - 2)!} m^{N-L} T_m^*(l_1, l_2; f, g), \tag{7.4}$$

for $m \geq 1$ and $c_0 = (-4\pi)^N a_0 b_0 E(0, n_1) E(0, n_2)$. Hence $c_0 = 0$ unless $n_1 = n_2 = 0$ and in that case $c_0 = a_0 b_0$. So we see that the differential operator $\mathcal{P}_{n_1, n_2}$ defined by

$$\mathcal{P}_{n_1, n_2}(f, g) = (4\pi)^{-N} \pi_{\mathrm{hol}} \left(R^{n_1} y^{k_1} f(z) \cdot R^{n_2} y^{k_2} g(z) \right) \tag{7.5}$$

gives a map $\mathcal{P}_{n_1, n_2} : \mathcal{M}_{2k_1} \times \mathcal{M}_{2k_2} \to \mathcal{S}_{2(N+K)}$ for $(n_1, n_2) \neq (0, 0)$. For $n_1 = n_2 = 0$ we have $\mathcal{P}_{0,0}(f, g) = fg$. Examples in Section 10 show that this map is not identically zero.

8 Rankin-Cohen Differential Operators

The map $\mathcal{P}$ above is similar to a construction of Cohen in [3]. For $f \in \mathcal{M}_{2k_1}$, $g \in \mathcal{M}_{2k_2}$ he shows that

$$\mathcal{F}_N(f,g) = (2\pi i)^{-N} \sum_{i=0}^{N} (-1)^i \binom{2k_1 - 1 + N}{N - i} \binom{2k_2 - 1 + N}{i} \partial_z^i f \partial_z^{N-i} g$$
(8.1)

is an element of $\mathcal{M}_{2(K+N)}$, where ∂_z^i means $\frac{d^i}{dz^i}$.

How are $\mathcal{F}$ and $\mathcal{P}$ related? In fact it is not hard to show that $\mathcal{F}_N$ is a certain average of the $\mathcal{P}_{n_1,n_2}$'s.

Proposition 8.1. *For every* $f \in \mathcal{M}_{2k_1}$, $g \in \mathcal{M}_{2k_2}$

$$(-1)^N \sum_{i=0}^{N} (-1)^i \binom{2k_1 - 1 + N}{N - i} \binom{2k_2 - 1 + N}{i} \mathcal{P}_{i,N-i}(f,g) = \mathcal{F}_N(f,g).$$
(8.2)

Proof. If $N = 0$ then the proposition is true. For $N \geq 1$ we may write the left hand side of (8.2) as $\sum_{m=1}^{\infty} d_m q^m$ with, after simplifying,

$$d_m = \sum_{i=0}^{N} (-1)^i \binom{2k_1 - 1 + N}{N - i} \binom{2k_2 - 1 + N}{i} T_m^*(i, N - i; f, g).$$

The Fourier coefficients of the right hand side are identical. □

As with $\mathcal{F}$, $\mathcal{P}$ may be expressed in terms of the derivatives of the modular forms.

Proposition 8.2. *For every* $f \in \mathcal{M}_{2k_1}$, $g \in \mathcal{M}_{2k_2}$

$$\mathcal{P}_{n_1,n_2}(f,g) = (2\pi i)^{-N} \sum_{l_1=0}^{n_1} \sum_{l_2=0}^{n_2} (-1)^L \frac{n_1! n_2!}{l_1! l_2!} \binom{2k_1 - 1 + n_1}{n_1 - l_1} \binom{2k_2 - 1 + n_2}{n_2 - l_2}$$
$$\times \frac{(2K + N + L - 2)!}{(2K + 2N - 2)!} \partial_z^{N-l_1-l_2} \left(\partial_z^{l_1} f \partial_z^{l_2} g \right).$$
(8.3)

Proof. Compare Fourier coefficients. □

Rankin in [10],[11] considers the general question of which polynomials in the derivatives of modular forms are again modular forms. His operator in [12] includes $\mathcal{F}$ as a special case. It is formulated as follows. Set $r \geq 2$ and label r modular forms $f_i \in \mathcal{M}_{2k_i}$ for $1 \leq i \leq r$. Also define

$$V(r, N) = \{(v_1, v_2, \ldots, v_r) \colon v_i \in \mathbb{N}, \sum_i v_i = N\},$$

$$U(r) = \{(u_1, u_2, \ldots, u_r) \colon u_i \in \mathbb{C}, \sum_i u_i = 0\}.$$

Then, for a fixed $u \in U(r)$,

$$(2\pi i)^{-N} \sum_{v \in V(r,N)} \frac{\partial^{v_1} f_1}{(2k_1 - 1 + v_1)! v_1!} \cdots \frac{\partial^{v_r} f_r}{(2k_r - 1 + v_r)! v_r!} u_1^{v_1} \ldots u_r^{v_r} \quad (8.4)$$

which we will denote by $\mathcal{G}_N(f_1, f_2, \ldots, f_r)$ is an element of $\mathcal{M}_{2(K+N)}$. This operator may also be expressed as an average, this time of

$$\mathcal{P}_{n_1,\ldots,n_r}(f_1, \ldots f_r) = (4\pi)^{-N} \pi_{\mathrm{hol}}(R^{n_1} y^{k_1} f_1(z) \ldots R^{n_r} y^{k_r} f_r(z)). \quad (8.5)$$

Proposition 8.3. *With the above notation* $\mathcal{G}_N(f_1, \ldots, f_r)$ *equals*

$$(-1)^N \sum_{v \in V(r,N)} \frac{u_1^{v_1}}{(2k_1 - 1 + v_1)! v_1!} \cdots \frac{u_r^{v_r}}{(2k_r - 1 + v_r)! v_r!} \mathcal{P}_{v_1,\ldots,v_r}(f_1, \ldots, f_r).$$

Proof. Same as Proposition 8.1.   □

For more information on Rankin-Cohen differential operators see [15], [16]. Similar operators for Siegel modular forms are constructed in [4].

9 Convolution Sums Involving the Divisor Function

We give a straightforward application of this material to finding explicit formulas for the sums

$$S_m(n_1, n_2; r_1, r_2) = \sum_{j=1}^{m-1} j^{n_1}(m - j)^{n_2} \sigma_{r_1}(j) \sigma_{r_2}(m - j). \quad (9.1)$$

Glaisher [5], Ramanujan [9, pp. 136–162] and Lahiri [8] were the first to systematically evaluate S_m for small values of n_1, n_2, r_1, r_2. Ramanujan manipulated the expressions

$$P = 1 - 24 \sum_{m=1}^{\infty} \frac{mq^m}{1-q^m},$$

$$Q = 1 + 240 \sum_{m=1}^{\infty} \frac{m^3 q^m}{1-q^m},$$

$$R = 1 - 504 \sum_{m=1}^{\infty} \frac{m^5 q^m}{1-q^m}, \tag{9.2}$$

to obtain his identities, and this work was extended in [8]. Ramanujan's series P, Q, R are none other than E_2, E_4, E_6; see [1]. For an interesting elementary method employing a generalization of Liouville's identity to find explicit formulas for S_m and other sums, see the paper [6] in these Proceedings.

Set

$$G_{2k} = \frac{-B_{2k}}{4k} E_{2k} = \frac{-B_{2k}}{4k} + \sum_{m=1}^{\infty} \sigma_{2k-1}(m) q^m. \tag{9.3}$$

While G_{2k} is in $\mathcal{M}_{2k}$ for $k \geq 2$ the series G_2 is not in the zero space $\mathcal{M}_2$. If we let $G_2^*(z) = G_2(z) + (8\pi y)^{-1}$, then G_2^* does transform correctly under the action of Γ and it has weight 2. Although it is no longer holomorphic, we do have $yG_2^*(z) \in C^{\infty}(\Gamma \backslash \mathfrak{H}, 2)$.

Set

$$S_m^*(l_1, l_2; r_1, r_2) = T_m^*(l_1, l_2; G_{r_1+1}, G_{r_2+1})$$

$$= S_m(l_1, l_2; r_1, r_2) - \frac{B_{r_1+1}}{2(r_1+1)} \sigma_{r_2}(m) E(0, l_1) E(m, l_2)$$

$$- \frac{B_{r_2+1}}{2(r_2+1)} \sigma_{r_1}(m) E(0, l_2) E(m, l_1). \tag{9.4}$$

Thus the sum we are interested in, (9.1), arises naturally in the Fourier coefficients of $\mathcal{P}_{l_1, l_2}(G_{r_1+1}, G_{r_2+1})$. Then it can be seen that (7.4) implies that the expressions S_m^* satisfy the relation

$$\sum_{l_1=0}^{n_1} \sum_{l_2=0}^{n_2} (-1)^L \frac{n_1! n_2!}{l_1! l_2!} \binom{r_1+n_1}{r_1+l_1} \binom{r_2+n_2}{r_2+l_2}$$

$$\times \frac{(r_1+r_2+N+L)!}{(r_1+r_2+2N)!} m^{N-L} S_m^*(l_1, l_2; r_1, r_2)$$

$$= \alpha_m \quad \text{for } r_1, r_2 \geq 3 \text{ and odd}, \tag{9.5a}$$

where $\sum_{m=1}^{\infty} \alpha_m q^m$ is in $\mathcal{S}_{2N+2+r_1+r_2}$ if $(n_1, n_2) \neq (0,0)$. If $(n_1, n_2) = (0,0)$ then $\sum_{m=0}^{\infty} \alpha_m q^m$ is in $\mathcal{M}_{2N+2+r_1+r_2}$ and $\alpha_0 = B_{r_1+1} B_{r_2+1} (4(r_1+1)(r_2+1))^{-1}$.

This means that we can express $S_m^*(n_1, n_2; r_1, r_2)$ (and hence, by (9.4), $S_m(n_1, n_2; r_1, r_2)$) in terms of the sums $S_m^*(l_1, l_2; r_1, r_2)$ for $0 \leq l_1 < n_1$ and $0 \leq l_2 < n_2$ and the coefficients of a cusp form in $\mathcal{S}_{r_1+r_2+2+2N}$. This cusp form may be identified by calculating its first few terms.

If $r_1 = 1$ or $r_2 = 1$ then the recurrence relation is slightly different to take into account the extra factor in the constant term of G_2^*. We have (each for $m \geq 1$)

$$S_m^*(0, 0; 1, 1) = \frac{-m}{2}\sigma_1(m) + \frac{5}{12}\sigma_3(m), \tag{9.5b}$$

$$\sum_{l_1=0}^{n_1}\sum_{l_2=0}^{n_2}(-1)^L \frac{n_1! n_2!}{l_1! l_2!}\left(\frac{1+n_1}{1+l_1}\right)\left(\frac{1+n_2}{1+l_2}\right)$$
$$\times \frac{(2+N+L)!}{(2+2N)!}m^{N-L}S_m^*(l_1, l_2; 1, 1) + \left((-1)^{n_1}\right.$$
$$\left.+ (-1)^{n_2}\right)\frac{\sigma_1(m)}{2}m^{N+1}\frac{N!(N+1)!}{(2N+2)!} = \beta_m \quad \text{for } (n_1, n_2) \neq (0, 0), \tag{9.5c}$$

$$\sum_{l_1=0}^{n_1}\sum_{l_2=0}^{n_2}(-1)^L \frac{n_1! n_2!}{l_1! l_2!}\left(\frac{1+n_1}{1+l_1}\right)\left(\frac{r+n_2}{r+l_2}\right)$$
$$\times \frac{(1+r+N+L)!}{(1+r+2N)!}m^{N-L}S_m^*(l_1, l_2; 1, r)$$
$$+ (-1)^{n_2}\frac{\sigma_r(m)}{2}m^{N+1}\frac{N!(N+r)!}{(2N+r+1)!} = \gamma_m \quad \text{for } r \geq 3 \text{ odd}, \tag{9.5d}$$

where $\sum_{m=1}^{\infty}\beta_m q^m$ is in $\mathcal{S}_{2N+4}$. Also $\sum_{m=1}^{\infty}\gamma_m q^m$ is in $\mathcal{S}_{2N+r+3}$ if (n_1, n_2) $\neq (0, 0)$. If $(n_1, n_2) = (0, 0)$ then $\sum_{m=0}^{\infty}\gamma_m q^m$ is in $\mathcal{M}_{r+3}$ and $\gamma_0 = B_{r+1}(48(r+1))^{-1}$.

Recall that $S_m^*(n_1, n_2; r_1, r_2) = S_m(n_1, n_2; r_1, r_2)$ unless $n_1 n_2 = 0$, and in that case S_m^* has the two extra terms given by (9.4). Note that (9.5b) appears in [15], where it is proved by finding $\pi_{\text{hol}}((yG_2^*)^2)$. The relations (9.5a), (9.5c) and (9.5d) generalize this idea to cover all the other cases.

10 Examples

To illustrate these ideas (and check the equations) we will give some examples. For $S_m^*(0, 0; 3, 3)$ use relation (9.5a) with $n_1 = n_2 = 0$ to see that $S_m^*(0, 0; 3, 3) = \alpha_m$ with $\sum_{m=0}^{\infty}\alpha_m q^m$ in $\mathcal{M}_8$ and $\alpha_0 = B_4^2/64 =$

$1/(64 \cdot 900)$. Thus $\alpha_m = \sigma_7(m)/120$ and

$$S_m^*(0,0;3,3) = \frac{1}{120}\sigma_7(m). \tag{10.1}$$

When $n_1 = 1$ and $n_2 = 0$, (9.5a) implies that

$$\frac{m}{2}S_m^*(0,0;3,3) - S_m^*(1,0;3,3) = \alpha_m$$

with $\sum_{m=1}^{\infty} \alpha_m q^m$ in $\mathcal{S}_{10}$. Consequently,

$$S_m^*(1,0;3,3) = S_m^*(0,1;3,3) = \frac{m}{2}S_m^*(0,0;3,3) = \frac{1}{240}m\sigma_7(m). \tag{10.2}$$

When $n_1 = 1$ and $n_2 = 1$ (9.5a) implies that

$$\frac{8}{45}m^2 S_m^*(0,0;3,3) - \frac{4}{5}mS_m^*(1,0;3,3) + S_m^*(1,1;3,3) = \alpha_m$$

with $\sum_{m=1}^{\infty} \alpha_m q^m$ in $\mathcal{S}_{12}$. The one dimensional space $\mathcal{S}_{12}$ contains the discriminant function $\sum_{m=1}^{\infty} \tau(m)q^m$. Since $\tau(1) = 1$ and $S_1^*(1,1;3,3) = 0$ we have $\alpha_m = -\tau(m)/540$ and

$$S_m^*(1,1;3,3) = \frac{1}{540}(\tau(m) + m^2\sigma_7(m)). \tag{10.3}$$

From (10.1), (10.2) and (10.3) we get

$$S_m(0,0;3,3) = \frac{1}{120}(\sigma_7(m) - \sigma_3(m)) \tag{10.4}$$

$$S_m(1,0;3,3) = S_m(0,1;3,3) = \frac{1}{240}m(\sigma_7(m) - \sigma_3(m)), \tag{10.5}$$

$$S_m(1,1;3,3) = \frac{1}{540}(\tau(m) + m^2\sigma_7(m)). \tag{10.6}$$

Continuing this procedure we obtain

$$S_m(2,0;3,3) = \frac{1}{2160}(4\tau(m) + 5m^2\sigma_7(m) - 9m^2\sigma_3(m)), \tag{10.7}$$

$$S_m(2,1;3,3) = \frac{1}{1080}(m^3\sigma_7(m) - m\tau(m)), \tag{10.8}$$

$$S_m(2,2;3,3) = \frac{1}{30888}(13m^4\sigma_7(m) - 22m^2\tau(m) + 9\tau(m) + 2160r(m)), \tag{10.9}$$

where $r(m) = \sum_{j=1}^{m-1} \sigma_3(m)\tau(m-j)$ comes from the cuspform $G_4\Delta \in \mathcal{S}_{16}$.

With $r_1 = r_2 = 1$ we have

$$S_m^*(0,0;1,1) = -m/2\sigma_1(m) + 5/12\sigma_3(m)$$

by (9.5b). By (9.5c),

$$S_m^*(1,0;1,1) = \frac{m}{2}S_m^*(0,0;1,1) = \frac{1}{24}(-6m^2\sigma_1(m) + 5m\sigma_3(m)). \quad (10.10)$$

Also,

$$\frac{4}{30}m^2 S_m^*(0,0;1,1) - \frac{2}{3}mS_m^*(1,0;1,1) + S_m^*(1,1;1,1) - \frac{1}{60}m^3\sigma_1(m) = 0,$$

so that

$$S_m^*(1,1;1,1) = \frac{1}{12}(m^2\sigma_3(m) - m^3\sigma_1(m)). \quad (10.11)$$

Therefore

$$S_m(0,0;1,1) = \frac{1}{12}(\sigma_1(m) - 6m\sigma_1(m) + 5\sigma_3(m)), \quad (10.12)$$

$$S_m(1,0;1,1) = \frac{1}{24}(m\sigma_1(m) - 6m^2\sigma_1(m) + 5m\sigma_3(m)), \quad (10.13)$$

$$S_m(1,1;1,1) = \frac{1}{12}(m^2\sigma_3(m) - m^3\sigma_1(m)). \quad (10.14)$$

Finally we consider the case $r_1 = 1, r_2 = 3$. By (9.5d) we have

$$S_m^*(0,0;1,3) + \frac{1}{8}\sigma_3(m)m = \gamma_m$$

with $\sum \gamma_m q^m$ in $\mathcal{M}_6$ and $\gamma_0 = -1/5760$. Hence $\gamma_m = 7\sigma_5(m)/80$ for $m \geq 1$ and

$$S_m^*(0,0;1,3) = \frac{1}{80}(7\sigma_5(m) - 10m\sigma_3(m)). \quad (10.15)$$

Similarly,

$$\frac{m}{3}S_m^*(0,0;1,3) - S_m^*(1,0;1,3) + \frac{1}{60}m^2\sigma_3(m) = \gamma_m$$

with $\sum \gamma_m q^m$ in $\mathcal{S}_8$. This implies that

$$S_m^*(1,0;1,3) = \frac{1}{240}(7m\sigma_5(m) - 6m^2\sigma_3(m)). \quad (10.16)$$

We also have

$$S_m^*(0,1;1,3) = \frac{1}{120}(7m\sigma_5(m) - 3m^2\sigma_3(m)). \quad (10.17)$$

For the last calculation (9.5d) gives

$$\frac{m^2}{7} S_m^*(0,0;1,3) - \frac{m}{4} S_m^*(0,1;1,3) - \frac{m}{2} S_m^*(1,0;1,3)$$
$$+ S_m^*(1,1;1,3) - \frac{1}{336} m^3 \sigma_3(m) = \gamma_m$$

with $\sum \gamma_m q^m$ in $\mathcal{S}_{10}$. Thus, as before, $\gamma_m = 0$ and

$$S_m^*(1,1;1,3) = \frac{1}{60}(m^2 \sigma_5(m) - m^3 \sigma_3(m)). \tag{10.18}$$

Equations (10.15), (10.16), (10.17) and (10.18) imply that

$$S_m(0,0;1,3) = \frac{1}{240}(21\sigma_5(m) - 30m\sigma_3(m) + 10\sigma_3(m) - \sigma_1(m)), \tag{10.19}$$

$$S_m(1,0;1,3) = \frac{1}{240}(7m\sigma_5(m) - 6m\sigma_3(m) - m\sigma_1(m)), \tag{10.20}$$

$$S_m(0,1;1,3) = \frac{1}{120}(7m\sigma_5(m) - 12m^2\sigma_3(m) + 5m\sigma_3(m)), \tag{10.21}$$

$$S_m(1,1;1,3) = \frac{1}{60}(m^2\sigma_5(m) - m^3\sigma_3(m)). \tag{10.22}$$

The equations (10.4), (10.5), (10.12), (10.13), (10.14), (10.19), (10.20), (10.21) and (10.22) appear in [8] and [6].

References

[1] B. C. Berndt, *Ramanujan's Notebooks Part II*, Springer-Verlag, New York, 1989.

[2] D. Bump, *Automorphic Forms and Representations*, Cambridge Stud. Adv. Math., vol. 55, Cambridge Univ. Press, Cambridge, 1997.

[3] H. Cohen, *Sums involving the values at negative integers of L functions of quadratic characters*, Math. Ann. **217** (1975), 271–285.

[4] E. Eholzer and T. Ibukiyama, *Rankin-Cohen type differential operators for Siegel modular forms*, Internat. J. Math. **9** (1998), 443–463.

[5] J. Glaisher, *Expressions for the first five powers of the series in which the coefficients are the sums of the divisors of the exponents*, Mess. Math. **15** (1885), 33–36.

[6]   J. Huard, Z. Ou, B. Spearman, and K. Williams, *Elementary evalua-tion of certain convolution sums involving divisor functions*, Number Theory for the Millennium (Proc. Millennial Conf. Number Theory, Urbana, Illinois, 2000; M.A. Bennett et al., ed.), Vol. 2, A K Peters, Boston, 2002, pp. 229–274.

[7]   H. Iwaniec, *Introduction to the spectral theory of automorphic forms*, Bibl. Rev. Mat. Iber., Madrid, 1995.

[8]   D. Lahiri, *On Ramanujan's function $\tau(n)$ and the divisor function $\sigma(n)$, I*, Bull. Calcutta Math. Soc. **38** (1946), 193–206.

[9]   S. Ramanujan, *Collected Papers*, Chelsea, New York, 1962.

[10] R. Rankin, *The construction of automorphic forms from the derivatives of a given form*, J. Indian Math. Soc. **20** (1956), 103–116.

[11] ———, *The construction of automorphic forms from the derivatives of given forms*, Michigan Math. J. **4** (1957), 181–186.

[12] ———, *The construction of automorphic forms from the derivatives of a given form II*, Canad. Math. Bull. **28** (1985), 306–316.

[13] J. Riordan, *Combinatorial Identities*, John Wiley & Sons, New York, 1968.

[14] J. Sturm, *Projections of C^∞ automorphic forms*, Bull. Amer. Math. Soc. **2** (1980), 435–439.

[15] D. Zagier, *Introduction to modular forms*, From number theory to physics (Les Houches, 1989), Springer, Berlin, 1992, pp. 238–291.

[16] ———, *Modular forms and differential operators*, Proc. Indian Acad. Sci. (Math. Sci.) **104** (1994), 57–75.

Iterative Methods for Pairs of Additive Diophantine Equations

Scott T. Parsell

1 Introduction

Let k and n be integers with $k \geq n \geq 1$, and let $c_1, \ldots, c_s$ and $d_1, \ldots, d_s$ be non-zero integers. We consider the problem of determining conditions under which one can demonstrate that the system of equations

$$
\begin{aligned}
c_1 x_1^k + \cdots + c_s x_s^k &= 0 \\
d_1 x_1^n + \cdots + d_s x_s^n &= 0
\end{aligned}
\tag{1.1}
$$

possesses a non-trivial integral solution. One obvious requirement is that the system must have a non-trivial real solution and a non-trivial p-adic solution for every prime p. In fact, the success of the Hardy-Littlewood method depends explicitly upon good information concerning the density of such solutions, and hence one must typically show that there are non-singular solutions in each local field. Unfortunately, the local solubility problem for (1.1) tends to be quite hard. When $p > k^4 n^2$, Wooley [25] demonstrates the existence of non-trivial p-adic solutions to (1.1) provided only that $s > 2(k + n)$, but the number of variables required for smaller primes in the recent work of Knapp [14] is often significantly larger than what would be required to handle the minor arcs in an application of the circle method. Moreover, it is difficult to guarantee that the solutions produced by the p-normalization methods of [14] and [25] are non-singular when $k - n > 1$. We therefore focus our attention on determining how large s must be in order to establish a local-global principle for (1.1). Define $G^*(k, n)$ to be the least integer r such that, whenever $s \geq r$ and the system (1.1) has a non-singular real solution and a non-singular p-adic solution for every prime p, the system (1.1) has a non-trivial integral solution.

The situation in which $k = n$ has attracted interest for quite some time, beginning with work of Davenport and Lewis [11] on pairs of cubics in 18 variables. Cook [9] showed that $G^*(2, 2) \leq 9$, while the bound of Davenport and Lewis for $G^*(3, 3)$ has been steadily reduced over the years by work of Cook [10], Vaughan [19], Baker and Brüdern [4], and finally Brüdern [7], who obtained $G^*(3,3) \leq 14$. Obstructions to local solubility are already apparent here, as [11] shows that there are pairs of cubics in 15 variables possessing no non-trivial 7-adic solution. In attempting to bound $G^*(k,k)$, one considers exponential sums of the type

$$F(\gamma) = \sum_{|x| \leq P} e(\gamma x^k),$$

where we have written $e(z) = e^{2\pi i z}$. In practice, the summation is often further restricted to integers that are free of large prime factors, but we suppress this point for now. One then observes that the number of solutions to (1.1) lying in the box $[-P, P]^s$ is given by the integral

$$\int_{[0,1]^2} \prod_{i=1}^{s} F(c_i \alpha + d_i \beta) \, d\alpha \, d\beta.$$

To obtain an upper bound for the minor arc contribution to this integral, one can apply Hölder's inequality and then make a change of variables that factors the integral into a product of one-dimensional mean values to which the estimates of Vaughan and Wooley [22], [24] apply. Thus one typically expects bounds for $G^*(k,k)$ that are about twice the size of the corresponding bound for $G(k)$ in Waring's problem. However, the sharpest bounds for $G(k)$ are often obtained by employing a form of p-adic iteration to save an extra variable over what would result from a direct application of the mean value estimates and Weyl's inequality. As illustrated in Brüdern [7], the extension of such techniques to pairs of equations can present serious technical challenges, although the argument of [18], leading to the bound $G^*(5,5) \leq 34$, is actually somewhat manageable. For larger k, the methods of Brüdern and Cook [8] show that $G^*(k,k) \leq (2 + o(1))k \log k$, which compares as expected with the well-known bound of Wooley [28].

Pairs of equations of differing degree have received considerably less attention, as their study requires a distinctly two-dimensional approach involving exponential sums of the shape

$$F(\alpha, \beta) = \sum_{|x| \leq P} e(\alpha x^k + \beta x^n).$$

This situation was first tackled by Wooley [27], who developed a version of Vaughan's iterative method [20] suitable for generating mean value estimates for sums of this type restricted to smooth numbers. These estimates initially produced the bound $G^*(3,2) \leq 14$, which was later improved to $G^*(3,2) \leq 13$ in [30] using the results of [29]. In this particular case, Wooley [26] was actually able to establish p-adic solubility whenever $s \geq 11$, and hence one only needs to impose a real solubility hypothesis.

It is a straightforward exercise to generate bounds for other pairs of exponents using the general method of Wooley [29]. In particular, one should be able to demonstrate with little difficulty that $G^*(k, n) \leq (2+o(1))k \log k$, for large $k > n$. However, as in the current treatment of Waring's problem (see Vaughan and Wooley [22], [24]), there are various refinements that may be attempted in order to obtain good results for smaller exponents. As the amount of available technology associated with the Hardy-Littlewood method is nowadays quite substantial, one has many options for carrying out such refinements, and it is a non-trivial task to determine the optimal strategy in each case. Our goal here is mainly to provide an overview of the various possible approaches and their limitations, so we defer most of the technical details to [17]. In the following table, the entry appearing in row k and column n is the upper bound we obtain for $G^*(k, n)$.

	1	2	3	4	5	6	7
3	10	13	14				
4	17	20	24	24			
5	30	31	32	36	34		
6	49	50	49	47	50	49	
7	66	72	70	65	64	66	67

As mentioned above, the estimates for $G^*(3,3)$, $G^*(5,5)$, and $G^*(3,2)$ are obtained from [7], [18], and [30], respectively. The bounds for $G^*(k,k)$ in the remaining cases follow in relatively routine fashion from the mean value estimates of [20], [22], and [24], and it may be possible to save an additional variable in some of these instances by proceeding as in [18].

The estimates for mean values of exponential sums obtained in our analysis may be further applied to deal with the corresponding problem on pairs of diophantine inequalities, where one seeks to demonstrate that two forms with real coefficients take arbitrarily small values simultaneously at integral points. This problem has already been investigated by the author [15], [16] in the case of a cubic and quadratic form by employing the mean value estimates of Wooley [30], together with some new ideas of Bentkus and Götze [5] and Freeman [12]. We intend to pursue this application for other pairs of exponents in a later paper.

2   The Analytic Set-up

After possibly replacing some of the variables x_i by $-x_i$ and then changing
the signs of the corresponding coefficients in forms of odd degree, we may
suppose that the system (1.1) has a non-singular real solution with all
coordinates positive, and hence it suffices to consider solubility in positive
integers. Let

$$\mathcal{A}(P, R) = \{n \in [1, P] \cap \mathbb{Z} : p|n \Rightarrow p \le R\}$$

denote the set of R-smooth numbers of size at most P. As usual, we take
P to be a large positive number and R to be a small positive power of P,
so that one has $\mathrm{card}(\mathcal{A}(P, R)) \gg P$ (see for example Vaughan [21], §12.1).
We now write $\boldsymbol{\alpha} = (\alpha, \beta)$ and define the exponential sums

$$F(\boldsymbol{\alpha}) = \sum_{1 \le x \le P} e(\alpha x^k + \beta x^n)$$

and

$$f(\boldsymbol{\alpha}) = \sum_{x \in \mathcal{A}(P, R)} e(\alpha x^k + \beta x^n). \tag{2.1}$$

Further, write $F_i(\boldsymbol{\alpha}) = F(c_i\alpha, d_i\beta)$ and $f_i(\boldsymbol{\alpha}) = f(c_i\alpha, d_i\beta)$, and introduce
the decomposition $s = t + 2u$. Then one sees that

$$N(P) = \int_{[0,1]^2} \prod_{i=1}^{t} F_i(\boldsymbol{\alpha}) \prod_{i=t+1}^{s} f_i(\boldsymbol{\alpha}) \, d\boldsymbol{\alpha}$$

is the number of solutions of (1.1) with the variables satisfying

$$1 \le x_i \le P \quad (i = 1, \ldots, t) \quad \text{and} \quad x_i \in \mathcal{A}(P, R) \quad (i = t+1, \ldots, s).$$

We now describe our Hardy-Littlewood dissection. Write $t_k = \max |c_i|$
and $t_n = \max |d_i|$, and let $X_i = 2k^2 t_i P^{i-1}$ for $i = k, n$. Define the major
arcs $\mathfrak{M}$ to be the union of the rectangles

$$\mathfrak{M}(q, a, b) = \{\boldsymbol{\alpha} \in [0, 1)^2 : |q\alpha - a| \le X_k^{-1} \text{ and } |q\beta - b| \le X_n^{-1}\}$$

with $0 \le a, b \le q \le P$ and $(q, a, b) = 1$, and write

$$\mathfrak{m} = [0, 1]^2 \setminus \mathfrak{M}$$

for the minor arcs. Provided that s and t are not too small, the existence
of non-singular real and p-adic solutions allows one to show that

$$\int_{\mathfrak{M}} \prod_{i=1}^{t} F_i(\boldsymbol{\alpha}) \prod_{i=t+1}^{s} f_i(\boldsymbol{\alpha}) \, d\boldsymbol{\alpha} \gg P^{s-(k+n)} \tag{2.2}$$

by employing a straightforward extension of the argument of Wooley [27]. Here one uses the t variables ranging over a complete interval to prune back to a thinner set of major arcs, on which asymptotics for $f_i(\alpha)$ can be obtained. For the minor arcs, an application of Hölder's inequality, together with a consideration of the underlying diophantine equations, shows that

$$\int_{\mathfrak{m}} \prod_{i=1}^{t} F_i(\alpha) \prod_{i=t+1}^{s} f_i(\alpha) \, d\alpha \ll \sup_{\alpha \in \mathfrak{m}} |F_i(\alpha)|^t \int_{[0,1]^2} |f(\alpha)|^{2u} \, d\alpha$$

for some i. A minor arc bound for $F_i(\alpha)$ is provided by a generalization of Weyl's inequality due to Baker [1] (see also [2], [3]), while the estimation of the even moments of $f(\alpha)$ represents our main challenge. Baker's result allows us to save essentially P^σ per variable over the trivial estimate, where $\sigma = 2^{1-k}$, and our mean value estimates take the shape

$$\int_{[0,1]^2} |f(\alpha)|^{2u} d\alpha \ll P^{2u-(k+n)+\Delta_u+\varepsilon}, \tag{2.3}$$

where an estimate with $\Delta_u = 0$ would be essentially best possible. We therefore obtain the bound

$$\int_{\mathfrak{m}} \prod_{i=1}^{t} F_i(\alpha) \prod_{i=t+1}^{s} f_i(\alpha) \, d\alpha \ll P^{s-(k+n)-\delta}$$

for some $\delta > 0$, provided that $\sigma t > \Delta_u$. Thus on recalling (2.2), one has

$$G^*(k,n) \leq \min_{u \in \mathbb{N}} \left(2u + \left[\frac{\Delta_u}{\sigma} \right] + 1 \right). \tag{2.4}$$

We concentrate on obtaining bounds for Δ_u in the remaining sections.

3 Efficient Differencing

One sees from the above discussion that the number of variables required to establish a local-global principle for the system (1.1) is closely connected with the strength of the available estimates for mean values of the exponential sums (2.1). We now indicate a strategy, based on the method of Wooley [29], for obtaining such mean value estimates. Wooley [27] showed using elementary methods that the number of solutions of the system

$$\begin{aligned} x_1^k + x_2^k + x_3^k &= y_1^k + y_2^k + y_3^k \\ x_1^n + x_2^n + x_3^n &= y_1^n + y_2^n + y_3^n \end{aligned} \tag{3.1}$$

with $x_i, y_i \in [1, P]$ is $O(P^{3+\varepsilon})$ whenever $k > n \geq 1$. This provides an estimate for the 6th moment of $F(\alpha)$, and hence of $f(\alpha)$, that is essentially best possible in view of the diagonal solutions to (3.1). One only saves P^3 over the trivial estimate, however, so in the context of (2.3) one is forced to take $\Delta_3 = k + n - 3$. An examination of (2.4) with $u = 3$ therefore produces unimpressive bounds such as $G^*(3, 2) \leq 15$ and $G^*(5, 3) \leq 87$. To improve on these, one instead uses the information concerning the number of solutions to (3.1) as the basis for an iteration to higher moments via the method of Vaughan [20] and Wooley [28], [29].

A typical step in the iteration may be summarized as follows. Suppose that one has an estimate of the shape (2.3) for the mean value

$$S_s(P, R) = \int_{[0,1]^2} |f(\alpha)|^{2s} \, d\alpha,$$

and write $\lambda_s = 2s - (k + n) + \Delta_s$. Then $S_{s+2}(P, R)$ is bounded above by the number of solutions of the system

$$z_1^k - w_1^k + z_2^k - w_2^k = \sum_{i=1}^{s}(x_i^k - y_i^k)$$
$$z_1^n - w_1^n + z_2^n - w_2^n = \sum_{i=1}^{s}(x_i^n - y_i^n)$$
$$(3.2)$$

with $x_i, y_i \in \mathcal{A}(P, R)$ and $z_i, w_i \in [1, P]$. Now let $\theta \leq 1/k$ be a parameter at our disposal, and write

$$M = P^{\theta}, \quad Q = PM^{-1}, \quad \text{and} \quad H = PM^{-k}.$$

The solutions of (3.2) with some x_i or y_i smaller than M can be shown to contribute a negligible amount, and for the remaining solutions the fact that x_i and y_i are R-smooth implies that each has a divisor lying between M and MR. Then by applying Hölder's inequality, one reduces to the situation in which the divisors are all identical and is thus led to analyze the system

$$z_1^k - w_1^k + z_2^k - w_2^k = m^k \sum_{i=1}^{s}(u_i^k - v_i^k)$$
$$z_1^n - w_1^n + z_2^n - w_2^n = m^n \sum_{i=1}^{s}(u_i^n - v_i^n)$$
$$(3.3)$$

with $u_i, v_i \in \mathcal{A}(Q, R)$, $M < m \leq MR$, and $z_i, w_i \in [1, P]$. One may further suppose, after some effort, that m is coprime to any Jacobian determinant of z_i and w_j arising from the left-hand side.

Equipped with an estimate for the number of non-singular solutions in z_i, w_i, distinct modulo m^k, to the implicit pair of congruences in (3.3), one can further reduce, via Cauchy's inequality, to the situation in which

$$z_1 \equiv w_1 \pmod{m^k} \quad \text{and} \quad z_2 \equiv w_2 \pmod{m^k}. \tag{3.4}$$

These congruence conditions allow us to perform a differencing operation that is more efficient than the classical one of Weyl, since we can now write

$$w_1 = z_1 + h_1 m^k \quad \text{and} \quad w_2 = z_2 + h_2 m^k$$

with h_1 and h_2 of magnitude at most H. It is actually convenient to write $x_i = z_i + w_i$ and to consider the symmetric difference polynomials

$$\psi_i(x, h, m) = m^{-i}((x + hm^k)^i - (x - hm^k)^i).$$

One then finds (essentially) that

$$S_{s+2}(P, R) \ll P^\varepsilon M^{2s+k-n-1} \int_{[0,1]^2} F_1(\boldsymbol{\alpha}) |f_1(\boldsymbol{\alpha})|^{2s} \, d\boldsymbol{\alpha}, \tag{3.5}$$

where

$$F_1(\boldsymbol{\alpha}) = \sum_m \left| \sum_{h,z} e(\alpha_k \psi_k(z, h, m) + \alpha_n \psi_n(z, h, m)) \right|^2,$$

with the summations running over

$$M < m \le MR, \quad 1 \le h \le H, \quad \text{and} \quad 1 \le z \le 2P,$$

and where

$$f_1(\boldsymbol{\alpha}) = \sum_{x \in \mathcal{A}(2Q, R)} e(\alpha_k x^k + \alpha_n x^n).$$

Of course, one also needs to account for the terms with $h = 0$, but in practice one usually chooses θ in such a way that this diagonal contribution is of the same order of magnitude as the expression in (3.5). The factor of $M^{2s+k-n-1}$ appearing in (3.5) represents the cost of obtaining the uniform divisor m and imposing the strong congruence condition (3.4). One can now obtain an estimate for $S_{s+2}(P, R)$ by employing the trivial bound $F_1 \ll MH^2P^2$ together with the estimate $S_s(2Q, R) \ll Q^{\lambda_s+\varepsilon}$, which was obtained at the previous stage. Alternatively, one could extract divisors from the u_i and v_i in (3.3) and then repeat the differencing argument, perhaps multiple times, before making a final estimate. More creative strategies for dealing with (3.5), along the lines of Vaughan and Wooley [22], [24], are described in the next section.

4   End-game Strategies

If we apply the efficient differencing argument of the previous section j times, then we choose parameters $M_1, \ldots, M_j$ corresponding to the sizes of the divisors extracted at each difference. The variables u_i and v_i in the analogue of (3.3) are represented by an exponential sum f_j that is identical to f_1 except that the variables range only up to $2^j P(M_1 \cdots M_j)^{-1}$. The polynomial ψ_i is replaced by a polynomial of degree $i - j$ in z that is also a function of the divisors $m_1, \ldots, m_j$ and difference parameters $h_1, \ldots, h_j$. We can then write down an exponential sum $\mathcal{F}_j$ analogous to $\mathcal{F}_1$ and attempt to estimate an integral of the shape

$$\int_{[0,1]^2} \mathcal{F}_j(\boldsymbol{\alpha}) |f_j(\boldsymbol{\alpha})|^{2r} \, d\boldsymbol{\alpha}. \tag{4.1}$$

Here r is typically about $s - j$, where s is as in (3.5). Finally, we trace this estimate back through the differencing process, optimizing the parameters M_i along the way, until we reach (3.5). The fundamental decisions to be made are how many differences to take and how to estimate the integral (4.1) after the final difference.

One possible approach, useful near the beginning of the iteration, is to consider estimates for the various moments of $\mathcal{F}_j(\boldsymbol{\alpha})$. For example, it is fairly easy to demonstrate by considering the underlying diophantine equations that one has

$$\int_{[0,1]^2} \mathcal{F}_1(\boldsymbol{\alpha})^2 \, d\boldsymbol{\alpha} \ll P^{2+\varepsilon} M^2 H^3, \tag{4.2}$$

with similar results holding for $\mathcal{F}_j$ whenever $j \leq k - 2$. Higher moments can also be estimated, but the bounds are in most cases too weak to be useful in light of the other available methods. In order to use an estimate such as (4.2), we apply Hölder's inequality to (4.1), and this in turn may require us to estimate a higher moment of f_j that was not scheduled to be dealt with until later in the iteration. Thus we should first obtain preliminary bounds for all the relevant moments by applying (for example) the method described at the end of §3. It is sometimes effective when using mean values to employ a generalization of the differencing argument in which the congruence condition (3.4) is weakened in order to reduce the cost of imposing it. In bounding $G^*(5,3)$, for example, it is useful to consider the congruences modulo m_j^3 rather than modulo m_j^5 on the final difference when s is small.

Towards the end of the iteration, the use of mean values tends to become less effective. If we are sufficiently far from a diagonal situation, however, it is often possible to obtain major and minor arc estimates for the sums $\mathcal{F}_j$ and hence to treat the integral (4.1) by means of a Hardy-Littlewood dissection. For example, in the case $k = 5$ and $n = 3$ one has

$$\alpha_5\psi_5 + \alpha_3\psi_3 = 10h\alpha_5 z^4 + (20h^3m^{10}\alpha_5 + 6hm^2\alpha_3)z^2,$$

and one can deduce from Baker [1] that there are good rational approximations to the coefficients of z^4 and z^2 above (for many values of h and m) whenever the sum $\mathcal{F}_1$ is unusually large. However, the resulting approximations to α_5 and α_3 certainly depend on h and m, and it is not immediately clear that there is any fixed choice for which the approximations are of major arc quality. Fortunately, Baker's argument confronts essentially the same issue, in the context of ordinary Weyl differencing, and employs a result of Birch and Davenport [6] to obtain uniform approximations of the desired quality. While a version of Baker's argument works in principle for our situation, there are in fact several conditions to check, and these tend to fail unless j is relatively small in terms of k and n. In the case $k = 5$ and $n = 3$, for example, the method is currently only successful for first differences. A possible alternative for larger j is to attempt a one-dimensional dissection based solely on approximations to α_k, but this is usually inferior to the other available methods.

Finally, the constant interplay between exponential sums and underlying diophantine equations gives rise to a variety of ad-hoc methods for estimating the integral (4.1) more directly. For example, if we difference n times, then we reduce to a system analogous to (3.3), but where the left-hand side of the second equation is zero. For the quintic-cubic case, the resulting system is

$$480h_1h_2h_3(z_1^2 - z_2^2) = \sum_{i=1}^{r}(u_i^5 - v_i^5), \qquad 0 = \sum_{i=1}^{r}(u_i^3 - v_i^3).$$

To count the solutions with $z_1 = z_2$, it suffices to fix the variables z_1, h_1, h_2, and h_3 and to apply an estimate for $S_r(P, R)$. If $z_1 \neq z_2$, then we can instead fix the u_i and v_i using the second equation, at which point h_1, h_2, h_3, z_1, and z_2 are determined to $O(P^\varepsilon)$ by a divisor estimate. Note that in either case the variables m_1, m_2, and m_3 must be accounted for by a trivial estimate, since they do not appear explicitly in the system. This method turns out to be quite effective in the intermediate stages of the iteration and is noteworthy because it has no analogue in a one-dimensional situation like Waring's problem. When $k-n = 1$ and $j \leq n-1$, a variation employed by Wooley in bounding $G^*(3, 2)$ is often somewhat more effective. Here one deals with the non-diagonal solutions by taking a linear combination of the two equations and then applying a generalization of Hua's inequality ([13], Theorem 4). It is in some ways natural to view the type of analysis described in this paragraph as a more direct version of a Hardy-Littlewood dissection, with the non-diagonal and diagonal solutions corresponding respectively to the major and minor arcs (see [23]).

The whole iterative process can now be repeated, seeded with improved preliminary estimates, until the values of λ_s stabilize. Because of the number of possibilities at each stage, we use a computer to determine the optimal iterative scheme for each pair of exponents. In the quintic-cubic case, a combination of mean values, dissections, and ad-hoc analyses permits us to obtain $\Delta_{15} \leq 0.11321$, and the bound $G^*(5,3) \leq 32$ then follows immediately from (2.4).

It is generally not effective to difference more than n times, because the loss of the non-singularity condition on the implicit pair of congruences would then essentially force us into a one-dimensional analysis. It is therefore difficult to generate good estimates for $G^*(k,n)$ when n is small relative to k, since a large value of k would ordinarily call for a large number of differences. As a consequence, we often obtain better bounds for $G^*(k,n)$ when n is close to k than for $G^*(k,1)$ or $G^*(k,2)$. This is contrary to the expectation that systems of lower total degree should require fewer variables to solve, but it seems to be a fundamental difficulty associated with the method.

References

[1] R. C. Baker, *Weyl sums and diophantine approximation*, J. London Math. Soc. **25** (1982), 25–34.

[2] ———, *Diophantine inequalities*, Clarendon Press, Oxford, 1986.

[3] ———, *Correction to 'Weyl sums and diophantine approximation'*, J. London Math. Soc. **46** (1992), 202–204.

[4] R. C. Baker and J. Brüdern, *On pairs of additive cubic equations*, J. Reine Angew. Math. **391** (1988), 157–180.

[5] V. Bentkus and F. Götze, *Lattice point problems and distribution of values of quadratic forms*, Annals of Math. **150** (1999), 977–1027.

[6] B. J. Birch and H. Davenport, *On a theorem of Davenport and Heilbronn*, Acta Math. **100** (1958), 259–279.

[7] J. Brüdern, *On pairs of diagonal cubic forms*, Proc. London Math. Soc. **61** (1990), 273–343.

[8] J. Brüdern and R. J. Cook, *On simultaneous diagonal equations and inequalities*, Acta Arith. **62** (1992), 125–149.

[9] R. J. Cook, *Simultaneous quadratic equations*, J. London Math. Soc. **4** (1971), 319–326.

[10] _____, *Pairs of additive equations*, Michigan Math. J. **19** (1972), 325–331.

[11] H. Davenport and D. J. Lewis, *Cubic equations of additive type*, Philos. Trans. Roy. Soc. Ser. A **261** (1966), 97–136.

[12] D. E. Freeman, *Asymptotic lower bounds for diophantine inequalities*, Mathematika, to appear.

[13] L.-K. Hua, *Additive theory of prime numbers*, A.M.S., Providence, Rhode Island, 1965.

[14] M. P. Knapp, *Forms in many variables over p-adic fields*, Ph.D. thesis, University of Michigan, 2000.

[15] S. T. Parsell, *On simultaneous diagonal inequalities*, J. London Math. Soc. **60** (1999), 659–676.

[16] _____, *On simultaneous diagonal inequalities, II*, Mathematika, to appear.

[17] _____, *Pairs of additive equations of small degree*, Acta Arith., to appear.

[18] S. T. Parsell and T. D. Wooley, *On pairs of diagonal quintic forms*, Compositio Math., to appear.

[19] R. C. Vaughan, *On pairs of additive cubic equations*, Proc. London Math Soc. **34** (1977), 354–364.

[20] _____, *A new iterative method in Waring's problem*, Acta Math. **162** (1989), 1–71.

[21] _____, *The Hardy-Littlewood method*, 2nd ed., Cambridge University Press, Cambridge, 1997.

[22] R. C. Vaughan and T. D. Wooley, *Further improvements in Waring's problem*, Acta Math. **174** (1995), 147–240.

[23] _____, *On a certain nonary cubic form and related equations*, Duke Math. J. **80** (1995), 669–735.

[24] _____, *Further improvements in Waring's problem, IV: Higher powers*, Acta Arith. **94** (2000), 203–285.

[25] T. D. Wooley, *On simultaneous additive equations III*, Mathematika **37** (1990), 85–96.

[26] _____, *On simultaneous additive equations I*, Proc. London Math. Soc. **63** (1991), 1–34.

[27] _____, *On simultaneous additive equations II*, J. Reine Angew. Math. **419** (1991), 141–198.

[28] _____, *Large improvements in Waring's problem*, Annals of Math. **135** (1992), 131–164.

[29] _____, *On exponential sums over smooth numbers*, J. Reine Angew. Math. **488** (1997), 79–140.

[30] _____, *On simultaneous additive equations IV*, Mathematika **45** (1998), 319–335.

Quelques Remarques sur la Théorie d'Iwasawa des Courbes Elliptiques

Dans ce texte, nous avons cherché à donner un panorama partial et partiel[1] de la théorie d'Iwasawa des courbes elliptiques en utilisant les deux outils que sont les systèmes d'Euler construits par Kato et la théorie de l'exponentielle (ou du logarithme). Nous n'avons regardé ici que la partie interpolation p-adique des valeurs spéciales de la fonction L en 1 tordue par un caractère de Dirichlet[2]. L'idée importante pour nous que nous voudrions faire passer dans ce texte est que l'application logarithme élargi (ou régulateur[3] d'Iwasawa) qui généralise les constructions de Kummer, Iwasawa, Coates-Wiles et l'homomorphisme de Coleman, contient, par sa construction même d'une part et grâce à la loi de réciprocité de Colmez d'autre part, **tous** les renseignements nécessaires permettant le calcul de ses valeurs spéciales et leur lien avec les invariants arithmétiques. Les systèmes d'Euler-Kato permettant de voir la fonction L p-adique comme le régulateur d'Iwasawa d'un système compatible pour les normes construit de manière modulaire grâce au théorème de Kato, la plupart des formules spéciales sur la fonction L p-adique en un entier k s'en déduisent. Dans le cas des courbes elliptiques ayant bonne réduction en p, ces résultats se trouvent déjà dans [31]. Dans le cas des courbes elliptiques semi-stables en p, on obtient une (nouvelle) interprétation du "zéro trivial".

Soit E une courbe elliptique (modulaire) définie sur $\mathbb{Q}$, de conducteur N. La fonction L de Hasse-Weil associée $L(E, s)$ définie pour $Re(s) > 1$ se prolonge à tout le plan complexe en une fonction holomorphe. Si $L(E, s) = \sum_{n=1}^{\infty} a_n n^{-s}$ et si η est un caractère de Dirichlet prolongé par 0 sur les entiers non premiers à son conducteur, on pose $L(E, \eta, s) = \sum_{n=1}^{\infty} \eta(n) a_n n^{-s}$. Si M est un entier, on note $L_{\{M\}}(E, s)$ la fonction L incomplète : $L_{\{M\}}(E, s) = \sum_{(n,M)=1} a_n n^{-s}$.

---

[1] Pour d'autres panoramas, voir par exemple les articles de Greenberg, [15].

[2] Pour la partie arithmétique, voir [31] et [15], [20], [30], [32] dans le cas ordinaire.

[3] Le terme de régulateur désigne à l'origine un nombre et plus particulièrement un volume, l'application permettant de "mesurer" étant le logarithme : par exemple, le volume du tore, quotient du $\mathbb{R}$-espace vectoriel $E = \mathbb{R}^{r_1(F)+r_2(F)-1}$ par le $\mathbb{Z}$-module d'indice fini engendré par l'image des unités d'un corps de nombres F par l'application logarithme est le régulateur du corps F. Il semble que désormais "régulateur" désigne couramment l'application elle-même.

Soit $H^0(E, \Omega^1_{E/\mathbb{Q}})$ le $\mathbb{Q}$-espace vectoriel des formes différentielles invariantes de E définies sur $\mathbb{Q}$ et $\mathrm{Lie}(E)$ l'algèbre de Lie de E. Rappelons la suite exacte

$$0 \to H^0(E, \Omega^1_{E/\mathbb{Q}}) \to H^1_{\mathrm{dR}}(E) \to \mathrm{Lie}(E) \to 0$$

compatible avec l'accouplement canonique

$$H^1_{\mathrm{dR}}(E) \times H^1_{\mathrm{dR}}(E) \to \mathbb{Q}.$$

Si $\omega \in H^0(E, \Omega^1_{E/\mathbb{Q}})$, on note

$$\Omega^\pm_{E,\omega} = \Omega^\pm_E = \int_{E(\mathbb{C})^\pm} \omega$$

les périodes complexes associées ($E(\mathbb{C})^\pm$ est le $\pm$-espace propre de $E(\mathbb{C})$ relatif à l'action de la conjugaison complexe, une orientation est fixée). Soit $D_{\mathrm{dR}}(E)$ le $\mathbb{Q}$-espace vectoriel $H^1_{\mathrm{dR}}(E)^*$ muni de la filtration $\mathrm{Fil}^0 D_{\mathrm{dR}}(E) = \mathrm{Lie}(E)^* \cong H^0(E, \Omega^1_{E/\mathbb{Q}})$.

Fixons un nombre premier p impair et supposons que E a réduction semi-stable en p. Nous serons amenés à distinguer les cas suivants (non exclusifs)

(1) cas (br) : E a bonne réduction en p

    (a) cas (br-ord) : E a bonne réduction ordinaire

    (b) cas (br-ss) : E a bonne réduction supersingulière

(2) cas (mult) : E a mauvaise réduction semi-stable en p

    (a) cas (mult-dépl) : E a réduction multiplicative déployée

    (b) cas (mult-non dépl) : E a réduction multiplicative non déployée

(3) cas (ord) : E a réduction ordinaire ((br-ord)+(mult)).

Si $T_p(E)$ est le module de Tate p-adique de E, c'est-à-dire la limite projective des E_{p^n}, $T_p(E)$ et $V_p(E) = \mathbb{Q}_p \otimes_{\mathbb{Z}_p} T_p(E)$ sont munis d'une action continue des groupes de Galois absolu $G_{\mathbb{Q}} = \mathrm{Gal}(\overline{\mathbb{Q}}/\mathbb{Q})$ et $G_{\mathbb{Q}_p} = \mathrm{Gal}(\overline{\mathbb{Q}}_p/\mathbb{Q}_p)$. La représentation p-adique $V_p(E)$ est de poids de Hodge-Tate 0 et 1 : si $\mathbb{C}_p$ est la complétion p-adique de $\overline{\mathbb{Q}}_p$, $\mathbb{C}_p \otimes V_p(E)$ est isomorphe en tant que $G_{\mathbb{Q}_p}$-module à $\mathbb{C}_p \oplus \mathbb{C}_p(1)$ où $\mathbb{C}_p(1)$ est le twist de Tate de $\mathbb{C}_p$. D'autre part, le $\mathbb{Q}_p$-espace vectoriel $D_p(E) = \mathbb{Q}_p \otimes D_{\mathrm{dR}}(E)$ est muni d'une structure de (φ, N)-module filtré[4] sur $\mathbb{Q}_p$ se définissant uniquement à l'aide de la représentation p-adique $V_p(E)$ grâce aux isomorphismes de comparaison. La filtration de $D_p(E)$ est compatible avec celle de $D_{\mathrm{dR}}(E)$: $\mathrm{Fil}^0 D_p(E) = \mathbb{Q}_p \otimes \mathrm{Fil}^0 D_{\mathrm{dR}}(E)$. On a avec les conventions choisies

$$\det(1 - X\varphi | D_p(E)^{N=0}) = (1 - a_p p^{-1} X + \epsilon(p) p^{-1} X^2)$$

avec $\epsilon(p) = 0$ si $p | N$ et 1 sinon.

Dans le cas (ord), les valeurs propres de φ agissant sur $D_p(E)$ sont de valuation 0 et -1, on note $D_p(E)_{[0]}$ (resp. $D_p(E)_{[-1]}$) les espaces propres correspondants et $\pi = \pi_{[0]}$ (resp. $\pi_{[-1]}$) la projection de $D_p(E)$ sur $D_p(E)_{[0]}$ (resp. $D_p(E)_{[-1]}$) parallèlement à $D_p(E)_{[-1]}$ (resp. $D_p(E)_{[0]}$). Dans le cas (br-ss), les valeurs propres de φ sont de valuation $-1/2$. Dans le cas (mult), si e_0 est une base de $D_p(E)_{[0]}$, Ne_0 est une base de $D_p(E)_{[-1]}$ et (e_0, Ne_0) forme une base de $D_p(E)$. Lorsque E a réduction multiplicative déployée, on a $\varphi e_0 = e_0$, lorsque la réduction n'est pas déployée, $\varphi e_0 = -e_0$.

Soit μ_{p^∞} le groupe des racines de l'unité d'ordre une puissance de p (dans $\overline{\mathbb{Q}}$ ou $\overline{\mathbb{Q}}_p$). Le groupe de Galois $G_\infty = \mathrm{Gal}(\mathbb{Q}(\mu_{p^\infty})/\mathbb{Q})$ s'identifie naturellement à $\mathrm{Gal}(\mathbb{Q}_p(\mu_{p^\infty})/\mathbb{Q}_p)$. Le caractère cyclotomique $\chi \colon G_\infty \to \mathbb{Z}_p^*$ défini par $\sigma\zeta = \zeta^{\chi(\sigma)}$ pour $\zeta \in \mu_{p^\infty}$ est un isomorphisme. Si $\mathbb{Q}_\infty$ est la sous-$\mathbb{Z}_p$-extension de $\mathbb{Q}(\mu_{p^\infty})/\mathbb{Q}$, on note $\Delta = \mathrm{Gal}(\mathbb{Q}(\mu_p)/\mathbb{Q}) \cong \mathrm{Gal}(\mathbb{Q}(\mu_{p^\infty})/\mathbb{Q}_\infty)$ et $\Gamma = \mathrm{Gal}(\mathbb{Q}_p(\mu_{p^\infty})/\mathbb{Q}_p(\mu_p))$ (il est isomorphe à $\mathbb{Z}_p$). Si η est un caractère de Dirichlet de conducteur une puissance de p, nous notons de la même manière le caractère de G_∞ qui s'en déduit.

Notons $\mathcal{H}'$ l'algèbre des fonctions (strictement) analytiques en x dans la boule unité de $\mathbb{C}_p$: $|x| < 1$, ayant un développement en série entière dans $\mathbb{Q}_p[[x]]$ et $\mathcal{H}$ la sous-algèbre des éléments f de $\mathcal{H}'$ vérifiant une condition de "croissance logarithmique au bord" : les $(\log \rho)^r \|f\|_\rho$ sont bornés[5], lorsque ρ tend vers 1^- pour un réel $r \geq 0$; on dit alors que f est d'ordre $\leq r$.

On note alors $\mathcal{H}(\Gamma) \subset \mathbb{Q}_p[[\gamma - 1]]$ l'image de $\mathcal{H}$ par l'application qui envoie x sur $\gamma - 1$ pour γ un générateur topologique de Γ et $\mathcal{H}(G_\infty) = \mathbb{Z}_p[\Delta] \otimes_{\mathbb{Z}_p} \mathcal{H}(\Gamma)$. L'algèbre $\mathcal{H}(G_\infty)$ contient l'algèbre d'Iwasawa $\mathbb{Z}_p[[G_\infty]]$ (les éléments d'ordre 0 sont exactement les éléments de $\mathbb{Q}_p \otimes \mathbb{Z}_p[[G_\infty]]$).

Si η est un caractère continu de G_∞ dans $\mathbb{C}_p^*$, $\eta(f)$ a un sens et appar-

---

[4]Un (φ, N)-module filtré sur $\mathbb{Q}_p$ est un $\mathbb{Q}_p$-espace vectoriel D (de dimension finie) muni de deux opérateurs φ et N vérifiant $N\varphi = p\varphi N$ avec φ bijectif, et d'une filtration $\mathrm{Fil}^\cdot D$ décroissante exhaustive et séparée ([14]).

[5]Si $f = \sum_{n=0}^\infty a_n x^n \in \mathcal{H}'$ et $0 < \rho < 1$, $\|f\|_\rho = \sup_n |a_n| \rho^n = \sup_{|x| \leq \rho} |f(x)|$.

tient à $\mathbb{C}_p$:

$$\text{si } f = \sum_{\delta \in \Delta} \sum_{n=0}^{\infty} a_{n,\delta}(\gamma - 1)^n \delta,$$

$$\eta(f) = \sum_{\delta \in \Delta} \sum_{n=0}^{\infty} a_{n,\delta}(\eta(\gamma) - 1)^n \eta(\delta).$$

Un élément de $\mathcal{H}(G_\infty)$ peut donc être vu comme une fonction sur les caractères de G_∞ : on notera indifféremment $\eta(f)$ ou $f(\eta)$. D'autre part, l'action de G_∞ sur les racines de l'unité d'ordre une puissance de p se prolonge en une action de $\mathcal{H}(G_\infty)$: si $f \in \mathcal{H}(G_\infty)$ et $\zeta \in \mu_{p^\infty}$, on note le résultat $f \cdot \zeta \in \mathbb{Q}_p(\zeta) \subset \mathbb{C}_p$. De même, l'action de G_∞ sur $\mathcal{H}$ induite par $\tau(1+x) = (1+x)^{\chi(\tau)}$ induit une opération de $\mathcal{H}(G_\infty)$ sur $\mathcal{H}$ qui est notée $(f, g) \mapsto f \cdot g$.

Fixons un système compatible (ζ_n) de racines de l'unité d'ordre p^n, par exemple les $\exp(2i\pi/p^n)$ dans $\mathbb{C}$. Si η est un caractère de Dirichlet de conducteur p^n, on note $G(\eta)$ la somme de Gauss :

$$G(\eta) = \sum_{\substack{a \bmod p^n \\ (a,p)=1}} \eta(a)\zeta_n^a.$$

Si $\epsilon(\eta) = \eta(-1)$ est la signature de η, $\frac{G(\eta)L(E,\eta^{-1},1)}{\Omega_{E,\omega}^{\epsilon(\eta)}}$ appartient au corps $\mathbb{Q}(\eta)$ engendré sur $\mathbb{Q}$ par les valeurs de η. Ces valeurs spéciales pour η de conducteur une puissance de p ont des propriétés p-adiques très intéressantes qui impliquent le théorème suivant d'interpolation (la première version se trouve en 1974 dans [26] dans le cas de bonne réduction ordinaire, voir [41] et [1] dans le cas supersingulier et [27] dans le cas général).

Théorème (Interpolation). Cas (br-ss) : *Il existe un unique élément $L_p(E) \in \mathcal{H}(G_\infty) \otimes D_p(E)$ dont les composantes (dans une base de $D_p(E)$) sont d'ordre $\leq 1/2$ tel que pour tout caractère de Dirichlet non trivial η de conducteur p^n et de signature $\epsilon(\eta)$*

$$\eta(L_p(E)) = \frac{\epsilon(\eta)G(\eta)L(E,\eta^{-1},1)}{\Omega_{E,\omega}^{\epsilon(\eta)}}\varphi^n\omega. \tag{1}$$

Cas (ord) : *Il existe un unique élément $L_p^\pi(E) \in \mathbb{Z}_p[[G_\infty]] \otimes D_p(E)_{[0]}$ tel que pour tout caractère non trivial η de conducteur p^n*

$$\eta(L_p^\pi(E)) = \frac{\epsilon(\eta)G(\eta)L(E,\eta^{-1},1)}{\Omega_{E,\omega}^{\epsilon(\eta)}}\pi(\varphi^n\omega).$$

Pour être complet, donnons aussi la valeur sur le caractère trivial **1**. Dans le premier cas, il s'agit de

$$\mathbf{1}(L_p(E)) = \frac{L_{\{p\}}(E,1)}{\Omega_{E,\omega}^+}(1-\varphi)(1-p^{-1}\varphi^{-1})^{-1}\omega, \qquad (2)$$

dans le second cas[6]

$$\mathbf{1}(L_p^\pi(E)) = \frac{L_{\{p\}}(E,1)}{\Omega_{E,\omega}^+}(1-\varphi)(1-p^{-1}\varphi^{-1})^{-1}\pi(\omega).$$

Faisons quelques remarques.

(1) La somme de Gauss $G(\eta^{-1})$ peut s'interpréter comme $G(\eta^{-1}) = e_\eta(\zeta_n)$ avec e_η l'opérateur de projection sur la η-composante :

$$e_\eta = \sum_{\sigma \in \mathrm{Gal}(\mathbb{Q}(\mu_{p^n})/\mathbb{Q})} \eta^{-1}(\sigma)\sigma.$$

En utilisant la formule $p^n = G(\eta)G(\eta^{-1})\eta(-1)$ et le fait que $e_\eta(f \cdot \zeta_n) = \eta(f)e_\eta(\zeta_n) = \eta(f)G(\eta^{-1})$ pour $f \in \mathcal{H}(G_\infty)$, la formule (1) peut s'écrire

$$e_\eta(L_p(E) \cdot \zeta_n) = \frac{L(E,\eta^{-1},1)}{\Omega_{E,\omega}^{\epsilon(\eta)}}(p\varphi)^n\omega \qquad (3)$$

et de même pour la seconde.

(2) Si a est un entier premier à p, posons pour $n > 0$

$$L(E, a \bmod p^n, s) = \begin{cases} \displaystyle\sum_{\substack{m=1 \\ m \equiv a \bmod p^n}}^{\infty} a_m m^{-s} & \text{si } n > 0 \\ L_{\{p\}}(E,s) & \text{si } n = 0 \end{cases}$$

et

$$L^{\pm}(E, a \bmod p^n, s) = L(E, a \bmod p^n, s) \pm L(E, -a \bmod p^n, s).$$

---

[6]La formule est ainsi la même dans le cas de bonne et mauvaise réduction ordinaire, si α est la racine de $X^2 - a_p X + p\epsilon(p)$ qui est une unité p-adique, le second membre peut s'écrire eul $\frac{L(E,1)}{\Omega_{E,\omega}^+}$ avec

$$\mathrm{eul} = (1-\alpha^{-1})(1-p^{-1}\alpha)^{-1}(1-\alpha p^{-1})(1-\epsilon(p)\alpha^{-1}),$$

ce qui vaut $(1-\alpha^{-1})^2$ dans le cas de bonne réduction et $1-\alpha^{-1}$ dans le cas de mauvaise réduction.

On aimerait exprimer les formules définissant $L_p(E)$ sans caractères. Mais le comportement particulier pour le caractère trivial empêche que la formule "naturelle" qui vient à l'esprit soit vraie. Il faut donc faire plus compliqué. Faisons-le dans le cas (ss) (dans le cas (ord), il faudrait prendre une projection). Considérons l'opérateur continu φ sur $\mathcal{H}$ tel que $\varphi(x) = (1+x)^p - 1$. Résolvons l'équation

$$(1 - \varphi \otimes \varphi)\mathcal{G}_E = L_p(E) \cdot (1+x) \tag{4}$$

dans $\mathcal{H} \otimes D_p(E)$, ce qui est possible car les valeurs propres de φ ne sont pas une puissance de p. Alors, pour η caractère non trivial de conducteur p^n,

$$e_\eta(\mathcal{G}_E(\zeta_n - 1)) = e_\eta(L_p(E) \cdot \zeta_n)$$

et $(1 - \varphi)\mathcal{G}_E(0) = \mathbf{1}(L_p(E))$. Les relations "simples" s'expriment en termes de $\mathcal{G}_E^\pm = \mathcal{G}_E \pm \mathcal{G}_E^\iota$ (avec $f^\iota(x) = f((1+x)^{-1} - 1)$) et non de $L_p(E)$: pour a premier à p et n entier strictement positif,

$$\mathcal{G}_E^\pm(\zeta_n^a - 1) = \frac{L^\pm(E, a \bmod p^n, 1)}{\Omega_{E,\omega}^\pm}(p\varphi)^n \omega, \tag{5}$$

et

$$(1 - p^{-1}\varphi^{-1})\mathcal{G}_E^\pm(0) = \frac{L^\pm(E, 1)}{\Omega_{E,\omega}^\pm}\omega. \tag{6}$$

Faisons la démonstration rapidement. La relation (4) prise en $x = 0$ implique que $(1 - \varphi)\mathcal{G}_E^+(0)/2 = \mathbf{1}(L_p(E)) = \frac{L_{\{p\}}^+(E,1)}{\Omega_{E,\omega}^+}$, d'où

$$(1 - p^{-1}\varphi^{-1})\mathcal{G}_E^+(0) = \frac{L_{\{p\}}^+(E,1)}{\Omega_{E,\omega}^+}\omega = 2\frac{L(E, \mathbf{1}, 1)}{\Omega_{E,\omega}^+}\omega.$$

On déduit alors de l'équation $\psi(\mathcal{G}_E) = (1 \otimes \varphi)\mathcal{G}_E$ que

$$(1 - p^{-1}\varphi^{-1})\mathcal{G}_E(0) = \mathrm{Tr}_{\mathbb{Q}(\mu_p)/\mathbb{Q}}((p\varphi)^{-1}\mathcal{G}_E^+(\zeta_1 - 1))$$
$$= e_{\mathbf{1}}((p\varphi)^{-1}\mathcal{G}_E^+(\zeta_1 - 1)).$$

Ainsi, pour tout caractère η pair de $(\mathbb{Z}/p^n\mathbb{Z})^*$, on a

$$e_\eta((p\varphi)^{-n}\mathcal{G}_E^+(\zeta_n - 1)) = \frac{L^+(E, \eta^{-1}, 1)}{\Omega_{E,\omega}^+}$$

(cela ne dépend pas de n à cause de la relation $\psi(\mathcal{G}_E) = (1 \otimes \varphi)\mathcal{G}_E$). On en déduit la relation (5) par application de la formule $\mathbf{1}_{a \bmod p^n} =$

$\frac{1}{p^{n-1}(p-1)/2} \sum_\eta \eta(a)e_\eta$ où la somme est prise sur les caractères de Dirichlet pairs modulo p^n et où $1_{a \bmod p^n}$ est la fonction caractéristique de la classe $a \bmod p^n$. La partie $-$ se démontre de la même manière.

(3) Si D est un φ-module de dimension finie munie d'une norme et $f \in \mathcal{H}(G_\infty) \otimes D$, on dit que f est d'ordre $\leq r$ (ou pour être plus précis de φ-ordre $\leq r$) si la suite $p^{-nr}||(1 \otimes \varphi)^{-n} f||_{\rho^{1/p^n}}$ est bornée (pour un ρ avec $0 < \rho < 1$). Ici, r peut prendre des valeurs négatives. La condition sur l'ordre de croissance de $L_p(E)$ peut se dire alors de manière plus "intrinsèque" : $L_p(E)$ est d'ordre ≤ 0.

Soit F une extension finie de $\mathbb{Q}$. Par la théorie de Kummer de la courbe elliptique, la suite de G_F-modules

$$0 \to E_{p^n} \to E(\overline{\mathbb{Q}}) \xrightarrow{p^n} E(\overline{\mathbb{Q}}) \to 0$$

est exacte et induit la suite exacte de cohomologies

$$0 \to E(F)/p^n E(F) \to H^1(F, E_{p^n}) \to H^1(F, E(\overline{\mathbb{Q}}))_{p^n} \to 0.$$

Par passage à la limite projective du premier homomorphisme, on obtient des injections $\mathbb{Z}_p \otimes_{\mathbb{Z}} E(F) \to H^1(F, T_p(E))$ et

$$\mathbb{Q}_p \otimes_{\mathbb{Z}} E(F) \to H^1(F, V_p(E)).$$

Plus précisément l'image de $\mathbb{Q}_p \otimes_{\mathbb{Z}} E(F)$ est contenue dans $H^1_f(F, V_p(E))$ (groupe de Selmer-Bloch-Kato) qui est traditionnellement défini comme le noyau de

$$H^1(F, V_p(E)) \to \prod_v \mathbb{Q}_p \otimes \varprojlim_n H^1(F_v, E(\overline{\mathbb{Q}}))_{p^n}$$

et coïncide avec le noyau de

$$H^1(F, V_p(E)) \to \prod_{v \nmid p} H^1(F_v^{nr}, V_p(E)) \times \prod_{v|p} H^1(F_v, B_{\text{cris}} \otimes V_p(E))$$

où F_v^{nr} est la plus grande extension non ramifiée de F_v et B_{cris} l'anneau des périodes p-adiques de Fontaine en p ([13]). Autrement dit, si $H^1_f(F_v, V_p(E))$ est $H^1(F_v^{nr}/F_v, V_p(E)^{G_{F_v,nr}})$ pour v ne divisant pas p et le noyau de $H^1(F_v, V_p(E)) \to H^1(F_v, B_{\text{cris}} \otimes V_p(E))$ pour v divisant p, $H^1_f(F, V_p(E))$ est le sous-espace vectoriel de $H^1(F, V_p(E))$ formé des éléments dont l'image par localisation dans $H^1(F_v, V_p(E))$ est dans $H^1_f(F_v, V_p(E))$. Dans tous les cas,

$$H^1_f(F_v, V_p(E)) = \mathbb{Q}_p \otimes \varprojlim_n E(F_v)/p^n E(F_v).$$

Pour $v|p$, l'espace tangent $t_E(F_v)$ de E/F_v à l'origine s'identifie à $F_v \otimes \left(D_p(E)/\operatorname{Fil}^0 D_p(E) \right)$ et l'application exponentielle de groupe de Lie

$$\exp_{v,E} \colon t_E(F_v) \xrightarrow{\cong} \mathbb{Q}_p \otimes E(F_v)$$

s'identifie à l'application exponentielle de Bloch-Kato

$$\exp_{v,E} \colon t_E(F_v) \xrightarrow{\cong} H_f^1(F, V_p(E)).$$

Nous utiliserons aussi l'application logarithme inverse de l'exponentielle

$$\log_{v,E} \colon H_f^1(F, V_p(E)) \to t_E(F_v).$$

Par la dualité locale de Tate

$$H^1(F_v, V_p(E)) \times H^1(F_v, V_p(E)) \to H^2(F_v, \mathbb{Q}_p(1)) \cong \mathbb{Q}_p,$$

l'application duale de $\exp_{v,E}$ est

$$\exp_{v,E}^* \colon H^1(F_v, V_p(E)) \to F_v \otimes \operatorname{Fil}^0 D_p(E)$$

dont le noyau est exactement $H_f^1(F_v, V_p(E))$ (on utilise ici le fait que l'accouplement de Weil induit un isomorphisme entre $V_p(E)$ et $V_p(E)^*(1)$).

Le fait que le groupe de Selmer puisse s'exprimer uniquement en termes de la représentation p-adique associée à E a été remarqué indépendamment par Greenberg dans le cas ordinaire et par Bloch et Kato, dans les années 1987-88. Le travail de Bloch et Kato dans [5], qui s'appuie sur la théorie de Fontaine a marqué un tournant dans la théorie d'Iwasawa et a permis de sortir du cadre des représentations p-adiques ordinaires en p.

La construction d'éléments de $H^1(\mathbb{Q}(\mu_{p^n}), T_p(E))$ faite par Kato à partir des éléments de Beilinson ([2], eux-même très liés aux unités de Siegel, de Kubert-Lang ou de Robert ([24], [36]) selon le contexte) et le calcul de leur valeur par $\exp_E^*$ fait par Kato permettent d'aborder sous un jour nouveau le théorème d'interpolation en interprétant les valeurs spéciales de la fonction L en 1 comme "régulateurs p-adiques" d'éléments de la cohomologie galoisienne. Donnons une version un peu vague du théorème de Kato (sans introduire la notion pourtant fondamentale de système d'Euler-Kolyvagin). Les références sont [39], [38, §7] et les articles de Kato à venir.[7]

Théorème (Kato). *Il existe des $c_n' \in H^1(\mathbb{Q}(\mu_{p^n}), T_p(E))$ "construits de manière modulaire", compatibles pour les applications de corestriction et une constante c_E non nulle tels que pour tout caractère η de conducteur p^n*

$$\sum_{\sigma \in \operatorname{Gal}(\mathbb{Q}(\mu_{p^n})/\mathbb{Q})} \eta^{-1}(\sigma) \exp_{p,E}^*(\sigma c_n') = c_E \frac{L_{\{pN\}}(E, \eta^{-1}, 1)}{\Omega_{E,\omega}^{\epsilon(\eta)}} \omega.$$

---

[7]Note ajoutée : cet article existe désormais ([21]) et on peut enlever le N des fonctions $L_{\{pN\}}$.

Notons

$$H^1_\infty(\mathbb{Q}, V_p(E)) = \mathbb{Q}_p \otimes \varprojlim_n H^1(\mathbb{Q}(\mu_{p^n}), T_p(E)).$$

Les $c_n = c(E)^{-1} c'_n$ forment un système projectif $c_\infty \in H^1_\infty(\mathbb{Q}, V_p(E))$ et on a pour tout caractère de conducteur p^n

$$\sum_{\sigma \in \mathrm{Gal}(\mathbb{Q}(\mu_{p^n})/\mathbb{Q})} \eta^{-1}(\sigma) \exp^*_{p,E}(\sigma c_n) = \frac{L_{\{pN\}}(E, \eta^{-1}, 1)}{\Omega^{\epsilon(\eta)}_{E,\omega}} \omega.$$

Soit

$$Z^1_\infty(\mathbb{Q}_p, V_p(E)) = \mathbb{Q}_p \otimes \varprojlim_n H^1(\mathbb{Q}_p(\mu_{p^n}), T_p(E)).$$

L'homomorphisme de Coleman associé à un groupe formel[8] [8] se généralise en une application $\mathcal{L}_E$ de $Z^1_\infty(\mathbb{Q}_p, V_p(E))$ dans un $\mathcal{H}(G_\infty)$-module $\mathcal{D}_\infty(E)$; $\mathcal{L}_E$ joue le rôle de régulateur pour ce module d'Iwasawa, nous l'appellerons régulateur d'Iwasawa, quant à $\mathcal{D}_\infty(E)$, on peut le décrire en termes uniquement du (φ, N)-module $D_p(E)$. Dans le cas de bonne réduction, $\mathcal{D}_\infty(E)$ a une description très simple : il s'agit de $\mathcal{H}(G_\infty) \otimes D_p(E)$ qu'il est commode de voir comme sous-espace de $\mathcal{H} \otimes D_p(E)$ par la transformée de Mellin :

$$\mathcal{H}(G_\infty) \otimes D_p(E) \to \mathcal{H} \otimes D_p(E)$$
$$f \mapsto f \cdot (1 + x).$$

Ici $\tau \cdot (1 + x) = (1 + x)^{\chi(\tau)}$ pour $\tau \in G_\infty$ et cette action se prolonge naturellement par linéarité et continuité à $\mathcal{H}(G_\infty)$. L'image de $f \mapsto f \cdot (1 + x)$ est égale au noyau de l'opérateur ψ défini[9] par

$$\varphi \circ \psi(g)(x) = p^{-1} \sum_{\zeta \in \mu_p} g(\zeta(1 + x) - 1)$$

avec φ l'opérateur de Frobenius : $\varphi(g)(x) = g((1 + x)^p - 1)$.

Le régulateur d'Iwasawa $\mathcal{L}_E$ est un homomorphisme de G_∞-modules

$$\mathcal{L}_E \colon Z^1_\infty(\mathbb{Q}_p, V_p(E)) \to \mathcal{H}(G_\infty) \otimes D_p(E)$$

dont l'image est formée d'éléments d'ordre ≤ 0. Avec les notations de [35, prop. 5.4.5], [31, §1] au signe près, on a $\mathcal{L}_E = l_0 \Omega^{-1}_{V_p(E),1}$. La condition

---

[8]Il trouve d'ailleurs son origine dans les travaux d'Iwasawa [17] et de Coates et Wiles [7].

[9]Cet opérateur qui est devenu fondamental dans ces constructions se trouve dans les travaux de Dwork ([12], [6]) et dans les constructions de Coleman ([8] sous le nom de $\mathcal{S}$).

sur l'ordre se traduit dans le cas (br-ss) par le fait que les composantes de l'image de $\mathcal{L}_E$ dans une base de $D_p(E)$ sont d'ordre $\leq 1/2$. Dans le cas (br-ord), la composante de $\pi_{[0]}(\mathcal{L}_E(z))$ dans une base de $D_p(E)_{[0]}$ est d'ordre ≤ 0, c'est-à-dire appartient à $\mathbb{Q}_p \otimes \mathbb{Z}_p[[G_\infty]]$, la composante de $\pi_{[-1]}(\mathcal{L}_E(z))$ est d'ordre ≤ 1. Notons

$$\tilde{\mathcal{L}}_E \colon Z^1_\infty(\mathbb{Q}_p, V_p(E)) \to \mathcal{D}_\infty(E) = \mathcal{H}^{\psi=0} \otimes D_p(E)$$

pour éviter les confusions.

Dans le cas (mult), $\mathcal{D}_\infty(E)$ est plus compliqué. Commençons par introduire une algèbre contenant $\mathcal{H}$. Pour cela, considérons l'algèbre $\mathcal{B}$ des séries de Laurent $\sum_{n\in\mathbb{Z}} a_n x^n$, avec $a_n \in \mathbb{Q}_p$, analytiques sur la couronne $\{x \in \mathbb{C}_p, p^{-1/(p-1)} < |x| < 1\}$ vérifiant une condition de croissance comme précédemment, puis l'algèbre $\mathcal{B}[\log x]$ des polynômes en $\log x$ à coefficients dans $\mathcal{B}$. On peut ici voir $\log x$ soit comme une fonction localement analytique vérifiant les propriétés usuelles du logarithme, soit comme un élément formel auquel on impose des règles de calcul : ainsi, $\mathcal{B}[\log x]$ est muni (entre autres) d'opérateurs φ, ψ, d'un opérateur de monodromie N trivial sur $\mathcal{B}$ et tel que $N \log x = 1$, d'une action continue de G_∞, d'une dérivation $D(f) = (1+x)\frac{d}{dx}f$ pour $f \in \mathcal{B}$ et $D \log x = \frac{1+x}{x}$ (remarquons que l'on est obligé d'introduire $\mathcal{B}$ pour définir D et φ, par exemple :

$$\varphi \log x = p \log x + \log \frac{(1+x)^p - 1}{x^p}$$

et un moyen[10] de donner un sens à $\log \frac{(1+x)^p-1}{x^p}$ dans $\mathcal{B}$ est d'écrire

$$\frac{(1+x)^p - 1}{x^p} = 1 + \frac{p}{x} + \ldots + \frac{p}{x^{p-1}}$$

et d'utiliser le développement $\log(1+x) = \sum_{n=1}^\infty (-1)^{n-1}\frac{x^n}{n})$. On définit $\mathcal{B}(G_\infty)$ à partir de $\mathcal{B}$ comme pour $\mathcal{H}(G_\infty)$. On montre que $\mathcal{B}(G_\infty)$ agit naturellement sur $\mathcal{B}[\log x]^{\psi=0}$ et que $\left(\mathcal{B}[\log x]^{\psi=0} \otimes D_p(E)\right)^{N=0}$ est un $\mathcal{B}(G_\infty)$-module de rang[11] 2. Lorsque $N = 0$ sur $D_p(E)$, on a

$$\left(\mathcal{B}[\log x]^{\psi=0} \otimes D_p(E)\right)^{N=0} = \mathcal{B}^{\psi=0} \otimes D_p(E).$$

---

[10] Un autre moyen serait d'écrire

$$\log\left((1+x)^p - 1\right) = \log_p p + \log x + \log(1 + \frac{x}{p} + \ldots + \frac{x^{p-1}}{p}),$$

le troisième terme donne une série convergente sur $|x| < p^{-1/(p-1)}$, ce qui ne peut pas être évaluer sur $\zeta - 1$ pour ζ racine de l'unité non triviale.

[11] Ce qui signifie qu'une fois tensorisé par l'anneau total des fractions $\mathrm{Frac}(\mathcal{B}(G_\infty))$ de $\mathcal{B}(G_\infty)$, il est localement de rang 2 sur $\mathrm{Frac}(\mathcal{B}(G_\infty))$.

On peut décrire à l'intérieur un $\mathcal{H}(G_\infty)$-module $\mathcal{D}_\infty(E)$ de rang 2 : en gros, il s'agit des éléments g de $\left(\mathcal{B}[\log x]^{\psi=0} \otimes D_p(E)\right)^{N=0}$ tels que l'on puisse résoudre dans $\mathcal{B}[\log x] \otimes D_p(E)$ une équation du type $(1 - \varphi \otimes \varphi)G = g + \mathfrak{M}\log(1+x)$ (voir [35, §2.3], ici $\mathcal{D}_\infty(E)$ est le $\mathcal{D}_{\infty,f}$ de [35]). Il se trouve dans la suite exacte

$$0 \to \mathcal{H}^{\psi=0} \otimes \mathbb{Q}_p Ne_0 \to \mathcal{D}_\infty(E) \to \mathcal{H}^{\psi=0} \otimes \mathbb{Q}_p e_0 \to \mathbb{Q}_p \to 0 \qquad (7)$$

où la dernière flèche est l'évaluation en 0 dans le cas déployé et

$$0 \to \mathcal{H}^{\psi=0} \otimes \mathbb{Q}_p Ne_0 \to \mathcal{D}_\infty(E) \to \mathcal{H}^{\psi=0} \otimes \mathbb{Q}_p e_0 \to 0 \qquad (8)$$

dans le cas non déployé. Un élément fe_0 de $\mathcal{H}^{\psi=0} \otimes \mathbb{Q}_p e_0$ (avec $f(0) = 0$ dans le cas déployé) se relève de la manière suivante : on montre qu'il existe un élément F de $(\mathcal{B}[\log x] \otimes D_p(E))^{N=0}$ tel que $(1 - \varphi \otimes \varphi)F \equiv f \bmod \mathcal{B}[\log x] \otimes Ne_0$. Autrement dit, on choisit $F_1 \in \mathcal{H}$ tel que $(1 \mp \varphi)F_1 = f$ avec $\mp = -$ dans le cas déployé (on peut alors supposer que $F_1(0) = 0$) et $+$ dans le cas non déployé ; il existe $F_2 \in \mathcal{B}[\log x]$ tel que $\psi(F_2) = p^{-1}F_2$, $NF_2 = F_1$; un relèvement de fe_0 est alors $(1 - \varphi \otimes \varphi)(F_1e_0 - F_2Ne_0)$. Nous donnons dans l'appendice une manière de calculer un tel relèvement et en particulier ses valeurs sur $\zeta - 1$ avec $\zeta \in \mu_{p^\infty}$.

De manière générale, un élément f de $\mathcal{D}_\infty(E)$ vérifie (outre l'équation fonctionnelle $\psi(f) = 0$) la propriété qu'il existe $F \in (\mathcal{B}[\log x] \otimes D_p(E))^{N=0}$ et $\mathfrak{M} \in \mathbb{Q}_p$ tels que $(1 - \varphi \otimes \varphi)F = f + \mathfrak{M}\log(1+x)Ne_0$ (avec $\mathfrak{M} = 0$ dans le cas non déployé). Le F n'est pas unique dans le cas déployé, un choix possible est d'imposer la nullité de la composante sur e_0 en 0. Nous noterons $\mathcal{S}(f) = \mathcal{S}(f, \mathfrak{M})$ (pour $\mathfrak{M}$ convenable) la solution de $(1-\varphi\otimes\varphi)F = f + \mathfrak{M}\log(1+x)Ne_0$ dont la composante sur e_0 est nulle en 0 dans le cas déployé.

Remarquons que $(\tau - 1)(1+x)$ forme une base du noyau de $\mathcal{H}^{\psi=0} \to \mathbb{Q}_p$ pour τ un générateur de G_∞. Notons $\mathcal{T}_\tau$ un relèvement dans $\mathcal{D}_\infty(E)$ d'ordre minimal (d'ordre ≤ 1 modulo $\log x\, \mathcal{B} \otimes D_p(E)$). Enfin, $\mathcal{T}_0 = (\tau - 1)^{-1}\mathcal{T}_\tau$ a un sens dans $(\mathcal{B}[\log x]^{\psi=0} \otimes D_\infty(E))^{N=0}$ et a l'avantage de ne pas dépendre de τ. Nous écrirons $\mathcal{T}_0 = (1+x)e_0 - t_0Ne_0$, $\mathcal{T}_\tau = (\tau - 1)(1+x)e_0 - t_\tau Ne_0$ avec $t_\tau = (\tau - 1)t_0$ et $\mathcal{T}_{-1} = (1+x)Ne_0$.

Pour conclure cette description de $\mathcal{D}_\infty(E)$, disons que $(\mathcal{T}_{-1}, \mathcal{T}_\tau)$ (resp. $(\mathcal{T}_{-1}, \mathcal{T}_0)$) en forme une base dans le cas déployé (resp. non déployé). Rappelons que lorsque $N = 0$ sur $D_p(E)$, ce sont les $(\mathcal{T}_i = (1+x)e_i)_{i=0,-1}$ qui forment une base du $\mathcal{H}(G_\infty)$-module $\mathcal{D}_\infty(E)$.

Un résultat important ici et démontré dans [35] est qu'il existe un homomorphisme de G_∞-modules naturel $\tilde{\mathcal{L}}_E$ (appelé ici régulateur d'Iwasawa, ailleurs logarithme ou "exponentielle duale")

$$\tilde{\mathcal{L}}_E : Z^1_\infty(\mathbb{Q}_p, V_p(E)) \to \mathcal{D}_\infty(E) \subset \left(\mathcal{B}[\log x]^{\psi=0} \otimes D_p(E)\right)^{N=0}.$$

Avec les notations de [35, prop. 5.4.5], il s'agit de $l_0\Omega_{V_p(E),1}^{-1}$; $\Omega_{V_p(E),1}(g)$ est obtenu en interpolant les valeurs des exponentielles d'éléments construits à partir de G avec $(1 - \varphi \otimes \varphi)G = g - \mathfrak{M}\log(1 + x)$, les exponentielles étant les exponentielles de Bloch-Kato pour la représentation p-adique $V_p(E)$, mais aussi pour ses twists. Le fait que $l_0\Omega_{V_p(E),1}^{-1}$ soit à valeurs dans $\mathcal{D}_\infty(E) \subset \left(\mathcal{B}[\log x]^{\psi=0} \otimes D_p(E)\right)^{N=0}$ se démontre à partir de la loi de réciprocité (Réc) démontrée par Colmez ([11], voir aussi un article annoncé de Kato, Kurihara et Tsuji et [3]) et de calculs sur le déterminant de $\Omega_{V_p(E),1}$ ([35, prop. 5.4.5]). Par composé avec l'application naturelle de localisation

$$H_\infty^1(\mathbb{Q}, V_p(E)) \to Z_\infty^1(\mathbb{Q}_p, V_p(E)),$$

on obtient

$$\tilde{\mathcal{L}}_E \colon H_\infty^1(\mathbb{Q}, V_p(E)) \to \mathcal{D}_\infty(E).$$

Choisissons un logarithme $\log_p$ sur $\mathbb{C}_p - \{0\}$: il suffit pour cela de fixer $\log_p p$, par exemple $\log_p(p) = 0$. Si η est un caractère non trivial de G_∞ de conducteur p^n, l'application $f \in \mathcal{B}[\log x] \mapsto e_\eta(f(\zeta_n - 1))$ induit une application naturelle d'évaluation

$$\mathrm{ev}_\eta \colon \mathcal{B}[\log x] \otimes \mathcal{D} \to \mathbb{Q}_p(\eta) \otimes D_p(E).$$

Dans le cas (br), on a

$$\mathrm{ev}_\eta(\tilde{\mathcal{L}}_E(z)) = G(\eta^{-1})\eta(\mathcal{L}_E(z)) = \epsilon(\eta)p^n G(\eta)\eta(\mathcal{L}_E(z)).$$

En utilisant les propriétés de $\mathcal{L}_E$ (voir [35] par exemple) et en procédant comme dans [31], on peut écrire le théorème de Kato sous la forme suivante.

Théorème (Kato). *Pour tout caractère non trivial η de G_∞ de conducteur p^n,*

$$\mathrm{ev}_\eta(\tilde{\mathcal{L}}_E(c_\infty)) = \frac{L_{\{Np\}}(E, \eta^{-1}, 1)}{\Omega_{E,\omega}^{\epsilon(\eta)}}(p\varphi)^n \omega \tag{9}$$

On obtient un moyen détourné de retrouver le théorème d'interpolation et un peu plus (excepté les facteurs locaux aux places divisant N). Regardons d'abord le cas (br) :

Proposition. *Dans le cas* (br), *il existe un élément $L_{p,\{N\}}(E)$ de $\mathcal{H}(G_\infty)\otimes D_p(E)$ d'ordre ≤ 0 tel que, pour tout caractère non trivial η de G_∞ de conducteur p^n,*

$$\eta(L_{p,\{N\}}(E)) = \frac{\epsilon(\eta)G(\eta)L_{\{Np\}}(E, \eta^{-1}, 1)}{\Omega_{E,\omega}^{\epsilon(\eta)}}\varphi^n \omega.$$

Il est unique vérifiant ces conditions dans le cas (ss), *mais pas dans le cas* (br-ord).

On prend[12] $L_{p,\{N\}}(E) = \mathcal{L}_E(c_\infty)$. C'est la fonction L p-adique (incomplète) de $E/\mathbb{Q}$. Si l'on veut obtenir la fonction L p-adique complète, il faut diviser $L_{p,\{N\}}(E)$ par un élément de $\mathbb{Z}_p[[G_\infty]]$ qui n'est pas un diviseur de zéro[13] :

$$L_p(E) = \prod_{\substack{q|N \\ q \neq p}} (1 - a_q \sigma_q^{-1})^{-1} L_{p,\{N\}}(E)$$

où σ_q est l'élément de G_∞ vérifiant $\sigma_q \zeta = \zeta^q$ pour $\zeta \in \mu_{p^\infty}$. On a en particulier dans le cas (br-ord) $\pi_{[0]}(L_p(E)) = L_p^\pi(E)$.

Par définition, la fonction L p-adique est donc le régulateur d'Iwasawa d'un élément "spécial" de $H_\infty^1(\mathbb{Q}, V_p(E))$. Cet élément joue le rôle des unités cyclotomiques.

Traduisons ce que cela signifie dans le cas ordinaire. La racine de $X^2 - a_p X + p\epsilon(p)$ qui est une unité dans $\mathbb{Z}_p$ est notée α et on prend $\beta = p/\alpha$.

Proposition. *Dans le cas* (br-ord), *il existe un unique élément $L_{p,\alpha}$ de $\mathbb{Q}_p \otimes \mathbb{Z}_p[[G_\infty]]$ tel que pour tout caractère η non trivial de conducteur p^n*

$$\eta(L_{p,\alpha}) = \alpha^{-n} \frac{\epsilon(\eta) G(\eta) L_{\{Np\}}(E, \eta^{-1}, 1)}{\Omega_{E,\omega}^{\epsilon(\eta)}}$$

et il existe un élément $L_{p,\beta}$ de $\mathcal{H}(G_\infty)$ d'ordre ≤ 1 vérifiant pour tout caractère non trivial η de conducteur p^n

$$\eta(L_{p,\beta}) = \beta^{-n} \frac{\epsilon(\eta) G(\eta) L_{\{Np\}}(E, \eta^{-1}, 1)}{\Omega_{E,\omega}^{\epsilon(\eta)}}.$$

On passe de la proposition précédente à celle-ci en écrivant $\omega = e_\alpha + e_\beta$ avec $e_\alpha \in D_p(E)_{[0]}$ et $e_\beta \in D_p(E)_{[-1]}$ et en prenant les composantes dans la "base" (e_α, e_β).

---

[12]Il y a une autre définition possible de la fonction L p-adique de E comme élément $L_p^*(E)$ de $\mathcal{H}(G_\infty) \otimes \mathrm{Hom}(D_p(E), \mathbb{Q}_p)$ par

$$L_p^*(E)(n) = L_p(E)^\iota \wedge n$$

pour $n \in D_p(E)$ (ici $\iota\tau = \tau^{-1}$). On a alors

$$\chi^k(L_p^*(E))(n) \overset{\text{déf}}{=} \chi^k(L_p^*(E)(n)) = \chi^{-k}(L_p(E)) \wedge n.$$

C'est cette définition qui a été prise dans le cas général de [33] ; les valeurs spéciales correspondant au motif $M(k)$ sont alors liées aux invariants arithmétiques de $M^*(1-k)$ ce qui est plus conforme à ce que l'on fait dans le cas complexe.

[13]Remarquons que la valeur de $1 - a_q q^{-1} \sigma_q^{-1}$ sur $\chi^{-k}\eta$ avec η un caractère d'ordre fini est $1 - a_q q^{k-1} \eta^{-1}(q)$ avec $a_q = \pm 1$ ou 0 et n'est donc jamais nulle sauf peut-être si η est trivial et $k = 1$. Cela ne posera donc pas de problèmes particuliers dans les formules d'évaluation sauf peut-être sur le caractère trivial.

Supposons maintenant que E a réduction multiplicative et exprimons $\tilde{L}_{p,\{N\}}(E) = \tilde{\mathcal{L}}_E(c_\infty)$ sur le système $(\mathcal{T}_0, \mathcal{T}_{-1})$:

$$\tilde{L}_{p,\{N\}}(E) = L_{p,\alpha} \cdot \mathcal{T}_0 + L_{p,\beta} \cdot \mathcal{T}_{-1} \tag{10}$$

avec $L_{p,\alpha}$ et $L_{p,\beta} = L_{p,\beta}^{\mathcal{T}_0}$ appartenant à $\mathcal{H}(G_\infty)$. D'où,

$$\begin{aligned}
\tilde{L}_{p,\{N\}}(E) &= L_{p,\alpha} \cdot ((1+x)e_0 - t_0 N e_0) + L_{p,\beta} \cdot (1+x) N e_0 \\
&= L_{p,\alpha} \cdot (1+x)e_0 + (L_{p,\beta} \cdot (1+x) - L_{p,\alpha} \cdot t_0) N e_0.
\end{aligned}$$

Comme $\tilde{L}_{p,\{N\}}(E) \in D_\infty(E)$, lorsque E a réduction déployée en p, $L_{p,\alpha}$ appartient en fait à $(\tau - 1)\mathcal{H}(G_\infty)$ et même à $\mathbb{Q}_p \otimes (\tau - 1)\mathbb{Z}[[G_\infty]]$, puisque $((\tau - 1)\mathcal{T}_0, \mathcal{T}_1)$ est une base de $\mathcal{D}_\infty(E)$. Donc $\mathbf{1}(L_{p,\alpha})$ est nul. On trouve ainsi une interprétation intéressante du "zéro trivial" traditionnel de la fonction L p-adique dans le cas (mult-dépl). Il provient du fait que par la suite exacte (7), la suite

$$
\begin{array}{ccccccc}
0 & \to & \mathcal{D}_\infty(\mathbb{Q}_p(1)) & \to & \mathcal{D}_\infty(E) & \to & \mathcal{D}_\infty(\mathbb{Q}_p) & \to 0 \\
 & & \| & & & & \| \\
 & & \mathcal{H}^{\psi=0} & & & & \mathcal{H}^{\psi=0} &
\end{array}
$$

n'est pas exacte. Ce zéro est donc d'une autre nature que dans le cas du carré symétrique d'une courbe elliptique ayant bonne réduction où ce qui intervient est le fait que les groupes de Bloch-Kato H_e^1, H_f^1 et H_g^1 sont différents alors qu'ils sont ici égaux.

Question 1. *Que vaut $L_{p,\beta}$ en $\mathbf{1}$?*

Cette valeur ne dépend pas du choix de $\mathcal{T}_0$ puisque $\mathbf{1}(L_{p,\alpha}) = 0$. Je ne vois aucune raison à sa nullité mais ne sais pas le calculer! Pour l'instant, $\tilde{L}_p(E)$ ou plutôt $\tilde{L}_{p,\{N\}}(E)$ n'est pas définie en 0, mais nous allons voir que l'on peut donner un sens à $(1 - p\varphi \otimes \varphi)\tilde{L}_{p,\{N\}}(E)$ en 0. Et cela n'est pas nul. Le "zéro trivial" n'est donc pas un zéro de la fonction L p-adique mais d'une de ses composantes.

Écrivons ω dans la base $(e_0, N e_0)$: quitte à changer e_0 par un multiple, on a $\omega = e_0 - \mathcal{L}(E) N e_0$ avec $\mathcal{L}(E) \in \mathbb{Q}_p^*$. Si q_E est le paramètre de Tate de E, on a $\mathcal{L}(E) = \frac{\log_p q_E}{\mathrm{ord}_p q_E}$.

Proposition. *Dans le cas* (mult), *il existe un unique élément $L_{p,\alpha}$ de $\mathbb{Q}_p \otimes \mathbb{Z}_p[[G_\infty]]$, nul sur le caractère trivial dans le cas* (mult-dépl) *et tel que pour tout caractère η non trivial de conducteur p^n*

$$\eta(L_{p,\alpha}) = \alpha^{-n} \frac{\epsilon(\eta) G(\eta) L_{\{N\}}(E, \eta^{-1}, 1)}{\Omega_{E,\omega}^{\epsilon(\eta)}} \tag{11}$$

et il existe un élément $L_{p,\beta}$ de $\mathcal{H}(G_\infty)$ d'ordre ≤ 1 vérifiant pour tout caractère η non trivial de conducteur p^n

$$\eta(L_{p,\beta}) = v(\eta)\beta^{-n} \frac{\epsilon(\eta)G(\eta)L_{\{N\}}(E, \eta^{-1}, 1)}{\Omega_{E,\omega}^{\epsilon(\eta)}} \tag{12}$$

avec $v(\eta) = v^{\mathcal{T}_0}(\eta) = -(\mathcal{L}(E) + \epsilon(\eta)G(\eta) \operatorname{ev}_\eta(t_0))$.

Rappelons que l'on peut changer $\mathcal{T}_0$ en $\mathcal{T}_0 + \hat{t} \cdot (1+x)Ne_0$ avec $\hat{t} \in \mathcal{H}(G_\infty)$, ce qui change alors $L_{p,\beta}$ en $L_{p,\beta} - \hat{t}L_{p,\alpha}$. Le terme $v^{\mathcal{T}_0}(\eta)$ est un terme p-adique un peu mystérieux. Il dépend du choix de $\log_p p$. Il dépend surtout de $\mathcal{T}_0$ de même que $L_{p,\beta} = L_{p,\beta}^{\mathcal{T}_0}$.

Remarque. Le fait que les valeurs de $L_p(E)$ sur un caractère non trivial de conducteur p^n appartiennent à $\mathbb{Q}_p(\mu_{p^n}) \otimes \varphi^n \operatorname{Fil}^0 D_p(E)$ (et qui se déduit du fait que l'image de $\exp^*$ est contenue dans $K_n \otimes \operatorname{Fil}^0 D_p(E)$, c'est-à-dire dans une "droite fixe") joue un rôle très important et permet de déduire les formules concernant $L_{p,\beta}^{\mathcal{T}_0}$ de celles concernant $L_{p,\alpha}$. Remarquons que c'est ce genre de propriétés qui permet de donner des résultats locaux sur les normes universelles ([34]) ou sur la croissance du groupe de Tate-Shafarevich ([25]).

Les formules peuvent se réécrire sans caractères comme en (5). Pour cela, on résout l'équation $(1 - \varphi \otimes \varphi)\mathcal{G}_E = \tilde{L}_p(E) + \mathfrak{M}_E \log(1+x)$. Il est déterminé à un élément de $D(E)^{\varphi=1}$ près, mais on peut le choisir de manière à ce que pour $n > 0$,

$$\mathcal{G}_E^{\pm}(\zeta_n^a - 1) = \frac{L_{\{p\}}^{\pm}(E, a \bmod p^n, 1)}{\Omega_{E,\omega}^{\pm}}(p\varphi)^n \omega \tag{13}$$

avec $\mathcal{G}_E^{\pm} = \mathcal{G}_E \pm \mathcal{G}_E^\iota$.

Le passage à $n = 0$ est un peu plus délicat car les fonctions $\mathcal{S}(\mathcal{T}_\tau)$ (ou $\mathcal{S}(\mathcal{T}_0)$ dans le cas (mult-non dépl)) et $\mathcal{G}_E$ ne sont pas a priori définies en 0. Par contre, elles vérifient des équation fonctionnelles du type $\psi(G) = 1 \otimes \varphi(G) + m \log(1+x)$, c'est-à-dire

$$\sum_{\zeta \in \mu_p} G(\zeta(1+x) - 1) = p(1 \otimes \varphi)G(x) + pm \log(1+x).$$

Cette équation fonctionnelle implique que G est définie sur $|x| \geq p^{-1/(p-1)}$. On définit alors $[(1 - p\varphi \otimes \varphi)G]$ pour $|x| < p^{-1/(p-1)}$ par

$$[(1 - p\varphi \otimes \varphi)G](x) = -\sum_{\substack{\zeta \in \mu_p \\ \zeta \neq 1}} G(\zeta(1+x) - 1) + pm \log(1+x)$$

et

$$[(1 - p\varphi \otimes \varphi)G](0) = -\sum_{\substack{\zeta \in \mu_p \\ \zeta \neq 1}} G(\zeta - 1).$$

Il est facile de voir que si $\tau \in G_\infty$, $[(1 - p\varphi \otimes \varphi)\tau(G)](0) = [(1 - p\varphi \otimes \varphi)G](0)$.

Remarque. Comme $\psi(\log x) = p^{-1} \log x$, on trouve de même que l'on peut définir

$$[(1 - \varphi) \log](0) = -\log(\prod_{\zeta \in \mu_p, \zeta \neq 1} \zeta - 1) = -\log_p p,$$

ce qui est compatible avec la formule (10).

Faisons le calcul dans le cas (mult-dépl). Rappelons que l'on peut écrire $\mathcal{S}(\mathcal{T}_\tau) = (\tau - 1)T_0 e_0 - S_\tau N e_0$ avec $T_0 \in \mathcal{H}$ vérifiant $(1 - \varphi)T_0 = (1 + x) - 1$, $T_0(0) = 0$ et $(1 - p^{-1}\varphi)S_\tau = t_\tau$ avec $S_\tau \in \mathcal{B}[\log x]$ et $NS_\tau = T_0$. Écrivons aussi $\mathcal{S}(\mathcal{T}_1) = T_1 N e_0$ avec $(1 - p^{-1}\varphi)T_1 = (1 + x) - \log(1 + x)$. De la formule (10), on déduit que

$$\begin{aligned}
\mathcal{G}_E &= L_{p,\beta} \cdot T_1 N e_0 + L_{p,\alpha} \cdot \mathcal{S}(\mathcal{T}_0) + \lambda e_0 \\
&= L_{p,\beta} \cdot T_1 N e_0 + L_{p,\alpha} \cdot T_0 e_0 - (\tau - 1)^{-1} L_{p,\alpha} \cdot S_\tau N e_0 + \lambda e_0 \qquad (14) \\
&= (L_{p,\alpha} \cdot T_0 + \lambda)e_0 + (L_{p,\beta} \cdot T_1 - (\tau - 1)^{-1} L_{p,\alpha} \cdot S_\tau)N e_0
\end{aligned}$$

avec $\lambda \in \mathbb{Q}_p$. Rappelons que $\mathbf{1}(L_{p,\alpha}) = 0$ et posons $L'_{p,\alpha} = \frac{d}{ds}\langle \chi \rangle^s (L_{p,\alpha})_{|s=0}$ avec $\langle \chi \rangle$ le composé de χ avec la projection $\mathbb{Z}_p^* \to 1 + p\mathbb{Z}_p$, d'où

$$\mathbf{1}(L'_{p,\alpha}) = \log \chi(\tau)\mathbf{1}((\tau - 1)^{-1} L_{p,\alpha}).$$

On obtient en prenant la valeur en 0 :

$$\begin{aligned}
[(1 - p\varphi \otimes \varphi)(\mathcal{G}_E)](0) &= \mathbf{1}(L_{p,\beta}) T_1(0)(1 - p\varphi)N e_0 \\
&\quad - \mathbf{1}(L'_{p,\alpha})\frac{[(1 - p\varphi \otimes \varphi)S_\tau](0)}{\log \chi(\tau)}N e_0 + (1 - p)\lambda e_0 \\
&= -(p - 1)\lambda e_0 + c\mathbf{1}(L'_{p,\alpha})N e_0
\end{aligned}$$

avec $c = -\frac{[(1-p\varphi\otimes\varphi)S_\tau](0)}{\log \chi(\tau)} \in \mathbb{Q}_p$ indépendant de τ, de E et même du choix de S_τ. Nous démontrons dans l'appendice que $c = 1$. On a d'autre part en utilisant (13) pour $n = 1$

$$(p\varphi)^{-1}[(1 - p\varphi \otimes \varphi)(\mathcal{G}_E)(0)] = -\frac{L_{\{N\}}(E, 1)}{\Omega_{E,\omega}}\omega.$$

On en déduit que

$$(1 - \frac{1}{p})\lambda e_0 - \mathbf{1}(L'_{p,\alpha})N e_0 = \frac{L_{\{N\}}(E, 1)}{\Omega_{E,\omega}}\omega \qquad (15)$$

D'où, en écrivant $\omega = e_0 - \mathcal{L}(E)Ne_0$, la formule

$$\mathbf{1}(L'_{p,\alpha}) = \mathcal{L}(E)\frac{L_{\{N\}}(E,1)}{\Omega_{E,\omega}}. \tag{16}$$

La formule (16) est une conjecture de Mazur-Tate-Teitelbaum ([27]) et a été démontrée par Greenberg et Stevens ([16]) (le passage de N à p est immédiat). Elle a été redémontrée par Kato-Kurihara-Tsuji, il y a quelques années. Redisons que nous utilisons la construction de Kato de systèmes d'Euler modulaires et qu'il n'y a donc peut-être rien de nouveau. Nous voulons simplement mettre en valeur la manière dont ces résultats sur les valeurs spéciales peuvent se montrer à partir de l'application régulateur $\mathcal{L}_E$, osons rajouter, de manière simple, c'est-à-dire en faisant des calculs sur des éléments de l'algèbre $\mathcal{B}[\log x]$. Autrement dit, pour $a = (a_n) \in Z^1_\infty(\mathbb{Q}_p, V)$, le régulateur $\mathcal{L}_E(a)$ contient tous les renseignements sur les régulateurs des a_n, et même de leur twist comme nous allons le voir un peu plus loin. Il y a derrière ces calculs l'étude du module $\mathcal{D}_\infty(E)$ et son lien avec $Z^1_\infty(\mathbb{Q}_p, V_p(E))$ qui utilise en particulier la loi de réciprocité. Mais la généralité est finalement grande!

Remarque. Il est à remarquer que les formules s'expriment mieux et sont plus précises en termes de $\mathcal{G}_E$ que de $L_{p,\{N\}}(E)$. Mais il y a d'autre raisons pour préférer $L_p(E)$ puisque c'est sur l'ordre de cette fonction que l'on a des renseignements. Il faut donc peut-être prendre l'habitude de considérer le couple $(L_{p,\{N\}}(E), \mathcal{G}_E)$. Lorsque φ n'a pas de valeurs propres égales à une puissance de p, la première détermine la seconde. Ce qui n'est pas le cas dans le cas (mult-dépl). L'équation (15) est plus précise que (16). On déduit ainsi de (15) que

$$\lambda = (1 - \frac{1}{p})^{-1}\frac{L_{\{N\}}(E,1)}{\Omega_{E,\omega}} \ .$$

D'où en remarquant que $N\mathcal{G}_E$ est bien définie en 0 et que le facteur d'Euler en p est $(1 - p^{-s})$,

$$(N\mathcal{G}_E)(0) = \frac{L_{\{N/p\}}(E,1)}{\Omega_{E,\omega}}Ne_0 \tag{17}$$

Le module $\mathcal{D}_\infty(E)$ ne dépend que de la structure de (φ, N)-module de $D_p(E)$ et absolument pas de sa filtration. C'est dans le régulateur d'Iwasawa que cette filtration se manifeste. En particulier, soit f une forme modulaire de poids k pour $\Gamma_0(N)$ avec p divisant exactement N et $a_p(f) = p^{(k-2)/2}$. Soit $D_p(f)$ le (φ, N)-module filtré associé à f en p. Comme plusieurs normalisations sont possibles, précisons que la filtration de $D_p(f)$ a

comme poids de Hodge 0 et $-k + 1$. Le (φ, N)-module $D_p(f)[(k - 2)/2]$
admet une base de vecteurs propres (e_0, Ne_0) avec $\varphi e_0 = e_0$. En tant que
(φ, N)-module filtré, il est irréductible pour $k \neq 2$. Par contre, en tant que
(φ, N)-module, il est réductible, le $\mathcal{H}(G_\infty)$-module $\mathcal{D}_\infty(D_p(f)[(k - 2)/2])$
est isomorphe à celui construit pour une courbe elliptique ayant réduction
multiplicative déployée et on a encore une suite exacte de $\mathcal{H}(G_\infty)$-modules

$$0 \to \mathcal{H}^{\psi=0} \to \mathcal{D}_\infty(D_p(f)[(k - 2)/2]) \to \mathcal{H}^{\psi=0} \to \mathbb{Q}_p \to 0.$$

En particulier, tous les calculs faits précédemment sont valables. Mis avec
la construction d'un système d'Euler-Kolyvagin à partir des éléments de
Beilinson, qui est faite par Kato, on a ainsi démontré la **conjecture de
Mazur-Tate-Teitelbaum** [27] pour les formes modulaires de poids k, de
conducteur divisible par p (et non par p^2) telles que $a_p = p^{(k-2)/2}$, avec
$\mathcal{L}(E)$ la pente de la filtration de $D_p(f)[(k - 2)/2]$ dans la base $e_0, -Ne_0$.
Stevens vient de démontrer une formule de ce type avec $\mathcal{L}(E)$ l'invariant
défini par Coleman dans [9]. Les deux résultats ensemble montrent l'égalité
de ces deux invariants (au moins lorsque $L(f, k/2)$ est non nul!).

Nous allons maintenant nous intéresser aux twists à la Tate de $V_p(E)$.
Soit k un entier strictement positif. Soit $H^2_{\mathcal{M}}(E, k) = (\mathbb{Q} \otimes K_{2k-2}(E))^{(k)}$ le
k-ième espace propre pour l'opérateur d'Adams de la K-théorie de Quillen
de E. Nous ne rappelons pas la définition des régulateurs complexes par
Beilinson. Du point de vue p-adique, on définit l'application régulateur
comme le composé Reg des classes de Chern étales p-adiques :

$$H^2_{\mathcal{M}}(E, k + 1) \to H^1(G_{S,\mathbb{Q}}, V_p(E)(k)) \to H^1(\mathbb{Q}_p, V_p(E)(k))$$

avec $r_k = \log_{E,k}$ le logarithme de Bloch-Kato :

$$r_k \colon H^1(\mathbb{Q}_p, V_p(E)(k)) = H^1_f(\mathbb{Q}_p, V_p(E)(k)) \overset{\log_{V_p(E)(k)}}{\to} D_p(E)[-k] \cong D_p(E).$$

Ici, S est un ensemble fini de places contenant les places de mauvaise
réduction de E et les places divisant p et $G_{S,\mathbb{Q}}$ le groupe de Galois su r
$\mathbb{Q}$ de la plus grande extension non ramifiée en dehors de S. Un autre point
de vue est de définir directement l'application Reg : $H^2_{\mathcal{M}}(E, k+1) \to D_p(E)$
(Niziol [29], Nekovář [28], Besser [4]). Nous n'en dirons pas plus ici. Nous
notons aussi Reg l'application $H^1(G_{S,\mathbb{Q}}, V_p(E)(k)) \to D_p(E)$. De même, si
F est une extension finie de $\mathbb{Q}$, on définit Reg

$$H^2_{\mathcal{M}}(E/F, k + 1) \to H^1(G_{S,F}, V_p(E)(k))$$
$$\to \prod_{v|p} H^1(F_v, V_p(E)(k)) \to F \otimes_{\mathbb{Q}} D_p(E).$$

Lorsque k est un entier négatif ou nul, on dispose d'une application

$$r_k \colon \oplus_{v|p} H^1(F_v, V_p(E)(k)) \to H^1_{/f}(F_v, V_p(E)(k))$$

$$\overset{\exp^*_{V_p(E)(1-k)}}{\to} \quad F \otimes_{\mathbb{Q}} D_p(E)[-k] = F \otimes_{\mathbb{Q}} D_p(E)$$

avec $H^1_{/f}(F_v, V_p(E)(k)) = H^1(F_v, V_p(E)(k))/H^1_f(F_v, V_p(E)(k))$ et on note encore Reg le composé : $H^1(G_{S,F}, V_p(E)(k)) \to F \otimes_{\mathbb{Q}} D_p(E)$.

Jannsen ([18], [19]) conjecture que $H^1(G_{S,F}, V_p(E)(k))$ est de dimension $[F \colon \mathbb{Q}]$, pour $k \neq 0$ et $k \neq \pm 1$ et pour F une extension finie de $\mathbb{Q}$, et plus précisément que si F est une extension abélienne de $\mathbb{Q}$ et η un caractère de $\mathrm{Gal}(F/\mathbb{Q})$, $H^1(G_{S,F}, V_p(E)(k))^{(\eta)}$ (image du projecteur e_η) est de dimension 1 sur $\mathbb{Q}(\eta)$. Il montre l'équivalence avec la nullité de $H^2(G_{S,F}, V_p(E)(k))$ pour ces valeurs de k par application de la formule de caractéristique d'Euler-Poincaré de ces groupes de cohomologie. Lorsque E a bonne réduction en p, il semble raisonnable d'inclure aussi le cas où $k = \pm 1$. Par contre, dans le cas (mult-dépl), on dispose d'une surjection

$$H^2(G_{S,\mathbb{Q}}, V_p(E)(1)) \to H^2(\mathbb{Q}_p, V_p(E)(1)) \cong H^0(\mathbb{Q}_p, V_p(E)^*) = \mathbb{Q}_p.$$

La conjecture est alors que le noyau de cette application est nul. Remarquons à ce propos que, dans ce cas, les groupes $H^1(\mathbb{Q}_p, V_p(E)(1)) = H^1_g(\mathbb{Q}_p, V_p(E)(1))$ et $H^1_f(\mathbb{Q}_p, V_p(E)(1))$ sont différents et diffèrent de

$$(D_p(E)[-1]/ND_p(E)[-1])^{\varphi=1} = (D_p(E)/ND_p(E))^{\varphi=p^{-1}} \cong \mathbb{Q}_p.$$

Revenons à la théorie d'Iwasawa. Elle permet à partir d'éléments compatibles pour la norme des $H^1(G_{S,\mathbb{Q}(\mu_{p^n})}, T_p(E))$ de construire des éléments de $H^1(G_{S,\mathbb{Q}(\mu_{p^n})}, T_p(E)(k))$ pour tout entier k. Les premiers exemples de telles constructions ont été donnés par Soulé (par exemple [40]) et c'est un outil extrêmement puissant. Si c appartient à[14] $H^1_\infty(\mathbb{Q}, V_p(E))$ (resp. à $Z^1_\infty(\mathbb{Q}_p, V_p(E)))$, on note $c(k)$ son image dans

$$H^1_\infty(\mathbb{Q}, V_p(E)(k)) = \mathbb{Q}_p \otimes \varprojlim_n H^1(G_{S,\mathbb{Q}(\mu_{p^n})}, T_p(E)(k))$$

(resp. dans $Z^1_\infty(\mathbb{Q}_p, V_p(E)(k)))$. Si η est un caractère de conducteur p^n, on note encore ev_η le composé de la projection de $H^1_\infty(\mathbb{Q}, V_p(E)(k))$ sur $H^1(G_{S,\mathbb{Q}(\mu_{p^n})}, T_p(E)(k))$ avec le projecteur e_η (et de même pour les groupes de cohomologie locaux). Ainsi, $\mathrm{ev}_{\mathbf{1}}$ est simplement la projection dans $H^1(G_{S,\mathbb{Q}}, T_p(E)(k))$. On note d'autre part D la dérivation $(1+x)\frac{d}{dx}$ et l'opérateur qu'elle induit sur $\mathcal{D}_\infty(E)$.

---

[14]Rappelons que $\mathbb{Q}_p \otimes \varprojlim_n H^1(G_{S,\mathbb{Q}(\mu_{p^n})}, T_p(E))$ est indépendant de S et égal à $H^1_\infty(\mathbb{Q}, V_p(E))$.

La construction du régulateur d'Iwasawa $\tilde{\mathcal{L}}_E$ utilise de façon essentielle cette opération de twist et ses propriétés impliquent (et même sont) la proposition suivante (ici $\Gamma^*(j)$ est $(j-1)!$ si $j \geq 1$ et $(-1)^j/j!$ si $j \leq 0$) :

Proposition. *Pour tout entier k et tout caractère η de conducteur p^n non trivial, on a*

$$\mathrm{ev}_\eta(D^{-k}(\tilde{\mathcal{L}}_E)(c_\infty)) = \Gamma^*(-k)(p^{1-k}\varphi)^n \, \mathrm{Reg}(\mathrm{ev}_\eta(c_\infty(k))) \qquad (18)$$

Dans le cas (br), *cela s'écrit aussi*

$$\eta\chi^{-k}(\mathcal{L}_E(c_\infty)) = \Gamma^*(-k)\epsilon(\eta)G(\eta)(p^{-k}\varphi)^n \, \mathrm{Reg}(\mathrm{ev}_\eta(c_\infty(k))) \qquad (19)$$

On peut alors traduire cela comme un énoncé sur les fonction L p-adiques classiques.

Proposition. *Pour tout entier k, on a*

(1) *dans le cas ordinaire,*

$$\eta\chi^{-k}(L^\pi_{p,\{N\}}(E)) = \Gamma^*(-k)\epsilon(\eta)G(\eta)(p^{-k}\varphi)^n \pi_{[0]}(\mathrm{Reg}(\mathrm{ev}_\eta(c_\infty(k))))$$

(2) *dans le cas supersingulier,*

$$\eta\chi^{-k}(L_{p,\{N\}}(E)) = \Gamma^*(-k)\epsilon(\eta)G(\eta)(p^{-k}\varphi)^n \, \mathrm{Reg}(\mathrm{ev}_\eta(c_\infty(k))).$$

Mais la formule (18) est plus complète et en termes des fonctions $\tilde{L}_{p,\{N\}}(E)$ ou $L_{p,\{N\}}(E)$, elle s'écrit

$$\mathrm{ev}_\eta(D^{-k}(\tilde{L}_{p,\{N\}}(E))) = \Gamma^*(-k)(p^{1-k}\varphi)^n \, \mathrm{Reg}(\mathrm{ev}_\eta(c_\infty(k))) \qquad (20)$$

ou

$$\eta\chi^{-k}(L_{p,\{N\}}(E)) = \Gamma^*(-k)\epsilon(\eta)G(\eta)(p^{-k}\varphi)^n \, \mathrm{Reg}(\mathrm{ev}_\eta(c_\infty(k))). \qquad (21)$$

Des formules analogues sont valables pour le caractère trivial (au moins dans le cas (br)). Par exemple pour $k = 1$, on peut interpréter la valeur de la fonction p-adique en $s = 0$ (avec $L_p(E,s) = \langle\chi\rangle^{1-s}(L_p(E))$) comme le régulateur d'un point C_2 de $K_2(E, \mathbb{Z}_p)$:

$$L_p(E,0) = (1 - p^{-2}\varphi^{-1})^{-1}(1 - p^{-1}\varphi) \, \mathrm{Reg} \, C_2. \qquad (22)$$

Une formule de ce type a été montrée par Coleman et de Shalit [10] dans le cas de multiplication complexe et par Kings [22] dans le même cadre pour tout entier k. L'intérêt supplémentaire qu'il y a dans la formule (22) (outre le fait qu'elle se généralise à tout entier k) est qu'elle tient compte du régulateur complet et non seulement de sa projection sur une droite spéciale.

Restons avec le cas $k = 1$. Par définition même de c_∞, C_2 provient d'un élément de $H^2_{\mathcal{M}}(E, 2)$. Pour obtenir l'équation (22), nous avons ici appliqué deux fois la loi de réciprocité (une fois celle de Kato et une fois celle que j'ai introduite) et twisté une fois vers la gauche et une fois vers la droite. Il devrait donc être possible de s'en passer. Et en effet, il semble possible de montrer directement la formule (22) et ses analogues pour un caractère η. Grâce à la loi de réciprocité de Colmez, on en déduit alors les formules (9) de Kato pour $c_\infty(1)$ au moins dans le cas (br) (article en préparation).

Question 2. *Dans le cas (mult-dépl), $C_2 = \mathrm{ev}_1\,(c_\infty(1))$ appartient-il à $H^1_f(\mathbb{Q}_p, V_p(E)(1))$? dans le cas contraire, que vaut son image dans*

$$H^1_{g/f}(\mathbb{Q}_p, V_p(E)(1)) \cong (D_p(E)/ND_p(E))^{p^{-1}\varphi = p^{-1}} \cong \mathbb{Q}_p e_0?$$

Comme $c_\infty(1)$ est très concret, une réponse est envisageable!

Question 3. *Dans le cas (mult-dépl), $C_0 = \mathrm{ev}_1\,(c_\infty(-1))$ appartient-il à $H^1_f(\mathbb{Q}_p, V_p(E)(-1))$? dans ce cas, que vaut son image dans*

$$H^1_{f/e}(\mathbb{Q}_p, V_p(E)(-1)) \cong D_p(E)^{N=0}/(1 - p\varphi) \cong \mathbb{Q}_p Ne_0?$$

Question 4. *Pourquoi ne semble-t-on envisager la possibilité d'un zéro trivial pour $L_{p,\alpha}$ en χ^{-1} ou χ alors que le facteur d'Euler de la fonction L s'annule ?*

Peut-être parce que ce ne serait pas ici la projection sur $\mathbb{Q}_p e_0$ qui en ferait apparaître un.

L'utilisation des techniques extrêmement puissantes introduites par Kolyvagin dans [23] et relatives à ce qu'il a appelé systèmes d'Euler ont permis à Kato de montrer la divisibilité de la série caractéristique de

$$H^2_{\infty,S}(\mathbb{Q}, V_p(E)) = \mathbb{Q}_p \otimes \varprojlim_n H^2(G_{\mathbb{Q}(\mu_{p^n}),S}, T_p(E))$$

par une fonction liée au système d'Euler. Ce résultat peut se dire de manière simple en utilisant l'idéal arithmétique $I_{\mathrm{arith}}(E)$ introduit dans [31] dans le cas des courbes elliptiques : lorsque $H^1_\infty(\mathbb{Q}, V_p(E))$ est de rang 1, $I_{\mathrm{arith}}(E)$ est simplement le $\mathbb{Z}_p[[G_\infty]]$-module, image dans $\mathcal{D}_\infty(E)$ de $H^1_\infty(\mathbb{Q}, T_p(E))$ multiplié par la série caractéristique de $H^2_{\infty,p}(\mathbb{Q}, T_p(E))$ défini comme le noyau

$$H^2_{\infty,S}(\mathbb{Q}, T_p(E)) \to \varprojlim_n \prod_{v \in S} H^2(\mathbb{Q}(\mu_{p^n})_v, T_p(E))$$

autrement dit, en utilisant le langage des déterminants, l'image par le régulateur d'Iwasawa $\mathcal{L}_E$ de $\det(H^2_{\infty,p}(\mathbb{Q}, T_p(E)))^{-1} \cdot H^1_\infty(\mathbb{Q}, T_p(E))$ dans $\mathcal{D}_\infty(E)$. Par un théorème de Rohrlich [37] sur les valeurs de la fonction L tordue par des caractères, $\tilde{L}_p(E)$ est non nul. Le théorème de Kato peut alors s'énoncer ainsi (bien qu'il ne le fasse pas explicitement)

Théorème (Kato). *Le $\mathbb{Z}_p[[G_\infty]]$-module $H^1_\infty(\mathbb{Q}, V_p(E))$ est de rang 1 et le $\mathbb{Z}_p[[G_\infty]]$-module engendré par $\tilde{L}_p(E)$ est contenu dans $\mathbb{Q}_p \otimes \mathcal{I}_{arith}(E)$.*

Ainsi, la "conjecture principale" de [31] qui prédit que $\mathcal{I}_{arith}(E) = (\tilde{L}_p(E))$ est à moitié vérifiée.

Nous nous limitons par prudence et manque de temps au cas (br) dans la proposition suivante :

Proposition. (1) (Kato) *Si $L(E, \eta^{-1}, 1)$ est non nul pour un caractère η de conducteur p^n, les η-composantes de $E(\mathbb{Q}(\mu_{p^n}))$ et du groupe de Shafarevich-Tate $\text{III}(E/\mathbb{Q}(\mu_{p^n}))$ sont finis.*

(2) Soit $k \neq 0$ tel que $\chi^{-k}\eta(L_p(E)) \neq 0$. Alors $H^1(G_{S,\mathbb{Q}(\mu_{p^n})}, V_p(E)(k))^{(\eta)}$ est de rang 1, engendré par $\text{ev}_\eta(c_\infty(k))$ et la conjecture de Jannsen est vraie.

Sous l'hypothèse faite, les facteurs d'Euler pour les places divisant N ne s'annulent pas. Il serait peut-être quand même raisonnable de vérifier numériquement que $\chi^{-k}(\tilde{L}_p(E))$ n'est pas nul!

Donnons rapidement la démonstration de 2). Comme $\chi^{-k}\eta(L_p(E)) \neq 0$, il en est de même de $\chi^{-k}\eta(\mathcal{I}_{arith}(E))$. Cela implique en particulier que la série caractéristique de $H^2_{\infty,\{p\}}(\mathbb{Q}, V_p(E)(k))$ ne s'annule pas en η^{-1} et par des arguments classiques que $H^2(G_{S,\mathbb{Q}(\mu_{p^n})}, V_p(E)(k))^{(\eta)}$ est fini et que $H^1(G_{S,\mathbb{Q}(\mu_{p^n})}, V_p(E)(k))^{(\eta)}$ est de rang 1 (formule de caractéristique de Tate-Poitou). On peut en fait calculer les valeurs de $\chi^{-k}\eta(\mathcal{I}_{arith}(E))$ pour $I_{arith}(E)$ un générateur de $\mathcal{I}_{arith}(E)$. Le calcul a été fait dans [33] dans le cas de bonne réduction et pour η caractère trivial (voir aussi [31] dans le cas des courbes elliptiques), et n'est pas plus difficile pour un caractère non trivial.

Finissons par une question. Posons $C_k = \text{ev}_1(c_\infty(k-1))$ et supposons $\text{Reg}\, C_k$ non nul, ce qui est équivalent à ce que $\chi^{-k}(L_p(E))$ soit non nul. En particulier, C_k est non nul et est une base de $H^1(G_{S,\mathbb{Q}}, V_p(E)(k-1))$ pour $k \neq 1$.

Notons ici D_α et D_β les espaces propres de $D_p(E)$ pour les valeurs propres α^{-1} et β^{-1} de φ et supposons que $\text{Fil}^0 D(E)$ n'est pas stable par φ. Lorsque $\text{Reg}\, C_k$ est non nul, pour le repérer dans le plan $D_p(E)$, on peut utiliser le birapport $\ell_k(E)$ des quatre droites D_α, D_β, $\text{Fil}^0 D_p(E)$ et

$\mathbb{Q}_p \operatorname{Reg} C_k = \operatorname{Reg}(H^1(G_{S,\mathbb{Q}}, V_p(E)(k-1)))$. Ainsi, si on écrit $\omega = e_\alpha + e_\beta$ avec $e_\alpha \in D_\alpha$, $e_\beta \in D_\beta$,

$$\operatorname{Reg} C_k \in \mathbb{Q}_p (e_\alpha + \ell_k(E) e_\beta).$$

Lorsque $k = 1$, les deux droites $\operatorname{Fil}^0 D_p(E)$ et $\operatorname{Reg} C_1$ coïncident et $\ell_1(E) = 1$. D'autre part, dans le cas supersingulier, il est facile de voir que $\ell_k(E)$ est de norme 1 dans l'extension quadratique $\mathbb{Q}_p(\alpha)/\mathbb{Q}_p$. On peut aussi écrire

$$\mathbb{Q}_p \operatorname{Reg} C_k = \mathbb{Q}_p (\omega - \lambda_k(E) p \varphi \omega)$$

et $\lambda_k(E) \in \mathbb{Q}_p \cup \infty$ ne dépend pas du choix de $\omega \in \operatorname{Fil}^0 D_p(E)$. Il est facile de voir qu'en fait $\ell_k(E) = (1 + \lambda_k(E)\alpha)/(1 + \lambda_k(E)\beta)$ si $\lambda_k(E) \neq \infty$ et -1 si $\lambda_k(E) = \infty$. On a $\lambda_1(E) = 0$. On peut donner des définitions analogues pour un caractère η de conducteur une puissance $p^{n(\eta)}$ de p et obtenir ainsi des invariants $\ell_{k,\eta}(E)$ et $\lambda_{k,\eta}(E)$. Par exemple, $\ell_{1,\eta}(E) = 1$.

Question 5. *Que peut-on dire de la répartition des $\lambda_k(E)$ dans $\mathbb{P}^1(\mathbb{Q}_p)$? Peut-il exister une formule simple pour $\lambda_{k,\eta}(E)$ pour $k \neq 1$ fixé ?*

Appendice

Dans cet appendice, nous expliquons comment peuvent se calculer des relèvements d'éléments de $\mathcal{D}_\infty(\mathbb{Q}_p)$ dans $\mathcal{D}_\infty(E)$ et en particulier nous calculons la constante c qui intervient dans la démonstration de la formule de Mazur-Tate-Teitelbaum.

Plaçons-nous dans le cas (mult-dépl). Soit $f \in \mathcal{H}^{\psi=0}$ tel que $f(0) = 0$. Soit $F_1 \in \mathcal{H}$ tel que $(1 - \varphi)F_1 = f$ et $F_1(0) = 0$. Nous allons montrer que si $f \in \mathbb{Q}_p \otimes \mathbb{Z}_p[[x]]$, la suite $p^n \psi^n(F_1 \log x)$ converge dans $\mathcal{B}[\log x]$ vers un élément F_2 tel que $\psi(F_2) = p^{-1} F_2$ et $N F_2 = F_1$; $(1 - \varphi \otimes \varphi)(F_1 e_0 - F_2 N e_0)$ est alors un élément de $\mathcal{D}_\infty(E)$.

Posons donc $U_n = p^n \psi^n(F_1 \log x)$. On a

$$
\begin{aligned}
U_{n+1} - U_n &= p^{n+1}\psi^{n+1}\left((1 - p^{-1}\varphi)(F_1 \log x)\right) \\
&= p^{n+1}\psi^{n+1}\left((1-\varphi)(F_1)\log x + \varphi(F_1)(1 - p^{-1}\varphi)(\log x)\right) \\
&= p^{n+1}\psi^{n+1}\left(f \log x\right) + p^{n+1}\psi^n\left(F_1\psi((1 - p^{-1}\varphi)\log x)\right) \\
&= p^{n+1}\psi^{n+1}\left(f \log x\right)
\end{aligned}
$$

car

$$
\begin{aligned}
\psi((1 - p^{-1}\varphi)\log x) &= 0 \\
&= p^{n+1}\psi^{n+1}\left(f(1 - p^{-1}\varphi)\log x\right) + p^n\psi^{n+1}\left(f\varphi \log x\right) \\
&= p^{n+1}\psi^{n+1}\left(f(1 - p^{-1}\varphi)\log x\right)
\end{aligned}
$$

car $\psi(f) = 0$ et $\psi(f \varphi g) = \psi(f)g$. On remarque alors que $(1 - p^{-1}\varphi) \log x$ est un élément de $\mathbb{Q}_p \otimes \mathbb{Z}_p[[1/x]]$ convergeant pour $|x| > p^{-1/(p-1)}$. Comme $f \in \mathbb{Q}_p \otimes \mathbb{Z}_p[[x]]$, $f(1 - p^{-1}\varphi) \log x$ est la somme d'un tel élément et d'un élément de $\mathbb{Q}_p \otimes \mathbb{Z}_p[[x]]$ et si $p^{-1/(p-1)} < \rho < 1$, les $\|\psi^n(f(1 - p^{-1}\varphi) \log x)\|_\rho$ sont bornés par rapport à n. On en déduit la convergence de la suite U_n (comme $V_n = U_n - F_1 \log x \in \mathcal{B}$, cela signifie simplement la convergence de V_n sur les couronnes $\rho_1 \le |x| \le \rho_2$ pour $p^{-1/(p-1)} < \rho_1 \le \rho_2 < 1$). Si F_2 est sa limite, il est clair que $\psi(F_2) = p^{-1}F_2$ et que le coefficient de $\log x$ est F_1, c'est-à-dire que $N f_2 = F_1$.

Pour $\zeta - 1$ avec $\zeta \in \mu_{p^\infty}$, les valeurs de F_1 ne sont pas difficiles à écrire en fonction de f :

$$F_1(x) = \sum_{m=0}^{n-1} f(\varphi^m(x)) + F_1(\varphi^n(x))$$

et donc pour $\zeta \in \mu_{p^n}$ avec $n \ge 1$

$$F_1(\zeta - 1) = \sum_{m=0}^{n-1} f(\zeta^{p^m} - 1).$$

Passons à F_2. On a alors en fonction de F_1

$$F_2(\zeta_n - 1) = \lim_{s \to \infty} \mathrm{Tr}_{s/n} \left(F_1(\zeta_s - 1) \log(\zeta_s - 1) \right)$$

$$= \lim_{s \to \infty} \sum_{\substack{\zeta \in \mu_{p^s} \\ \zeta \mapsto \zeta_n}} F_1(\zeta - 1) \log(\zeta - 1)$$

où $\zeta \mapsto \zeta_n$ signifie qu'il existe un entier k tel que $\zeta^{p^k} = \zeta_n$ et $\mathrm{Tr}_{m/n}$ désigne la trace de $\mathbb{Q}(\mu_{p^m})$ à $\mathbb{Q}(\mu_{p^n})$, d'où

$$F_2(\zeta_n - 1) = \lim_{s \to \infty} \sum_{m=0}^{n-1} \sum_{\substack{\zeta \in \mu_{p^s} \\ \zeta \mapsto \zeta_n}} f(\zeta^{p^m} - 1) \log(\zeta - 1).$$

Prenons maintenant $f = (\tau - 1) \cdot (1 + x)$, pour $S_\tau = F_2$, calculons

$$c_\tau = -[(1 - p\varphi \otimes \varphi)S_\tau](0) = \sum_{\substack{\zeta \in \mu_p \\ \zeta \neq 1}} S_\tau(\zeta - 1)$$

$$= \lim_{n \to \infty} \mathrm{Tr}_{n/0}(F_1(\zeta_n - 1) \log(\zeta_n - 1)).$$

Posons $G_n = \mathrm{Gal}(\mathbb{Q}_p(\mu_{p^n})/\mathbb{Q}_p)$ et calculons donc

$$X_n = \sum_{\sigma \in G_n} \sum_{m=0}^{n-1} (\tau - 1) \cdot \zeta_{n-m}^\sigma \log(\zeta_n^\sigma - 1).$$

On remarque d'abord que

$$\sum_{\sigma \in G_n} (\tau - 1) \cdot \zeta_{n-m}^\sigma \log(\zeta_n^\sigma - 1) = \sum_{\sigma \in G_m} (\tau - 1) \cdot \zeta_{n-m}^\sigma \log(\zeta_{n-m}^\sigma - 1)$$

car $p\psi(\log x) = \log x$. D'où,

$$X_n = \sum_{m=0}^{n-1} \sum_{\sigma \in G_{n-m}} \zeta_{n-m}^\sigma \log \frac{\zeta_{n-m}^{\tau^{-1}\sigma} - 1}{\zeta_{n-m}^\sigma - 1}.$$

En posant $a_{\tau-1} = (\tau^{-1} - 1) \log x = \log \frac{(1+x)^{\chi(\tau)^{-1}} - 1}{x} \in \mathcal{H}^{\psi=0}$ et en remarquant que les ζ_{n-m}^σ parcourent toutes les racines p^n-ièmes de l'unité sauf 1, on obtient que

$$X_n = \sum_{\zeta \in \mu_{p^n} - \{1\}} \zeta a_{\tau-1}(\zeta - 1).$$

Ecrivons $a_{\tau-1} \equiv \sum_{\substack{0 < k < p^n \\ (k,p)=1}} a^{k,n} (1+x)^k \mod (1+x)^{p^n} - 1$ et remarquons que

$$\sum_{\zeta \in \mu_{p^n}} \zeta^j = \begin{cases} 0 & \text{si } j \not\equiv 0 \bmod p^n \\ p^n & \text{si } j \equiv 0 \bmod p^n \end{cases} \equiv 0 \bmod p^n.$$

Donc

$$X_n \equiv - \sum_{\substack{0 < k < p^n \\ (k,p)=1}} a^{k,n} \bmod p^n$$

$$\equiv -a_{\tau-1}(0) \bmod p^n$$

et la limite c_τ de X_n lorsque $n \to \infty$ est $-a_{\tau-1}(0) = \log \chi(\tau)$. Ce qui démontre l'égalité désirée

$$c = -\frac{[(1 - p\varphi \otimes \varphi)S_\tau](0)}{\log \chi(\tau)} = 1.$$

On peut de la même manière écrire une formule pour $\mathrm{ev}_\eta(t_0) = (\eta(\tau) - 1)^{-1}\,\mathrm{ev}_\eta(S_\tau)$ pour η caractère de conducteur p^n avec ce choix de S_τ :

$$(\eta(\tau) - 1)\,\mathrm{ev}_\eta(t_0) = \sum_{\sigma \in G_n} \eta^{-1}(\sigma) \lim_{s \to \infty} \mathrm{Tr}_{s/n}\left(\sum_{j=0}^{s-1} (\tau - 1)\zeta_s^{j\sigma} \log(\zeta_s^\sigma - 1) \right).$$

On peut la transformer, mais je n'ai pas trouvé de manière de bien la présenter! Il est à remarquer cependant qu'interviennent des traces d'éléments du type $\zeta_s \log(\zeta_s - 1)$. Je ne sais pas si de telles expressions interviennent ailleurs.

Références

[1] Y. Amice and J. Vélu, *Distributions p-adiques associées aux séries de Hecke*, Astérisque (1975), no. 24–25, 119–131.

[2] A. A. Beilinson, *Higher regulators of modular curves*, Applications of algebraic K-theory to algebraic geometry and number theory, Part I, II, (Boulder, Colo., 1983), Amer. Math. Soc., Providence, R.I., 1986, pp. 1–34.

[3] D. Benois, *On Iwasawa theory of crystalline representations*, Duke Math. J. **104** (2000), 211–267.

[4] A. Besser, *Syntomic regulators and p-adic integration. II. K_2 of curves*, Proceedings of the Conference on p-adic Aspects of the Theory of Automorphic Representations (Jerusalem, 1998), vol. 120, 2000, pp. 335–359.

[5] S. Bloch and K. Kato, *L-functions and Tamagawa numbers of motives*, The Grothendieck Festschrift, vol. I, Birkhäuser Boston, Boston, MA, 1990, pp. 333–400.

[6] G. Christol, *Systèmes différentiels linéaires p-adiques, structure de Frobenius faible*, Bull. Soc. Math. France **109** (1981), 83–122.

[7] J. Coates and A. Wiles, *On p-adic L-functions and elliptic units*, J. Austral. Math. Soc. Ser. A **26** (1978), 1–25.

[8] R. F. Coleman, *Division values in local fields*, Invent. Math. **53** (1979), 91–116.

[9] _____, *A p-adic Shimura isomorphism and p-adic periods of modular forms*, p-adic monodromy and the Birch and Swinnerton-Dyer conjecture, (Boston, MA, 1991), Amer. Math. Soc., Providence, RI, 1994, pp. 21–51.

[10] R. F. Coleman and E. de Shalit, *p-adic regulators on curves and special values of p-adic L-functions*, Invent. Math. **93** (1988), 239–266.

[11] P. Colmez, *Théorie d'Iwasawa des représentations de de Rham d'un corps local*, Ann. of Math. **148** (1998), 485–571.

[12] B. Dwork, *p-adic cycles*, Inst. Hautes Études Sci. Publ. Math. **37** (1969), 27–115.

[13] J.-M. Fontaine, *Le corps des périodes p-adiques*, Astérisque **223** (1994), 59–111.

[14] _____, *Représentations p-adiques semi-stables*, Astérisque **223** (1994), 113–184.

[15] R. Greenberg, *Iwasawa theory for elliptic curves*, Arithmetic theory of elliptic curves (Cetraro, 1997), Springer, Berlin, 1999, pp. 51–144.

[16] R. Greenberg and G. Stevens, *On the conjecture of Mazur, Tate, and Teitelbaum*, p-adic monodromy and the Birch and Swinnerton-Dyer conjecture (Boston, MA, 1991), Amer. Math. Soc., Providence, RI, 1994, pp. 183–211.

[17] K. Iwasawa, *Explicit formulas for the norm residue symbol*, J. Math. Soc. Japan **20** (1968), 151–164.

[18] U. Jannsen, *On the Galois cohomology of l-adic representations attached to varieties over local or global fields*, Séminaire de Théorie des Nombres, Paris 1986–87, Birkhäuser Boston, Boston, MA, 1988, pp. 165–182.

[19] _____, *On the l-adic cohomology of varieties over number fields and its Galois cohomology*, Galois groups over **Q** (Berkeley, CA, 1987), Springer, New York, 1989, pp. 315–360.

[20] J. W. Jones, *Iwasawa L-functions for multiplicative abelian varieties*, Duke Math. J. **59** (1989), 399–420.

[21] K. Kato, *p-adic Hodge theory and values of zeta functions of modular forms*, prépublication, 2001.

[22] G. Kings, *The Tamagawa number conjecture for CM elliptic curves*, Invent. Math. **143** (2001), 571–627.

[23] V. A. Kolyvagin, *Euler systems*, The Grothendieck Festschrift, vol. II, Birkhäuser Boston, Boston, MA, 1990, pp. 435–483.

[24] D. S. Kubert and S. Lang, *Modular units*, Springer-Verlag, New York, 1981.

[25] M. Kurihara, *On the Tate-Shafarevich groups over the cyclotomic fields of an elliptic curve with supersingular reduction I*, prépublication, 2000.

[26] B. Mazur and P. Swinnerton-Dyer, *Arithmetic of Weil curves*, Invent. Math. **25** (1974), 1–61.

[27] B. Mazur, J. Tate, and J. Teitelbaum, *On p-adic analogues of the conjectures of Birch and Swinnerton-Dyer*, Invent. Math. **84** (1986), 1–48.

[28] J. Nekovář, *p-adic Abel-Jacobi maps and p-adic heights*, The arithmetic and geometry of algebraic cycles (Banff, AB, 1998), Amer. Math. Soc., Providence, RI, 2000, pp. 367–379.

[29] W. Niziol, *On the image of p-adic regulators*, Invent. Math. **127** (1997), 375–400.

[30] B. Perrin-Riou, *Théorie d'Iwasawa et hauteurs p-adiques*, Invent. Math. **109** (1992), 137–185.

[31] ———, *Fonctions L p-adiques d'une courbe elliptique et points rationnels*, Ann. Inst. Fourier (Grenoble) **43** (1993), 945–995.

[32] ———, *Théorie d'Iwasawa et hauteurs p-adiques (cas des variétés abéliennes)*, Séminaire de Théorie des Nombres, Paris, 1990–91, Birkhäuser Boston, Boston, MA, 1993, pp. 203–220.

[33] ———, *p-adic L-functions and p-adic representations*, Amer. Math. Soc., Providence, RI, 2000, Translated from the 1995 French original by Leila Schneps and revised by the author.

[34] ———, *Représentations p-adiques et normes universelles. I. Le cas cristallin*, J. Amer. Math. Soc. **13** (2000), 533–551.

[35] ———, *Théorie d'Iwasawa des représentations p-adiques semi-stables*, Mém. Soc. Math. Fr. (N.S.) **84** (2001).

[36] G. Robert, *Unités elliptiques*, Bull. Soc. Math. France, Mém. No. 36, vol. 101, Société Mathématique de France, Paris, 1973.

[37] D. E. Rohrlich, *On L-functions of elliptic curves and cyclotomic towers*, Invent. Math. **75** (1984), 409–423.

[38] K. Rubin, *Euler systems and modular elliptic curves*, Galois representations in arithmetic algebraic geometry (Durham, 1996), Cambridge Univ. Press, Cambridge, 1998, pp. 351–367.

[39] A. J. Scholl, *An introduction to Kato's Euler systems*, Galois representations in arithmetic algebraic geometry (Durham, 1996), Cambridge Univ. Press, Cambridge, 1998, pp. 379–460.

[40] C. Soulé, *p-adic K-theory of elliptic curves*, Duke Math. J. **54** (1987), 249–269.

[41] M. M. Vishik, *Non-archimedian measures connected with Dirichlet series*, Math. USSR Sbornik **28** (1976), 216–228.

Computing Rational Points on Curves

Bjorn Poonen[1]

1 Introduction

The solution of diophantine equations (such as $x^{13} + y^{13} = z^{13}$) over the integers often reduces to the problem of determining the *rational* number solutions to a single polynomial equation in two variables. Such an equation describes a curve, and the problem of finding rational number solutions can be interpreted geometrically as finding the rational points on the curve, i.e., the points on the curve with rational coordinates.

Despite centuries of effort, we still do not know if there is a general algorithm that takes the equation of a curve, and outputs a list of its rational points, in the cases where the list is finite.[2] On the other hand, qualitative results such as Faltings' Theorem [26] on the finiteness of the number of rational points on curves of genus at least 2, the wide variety of conjectural effective approaches, and the practical success of recent efforts in determining the rational points on individual curves, have led many to believe that such an algorithm exists.

The reader expecting a thorough introduction or a comprehensive survey of known results may be disappointed by this article. We have selected only a few of the many aspects of the subject that we could have discussed. On the other hand, as we go along, we provide pointers to the literature for the reader wishing to delve more deeply in some particular direction. Finally, most of what we say for the field $\mathbf{Q}$ of rational numbers can be generalized easily to arbitrary number fields.

2 Hilbert's 10th Problem and Undecidability

First,[3] let us consider the problem of what can be computed *in theory*, and let us broaden the perspective to millennial proportions by considering

---

[1]This research was supported by NSF grant DMS-9801104, a Sloan Fellowship, and a Packard Fellowship.

[2]In fact, we do not even know if there is an algorithm that can always decide whether the list *is* finite.

[3]This section is not prerequisite for the rest of the article.

not only curves, but also higher dimensional varieties. Also, before asking whether we can compute all rational points, let's ask first whether we can determine whether a variety *has* a rational point. This leads to "Hilbert's 10th Problem over **Q**":[4]

Is there an algorithm for deciding whether a system of polynomial equations with integer coefficients

$$f_1(x_1, \ldots, x_n) = 0$$
$$f_2(x_1, \ldots, x_n) = 0$$
$$\vdots$$
$$f_m(x_1, \ldots, x_n) = 0$$

has a solution with $x_1, x_2, \ldots, x_n \in \mathbf{Q}$?

The system of equations defines a variety[5] over **Q**, so equivalently we may ask, does there exist an algorithm for deciding whether a variety over **Q** has a rational point?

By "algorithm" we mean Turing machine: see [36] for a definition. The machine is to be fed (for instance) a finite stream of characters containing the TEX code for a system of polynomial equations over **Q**, and is supposed to output yes or no in a finite amount of time, according to whether there is a rational solution or not. There is no insistence that the running time of the algorithm be bounded by a fixed polynomial in the length of the input stream; in Hilbert's 10th Problem, we are happy as long as the algorithm terminates after some unspecified number of steps on each input.

The answer to Hilbert's 10th Problem over **Q** is not known. This can be stated in logical terms as follows: we do not know whether there exists an algorithm for deciding the truth of all sentences such as

$$(\exists x)(\exists y)((2 * x * x + y = 0) \wedge (x + y + 3 = 0))$$

involving only rational numbers, the symbols $+, *, =, \exists$, logical relations $\wedge$ ("and"), $\vee$ ("or"), and variables x, y, ... bound by existential quantifiers. One can try asking for more, namely, for an algorithm to decide the entire *first order theory* of $(\mathbf{Q}, 0, 1, +, *)$; this would mean an algorithm for deciding the truth of sentences such as the one above, but in which in addition the symbols $\forall$ ("for all") and $\neg$ ("not") are allowed to appear. For this

---

[4]The analogous problem with **Q** replaced by **Z** was Problem 10 in the list of 23 problems that Hilbert presented to the mathematical community in 1900. This question over **Z** was settled in the negative [49] around 1970.

[5]In this article, varieties will not be assumed to be irreducible or reduced unless so specified.

more general problem, it is known that there is no algorithm that solves it [67].

To put the situation in perspective, we list the answers for the analogous questions about the existence of algorithms for deciding Hilbert's 10th Problem (existence of solutions to a polynomial system) or for deciding the first order theory, over other commutative rings.[6] Here YES means that there is an algorithm, NO means that no algorithm exists (i.e., Hilbert's 10th Problem of the first order theory is undecidable), and ? means that it is not known whether an algorithm exists.

Ring	Hilbert's 10th Problem	First order theory
$\mathbf{C}$	YES	YES
$\mathbf{R}$	YES	YES
$\mathbf{F}_p$	YES	YES
$\mathbf{Q}_p$	YES	YES
$\mathbf{Q}$	?	NO
$\mathbf{F}_p(t)$	NO	NO
$\mathbf{Z}$	NO	NO

The rings are listed approximately in order of increasing "arithmetic complexity." There is no formal definition of arithmetic complexity, but roughly we can measure the complexity of fields k by the "size" of the absolute Galois group, i.e., the Galois group of the algebraic closure $\bar{k}$ over k. And nonfields can be thought of as more complex than their fields of fractions, for instance because there is "extra structure" coming from the nontriviality of the divisibility relation. Whether or not $\mathbf{F}_p(t)$ is more complex than $\mathbf{Q}$ is debatable, but it is the extra structure coming from the p-th power map on the former that enabled the proof of undecidability of

---

[6]A technical point: in the cases where the commutative ring R is uncountable ($\mathbf{C}$, $\mathbf{R}$, $\mathbf{Q}_p$), we must be careful with our statement of the problem, because for instance, a classical Turing machine cannot examine the entirety of an infinite precision real number in a finite number of steps, and hence cannot even decide equality of two real numbers if fed the strings of their decimal digits on two infinite input tapes. To circumvent the problem, in the uncountable cases we restrict attention to decidability questions in which the constants appearing in the input polynomial system or first order sentence are *integers*. We still, however, require the machine to decide the existence of solutions or truth of the sentence with the variables ranging over all of R. It makes sense to ask this, since the output is to be simply yes or no.

Hilbert's 10th Problem for it.

For the complex numbers $\mathbf{C}$, the fact that the first order theory (and hence also Hilbert's 10th Problem) is decidable is a consequence of classical elimination theory. The first order theory has elimination of quantifiers: this means that a first order sentence involving $n \geq 1$ quantifiers $(\exists, \forall)$ can be transformed into a sentence with $n - 1$ quantifiers, in an algorithmic way, such that the latter sentence is true if and only if the former is. Algebraically, this corresponds to the elimination of a single variable from a system of equations, and geometrically it amounts to the fact that the projection from $\mathbf{C}^n$ to $\mathbf{C}^{n-1}$ of a Boolean combination of algebraic subsets of $\mathbf{C}^n$ can be written as a Boolean combination of algebraic subsets of $\mathbf{C}^{n-1}$. See [34, Exercise II.3.19] and the references [13, Exposé 7] and [50, Chapter 2, §6] listed there for a generalization due to Chevalley, which shows that for the same reasons, the first order theory of any algebraically closed field is decidable.

The analogous statement about the decidability of the first order theory of the real numbers $\mathbf{R}$ was proved by Tarski [76] using the theory of Sturm sequences. Again there is an elimination of quantifiers, provided that one augments the language by adding a symbol for $\leq$. The proof generalizes to real closed fields: see [37].

For the finite field $\mathbf{F}_p$ of p elements, the decidability results are obvious, since a Turing machine can simply loop over all possible values of the variables.

Tarski conjectured that the only fields with a decidable first order theory were the algebraically closed, real closed, and finite fields. This turned out to be false: Ax and Kochen [1] proved decidability for the field $\mathbf{Q}_p$ of p-adic numbers, which is the completion of $\mathbf{Q}$ with respect to the p-adic absolute value. (See [39] for the definition and basic properties of $\mathbf{Q}_p$.) They [2] gave several other examples of decidable fields. Macintyre [46] showed that there is an elimination of quantifiers for $\mathbf{Q}_p$ analogous to that for $\mathbf{R}$.

It is not known whether Hilbert's 10th Problem over the field $\mathbf{Q}$ of rational numbers is decidable or not; see [52] for a survey. On the other hand, Robinson [67] proved the undecidability of the first order theory of $\mathbf{Q}$, using the Hasse principle for quadratic forms. For a statement and proof of the latter, see [69, Chapter IV, §3, Theorem 8].

For the field $\mathbf{F}_p(t)$ of rational functions with coefficients in $\mathbf{F}_p$, Pheidas [60] proved the undecidability of Hilbert's 10th Problem, at least for $p \neq 2$. The $p = 2$ case was settled shortly thereafter by Videla [78]. The simpler problem of proving undecidability of the first order theory was done earlier, by Ershov [24] and Penzin [59] for $p \neq 2$ and $p = 2$, respectively.

Undecidability of Hilbert's 10th Problem itself (over the ring $\mathbf{Z}$ of integers) was proved by Matiyasevich [49]. The undecidability of the first order theory followed earlier from the fundamental work of Gödel [31]. Hilbert's 10th Problem for the ring of integers $\mathbf{Z}_K$ of a number field K is expected to be undecidable, but has been proved so only for certain K. For an up-to-date account of results in this direction, see [72]. For a survey of Hilbert's 10th Problem over commutative rings in general, see [61].

3 Rational Points on Varieties of Arbitrary Dimension

As discussed in the previous section, we do not know if there is an algorithm to decide in general whether a variety X over $\mathbf{Q}$ has a rational point. In order to pinpoint what is known and what is not, let us subdivide the problem according to the dimension of X. As usual, $X(\mathbf{Q})$ will denote the set of rational points of X, i.e., the set of rational solutions to the system of polynomials defining X.

dim X	$\exists$ algorithm to decide if $X(\mathbf{Q}) \neq \emptyset$?
0	YES
1	not known, but probably YES
≥ 2	?

If dim $X = 0$, then elimination theory lets us reduce to the case where X is a 0-dimensional subset of the affine line, and hence the problem becomes that of deciding whether a polynomial $f \in \mathbf{Q}[x]$ has a rational root. The latter can be done effectively, even in polynomial time [43].

The dim $X = 1$ case is the main subject of this article. Details follow in later sections.

For varieties of higher dimension, very little has been proved about $X(\mathbf{Q})$. On the other hand, below is a sample of some qualitative conjectures/questions that have been thrown around. All of these are known for dim $X \leq 1$, and for certain varieties of higher dimension. No counterexamples are known.

3.1 Bombieri, Lang (Independently)

Define the *special set* $S \subset X$ as the Zariski closure of the union of all positive dimensional images of morphisms of abelian varieties to X. Is it true that all but finitely many rational points of X lie in S?

The dim $X = 1$ case is equivalent to the Mordell conjecture [56], now Faltings' Theorem [26]: it states that a curve of genus at least 2 has at most finitely many rational points. Faltings [27] used diophantine approximation methods of Vojta to prove more generally that the answer is yes whenever X can be embedded in an abelian variety. For more conjectures along these lines, see [42, Chapter I, §3].

3.2   Colliot-Thélène and Sansuc

If X is smooth and projective, and X is birational to $\mathbf{P}^d$ over $\overline{\mathbf{Q}}$ (for instance, X could be a smooth cubic surface in $\mathbf{P}^3$), is the Brauer-Manin obstruction to the Hasse principle the only one? Actually, this was posed originally only for surfaces, as question (k1) in [25] but as Colliot-Thélène has pointed out, the answer could be yes in higher dimensions as well.

The Hasse principle is the statement that $X(\mathbf{Q}) \neq \emptyset$ if and only if $X(\mathbf{Q}_p) \neq \emptyset$ for all primes $p \leq \infty$. (By convention, $\mathbf{Q}_\infty = \mathbf{R}$.) This statement is proven for some varieties X (e.g., all degree 2 hypersurfaces in $\mathbf{P}^n$) and is known to be false for others (e.g., certain genus 1 curves). In 1970, Manin [48] discovered a possible obstruction to the Hasse principle coming from elements of the Brauer group of X, and he and others subsequently showed that this obstruction accounted for all violations of the Hasse principle known at the time. Much later, Skorobogatov [75] constructed an example of a surface X with no rational points, even though there was no Brauer-Manin obstruction; in other words, one could say that other nontrivial obstructions exist. Nevertheless, it is conceivable, and this is the point of the question of Colliot-Thélène and Sansuc, that for geometrically rational varieties, the nonexistence of a Brauer-Manin obstruction is a necessary and sufficient condition for the existence of rational points.

3.3   Mazur

Does the topological closure of $X(\mathbf{Q})$ in $X(\mathbf{R})$ have at most finitely many connected components?

See [51] and [53] for this and related questions. In [16] a counterexample is given to the following stronger version: If X is a smooth integral variety over $\mathbf{Q}$ such that $X(\mathbf{Q})$ is Zariski dense in X, then the topological closure of $X(\mathbf{Q})$ in $X(\mathbf{R})$ is a union of connected components of $X(\mathbf{R})$. The end of the paper [16] also suggests some other variants.

Even if the three questions above were answered tomorrow, we still would not know whether there exists an algorithm for deciding the existence of rational points on varieties.

4 Rational Points on Curves

For the rest of this article, we consider the problem of determining $X(\mathbf{Q})$ in the case dim $X = 1$, i.e., the case of curves. We begin with a few reductions, so that in the future we need consider only "nice" curves. Computational algebraic geometry provides algorithms for decomposing X into irreducible components over $\mathbf{Q}$ and over $\overline{\mathbf{Q}}$. Clearly then we can reduce to the case that X is irreducible over $\mathbf{Q}$. If X is irreducible over $\mathbf{Q}$, but not over $\overline{\mathbf{Q}}$, then the action of Galois acts transitively on the $\overline{\mathbf{Q}}$-irreducible components, but rational points are fixed by Galois, so $X(\mathbf{Q})$ is contained in the intersection of the $\overline{\mathbf{Q}}$-irreducible components; thus in this case we reduce to the 0-dimensional problem, which according to Section 3 is easily solved. Hence from now on, we will assume that X is geometrically integral.

Taking a projective closure and blowing up singularities changes X only by 0-dimensional sets, whose rational points we understand. Therefore, from now on, all our curves will be assumed to be smooth, projective, and geometrically integral. (Alternatively, at the expense of introducing nodes (violating smoothness), we could project X to a curve in $\mathbf{P}^2$ so that X would be described by a single polynomial equation. But in this article we prefer to talk about the smooth curves, except when presenting a curve explicitly, in which case we sometimes give an equation for an affine plane curve birational to the smooth projective curve that we are really interested in.)

The most important geometric invariant of a (smooth, projective, and geometrically integral) curve over $\mathbf{Q}$ is its genus g, which has several equivalent definitions:

(1) $g = \dim_{\mathbf{Q}} \Omega$ where Ω is the vector space of everywhere regular differentials on X. (Here regular means "no poles.") See [34, Chapter II, §8, p. 181] or [73, Chapter II, §5, p. 39] for more details.

(2) g is the topological genus of the compact Riemann surface $X(\mathbf{C})$.

(3) g is the dimension of the sheaf cohomology group $H^1(X, \mathcal{O}_X)$. See [34, Chapter III] for definitions.

(4) $g = \dfrac{(d-1)(d-2)}{2} - (\text{terms for singularities})$, where Y is a (possibly singular) plane curve of degree d birational to X (e.g., the image of X under a sufficiently generic projection to $\mathbf{P}^2$). In this formula one subtracts a computable positive integer for each singularity. The integer depends on the complexity of the singularity: for nodes (ordinary double points), the integer is 1. See [34, Chapter V, Example 3.9.2, p. 393] for more details.

(The equivalence of these is certainly not obvious.)

Although the genus is a measure of geometric complexity, it has been discovered over the years that the geometry also influences the rational points. Hence we subdivide the problem of determining $X(\mathbf{Q})$ according to the genus. We summarize the situation in the following table:

genus g	$\exists$ algorithm to determine $X(\mathbf{Q})$?
0	YES
1	YES, if $\text{III}(\text{Jac}\, X)$ is finite
≥ 2	Not known, but probably YES

4.1 Genus Zero

In this case, one shows using the Riemann-Roch Theorem [34, Chapter IV, §1] that the anticanonical divisor class on X induces an embedding of X as a degree 2 curve in $\mathbf{P}^2$. In other words, X is isomorphic to a conic, the zero locus in $\mathbf{P}^2$ of an absolutely irreducible homogeneous polynomial $f \in \mathbf{Q}[x, y, z]$ of degree 2. By a linear change of variables, we may assume that f has the form $ax^2 + by^2 + cz^2$ for some nonzero $a, b, c \in \mathbf{Z}$. Conversely, nonsingular degree 2 curves in $\mathbf{P}^2$ are curves of genus zero, by the fourth definition of g above. Over an algebraically closed field, we could say further that X is isomorphic to the projective line $\mathbf{P}^1$, but this is not necessarily the case for genus zero curves over $\mathbf{Q}$: for instance, the conic defined by $x^2 + y^2 + z^2 = 0$ in $\mathbf{P}^2$ is not isomorphic to $\mathbf{P}^1$ over $\mathbf{Q}$, because the latter has rational points, whereas the former does not.

As mentioned in Section 3.2, degree 2 curves in $\mathbf{P}^2$ (and more generally degree 2 hypersurfaces in $\mathbf{P}^2$) satisfy the Hasse principle, so we can decide the existence of a rational point by checking the existence of a $\mathbf{Q}_p$-point for each prime $p \leq \infty$. And the latter is in fact a finite problem since one can show a priori that $ax^2 + by^2 + cz^2 = 0$ has $\mathbf{Q}_p$-points for p not dividing $2abc$, and for each of the finitely many remaining p (including ∞) one has an algorithm for deciding the existence of a $\mathbf{Q}_p$-point, as explained in Section 2. For a more explicit criterion, see [57], for instance.

In the case where X does have a rational point P, there is an isomorphism $\mathbf{P}^1 \to X$ defined as follows: thinking of $\mathbf{P}^1$ as the set of lines in $\mathbf{P}^2$ through P, map the line L to the point $Q \in L \cap X$ not equal to P. (If L is the tangent line to X at P, take $Q = P$.) Hence we obtain an explicit parameterization of $X(\mathbf{Q})$.

4.2   Genus One

A genus one curve over $\mathbf{Q}$ with a rational point is called an *elliptic curve over* $\mathbf{Q}$. One shows using the Riemann-Roch Theorem that an elliptic curve E over $\mathbf{Q}$ is isomorphic to the projective closure of the affine curve $y^2 = x^3 + Ax + B$ for some $A, B \in \mathbf{Z}$ with $4A^3 + 27B^2 \neq 0$. More importantly, it can be shown that E can be given the structure of an algebraic group [73, Chapter 3]. Roughly, this means that there are rational functions that induce a group structure on the set $E(k)$ for any field k containing $\mathbf{Q}$. The Mordell-Weil Theorem[7] states that the set $E(\mathbf{Q})$ of rational points on an elliptic curve form a finitely generated abelian group [73, Chapter 8]. The group $E(\mathbf{Q})$ is called the Mordell-Weil group, and its rank is called the Mordell-Weil rank or simply the rank of E. The torsion subgroup of $E(\mathbf{Q})$ is easy to compute. But the equivalent problems of determining the rank of $E(\mathbf{Q})$ and of determining a list of generators have not yet been solved. There is a proposed method, "descent," for solving these problems, which is a generalization of the infinite descent method used by Fermat. Descent usually works well in practice when A and B are not too large (see [19]), but its success in general relies upon the conjecture that the Shafarevich-Tate group $\mathrm{III}(E)$, a certain abelian group associated to E, is finite, or at least that the p-primary part of $\mathrm{III}(E)$ is finite for some prime p. Using the modularity of elliptic curves over $\mathbf{Q}$ [7] and the work of Kolyvagin [40] supplemented by [12] or [58], we know $\#\mathrm{III}(E) < \infty$ for infinitely many E, namely, those for which $\mathrm{ord}_{s=1} L_E(s) \leq 1$, where $L_E(s)$ is the L-function of E. (See [73, Appendix C, §16] for a definition of $L_E(s)$.)

A general genus one curve X over $\mathbf{Q}$ need not have a rational point. In fact, Lind [44] and Reichardt [64] independently discovered examples where X does not satisfy the Hasse principle. Explicitly, the smooth projective models of the affine curves $2y^2 = 1 - 17x^4$ and $3x^3 + 4y^3 = 5$ (from [64] and [68], respectively) are curves having $\mathbf{Q}_p$-points for all $p \leq \infty$, but no rational points. To any genus one curve X one can associate an elliptic curve E, namely the *Jacobian* of X. The Jacobian $\mathrm{Jac}\,X$ of a curve X of genus g is an abelian variety (irreducible projective algebraic group) of dimension g, whose geometric points correspond to elements of $\mathrm{Pic}^0(X_{\overline{\mathbf{Q}}})$, i.e., to divisor classes of degree zero on $X_{\overline{\mathbf{Q}}}$. (Note: $X_{\overline{\mathbf{Q}}}$ denotes the same variety as X except where the defining polynomials are viewed as having coefficients in $\overline{\mathbf{Q}}$, even though the coefficients actually are in the subfield $\mathbf{Q}$. See [73, Chapter 2] for a definition of Pic^0, and see [55] for details about Jacobians.) In the case where $g = 1$, X is a *principal homogeneous*

---

[7]Actually Mordell alone proved this fact. Weil generalized Mordell's result in two directions: elliptic curves were replaced by abelian varieties of arbitrary dimension, and $\mathbf{Q}$ was replaced by an arbitrary number field.

space [73, Chapter 10], or *torsor*, of its Jacobian E. This means that there is an isomorphism $E_{\overline{\mathbf{Q}}} \simeq X_{\overline{\mathbf{Q}}}$ over $\overline{\mathbf{Q}}$, and a morphism of varieties $E \times X \to X$ over $\mathbf{Q}$, which when considered over $\overline{\mathbf{Q}}$ becomes equivalent (after identifying $X_{\overline{\mathbf{Q}}}$ with $E_{\overline{\mathbf{Q}}}$) to the addition morphism $E_{\overline{\mathbf{Q}}} \times E_{\overline{\mathbf{Q}}} \to E_{\overline{\mathbf{Q}}}$. Then $\text{III}(E)$ is defined as the set of principal homogeneous spaces for E that have $\mathbf{Q}_p$-points for all $p \leq \infty$, up to isomorphism as principal homogeneous spaces of E over $\mathbf{Q}$. It is known that if $\text{III}(E)$ is finite, then in principle, there is an algorithm for determining whether $X(\mathbf{Q})$ is nonempty: if $X(\mathbf{Q})$ is nonempty, a rational point can be found by search; if $X(\mathbf{Q}_p) = \emptyset$ for some $p \leq \infty$, then $X(\mathbf{Q}) = \emptyset$; if neither holds, then X represents a nonzero element of $\text{III}(E)$, and this can proved by finding another element Y of $\text{III}(E)$ such that the Cassels-Tate pairing of X and Y is nonzero in $\mathbf{Q}/\mathbf{Z}$. See [63] for some definitions of the Cassels-Tate pairing.

4.3 Genus At Least Two

Let X be a curve over $\mathbf{Q}$ of genus at least 2. Mordell [56] conjectured in 1922 that $X(\mathbf{Q})$ is finite, and this was finally proved in 1983 by Faltings [26]. A new proof based on diophantine approximation was found by Vojta [79]. Simplifications of Vojta's argument were found by Faltings and by Bombieri, who presented a relatively elementary proof in [5].

These proofs let one calculate a bound on the *number* of rational points on X given the equations defining X. But they are ineffective in that they do not provide an upper bound on the numerators and denominators of the coordinates of the rational points, so they cannot be used to determine $X(\mathbf{Q})$ rigorously. They are unable to decide even whether $X(\mathbf{Q})$ is empty.

Ironically, certain other methods (mostly older), which so far have failed to prove the Mordell conjecture in full generality, are the ones that have succeeded in determining $X(\mathbf{Q})$ in many examples. In the next section, we discuss these methods, and the ways in which they have been developed recently into practical algorithms.

5 Methods for Curves of Genus At Least Two

We use the following brief names to refer to the various methods that are used to determine $X(\mathbf{Q})$ for a curve X over $\mathbf{Q}$ of genus at least 2:
   (1) Local points
   (2) Dem'yanenko-Manin
   (3) Chabauty
   (4) Going-down
   (5) Going-up (Chevalley-Weil)

The last two are transitional in the sense that they by themselves do not determine $X(\mathbf{Q})$ directly, but instead reduce the problem of determining $X(\mathbf{Q})$ to the problem of determining the rational points on certain auxiliary varieties. In addition to the methods listed, there is a method based on the modularity of elliptic curves, and another method called "elliptic Chabauty." We will discuss these too, but in fact they can be interpreted as combinations of the methods already listed.

5.1 Local Points

This method attempts to prove that $X(\mathbf{Q})$ is empty without much work, by showing that $X(\mathbf{Q}_p)$ is empty for some $p \leq \infty$.

For a curve X over $\mathbf{Q}$ of any genus g, it is possible to compute a finite set S of primes such that $X(\mathbf{Q}_p) \neq \emptyset$ for all $p \notin S$. (Because of Hensel's lemma, S can be taken as the set of primes of bad reduction, together with ∞ and the primes p for which the Weil lower bound $p+1-2g\sqrt{p}$ for $X(\mathbf{F}_p)$ is nonpositive.) Then for each $p \in S$, it is possible to check whether $X(\mathbf{Q}_p)$ is empty, since according to Section 2 this problem is decidable for fixed p.

If we find p for which $X(\mathbf{Q}_p)$ is empty, then we know that $X(\mathbf{Q})$ is empty, and we are done. Otherwise we have learned nothing: the Hasse principle, the statement that $X(\mathbf{Q}_p) \neq \emptyset$ for all p implies $X(\mathbf{Q}) \neq \emptyset$, often fails for curves of positive genus.

5.2 Dem'yanenko-Manin

This method applies to certain special curves X. If A is an abelian variety, the group structure on A induces a group structure on the set of morphisms $X \to A$ over $\mathbf{Q}$. Let $\mathrm{Mor}(X, A)$ denote the quotient of this group by the subgroup of constant morphisms. If there is an abelian variety A over $\mathbf{Q}$ such that we can prove $\mathrm{rank}\,\mathrm{Mor}(X, A) > \mathrm{rank}\,A(\mathbf{Q})$, then the Dem'yanenko-Manin method [23], [47] provides an explicit upper bound on the sizes of the numerators and denominators of the coordinates of the rational points on X, so that $X(\mathbf{Q})$ can be computed by a finite search. Note that it is not necessary to know $\mathrm{rank}\,A(\mathbf{Q})$ exactly. For $\mathbf{Q}$-simple abelian varieties A, the needed inequality is equivalent to the condition that A^m appears in the decomposition of $\mathrm{Jac}\,X$ into $\mathbf{Q}$-simple abelian varieties up to isogeny, with

$$m > \frac{\mathrm{rank}\,A(\mathbf{Q})}{\mathrm{rank}\,\mathrm{End}_{\mathbf{Q}}\,A},$$

where $\mathrm{End}_{\mathbf{Q}}\,A$ denotes the ring of endomorphisms of A that are defined over $\mathbf{Q}$. For a fuller exposition of this method, see [70], and for some

explicit applications of it; see [74], [41], and [33]. Its main disadvantage is
that the condition necessary for its application fails for most curves.

5.3 Chabauty

This is a method based on p-adic geometry. Suppose that X embeds in
an abelian variety A such that $\operatorname{rank} A(\mathbf{Q}) < \dim A$. Suppose also that X
generates A in the sense that the differences of points $P - Q$ of $X(\overline{\mathbf{Q}})$
generate the group $A(\overline{\mathbf{Q}})$. For example, A might be $\operatorname{Jac} X$, in which
case the condition becomes $\operatorname{rank} A(\mathbf{Q}) < g$, where g is the genus of X.
Then the p-adic closure $\overline{A(\mathbf{Q})}$ of $A(\mathbf{Q})$ in $A(\mathbf{Q}_p)$ can be shown to be an
"analytic subvariety" of dimension at most $\operatorname{rank} A(\mathbf{Q})$; as a topological
group, it is an extension of a finite abelian group by a free $\mathbf{Z}_p$-module of
finite rank. The inequality hypothesis guarantees that $\overline{A(\mathbf{Q})}$ has positive
codimension in $A(\mathbf{Q}_p)$. Hence by dimension counting, one expects that
$X(\mathbf{Q}_p) \cap \overline{A(\mathbf{Q})}$ is at most a zero-dimensional closed subset of the compact
group $A(\mathbf{Q}_p)$, hence finite. Chabauty [14] proved this finiteness statement.
But $X(\mathbf{Q}) \subseteq X(\mathbf{Q}_p) \cap \overline{A(\mathbf{Q})}$, and by computing the intersection to a given
p-adic precision, one can bound the number of rational points on $X(\mathbf{Q})$,
and obtain p-adic approximations to their possible locations. Coleman [15]
gave an explicit upper bound on the size of this intersection. The intersection can be computed to some p-adic precision either by working with
the formal group of J, or by looking at the p-adic integrals of regular differentials on X. The latter seems to be easier, especially for higher genus
curves.

   Unfortunately, $X(\mathbf{Q})$ may be strictly smaller than $X(\mathbf{Q}_p) \cap \overline{A(\mathbf{Q})}$, although heuristically this may be rare when $\operatorname{rank} A(\mathbf{Q}) \leq \dim A - 2$: in that
case the naive dimension count suggests that the intersection is empty so
perhaps, if there are points in the intersection, they are there for a rea-
son! Because of the possibility that $X(\mathbf{Q}) \neq X(\mathbf{Q}_p) \cap \overline{A(\mathbf{Q})}$, the condition
$\operatorname{rank} A(\mathbf{Q}) < \dim A$ alone is not sufficient for success. In practice, however,
Chabauty's method has proved successful, especially in conjunction with
the transitional methods to be discussed below. See for example, [32], [54],
[28], [3], and [45].

5.4 Going-Down

If X admits a nonconstant morphism $X \to Y$ over $\mathbf{Q}$ to another variety
Y over $\mathbf{Q}$, where $Y(\mathbf{Q})$ is finite and computable, then one can determine
$X(\mathbf{Q})$, since it suffices to examine the finitely many points in $X(\overline{\mathbf{Q}})$ mapping to the points in $Y(\mathbf{Q})$: every point in $X(\mathbf{Q})$ must map to some point

in $Y(\mathbf{Q})$. In practice, Y is usually an abelian variety with $Y(\mathbf{Q})$ finite, or Y is another curve.

5.5 Going-Up

If $f\colon Y \to X$ is an *unramified* morphism of curves over $\mathbf{Q}$, Chevalley and Weil proved that there is a computable finite extension k of $\mathbf{Q}$ such that $f^{-1}(X(\mathbf{Q})) \subseteq Y(k)$. If X has genus at least 2, then Y does too, so $Y(k)$ is known to be finite. Hence one can reduce the problem of computing $X(\mathbf{Q})$ to that of computing $Y(k)$.

One difficulty with this method is that k may be much larger than $\mathbf{Q}$. Fortunately, Coombes and Grant [17] and Wetherell [80] found variants of the method that instead gave a finite set of unramified covering curves $Y_i \to X$ over $\mathbf{Q}$, $1 \le i \le n$, all isomorphic over $\overline{\mathbf{Q}}$, such that $X(\mathbf{Q}) \subseteq \bigcup_{i=1}^{n} f_i(Y_i(\mathbf{Q}))$. Given such a covering collection, one can determine $X(\mathbf{Q})$ if one can determine $Y_i(\mathbf{Q})$ for all i. Specialized to the case where X is an elliptic curve, this becomes the method of descent used to prove the Mordell-Weil Theorem.

Let us give one example of this method, taken from [8]. Suppose that we want to find $X(\mathbf{Q})$, where X is the curve of genus 2 that is the smooth projective model of the affine hyperelliptic curve $y^2 = 6x(x^4 + 12)$. (This is one of the curves that comes up when one studies the integer solutions to the equation $x^8 + y^3 = z^2$.) Suppose we have an affine point $(x_0, y_0) \in X(\mathbf{Q})$ with $x_0, y_0 \ne 0$. If we write $x_0 = X/Z$ in lowest terms with $X, Z \in \mathbf{Z}$, $Z \ne 0$, we obtain $y_0^2 Z^6 = 6XZ(X^4 + 12Z^4)$. Setting $Y = y_0 Z^3$, which must be an integer, since its square equals the right hand side, we obtain $Y^2 = 6XZ(X^4 + 12Z^4)$. If a prime $p \ge 5$ divides both $6XZ$ and $X^4 + 12Z^4$, then it divides either X or Z, and then in order to divide $X^4 + 12Z^4$ it must divide *both* X and Z, contradicting the assumption that X/Z is in lowest terms. (More generally, one would argue using primes not dividing the resultant of the two homogeneous polynomial factors.) Thus each prime $p \ge 5$ divides at most one of $6XZ$ and $X^4 + 12Z^4$. But their product is a square, so if p divides one of $6XZ$ and $X^4 + 12Z^4$, the exponent of p in that factor is even. Since this holds for all $p \ge 5$, we have $X^4 + 12Z^4 = \delta W^2$ for some $\delta \in \{\pm 1, \pm 2, \pm 3, \pm 6\}$ and $W \in \mathbf{Z}$. Dividing by Z^4, we obtain a rational point (u, v) on the curve $E_\delta\colon \delta v^2 = u^4 + 12$. We may assume $\delta > 0$ (since otherwise $E_\delta(\mathbf{R}) = \emptyset$) and $2 \mid \delta$ (since otherwise $E_\delta(\mathbf{Q}_2) = \emptyset$). Therefore we need only search for rational points on

$$E_1\colon v^2 = u^4 + 12, \quad \text{and} \quad E_3\colon 3v^2 = u^4 + 12.$$

These are curves of genus one, and with a little work, using descent, one can show that $E_1(\mathbf{Q})$ and $E_3(\mathbf{Q})$ are finite, each of size 2, counting points on the nonsingular projective models. Checking to see what points on X these give rise to, we find that $X(\mathbf{Q})$ consists of $(0,0)$ and a point at infinity on the projective model.

Geometrically what has happened here is that for each $\delta \in \mathbf{Q}^*$, the genus 3 curve Y_δ defined by the *system* of equations

$$y^2 = 6x(x^4 + 12), \quad \delta z^2 = x^4 + 12$$

in (x, y, z)-space maps to X via $(x, y, z) \mapsto (x, y)$, and Y_δ is an unramified cover of X. Moreover, the union of the images of $Y_\delta(\mathbf{Q})$ in X equals $X(\mathbf{Q})$. Since the isomorphism class of Y_δ depends only on the image of δ in $\mathbf{Q}^*/\mathbf{Q}^{*2}$, we may assume that $\delta \in \mathbf{Z}$ is nonzero and squarefree. If a prime $p \geq 5$ divides δ, then $Y_\delta(\mathbf{Q}_p) = \emptyset$, so we may discard Y_δ. By also demanding the existence of local points over $\mathbf{R}$ and $\mathbf{Q}_2$, we may discard all but Y_1 and Y_3. (As it turns out, $\mathbf{Q}_3$ gives no further restriction.) Finally, $(x, y, z) \mapsto (x, z)$ gives a nonconstant morphism $Y_\delta \to E_\delta$, so by "going down" it suffices to find $E_1(\mathbf{Q})$ and $E_3(\mathbf{Q})$, if these are finite. We are lucky: both E_1 and E_3 turn out to be elliptic curves of rank zero.

In general, covering collections by geometrically *abelian* covers are described by geometric class field theory. They arise as follows. Suppose that X is embedded in its Jacobian J using a basepoint $P_0 \in X(\mathbf{Q})$. Choose an isogeny $\phi \colon A \to J$, i.e., a surjective homomorphism between abelian varieties with finite kernel. Choose a representative $R \in J(\mathbf{Q})$ of each element of $J(\mathbf{Q})/\phi(A(\mathbf{Q}))$, let $\phi_R \colon A \to J$ be the composition ϕ followed by translation-by-R on J, and let Y_R denote the following fiber product:

In other words, Y_R is the inverse image $\phi_R^{-1}(X)$ in A. Then the Y_R form a finite set of unramified covers of X, and the union of the images of $Y_R(\mathbf{Q})$ in X equals $X(\mathbf{Q})$, since $\bigcup_R \phi_R(A(\mathbf{Q})) = J(\mathbf{Q})$.

Note that given X and hence J, there are many pairs (A, ϕ) where $\phi \colon A \to J$ is an isogeny over $\mathbf{Q}$: if nothing else is available, one can take ϕ as the multiplication-by-m map $[m] \colon J \to J$ for some $m \geq 2$. Hence going up is always possible. On the other hand, for $\phi = [m]$, the genus of each Y_R equals $m^{2g}(g - 1) + 1$ by the Riemann-Hurwitz formula [34, IV.2.4] so if m is too large, the Y_R may be difficult to work with. In fact, it might be preferable to use an isogeny ϕ of degree lower than $\deg[2] = 2^{2g}$ if one exists.

5.6 Modularity

Through the work of Wiles, Taylor, Breuil, Conrad, and Diamond [81], [77], [7], it is now known that every elliptic curve E over $\mathbf{Q}$ is modular, meaning that there exists a nonconstant morphism from the modular curve $X_0(N)$ to E, for some integer $N \geq 1$. See [42, Chapter V] for an introduction to this concept, and see [71] and [38] for more details on modular functions and curves. On the other hand, work of Frey, Serre, and Ribet [65] showed that a nontrivial rational point on $x^p + y^p = 1$ for prime $p > 2$ would give rise to an elliptic curve over $\mathbf{Q}$ that could not be modular. Together, these results proved Fermat's Last Theorem. For an overview of the whole proof, see [20] or the book [18]. In the past few years, methods based on modularity have been adapted to solve certain other Fermat-like diophantine equations. See [22] and [66], for instance.

Darmon has pointed out that these proofs can be interpreted as instances of the going-up method. Recall that the geometric class field theory construction at the end of Section 5.5 produces only unramified covers Y that when considered over $\overline{\mathbf{Q}}$ are Galois and abelian over the base curve X. These abelian covers are the easiest unramified covers to work with, but they form only a small subset of *all* the unramified covers. Kummer's partial results on Fermat's Last Theorem can be reinterpreted as a study of the abelian covers of the Fermat curve $X : x^p + y^p = 1$ with Galois group $\mathbf{Z}/p$. The modularity proof of Fermat's Last Theorem is equivalent to a study of certain unramified covers with nonabelian Galois group. More precisely, the map $(x, y) \mapsto x^p$ from $X \to \mathbf{P}^1$ exhibits X as a cover of $\mathbf{P}^1$ ramified above $\{0, 1, \infty\}$, and the latter can be thought of as the modular curve $X(2)$ with its three cusps. The fiber product

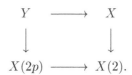

is an unramified cover Y of X with Galois group equal to that of $X(2p)$ over $X(2)$, namely the nonabelian group $\mathrm{PSL}_2(\mathbf{F}_p)$. One can then study the rational points on Y and its relevant twists by going down to $X(2p)$ and its twists. Thus one reduces to questions about rational points on modular curves, to which one can apply the work of Mazur.

5.7 Elliptic Chabauty

This is not so much a separate method as it is a clever way to combine the going-up and Chabauty methods. It was discovered independently by

Bruin [9] and by Flynn and Wetherell [29].

Suppose that X is a curve of genus 2 over $\mathbf{Q}$ embedded in its Jacobian J_X using a basepoint $P_0 \in X(\mathbf{Q})$. Consider the curves Y obtained as in Section 5.5 by pulling back X under the multiplication-by-2 isogeny $J_X \to J_X$, or its translates. Then Y is an unramified cover of X, and when considered over $\overline{\mathbf{Q}}$, it is an abelian cover with Galois group $J_X[2](\overline{\mathbf{Q}}) \simeq (\mathbf{Z}/2)^4$. As mentioned in Section 5.5, the Riemann-Hurwitz formula shows that the genus of Y is 17. By Galois theory, there are 15 intermediate covers Z_i of $X_{\overline{\mathbf{Q}}}$ of degree 2, and these have genus 3. Hence each Jacobian J_{Z_i} is isogenous over $\overline{\mathbf{Q}}$ to $J_X \times E_i$ for some elliptic curve E_i, and it follows that J_Y is isogenous to $J_X \times A$ where A is a 15-dimensional abelian variety isomorphic to $\prod_{i=1}^{15} E_i$ over $\overline{\mathbf{Q}}$. More precisely, one can show that one can take A to the Weil restriction of scalars $\mathrm{Res}_{K/\mathbf{Q}} E$ of an elliptic curve E over the 15-dimensional $\mathbf{Q}$-algebra K of global sections of the structure sheaf on $J_X[2] - \{0\}$. (See [6, Section 7.6] for the definition of the Weil restriction of scalars.) More concretely, K is the product of the fields of definition of representatives for the Galois orbits of nontrivial 2-torsion points of J; often there is just one orbit and K is a number field of degree 15 over $\mathbf{Q}$. One then can attempt to apply Chabauty to $Y \to A$ for each Y.

Whereas success of the direct Chabauty method required rank $J_X(\mathbf{Q}) < 2$, elliptic Chabauty requires the (independent?) condition rank $A(\mathbf{Q}) < \dim A$ for each A that arises. One of the properties of restriction of scalars is that $A(\mathbf{Q}) \simeq E(K)$, and we also know $\dim A = 15$, so the latter condition is equivalent to rank $E(K) < 15$. Perhaps this is more likely than rank $J_X(\mathbf{Q}) < 2$. In the worst case, when K is a number field and each nonzero 2-torsion point of E is defined only over a cubic extension of K, we apparently need to compute the 2-part of the class group of a degree 45 number field to complete the descent to compute rank $E(K)$. But in favorable cases, the number fields are much smaller, and the method is practical.

Elliptic Chabauty has the advantage over the original Chabauty method that one can do all the computations with the group law of E over K, instead of a Jacobian over $\mathbf{Q}$. The former is usually easier from the computational point of view: as Bruin says, "simple geometry over a field with complicated arithmetic is to be preferred over complicated geometry over a field with simple arithmetic."

6   The Mordell-Weil Race

If faced with the problem of determining $X(\mathbf{Q})$ for a specific curve X over $\mathbf{Q}$ of genus $g \geq 2$, it is probably best first to try the method of local points. If this fails, next one can try to see if X admits a nonconstant morphism to a curve Y where Y has genus at least 2, or where Y is a curve of genus 1 for which $Y(\mathbf{Q})$ is finite and can be determined. If not, one can attempt the Dem'yanenko-Manin method, and the method of Chabauty.

If all of these fail, one can go up to a set of unramified covering curves, and then recursively apply the above methods to each of these.

It seems plausible that iteration of going-up and Chabauty alone are sufficient to resolve $X(\mathbf{Q})$ for every curve X of genus $g \geq 2$ over $\mathbf{Q}$! Starting from X, one replaces the problem on X with the problem on a finite set of covering curves Y_R. For each Y_R whose rational points are not resolved by Chabauty, one must replace Y_R by a finite set of its covering curves, and so on. We obtain a tree and hope that all the branches will eventually be terminated by Chabauty. If one applies Chabauty to the Jacobians alone, the issue is whether the genus eventually outpaces the rank of the Jacobian as one goes up along any branch of the tree. If so, Chabauty is likely to succeed in terminating all the branches.

In practice, one can apply Chabauty to morphisms from the curve into *quotients* of its Jacobian at each node of the tree, as in the elliptic Chabauty method of Section 5.7. This is preferable especially if the original X has large rank: if Y covers X, the Jacobian J_Y is isogenous over $\mathbf{Q}$ to $J_X \times A$ for some abelian variety A over $\mathbf{Q}$, so rank $J_Y(\mathbf{Q}) = \text{rank } J_X(\mathbf{Q}) + \text{rank } A(\mathbf{Q})$, which will still be large, if rank $J_X(\mathbf{Q})$ was large to begin with. On the other hand, there seems to be no direct correlation between rank $A(\mathbf{Q})$ and rank $J_X(\mathbf{Q})$.

Unfortunately, virtually nothing is known about the growth of Mordell-Weil ranks as one ascends an unramified tower of curves, and hence it seems impossible to prove anything about the success of this tree method. All we have now are a few isolated examples in which the method has been successful. For a genus g curve X, is rank $J_X(\mathbf{Q})$ typically of size around g? Or is it typically $O(1)$ as $g \to \infty$? It seems difficult even to find a heuristic that predicts the answers. Perhaps analytic methods generalizing [11] will provide hints.

Even if the truth is that the going-up and Chabauty methods are *not* always enough to determine $X(\mathbf{Q})$, there are several other conjectural approaches towards an effective algorithm. See Section F.4.2 of [35] for a survey of some of these.

7 Some Success Stories

7.1 Diophantus

About 1700 years ago, Diophantus challenged his readers to find a solution to

$$y^2 = x^8 + x^4 + x^2$$

in positive rational numbers. This is not hard, but these days one wants to know *all* the solutions. Wetherell [80] combined going-up, going-down, and Chabauty to prove that $(1/2, 9/16)$ is the only positive rational solution.

7.2 Serre

Over 15 years ago, Serre challenged the mathematical community to find the rational points on $x^4 + y^4 = 17$, a genus 3 curve whose Jacobian is isogenous to $E \times E \times E'$ where E, E' are elliptic curves over $\mathbf{Q}$, each of rank 2. This past year, Flynn and Wetherell [30] found suitable unramified covers and applied a version of elliptic Chabauty to them to prove that the only rational points are the obvious eight with

$$\{|x|, |y|\} = \{1, 2\}.$$

7.3 Generalized Fermat

Work of Beukers [4], and of Darmon and Granville [21] reduces solving $x^p + y^q = z^r$ in relatively prime integers x, y, z for fixed $p, q, r > 1$ to computing $X(\mathbf{Q})$ for a finite set of curves X over $\mathbf{Q}$. As has already been mentioned, methods based on modularity of elliptic curves solve the equation for many (p, q, r). Some of the small exponent cases, too small for modularity methods to work easily, are worked out in [9], [8], [10], and [62]. For example, [8] applies elliptic Chabauty and other methods to prove that the only solutions to

$$x^8 + y^3 = z^2$$

in nonzero relatively prime integers are

$$(\pm 1, 2, \pm 3) \quad \text{and} \quad (\pm 43, 96222, \pm 30042907).$$

References

[1] J. Ax and S. Kochen, *Diophantine problems over local fields. II. A complete set of axioms for p-adic number theory*, Amer. J. Math. **87** (1965), 631–648.

[2] ———, *Diophantine problems over local fields. III. Decidable fields*, Annals of Math. **83** (1966), 437–456.

[3] M. H. Baker, *Kamienny's criterion and the method of Coleman and Chabauty*, Proc. Amer. Math. Soc. **127** (1999), 2851–2856.

[4] F. Beukers, *The Diophantine equation $Ax^p + By^q = Cz^r$*, Duke Math. J. **91** (1998), 61–88.

[5] E. Bombieri, *The Mordell conjecture revisited*, Ann. Scuola Norm. Sup. Pisa Cl. Sci. (4) **17** (1990), 615–640, Errata-corrige: ibid. **18** (1991), 473.

[6] S. Bosch, W. Lütkebohmert, and M. Raynaud, *Néron models*, Springer-Verlag, Berlin, 1990.

[7] C. Breuil, B. Conrad, F. Diamond, and R. Taylor, *On the modularity of elliptic curves over* **Q**, J. Amer. Math. Soc. **14** (2001), 843–939.

[8] N. Bruin, *Chabauty methods using covers on curves of genus 2*, preprint, 1999.

[9] ――――, *Chabauty methods using elliptic curves*, preprint, 1999.

[10] ――――, *The diophantine equations $x^2 \pm y^4 = \pm z^6$ and $x^2 + y^8 = z^3$*, Compositio Math. **118** (1999), 305–321.

[11] A. Brumer, *The average rank of elliptic curves. I*, Invent. Math. **109** (1992), 445–472.

[12] D. Bump, S. Friedberg, and J. Hoffstein, *Nonvanishing theorems for L-functions of modular forms and their derivatives*, Invent. Math. **102** (1990), 543–618.

[13] H. Cartan and C. Chevalley, *Géometrie algébrique*, Séminaire Cartan-Chevalley (Paris), Secrétariat Math., 1955–1956.

[14] C. Chabauty, *Sur les points rationnels des courbes algébriques de genre supérieur à l'unité*, C.R. Acad. Sci. Paris **212** (1941), 882–885.

[15] R. F. Coleman, *Effective Chabauty*, Duke Math. J. **52** (1985), 765–780.

[16] J.-L. Colliot-Thélène, A. N. Skorobogatov, and P. Swinnerton-Dyer, *Double fibres and double covers: paucity of rational points*, Acta Arith. **79** (1997), 113–135.

[17] K. R. Coombes and D. R. Grant, *On heterogeneous spaces*, J. London Math. Soc. (2) **40** (1989), 385–397.

[18] G. Cornell, J. H. Silverman, and G. Stevens, eds., *Modular forms and Fermat's last theorem*, Springer-Verlag, New York, 1997, Papers from the Instructional Conference on Number Theory and Arithmetic Geometry held at Boston University, Boston, MA, August 9–18, 1995.

[19] J. E. Cremona, *Algorithms for modular elliptic curves*, second ed., Cambridge University Press, Cambridge, 1997.

[20] H. Darmon, F. Diamond, and R. Taylor, *Fermat's last theorem*, Elliptic curves, modular forms, & Fermat's last theorem (Hong Kong, 1993) (John Coates and S. T. Yau, eds.), International Press, Cambridge, MA, second ed., 1997, pp. 2–140.

[21] H. Darmon and A. Granville, *On the equations $z^m = F(x, y)$ and $Ax^p + By^q = Cz^r$*, Bull. London Math. Soc. **27** (1995), 513–543.

[22] H. Darmon and L. Merel, *Winding quotients and some variants of Fermat's last theorem*, J. Reine Angew. Math. **490** (1997), 81–100.

[23] V. Dem'yanenko, *Rational points on a class of algebraic curves*, Amer. Math. Soc. Transl. **66** (1968), 246–272.

[24] Yu. Ershov, *Undecidability of certain fields (Russian)*, Dokl. Akad. Nauk SSSR **161** (1965), 349–352.

[25] J.-L. Colliot-Thélène et J.-J. Sansuc, *La descente sur les variétés rationnelles*, Journées de Géométrie Algébrique d'Angers (Angers, 1979), Sijthoff & Noordhoff, Alphen aan den Rijn, 1980, pp. 223–237.

[26] G. Faltings, *Endlichkeitssätze für abelsche Varietäten über Zahlkörpern*, Invent. Math. **73** (1983), 349–366, English translation: Arithmetic geometry (G. Cornell and J. Silverman, eds.), Springer-Verlag, New York-Berlin, 1986, pp. 2–27.

[27] ———, *The general case of S. Lang's conjecture*, Barsotti Symposium in Algebraic Geometry (Abano Terme, 1991), Perspect. Math., vol. 15, Academic Press, San Diego, CA, 1994, pp. 175–182.

[28] E. V. Flynn, *A flexible method for applying Chabauty's theorem*, Compositio Math. **105** (1997), 79–94.

[29] E. V. Flynn and J. L. Wetherell, *Finding rational points on bielliptic genus 2 curves*, Manuscripta Math. **100** (1999), 519–533.

[30] ———, *Covering collections and a challenge problem of Serre*, Acta Arith. **98** (2001), 197–205.

[31] K. Gödel, *Über formal unentscheidbare Sätze der Principia Mathematica und verwandter System I*, Monatshefte für Math. und Physik **38** (1931), 173–198, English translation by Elliot Mendelson: "On formally undecidable propositions of Principia Mathematica and related systems I" in M. Davis, The undecidable, Raven Press, 1965.

[32] D. Grant, *A curve for which Coleman's effective Chabauty bound is sharp*, Proc. Amer. Math. Soc. **122** (1994), 317–319.

[33] G. Grigorov and J. Rizov, *Heights on elliptic curves and the equation $x^4 + y^4 = cz^4$*, preprint, 1998.

[34] R. Hartshorne, *Algebraic geometry*, Graduate Texts in Mathematics, vol. 52, Springer-Verlag, New York-Heidelberg, 1977.

[35] M. Hindry and J. H. Silverman, *Diophantine geometry. An introduction*, Graduate Texts in Mathematics, vol. 201, Springer-Verlag, New York, 2000.

[36] J. E. Hopcroft and J. D. Ullman, *Formal languages and their relation to automata*, Addison-Wesley Publishing Co., Reading, Mass, 1969.

[37] N. Jacobson, *Basic algebra I*, second ed., W. H. Freeman and Company, New York, 1985.

[38] N. M. Katz and B. Mazur, *Arithmetic moduli of elliptic curves*, Annals of Mathematics Studies, vol. 108, Princeton University Press, Princeton, N.J., 1985.

[39] N. Koblitz, *p-adic numbers, p-adic analysis, and zeta-functions*, Graduate Texts in Mathematics, vol. 58, Springer-Verlag, New York-Berlin, second ed., 1984.

[40] V. Kolyvagin, *Finiteness of $E(\mathbf{Q})$ and $\text{Ш}(E, \mathbf{Q})$ for a subclass of Weil curves (Russian)*, Izv. Akad. Nauk SSSR Ser. Mat. **52** (1988), 522–540, 670–671, English translation: Math. USSR-Izv. **32** (1989), no. 3, 523–541.

[41] L. Kulesz, *Application de la méthode de Dem'janenko-Manin à certaines familles de courbes de genre 2 et 3*, J. Number Theory **76** (1999), 130–146.

[42] S. Lang, *Number theory. III. Diophantine geometry*, Encyclopaedia of Mathematical Sciences, vol. 60, Springer-Verlag, Berlin, 1991.

[43] A. K. Lenstra, H. W. Lenstra Jr., and L. Lovász, *Factoring polynomials with rational coefficients*, Math. Ann. **261** (1982), 515–534.

[44] C.-E. Lind, *Untersuchungen über die rationalen Punkte der ebenen kubischen Kurven vom Geschlecht Eins*, Ph.D. thesis, University of Uppsala, 1940.

[45] D. Lorenzini and T. J. Tucker, *Thue equations and the method of Chabauty-Coleman*, preprint, May 2000.

[46] A. Macintyre, *On definable subsets of p-adic fields*, J. Symbolic Logic **41** (1976), 605–610.

[47] Yu. I. Manin, *The p-torsion of elliptic curves is uniformly bounded (Russian)*, Izv. Akad. Nauk. SSSR Ser. Mat. **33** (1969), 459–465, English translation: *Amer. Math. Soc. Transl.*, 433–438.

[48] ———, *Le groupe de Brauer-Grothendieck en géométrie diophantienne*, Actes du Congrès International des Mathématiciens (Nice, 1970), vol. 1, Gauthier-Villars, Paris, 1971, pp. 401–411.

[49] Yu. Matiyasevich, *The Diophantineness of enumerable sets (Russian)*, Dokl. Akad. Nauk SSSR **191** (1970), 279–282.

[50] H. Matsumura, *Commutative algebra*, W. A. Benjamin, Inc., New York, 1970.

[51] B. Mazur, *The topology of rational points*, Experiment. Math. **1** (1992), 35–45.

[52] ———, *Questions of decidability and undecidability in number theory*, J. Symbolic Logic **59** (1994), 353–371.

[53] ———, *Speculations about the topology of rational points: an update*, Astérisque **4** (1995), 165–182, Columbia University Number Theory Seminar (New York, 1992).

[54] W. McCallum, *On the method of Coleman and Chabauty*, Math. Ann. **299** (1994), 565–596.

[55] J. S. Milne, *Jacobian varieties*, Arithmetic geometry (G. Cornell and J. H. Silverman, eds.), Springer-Verlag, New York, 1986, pp. 167–212.

[56] L. J. Mordell, *On the rational solutions of the indeterminate equations of the third and fourth degrees*, Proc. Cambridge Phil. Soc. **21** (1922), 179–192.

[57] ———, *On the magnitude of the integer solutions of the equation $ax^2 + by^2 + cz^2 = 0$*, J. Number Theory **1** (1969), 1–3.

[58] M. Ram Murty and V. Kumar Murty, *Mean values of derivatives of modular L-series*, Annals of Math. (2) **133** (1991), 447–475.

[59] Yu. Penzin, *Undecidability of fields of rational functions over fields of characteristic* 2 *(Russian)*, Algebra i Logika **12** (1973), 205–210, 244.

[60] T. Pheidas, *Hilbert's tenth problem for fields of rational functions over finite fields*, Invent. Math. **103** (1991), 1–8.

[61] T. Pheidas and K. Zahidi, *Undecidability of existential theories of rings and fields: a survey*, Hilbert's tenth problem: relations with arithmetic and algebraic geometry (Ghent, 1999), Contemporary Mathematics, vol. 270, Amer. Math. Soc., 2000, pp. 49–105.

[62] B. Poonen, *Some Diophantine equations of the form $x^n + y^n = z^m$*, Acta Arith. **86** (1998), 193–205.

[63] B. Poonen and M. Stoll, *The Cassels-Tate pairing on polarized abelian varieties*, Annals of Math. (2) **150** (1999), 1109–1149.

[64] H. Reichardt, *Einige im Kleinen überall lösbare im Grossen unlösbare diophantische Gleichungen*, J. Reine Angew. Math. **184** (1942), 12–18.

[65] K. A. Ribet, *On modular representations of* $\mathrm{Gal}(\overline{\mathbf{Q}}/\mathbf{Q})$ *arising from modular forms*, Invent. Math. **100** (1990), 431–476.

[66] _____, *On the equation $a^p + 2^\alpha b^p + c^p = 0$*, Acta Arith. **79** (1997), 7–16.

[67] J. Robinson, *Definability and decision problems in arithmetic*, J. Symbolic Logic **14** (1949), 98–114.

[68] E. Selmer, *The diophantine equation $ax^3 + by^3 + cz^3 = 0$*, Acta Math. **85** (1951), 203–362, and **92** (1954) 191–197.

[69] J.-P. Serre, *A course in arithmetic*, Graduate Texts in Mathematics, vol. 7, Springer-Verlag, New York-Heidelberg, 1973.

[70] _____, *Lectures on the Mordell-Weil theorem*, Aspects of Mathematics, vol. E15, Friedr. Vieweg & Sohn, Braunschweig, 1989.

[71] G. Shimura, *Introduction to the arithmetic theory of automorphic functions*, Publications of the Mathematical Society of Japan, vol. 11, Princeton University Press, Princeton, NJ, 1994, Reprint of the 1971 original.

[72] A. Shlapentokh, *Hilbert's Tenth Problem over number fields, a survey*, Hilbert's tenth problem: relations with arithmetic and algebraic geometry (Ghent, 1999), Contemporary Mathematics, vol. 270, Amer. Math. Soc., 2000, pp. 107–137.

[73] J. H. Silverman, *The arithmetic of elliptic curves*, Graduate Texts in Mathematics, vol. 106, Springer-Verlag, New York-Berlin, 1986.

[74] ———, *Rational points on certain families of curves of genus at least 2*, Proc. London Math. Soc. (3) **55** (1987), 465–481.

[75] A. N. Skorobogatov, *Beyond the Manin obstruction*, Invent. Math. **135** (1999), 399–424.

[76] A. Tarski, *A decision method for elementary algebra and geometry*, 2nd ed., University of California Press, Berkeley and Los Angeles, Calif., 1951.

[77] R. Taylor and A. Wiles, *Ring-theoretic properties of certain Hecke algebras*, Annals of Math. (2) **141** (1995), 553–572.

[78] C. Videla, *Hilbert's tenth problem for rational function fields in characteristic 2*, Proc. Amer. Math. Soc. **120** (1994), 249–253.

[79] P. Vojta, *Siegel's theorem in the compact case*, Annals of Math. (2) **133** (1991), 509–548.

[80] J. L. Wetherell, *Bounding the number of rational points on certain curves of high rank*, Ph.D. thesis, University of California at Berkeley, 1997.

[81] A. Wiles, *Modular elliptic curves and Fermat's last theorem*, Annals of Math. (2) **141** (1995), 443–551.

Norms of Products and Factors of Polynomials

Igor E. Pritsker[1]

1 Introduction

Let E be a compact set in the complex plane $\mathbb{C}$. Define the uniform (sup) norm on E as follows:

$$\|f\|_E = \sup_{z \in E} |f(z)|.$$

Consider algebraic polynomials $\{p_k(z)\}_{k=1}^m$ and their product

$$p(z) := \prod_{k=1}^m p_k(z).$$

We are interested here in polynomial inequalities of the form

$$\prod_{k=1}^m \|p_k\|_E \le C \|p\|_E. \tag{1.1}$$

One of the first results in this direction is due to Kneser [15], for $E = [-1, 1]$ and $m = 2$ (see also Aumann [1]), who proved that

$$\|p_1\|_{[-1,1]} \|p_2\|_{[-1,1]} \le K_{\ell,n} \|p_1 p_2\|_{[-1,1]}, \tag{1.2}$$

where

$$K_{\ell,n} := 2^{n-1} \prod_{k=1}^{\ell} \left(1 + \cos \frac{2k-1}{2n}\pi\right) \prod_{k=1}^{n-\ell} \left(1 + \cos \frac{2k-1}{2n}\pi\right), \tag{1.3}$$

$\deg p_1 = \ell$ and $\deg(p_1 p_2) = n$. Note that (1.2) becomes an equality for the Chebyshev polynomial $t(z) = \cos n \arccos z = p_1(z) p_2(z)$, with a proper

---

[1] Research supported in part by the National Science Foundation grants DMS-9996410 and DMS-9707359.

choice of the factors $p_1(z)$ and $p_2(z)$. P. B. Borwein [6] gave a new proof of (1.2)–(1.3) and generalized this to the multifactor inequality

$$\prod_{k=1}^{m} \|p_k\|_{[-1,1]} \leq 2^{n-1} \prod_{k=1}^{\left[\frac{n}{2}\right]} \left(1 + \cos \frac{2k-1}{2n}\pi\right)^2 \|p\|_{[-1,1]}. \qquad (1.4)$$

He has also showed that

$$2^{n-1} \prod_{k=1}^{\left[\frac{n}{2}\right]} \left(1 + \cos \frac{2k-1}{2n}\pi\right)^2 \sim (3.20991\ldots)^n, \text{ as } n \to \infty. \qquad (1.5)$$

Another case of the inequality (1.1) was considered by Gel'fond [12, p. 135] in connection with the theory of transcendental numbers, for $E = \overline{D}$, where $D := \{w \colon |w| < 1\}$ is the unit disk:

$$\prod_{k=1}^{m} \|p_k\|_{\overline{D}} \leq e^n \|p\|_{\overline{D}}. \qquad (1.6)$$

The latter inequality was improved by Mahler [18], who replaced e by 2:

$$\prod_{k=1}^{m} \|p_k\|_{\overline{D}} \leq 2^n \|p\|_{\overline{D}}. \qquad (1.7)$$

It is easy to see that the base 2 cannot be decreased, if $m = n$ and $n \to \infty$. However, (1.7) has recently been further improved in two directions. D. W. Boyd [7] [11] showed that, by taking into account the number of factors m in (1.7), one has

$$\prod_{k=1}^{m} \|p_k\|_{\overline{D}} \leq (C_m)^n \|p\|_{\overline{D}}, \qquad (1.8)$$

where

$$C_m := \exp\left(\frac{m}{\pi} \int_0^{\pi/m} \log\left(2\cos\frac{t}{2}\right) dt\right) \qquad (1.9)$$

is asymptotically best possible for each fixed m, as $n \to \infty$. Kroó and Pritsker [16] showed that, for any $m \leq n$,

$$\prod_{k=1}^{m} \|p_k\|_{\overline{D}} \leq 2^{n-1} \|p\|_{\overline{D}}, \qquad (1.10)$$

where equality holds in (1.10) for *each* $n \in \mathbb{N}$, with $m = n$ and $p(z) = z^n - 1$. In Section 2 we give an asymptotically sharp inequality for the norm of products of polynomials on arbitrary compact set, which generalizes the results of Mahler, Kneser and Borwein. This inequality and other connected to it results were originally obtained in [23].

A closely related problem is to estimate the norm of a single factor via the norm of the whole polynomial. Clearly, we have to normalize the problem by assuming that $p(z)$ is a monic polynomial of degree n, with a monic factor $q(z)$, so that

$$p(z) = q(z) r(z).$$

In the case of the unit disk, Boyd [7] proved an asymptotically sharp inequality

$$\|q\|_{\overline{D}} \leq \beta^n \|p\|_{\overline{D}}, \tag{1.11}$$

with

$$\beta := \exp\left(\frac{1}{\pi} \int_0^{2\pi/3} \log\left(2\cos\frac{t}{2}\right) dt\right). \tag{1.12}$$

This inequality improved upon a series of results by Mignotte [20], Granville [14] and Glesser [13].

Further progress was made by Borwein [6] for the segment $[-a, a]$, $a > 0$ (see Theorems 2 and 5 there or see Section 5.3 in [5]). In particular, Borwein proved that if $\deg q = m$ then

$$|q(-a)| \leq \|p\|_{[-a,a]} a^{m-n} 2^{n-1} \prod_{k=1}^{m}\left(1 + \cos\frac{2k-1}{2n}\pi\right), \tag{1.13}$$

where the bound is attained for a monic Chebyshev polynomial of degree n on $[-a, a]$ and a factor q. He also showed that, for $E = [-2, 2]$, the constant in the above inequality satisfies

$$\limsup_{n\to\infty}\left(2^{m-1}\prod_{k=1}^{m}\left(1 + \cos\frac{2k-1}{2n}\pi\right)\right)^{1/n}$$

$$\leq \lim_{n\to\infty}\left(2^{[2n/3]-1}\prod_{k=1}^{[2n/3]}\left(1 + \cos\frac{2k-1}{2n}\pi\right)\right)^{1/n}$$

$$= \exp\left(\int_0^{2/3}\log\left(2 + 2\cos\pi x\right) dx\right) = 1.9081\ldots,$$

which suggests that an analog of (1.11)–(1.12) for $[-2, 2]$ holds with the constant

$$C_{[-2,2]} = \exp\left(\int_0^{2/3}\log\left(2 + 2\cos\pi x\right) dx\right) = 1.9081\ldots. \tag{1.14}$$

In Section 3, we find an asymptotically sharp inequality of this type for a rather arbitrary compact set E. The general result is then applied to the cases of a disk and a line segment, so that we recover (1.11)–(1.12) and confirm (1.14). See also [22] for these results.

The above problems have applications in transcendence theory (see [12]) and in designing algorithms for factoring polynomials (see [10] and [17]). We confine ourselves to studying the sup norms for polynomials of only one variable. A survey of the results involving other norms (e.g., Bombieri norms) can be found in [10]. These inequalities are also of considerable interest for polynomials in several variables, where very little is known about sharp constants (cf. [2], [3], [4] and [19]).

2   Products of Polynomials in Uniform Norms

Inequalities (1.2)–(1.10) clearly indicate that the constant C in (1.1) grows exponentially fast with n, with the base for the exponential depending on the set E. A natural general problem arising here is to find the *smallest* constant $M_E > 0$, such that

$$\prod_{k=1}^{m} \|p_k\|_E \leq M_E^n \|p\|_E \tag{2.1}$$

for arbitrary algebraic polynomials $\{p_k(z)\}_{k=1}^{m}$ with complex coefficients, where $p(z) = \prod_{k=1}^{m} p_k(z)$ and $n = \deg p$. The solution of this problem is based on the logarithmic potential theory (cf. [25] and [24]). Let $\mathrm{cap}(E)$ be the *logarithmic capacity* of a compact set $E \subset \mathbb{C}$. For E with $\mathrm{cap}(E) > 0$, denote the *equilibrium measure* of E by μ_E. We remark that μ_E is a positive unit Borel measure supported on E (see [25, p. 55]). Define

$$d_E(z) := \max_{t \in E} |z - t|, \qquad z \in \mathbb{C}, \tag{2.2}$$

which is clearly a positive and continuous function on $\mathbb{C}$.

Theorem 2.1. *Let $E \subset \mathbb{C}$ be a compact set, $\mathrm{cap}(E) > 0$. Then the best constant M_E in (2.1) is given by*

$$M_E = \frac{\exp\left(\int \log d_E(z) d\mu_E(z)\right)}{\mathrm{cap}(E)}. \tag{2.3}$$

One can see from (2.1) or (2.3) that M_E is invariant with respect to rigid motions and dilations of the set E in the plane.

Note that the restriction $\mathrm{cap}(E) > 0$ excludes only very *thin* sets from our consideration (see [25, pp. 63–66]), e.g., finite sets in the plane. On the other hand, Theorem 2.1 is applicable to any compact set with a connected component consisting of more than one point (cf. [25, p. 56]). In particular, if E is a continuum, i.e., a connected set, then we obtain a simple universal bound for M_E.

Corollary 2.2. *Let $E \subset \mathbb{C}$ be a bounded continuum (not a single point). Then we have*

$$M_E \leq \frac{\text{diam}(E)}{\text{cap}(E)} \leq 4, \tag{2.4}$$

where $\text{diam}(E)$ *is the Euclidean diameter of the set E.*

For the unit disk $D = \{w \colon |w| < 1\}$, we have that $\text{cap}(\overline{D}) = 1$ [25, p. 84] and that

$$\mu_{\overline{D}} = \frac{1}{2\pi} d\theta, \tag{2.5}$$

where $d\theta$ is the arclength on ∂D. Thus Theorem 2.1 yields

$$M_{\overline{D}} = \exp\left(\frac{1}{2\pi} \int_0^{2\pi} \log d_{\overline{D}}(e^{i\theta}) \, d\theta\right) = \exp\left(\frac{1}{2\pi} \int_0^{2\pi} \log 2 \, d\theta\right) = 2, \tag{2.6}$$

so that we immediately obtain Mahler's inequality (1.7).

If $E = [-1, 1]$ then $\text{cap}([-1, 1]) = 1/2$ and

$$\mu_{[-1,1]} = \frac{dx}{\pi\sqrt{1 - x^2}}, \quad x \in [-1, 1], \tag{2.7}$$

which is the Chebyshev (or arcsin) distribution (see [25, p. 84]). Using Theorem 2.1, we obtain

$$M_{[-1,1]} = 2 \exp\left(\frac{1}{\pi} \int_{-1}^{1} \frac{\log d_{[-1,1]}(x)}{\sqrt{1 - x^2}} \, dx\right) \tag{2.8}$$

$$= 2 \exp\left(\frac{2}{\pi} \int_0^1 \frac{\log(1 + x)}{\sqrt{1 - x^2}} \, dx\right)$$

$$= 2 \exp\left(\frac{2}{\pi} \int_0^{\pi/2} \log(1 + \sin t) \, dt\right) \approx 3.2099123. \tag{2.9}$$

This gives the asymptotic version of Borwein's inequality (1.4)–(1.5).

It appears that the upper bound 4 in Corollary 2.2 is not the best possible. One might conjecture that the sharp universal bounds are

$$2 = M_{\overline{D}} \leq M_E \leq M_{[-1,1]} \approx 3.2099123, \tag{2.10}$$

for any bounded non-degenerate continuum E.

It is of interest to determine the nature of the extremal polynomials for (2.1). We characterized the asymptotically extremal polynomials for (2.1), i.e., those polynomials, for which (2.1) becomes an asymptotic equality as $n \to \infty$, by their asymptotic zero distributions. The precise statements of these results can be found in Theorems 2.3–2.5 and Corollaries 3.1–3.3 of [23].

3   Uniform Norm of a Single Factor

In the same way as in Section 2, we naturally arrive at the problem to find the best (smallest) constant C_E, such that

$$\|q\|_E \le C_E^n \|p\|_E, \quad \deg p = n, \tag{3.1}$$

is valid for *any* monic polynomial $p(z)$ and *any* monic factor $q(z)$. Our solution of this problem is based on similar ideas, involving the logarithmic capacity and the equilibrium measure of E.

Theorem 3.1. *Let $E \subset \mathbb{C}$ be a compact set, $\mathrm{cap}(E) > 0$. Then the best constant C_E in (3.1) is given by*

$$C_E = \frac{\max\limits_{u \in \partial E} \exp \left(\int_{|z-u| \ge 1} \log |z - u| d\mu_E(z) \right)}{\mathrm{cap}(E)}. \tag{3.2}$$

Furthermore, if E is regular then

$$C_E = \max_{u \in \partial E} \exp \left(-\int_{|z-u| \le 1} \log |z - u| d\mu_E(z) \right). \tag{3.3}$$

The above notion of regularity is to be understood in the sense of exterior Dirichlet problem (cf. [25, p. 7]).

One can readily see from (3.1) or (3.2) that the best constant C_E is invariant under the rigid motions of the set E in the plane. Therefore we consider applications of Theorem 3.1 to the family of disks $D_r := \{z : |z| < r\}$, which are centered at the origin, and to the family of segments $[-a, a]$, $a > 0$.

Corollary 3.2. *Let D_r be a disk of radius r. Then the best constant $C_{\overline{D}_r}$, for $E = \overline{D}_r$, is given by*

$$C_{\overline{D}_r} = \begin{cases} \dfrac{1}{r}, & 0 < r \le 1/2, \\[3mm] \dfrac{1}{r} \exp \left(\dfrac{1}{\pi} \displaystyle\int_0^{\pi - 2\arcsin \frac{1}{2r}} \log \left(2r \cos \dfrac{x}{2} \right) dx \right), & r > 1/2. \end{cases} \tag{3.4}$$

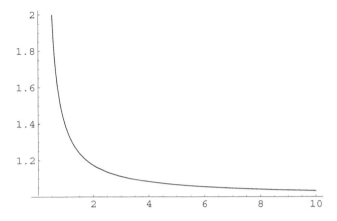

Figure 1. $C_{\overline{D}_r}$ as a function of r.

Note that (1.11)–(1.12) immediately follow from (3.4) for $r = 1$. The graph of $C_{\overline{D}_r}$ is in Figure 1.

Corollary 3.3. *If $E = [-a, a]$, $a > 0$, then*

$$
C_{[-a,a]} = \begin{cases} \dfrac{2}{a}, & 0 < a \leq 1/2, \\[2ex] \dfrac{2}{a} \exp\left(\displaystyle\int_{1-a}^{a} \dfrac{\log(t+a)}{\pi\sqrt{a^2 - t^2}} \, dt \right), & a > 1/2. \end{cases} \tag{3.5}
$$

Observe that (3.5), with $a = 2$, implies (1.14) by the change of variable $t = 2\cos \pi x$. We include the graph of $C_{[-a,a]}$ in Figure 2 below.

We now state two general consequences of Theorem 3.1. They explain some interesting features of C_E, which the reader may have noticed in Corollaries 3.2 and 3.3. Recall that the Euclidean diameter of E is defined by

$$
\mathrm{diam}(E) := \max_{z, \zeta \in E} |z - \zeta|.
$$

Corollary 3.4. *Suppose that $\mathrm{cap}(E) > 0$. If $\mathrm{diam}(E) \leq 1$ then*

$$
C_E = \frac{1}{\mathrm{cap}(E)}. \tag{3.6}
$$

It is well known that $\mathrm{cap}(D_r) = r$ and $\mathrm{cap}([-a, a]) = a/2$ (see [24, p. 135]), which clarifies the first lines of (3.4) and (3.5) by (3.6).

The next Corollary shows how the constant C_E behaves under dilations of the set E. Let αE be the dilation of E with a factor $\alpha > 0$.

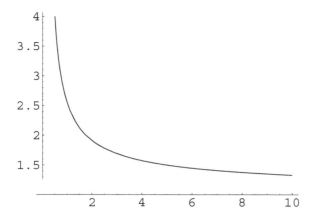

Figure 2. $C_{[-a,a]}$ as a function of a.

Corollary 3.5. *If E is regular then*

$$\lim_{a \to +\infty} C_{\alpha E} = 1. \qquad (3.7)$$

Thus Figures 1 and 2 clearly illustrate (3.7).

We conclude this section with two remarks.

Remark 3.6. One can deduce inequalities of the type (3.1), for various L_p norms, from Theorem 3.1, by using relations between L_p and L_∞ norms of polynomials on E (see, e.g., [21]).

Remark 3.7. Note that the inequalities considered in this section hold for any monic factor $q(z)$ of a monic polynomial $p(z)$, i.e., they hold for the *largest* factor in the terminology of [8]. However, if we are given the existence of a factoring $p(z) = q(z)r(z)$, then the norm of the *smallest* factor (cf. [9]) can be estimated from (2.1) as follows:

$$\|r\|_E \le M_E^{n/2} \|p\|_E^{1/2}, \quad \deg p = n, \qquad (3.8)$$

which may be better than (3.1) in some cases.

4   Proofs

The following lemma is a generalization of Lemma 2 in [11]. We refer to [23] for its proof.

Lemma 4.1. *Let $E \subset \mathbb{C}$ be a compact set (not a single point) and let*

$$d_E(z) := \max_{t \in E} |z - t|, \quad z \in \mathbb{C}.$$

Then $\log d_E(z)$ is a subharmonic function in $\mathbb{C}$ and

$$\log d_E(z) = \int \log |z - t| d\sigma_E(t), \quad z \in \mathbb{C}, \tag{4.1}$$

where σ_E is a positive unit Borel measure in $\mathbb{C}$ with unbounded support, i.e.,

$$\sigma_E(\mathbb{C}) = 1 \quad and \quad \infty \in \operatorname{supp} \sigma_E. \tag{4.2}$$

Lemma 4.2 (Bernstein-Walsh). *Let $E \subset \mathbb{C}$ be a compact set, $\operatorname{cap}(E) > 0$, with the unbounded component of $\overline{\mathbb{C}} \setminus E$ denoted by Ω. Then, for any polynomial $p(z)$ of degree n, we have*

$$|p(z)| \le \|p\|_E \, e^{n g_\Omega(z, \infty)}, \quad z \in \mathbb{C}, \tag{4.3}$$

where $g_\Omega(z, \infty)$ is the Green function of Ω, with pole at ∞, satisfying

$$g_\Omega(z, \infty) = \log \frac{1}{\operatorname{cap}(E)} + \int \log |z - t| d\mu_E(t), \quad z \in \mathbb{C}. \tag{4.4}$$

This is a well known result about the upper bound (4.3) for the growth of $p(z)$ off the set E (see [24, p. 156], for example). The representation (4.4) for $g_\Omega(z, \infty)$ is also classical (cf. Theorem III.37 in [25, p. 82]).

Consider the n-th Fekete points $\{a_{k,n}\}_{k=1}^n$ for a compact set $E \subset \mathbb{C}$ (cf. [24, p. 152]). Let

$$F_n(z) := \prod_{k=1}^n (z - a_{k,n}) \tag{4.5}$$

be the Fekete polynomial of degree n, and define the normalized counting measures in Fekete points by

$$\tau_n := \frac{1}{n} \sum_{k=1}^n \delta_{a_{k,n}}, \quad n \in \mathbb{N}, \tag{4.6}$$

where $\delta_{a_{k,n}}$ is a unit point-mass at $a_{k,n}$.

Lemma 4.3. *For a compact set $E \subset \mathbb{C}$, $\operatorname{cap}(E) > 0$, we have that*

$$\lim_{n \to \infty} \|F_n\|_E^{1/n} = \operatorname{cap}(E) \tag{4.7}$$

and

$$\tau_n \overset{*}{\to} \mu_E, \quad as \ n \to \infty. \tag{4.8}$$

Equation (4.7) is standard (see Theorems 5.5.4 and 5.5.2 in [24, pp. 153–155]), while (4.8) follows from (4.7) (see Ex. 5 on page 159 of [24]).

Proof of Theorem 2.1. First we show that the best constant M_E in (2.1) is at most the right hand side of (2.3). Clearly, it is sufficient to prove an inequality of the type (2.1) for monic polynomials only. Thus, we assume that $p_k(z), 1 \le k \le m$, are all monic, so that $p(z)$ is monic too. Let $\{z_{k,n}\}_{k=1}^n$ be the zeros of $p(z)$ and let ν_n be the normalized zero counting measure for $p(z)$. Then, we use (4.1), Fubini's theorem and Lemma 4.2 as follows:

$$\frac{1}{n} \log \frac{\prod_{k=1}^m \|p_k\|_E}{\|p\|_E}$$

$$\le \frac{1}{n} \log \frac{\prod_{k=1}^n \|z - z_{k,n}\|_E}{\|p\|_E} = \log \frac{1}{\|p\|_E^{1/n}} + \int \log d_E(z) d\nu_n(z)$$

$$= \log \frac{1}{\|p\|_E^{1/n}} + \iint \log |z - t| d\nu_n(z) d\sigma_E(t) = \int \log \frac{|p(t)|^{1/n}}{\|p\|_E^{1/n}} d\sigma_E(t)$$

$$\le \int g_\Omega(t, \infty) d\sigma_E(t) = \log \frac{1}{\operatorname{cap}(E)} + \iint \log |z - t| d\sigma_E(t) d\mu_E(z)$$

$$= \log \frac{1}{\operatorname{cap}(E)} + \int \log d_E(z) d\mu_E(z).$$

This gives that

$$M_E \le \frac{\exp(\int \log d_E(z) d\mu_E(z))}{\operatorname{cap}(E)}. \tag{4.9}$$

To show that equality holds in (4.9), we consider the n-th Fekete points $\{a_{k,n}\}_{k=1}^n$ for E and the Fekete polynomials $F_n(z), n \in \mathbb{N}$. Observe that

$$\|z - a_{k,n}\|_E = d_E(a_{k,n}), \quad 1 \le k \le n, \quad n \in \mathbb{N}.$$

Since $\operatorname{cap}(E) \ne 0$, the set E consists of more than one point and, therefore, $d_E(z)$ is a *strictly positive* continuous function in $\mathbb{C}$. Consequently, $\log d_E(z)$ is also continuous in $\mathbb{C}$, and we obtain by (4.8) of Lemma 4.3 that

$$\lim_{n \to \infty} \left(\prod_{k=1}^n \|z - a_{k,n}\|_E \right)^{1/n} = \lim_{n \to \infty} \exp \left(\frac{1}{n} \sum_{k=1}^n \log d_E(a_{k,n}) \right)$$

$$= \exp \left(\lim_{n \to \infty} \int \log d_E(z) d\tau_n(z) \right)$$

$$= \exp \left(\int \log d_E(z) d\mu_E(z) \right). \tag{4.10}$$

Finally, we have from the above and (4.7) that

$$M_E \geq \lim_{n\to\infty} \frac{(\prod_{k=1}^n \|z - a_{k,n}\|_E)^{1/n}}{\|F_n\|_E^{1/n}} = \frac{\exp(\int \log d_E(z) d\mu_E(z))}{\mathrm{cap}(E)}.$$

□

Proof of Corollary 2.2. Since E is a bounded continuum, we obtain from Theorem 5.3.2(a) of [24, p. 138] that

$$\mathrm{cap}(E) \geq \frac{\mathrm{diam}(E)}{4}.$$

Thus, the Corollary follows by combining this estimate with the obvious inequality

$$d_E(z) \leq \mathrm{diam}(E), \quad z \in E,$$

and by using that $\mu_E(\mathbb{C}) = 1$, $\mathrm{supp}\mu_E \subset E$.

□

Proof of Theorem 3.1. The proof of this result is quite similar to that of Theorem 2.1 (also see [7]). For $u \in \mathbb{C}$, consider a function

$$\rho_u(z) := \max(|z - u|, 1), \quad z \in \mathbb{C}.$$

One can immediately see that $\log \rho_u(z)$ is a subharmonic function in $z \in \mathbb{C}$, which has the integral representation (see [24, p. 29])

$$\log \rho_u(z) = \int \log |z - t| \, d\lambda_u(t), \quad z \in \mathbb{C}, \tag{4.11}$$

where $d\lambda_u(u + e^{i\theta}) = d\theta/(2\pi)$ is the normalized angular measure on $|t - u| = 1$.

Let $u \in \partial E$ be such that

$$\|q\|_E = |q(u)|.$$

If z_k, $k = 1, \ldots, n$, are the zeros of $p(z)$, arranged so that the first m zeros belong to $q(z)$, then

$$\log \|q\|_E = \sum_{k=1}^m \log |u - z_k| \leq \sum_{k=1}^m \log \rho_u(z_k) \leq \sum_{k=1}^n \log \rho_u(z_k)$$

$$= \sum_{k=1}^n \int \log |z_k - t| \, d\lambda_u(t) = \int \log |p(t)| \, d\lambda_u(t), \tag{4.12}$$

by (4.11).

It follows from (4.11)–(4.12), Lemma 4.2 and Fubini's theorem that

$$\frac{1}{n} \log \frac{\|q\|_E}{\|p\|_E} \leq \int \log \frac{|p(t)|^{1/n}}{\|p\|_E^{1/n}} \, d\lambda_u(t) \leq \int g_\Omega(t, \infty) \, d\lambda_u(t)$$

$$= \log \frac{1}{\mathrm{cap}(E)} + \iint \log |z - t| \, d\lambda_u(t) d\mu_E(z)$$

$$= \log \frac{1}{\mathrm{cap}(E)} + \int \log \rho_u(z) \, d\mu_E(z).$$

Using the definition of $\rho_u(z)$, we obtain from the above estimate that

$$\|q\|_E \leq \left(\frac{\max\limits_{u \in \partial E} \exp \left(\int \log \rho_u(z) \, d\mu_E(z) \right)}{\mathrm{cap}(E)} \right)^n \|p\|_E$$

$$= \left(\frac{\max\limits_{u \in \partial E} \exp \left(\int_{|z-u| \geq 1} \log |z - u| \, d\mu_E(z) \right)}{\mathrm{cap}(E)} \right)^n \|p\|_E.$$

Hence

$$C_E \leq \frac{\max\limits_{u \in \partial E} \exp \left(\int_{|z-u| \geq 1} \log |z - u| \, d\mu_E(z) \right)}{\mathrm{cap}(E)}. \tag{4.13}$$

In order to prove the inequality opposite to (4.13), we consider the n-th Fekete points $\{a_{k,n}\}_{k=1}^n$ for the set E and the Fekete polynomials $F_n(z)$, $n \in \mathbb{N}$. Let $u \in \partial E$ be a point, where the maximum of the right hand side of (4.13) is attained. Let $q_n(z)$ be the factor of $F_n(z)$, whose zeros are the n-th Fekete points satisfying $|a_{k,n} - u| \geq 1$. By (4.8) we have

$$\lim_{n \to \infty} \|q_n\|_E^{1/n} \geq \lim_{n \to \infty} |q_n(u)|^{1/n} = \lim_{n \to \infty} \exp \left(\frac{1}{n} \sum_{|a_{k,n} - u| \geq 1} \log |u - a_{k,n}| \right)$$

$$= \exp \left(\lim_{n \to \infty} \int_{|z-u| \geq 1} \log |u - z| \, d\tau_n(z) \right)$$

$$= \exp \left(\int_{|z-u| \geq 1} \log |u - z| d\mu_E(z) \right).$$

Combining the above inequality with (4.7) and the definition of C_E, we obtain that

$$C_E \geq \lim_{n \to \infty} \frac{\|q_n\|_E^{1/n}}{\|F_n\|_E^{1/n}} \geq \frac{\exp\left(\int_{|z-u| \geq 1} \log|z-u|\, d\mu_E(z)\right)}{\operatorname{cap}(E)}.$$

This shows that (3.2) holds true. Moreover, if $u \in \partial E$ is a regular point for Ω, then we obtain by Theorem III.36 of [25, p. 82]) and (4.4) that

$$\log \frac{1}{\operatorname{cap}(E)} + \int \log|u-t|\, d\mu_E(t) = g_\Omega(u, \infty) = 0.$$

Hence

$$\log \frac{1}{\operatorname{cap}(E)} + \int_{|z-u| \geq 1} \log|u-t|\, d\mu_E(t) = -\int_{|z-u| \leq 1} \log|u-t|\, d\mu_E(t),$$

which implies (3.3) by (3.2). □

Proof of Corollary 3.2. It is well known [25, p. 84] that $\operatorname{cap}(\overline{D_r}) = r$ and $d\mu_{\overline{D_r}}(re^{i\theta}) = d\theta/(2\pi)$, where $d\theta$ is the angular measure on ∂D_r. If $r \in (0, 1/2]$ then the numerator of (3.2) is equal to 1, so that

$$C_{\overline{D_r}} = \frac{1}{r}, \quad 0 < r \leq 1/2.$$

Assume that $r > 1/2$. We set $z = re^{i\theta}$ and let $u_0 = re^{i\theta_0}$ be a point where the maximum in (3.2) is attained. On writing

$$|z - u_0| = 2r \left|\sin \frac{\theta - \theta_0}{2}\right|,$$

we obtain that

$$C_{\overline{D_r}} = \frac{1}{r} \exp\left(\frac{1}{2\pi} \int_{\theta_0 + 2\arcsin\frac{1}{2r}}^{2\pi + \theta_0 - 2\arcsin\frac{1}{2r}} \log\left|2r \sin \frac{\theta - \theta_0}{2}\right| d\theta\right)$$

$$= \frac{1}{r} \exp\left(\frac{1}{2\pi} \int_{2\arcsin\frac{1}{2r} - \pi}^{\pi - 2\arcsin\frac{1}{2r}} \log\left(2r \cos \frac{x}{2}\right) dx\right)$$

$$= \frac{1}{r} \exp\left(\frac{1}{\pi} \int_0^{\pi - 2\arcsin\frac{1}{2r}} \log\left(2r \cos \frac{x}{2}\right) dx\right),$$

by the change of variable $\theta - \theta_0 = \pi - x$. □

Proof of Corollary 3.3. Recall that $\text{cap}([-a,a]) = a/2$ (see [25, p. 84]) and

$$d\mu_{[-a,a]}(t) = \frac{dt}{\pi\sqrt{a^2 - t^2}}, \quad t \in [-a, a].$$

It follows from (3.2) that

$$C_{[-a,a]} = \frac{2}{a}\exp\left(\max_{u\in[-a,a]}\int_{[-a,a]\setminus(u-1,u+1)}\frac{\log|t-u|}{\pi\sqrt{a^2-t^2}}\,dt\right). \tag{4.14}$$

If $a \in (0, 1/2]$ then the integral in (4.14) obviously vanishes, so that $C_{[-a,a]} = 2/a$. For $a > 1/2$, let

$$f(u) := \int_{[-a,a]\setminus(u-1,u+1)}\frac{\log|t-u|}{\pi\sqrt{a^2-t^2}}\,dt. \tag{4.15}$$

One can easily see from (4.15) that

$$f'(u) = \int_{u+1}^{a}\frac{dt}{\pi(u-t)\sqrt{a^2-t^2}} < 0, \quad u \in [-a, 1-a],$$

and

$$f'(u) = \int_{-a}^{u-1}\frac{dt}{\pi(u-t)\sqrt{a^2-t^2}} > 0, \quad u \in [a-1, a].$$

However, if $u \in (1-a, a-1)$ then

$$f'(u) = \int_{u+1}^{a}\frac{dt}{\pi(u-t)\sqrt{a^2-t^2}} + \int_{-a}^{u-1}\frac{dt}{\pi(u-t)\sqrt{a^2-t^2}}.$$

It is not difficult to verify directly that

$$\int\frac{dt}{\pi(u-t)\sqrt{a^2-t^2}} = \frac{1}{\pi\sqrt{a^2-u^2}}\log\left|\frac{a^2 - ut + \sqrt{a^2-t^2}\sqrt{a^2-u^2}}{t-u}\right| + C,$$

which implies that

$$f'(u) = \frac{1}{\pi\sqrt{a^2-u^2}}\log\left(\frac{a^2 - u^2 + u + \sqrt{a^2-(u-1)^2}\sqrt{a^2-u^2}}{a^2 - u^2 - u + \sqrt{a^2-(u+1)^2}\sqrt{a^2-u^2}}\right),$$

for $u \in (1-a, a-1)$. Hence

$$f'(u) < 0, u \in (1-a, 0), \quad \text{and} \quad f'(u) > 0, u \in (0, a-1).$$

Collecting all facts, we obtain that the maximum for $f(u)$ on $[-a, a]$ is attained at the endpoints $u = a$ and $u = -a$, and is equal to

$$\max_{u \in [-a,a]} f(u) = \int_{1-a}^{a} \frac{\log(t+a)}{\pi \sqrt{a^2 - t^2}} \, dt.$$

Thus (3.5) follows from (4.14) and the above equation. $\qquad\qquad\square$

Proof of Corollary 3.4. Note that the numerator of (3.2) is equal to 1, because $|z - u| \le 1$ for all $z \in E$ and all $u \in \partial E$. Thus (3.6) follows immediately. $\qquad\qquad\square$

Proof of Corollary 3.5. Observe that $C_E \ge 1$ for any $E \in \mathbb{C}$, so that $C_{\alpha E} \ge 1$. Since E is regular, we use the representation for C_E in (3.3). Let $T \colon E \to \alpha E$ be the dilation mapping. Then $|Tz - Tu| = \alpha |z - u|$, $z, u \in E$, and $d\mu_{\alpha E}(Tz) = d\mu_E(z)$. This gives that

$$C_{\alpha E} = \max_{Tu \in \partial(\alpha E)} \exp\left(- \int_{|Tz - Tu| \le 1} \log |Tz - Tu| \, d\mu_{\alpha E}(Tz) \right)$$

$$= \max_{u \in \partial E} \exp\left(- \int_{|z-u| \le 1/\alpha} \log(\alpha |z - u|) \, d\mu_E(z) \right)$$

$$= \max_{u \in \partial E} \exp\left(-\mu_E(\overline{D_{1/\alpha}(u)}) \log \alpha - \int_{|z-u| \le 1/\alpha} \log |z - u| \, d\mu_E(z) \right)$$

$$< \max_{u \in \partial E} \exp\left(- \int_{|z-u| \le 1/\alpha} \log |z - u| \, d\mu_E(z) \right),$$

where $\alpha \ge 1$. Using the absolute continuity of the integral, we have that

$$\lim_{\alpha \to +\infty} \int_{|z-u| \le 1/\alpha} \log |z - u| \, d\mu_E(z) = 0,$$

which implies (3.7). $\qquad\qquad\square$

References

[1]   G. Aumann, *Satz über das Verhalten von Polynomen auf Kontinuen*, Sitz. Preuss. Akad. Wiss. Phys.-Math. Kl. (1933), 926–931.

[2]   V. Avanissian and M. Mignotte, *A variant of an inequality of Gel'fond and Mahler*, Bull. London Math. Soc. **26** (1994), 64–68.

[3] B. Beauzamy, E. Bombieri, P. Enflo, and H. L. Montgomery, *Products of polynomials in many variables*, J. Number Theory **36** (1990), 219–245.

[4] C. Benitez, Y. Sarantopoulos, and A. Tonge, *Lower bounds for norms of products of polynomials*, Math. Proc. Cambridge Philos. Soc. **124** (1998), 395–408.

[5] P. Borwein and T. Erdélyi, *Polynomials and polynomial inequalities*, Springer-Verlag, New York, 1995.

[6] P. B. Borwein, *Exact inequalities for the norms of factors of polynomials*, Can. J. Math. **46** (1994), 687–698.

[7] D. W. Boyd, *Two sharp inequalities for the norm of a factor of a polynomial*, Mathematika **39** (1992), 341–349.

[8] ———, *Bounds for the height of a factor of a polynomial in terms of Bombieri's norms: I. The largest factor*, J. Symbolic Comp. **16** (1993), 115–130.

[9] ———, *Bounds for the height of a factor of a polynomial in terms of Bombieri's norms: II. The smallest factor*, J. Symbolic Comp. **16** (1993), 131–145.

[10] ———, *Large factors of small polynomials*, Contemp. Math. **166** (1994), 301–308.

[11] ———, *Sharp inequalities for the product of polynomials*, Bull. London Math. Soc. **26** (1994), 449–454.

[12] A. O. Gel'fond, *Transcendental and algebraic numbers*, Dover, New York, 1960.

[13] P. Glesser, *Nouvelle majoration de la norme des facteurs d'un polynôme*, C. R. Math. Rep. Acad. Sci. Canada **12** (1990), 224–228.

[14] A. Granville, *Bounding the coefficients of a divisor of a given polynomial*, Monatsh. Math. **109** (1990), 271–277.

[15] H. Kneser, *Das Maximum des Produkts zweies Polynome*, Sitz. Preuss. Akad. Wiss. Phys.-Math. Kl. (1934), 429–431.

[16] A. Kroó and I. E. Pritsker, *A sharp version of Mahler's inequality for products of polynomials*, Bull. London Math. Soc. **31** (1999), 269–278.

[17] S. Landau, *Factoring polynomials quickly*, Notices Amer. Math. Soc. **34** (1987), 3–8.

[18] K. Mahler, *An application of Jensen's formula to polynomials*, Mathematika **7** (1960), 98–100.

[19] _____, *On some inequalities for polynomials in several variables*, J. London Math. Soc. **37** (1962), 341–344.

[20] M. Mignotte, *Some useful bounds*, Computer Algebra, Symbolic and Algebraic Computation (B. Buchberger et al., ed.), Springer-Verlag, New York, 1982, pp. 259–263.

[21] I. E. Pritsker, *Comparing norms of polynomials in one and several variables*, J. Math. Anal. Appl. **216** (1997), 685–695.

[22] _____, *An inequality for the norm of a polynomial factor*, Proc. Amer. Math. Soc. **129** (2001), 2283–2291.

[23] _____, *Products of polynomials in uniform norms*, Trans. Amer. Math. Soc. **353** (2001), 3971–3993.

[24] T. Ransford, *Potential theory in the complex plane*, Cambridge University Press, Cambridge, 1995.

[25] M. Tsuji, *Potential theory in modern function theory*, Chelsea Publ. Co., New York, 1975.

G. H. Hardy As I Knew Him

Robert A. Rankin[1]

1 Introduction

This article stems from my desire to make a complete list of all the research students of G. H. Hardy (1877–1947) and that is still one of its objectives; see §§3–5. However, it occurs to me that it may be of some interest to set down some reminiscences of my association with Hardy during the last ten years of his life, and to add information about his family and ancestry.

After completing Part III of the Mathematical Tripos in 1937 I enrolled as a research student of A. E. Ingham (1900–1967). Ingham had done distinguished work on number theory, and, in particular, on Riemann's zeta-function, and was a very kind and considerate supervisor. I owe a lot to him; in particular, what is known as 'Rankin's trick' for the estimate of a Dirichlet series is really due to him. However, I was a very shy young man and he too was shy, so that I did not have the same rapport with him as I like to think I had later with Hardy, to whom in 1938 Ingham transferred me when he went on sabbatical leave as a Leverhulme Fellow. Hardy talked to one as an equal, which was flattering, although somewhat frightening. Also, although Hardy disliked small talk, he was at times quite chatty and his conversation was always interesting. He remains one of my heroes.

Like other research students in pure mathematics I attended Professor Hardy's seminar on Tuesday afternoons. This was advertised in the Cambridge University Reporter as the 'Conversation Class of Professors Hardy and Littlewood'; my memory is that in 1937 the word seminar was not as widely used as it is today, and that it tended to be restricted to its original German meaning of a course of lectures or study of an advanced topic. During my time Littlewood (1885–1977) never turned up at these meetings; the explanation of this I owe to Dame Mary Cartwright.

While Hardy was in Oxford, Littlewood held a regular seminar in his rooms in Trinity College and it was agreed that this should continue when Hardy returned to Cambridge and that the meetings should be chaired by them alternately. The first such meeting was held in Trinity as usual, with Littlewood giving the lecture. Hardy was present and asked a number

---

[1]The author died on January 27, 2001.

of questions. At this Littlewood took umbrage and said "I refuse to be heckled". The result was that thereafter they never appeared together at any of the meetings and eventually only Hardy came and the meetings were held elsewhere.

Both Hardy and Littlewood were admirable lecturers and were well prepared. Hardy had a curious mannerism of repeating the last six or so words in a sentence, then the last three, and ultimately converging to the last one. I think that Littlewood was marginally more stimulating than Hardy but foreign students could occasionally find him difficult to understand. I illustrate by referring to the English word *qua*, which is pronounced to rhyme with *way* but is originally the feminine ablative of the Latin relative pronoun *qui*; it is used to mean *in the capacity of*. Littlewood was lecturing on function theory and introduced a function f of two variables s and z. He was laying forth in his usual extravert style about f *qua* function of s and f *qua* function of z. A friend of mine, an Indian student called R. K. Rubugunday, who was classed in 1938 as Wrangler in the Mathematical Tripos, Part II, put up his hand and said "Please, sir, what is *qua*?" I do not recall that Littlewood gave any satisfactory reply, but shortly after poor Rubugunday was admitted to Fulbourn Mental Hospital, suffering, no doubt, from cultural shock at the strangeness of English *ways* (if not *quas*).

Hardy and Littlewood normally lectured in flannel trousers and sports jackets. I do not remember that their dress was particularly scruffy, but Hardy's sister Gertrude (1879–1963), who adored her older brother, told me that he was once taken for a beggar in the street and given sixpence. He was fond of wearing several woolen cricketing sweaters and would peel off one or two when he got too hot. However, once a month he appeared at his lecture wearing a beautiful dark crimson suit, and we knew it was the day when he went to London for the meeting of the London Mathematical Society. On these days the lecture finished early and he had a taxi waiting for him outside the Arts School to take him to the station.

He had a dislike of all formalities, including shaking hands. He told me once that when he was in Copenhagen at the invitation of Harald Bohr he was taken for a walk in the town and met several members of the mathematics staff, who all expected to shake hands; however, he kept his hands firmly in his pockets.

Hardy knew E. T. Whittaker (1873–1956) well as they were both Trinity men. Whittaker, who was knighted in 1945, was Professor of Mathematics in Edinburgh University from 1912 till 1946 and was, no doubt, responsible for Hardy's honorary LL.D. in Edinburgh. My aunt Dr. Mary Rankin, who was a Reader in Political Economy at that university and lived near the Whittakers in George Square, told me that Hardy created a bad impression by not attending one of the honorary degree parties, and going instead for a walk in the Pentland Hills with Whittaker.

In 1938 or 1939 Hardy told me that he had received an invitation from Whittaker to come to Edinburgh to give a course of lectures and was doubtful whether he should accept. I do not know why he consulted me on the subject, but he may have known that several members of my family had a strong connection with that university. He suspected that his conversation with Whitttker would involve too much discussion of religious matters, which he wanted to avoid. As a voting Fellow of Trinity Whittaker had been an elder in the Presbyterian church in Cambridge, but, over the years, he had turned to various other denominations, finally ending up as a convert to Roman Catholicism.

One of the most enjoyable occasions of my life was, surprisingly, my Ph.D. oral examination in June, 1940. My two examiners were supposed to be Hardy and Heilbronn, but Heilbronn had been interned in the Isle of Man, so that Littlewood took his place. Hardy invited me to dinner in Trinity, with strict instructions to meet him at 7:45 p.m. outside the door to the dining hall. As he would have been required to read the grace, if he had been the Senior Fellow present, we waited outside until grace had been said and then went in and had our dinner. After dinner Hardy, not being a drinking man, went back to his room, but I accompanied Littlewood to the Combination Room where we drank port and ate charcoal biscuits and salted almonds. We then went to Littlewood's room, Hardy joined us and we had, so far as I can remember, a very pleasant conversation; I suspected that Littlewood had not read the dissertation in detail, but had only glanced through it.

When I returned to Cambridge in 1945 after the war I served for a period of years as editor of the *Proceedings of the Cambridge Philosophical Society*. After reading Dorothy Sayers' book *The Nine Tailors*, I had written a paper [4] on campanology, which improved upon a result of W. H. Thomson [5] who had shown in 1886 that a complete peel of Grandshire Triples could not be rung using only plain and bob leads; this he had done without any knowledge of group theory. I was anxious that my paper should be published in Cambridge in the *Proceedings* and gave it to Hardy to send to a referee. Hardy, as I guessed, sent it to Philip Hall, who, fortunately, recommended publication. Hardy returned the paper to me with Hall's comments and his own letter to Hall. I was interested to see that he had written to Hall: "Fortunately the conclusions are negative, since anyone who had proved such feats possible would probably and deservedly be shot". Perhaps I should comment that at that time in Cambridge on Monday evenings those who lived in the centre of Cambridge had to endure a two hour practice by the change ringers in Great St. Mary's church.

I was greatly honoured, when sometime before the outbreak of war, Hardy asked me to read the manuscript and proofs of his book of lectures on the work of Ramanujan [2].

As is well known Hardy was very fond of intellectual games of various kinds. I remember his telling me that once, during a dull Royal Society Council meeting, he had before him a list of fellows divided into their different subjects. He wondered which group was the most aristocratic and assigned to the names 5 marks for a duke, 4 for an earl, etc. He told me that Mathematics came out better than he had expected, but that oceanography was the clear winner owing to the large number of admirals.

I very much regretted that I was unable to go and see him during his last illness when he had lost his interest in mathematics but retained his love of cricket. His sister used to visit him in the Evelyn Nursing Home and read to him out of the history of English county cricket, which to me seemed one of the most boring subjects imaginable. Unfortunately, I was no good at cricket at school, the summit of my achievements being made captain of a team below any eleven named Remnants, which ranked below all the other cricket teams fielded by my school at that time.

2   Family History

Most of the information in this section was obtained by my wife and me from searches in the Public Record Office and elsewhere. Some of it has been published in Kanigel's book [3], where further information can be found.

Hardy's father, Isaac Hardy (1842–1901), was born in Pinchbeck, a couple of miles north of Spalding in Lincolnshire. He taught Geography and Drawing at Cranleigh School in Surrey and later became bursar; he had a fine tenor voice and had been a keen footballer. Both his father, who was a labourer and foundryman, and his paternal grandfather were called Isaac. This must have been a popular name in Lincolnshire at the time, possibly deriving from Isaac Newton, who was a Lincolnshire man.

Hardy's mother, née Sophia Hall (1845–1917), was First Governess at Lincoln Diocesan College and was a remarkable woman; see [6]. At this school the teachers were called governesses and the pupils mistresses. She was born in Northampton. Her father Edward Hall was turnkey at Northampton County Gaol, and lived in the appropriately named Fetter Street nearby. He later became a baker in the village of Spratton, north of Northampton. Edward's wife was Charlotte Penn. Sophia Hall took up her employment at the Diocesan College in 1870. When she resigned her post in 1874 to get married, the Management Committee recorded their appreciation of the uniform excellence with which the responsible duties entrusted to her had been carried out and her wise combination of firmness and kindness.

Hardy's sister Gertrude Edith Hardy was unmarried and taught art and classics at St. Catherine's School, Bromley, a kind of sister school to Cranleigh. When her brother became ill after the war she came to Cambridge to look after him and lived in a Guest House at 5 West Road. It is possible that later on she was allowed by the College authorities to live in Hardy's rooms in Trinity in order to look after him better, and before he moved to the Evelyn Nursing home where he died. She later moved to Meadowcroft in Trumpington Road, where she died in 1963.

It is known that Hardy, possibly because of his atheism, never used his first name Godfrey. To his family and very close friends, such as Professor Donald Robertson, the Professor of Greek, he was known as Harold, and in the newspaper report of the death of his father he is listed among the mourners as Mr. Harold Hardy.

Hardy's atheism appears to have been partly a reaction against the strong religious views of his parents. At the time Cranleigh was known for its 'churchiness'. Gertrude's views were similar to her brother's, but probably not quite so strong. When she lived in Meadowcroft she was annoyed because the other old ladies living there kept on asking her where she went to church on Sundays. Her solution was to say that she was a Mohammedan, but that there was no mosque in Cambridge, and she asked her friend Marjorie Dibden, whose husband Kenneth was Secretary of the Cavendish Laboratory, to accompany her into town to purchase a prayer mat.

3 Early Days of Mathematical Research

The degree of Doctor of Philosophy was not introduced in the University of Cambridge until May 1920. A Board of Research Studies was then set up, which issued an Annual Report in the Cambridge University Reporter. In this report were listed the names and colleges of all research students, together with those of their supervisors, and a brief statement of the subjects of their research. Before that date arrangements were somewhat informal and supervisors' names can only occasionally be found in the records. From 1880 to 1913 Research Students were known as 'Advanced Students'. Of these there were two kinds, one proceeding to the Degree of B.A. or LL.B. by means of Tripos Examinations and the other by means of Certificates of Research. From 1913 to 1920 Research Students proceeded to the Degree of B.A. or LL.B. by Certificates of Research only.

1910	E. F. Clark, (b. 1887), Trin.
1900	B. Cookson, (1874–1909), Trin.
1910	W. T. David (1886–1948), Trin.
1912	G. G. Davidson, (1886–1959), Caius
1900	L. N. G. Filon (1875–1937), King's
1907	J. B. Hubrecht (1883–1978), Christ's
1916–17	E. L. Ince, (1891–1941), Trin.
1910	C. McNeil, (1886–1959), Jesus
1919–20	G. A. Newgass, (1889–1948), Trin.
1915–16	S. Ramanujan, (1887–1920), Trin.
1916–17	F. W. Richards, (b. 1890), Caius
1913–14	R. Rossi, (1888–1920), Trin.
1913–14	F. E. Rowett, (1889–1935), St. John's
1913–14	L. F. G. Simmons, (1890–1954), St. John's
1911	H. J. Swain, Emma.

Table 1. Holders of Research Certificates

In [1] a complete list is given of the colleges and Faculties of the 214 students who received Certificates of Research from 1899 to 1920. From this list I have excerpted, in Table 1, the names of the 15 students who gave Mathematics as their Faculty. Of these only three, namely, Louis Napoleon George Filon, a distinguished applied mathematician, Edward Lindsay Ince and Srinivasa Ramanujan were known to me because of their work. In this and the following tables I give the years of birth and death, where I have been able to find them, together with occasional comments on the students listed. This may enable interested readers to find biographical information from the *Dictionary of National Biography, Who was Who*, or other sources.

However, not every student who did research in Mathematics registered as a Research Student during this time. In Table 2 I list the names of students who, between 1911 and 1920, won a Smith's or Rayleigh Prize, and those who failed to do so, but were commended for their essays. With the exception of E. L. Ince, each person in the table had taken the Mathematical Tripos. It will be noticed that this is, on average, a much more distinguished list of students. The letters S, and R and C after the names denote Smith's Prize, Rayleigh Prize and Commended.

I now come to G. H. Hardy. From the time that he was elected a Fellow of Trinity College in 1900, his adult university career falls into three parts: (i) Cambridge 1900–1919, (ii) Oxford 1919–1931, (iii) Cambridge 1931–1947. During the first period it is known that he gave guidance and encouragement to K. Ananda Rau and S. Ramanujan, but there were, almost certainly others. The former had taken the Tripos, obtaining First Class Honours in both parts, and was a Smith's Prize winner, as may be seen from Table 2. As regards the latter, Hardy stated that he learnt more from Ramanujan than Ramanujan did from him. To illustrate that supervision arrangements were not as formalised as they became after 1920, it may be mentioned that E. L. Ince (an Edinburgh graduate) paid tribute to E. T. Whittaker, his Edinburgh professor, for help and advice while he was doing research in Cambridge; he may have had a nominal Cambridge supervisor, although that was, perhaps, unlikely to have been Hardy. I had hoped that the Minutes of the Degree Committee of the Faculty of Mathematics for the years before 1920 might have revealed the name of supervisors. These Minutes, which were lost, have only recently been found, but unfortunately they do not do so.

4   The Oxford Years

In the 1920's research students at the University of Oxford were admitted either as Advanced Students studying for the D.Phil. or as B.Sc. students. This second category was of a lower level, being approximately equivalent to a present University M.Sc. course. Table 3 lists the names of students supervised by Hardy who were B.Sc. students. The first column gives the year of admission and the last year of graduation, where relevant. It may be noted that T. Vijayaraghavan did not graduate, but received a Certificate entitling him to do so. Also, W. L. Ferrar (1893–1990) succeeded Hardy as supervisor for J. G. Nicholas.

Duminy later became Vice-Chancellor of the University of Capetown. Sutton was a distinguished meteorologist and Gertrude Stanley became head of the Mathematics Department at Westfield College, London.

I now turn to Hardy's Advanced Students reading for the Degree of D.Phil. They are listed in Table 4, arranged similarly to Table 3. It is interesting to note that E. C. Titchmarsh, although qualified to graduate, did not do so, and the same is true again for T. Vijayaraghavan. The asterisk against the last four names indicates that they were transferred to Titchmarsh as supervisor on Hardy's return to Cambridge.

1911	W. E. H. Berwick (S) (1888–1944)	1912	L. J. Mordell (S) (1888–1972)
1913	S. Chapman (S) (1888–1970)	1912	E. H. Neville (S) (1889–1961)
1912	P. J. Daniell (R) (1889–1946)	1920	S. Pollard (S) (1894–1925)
1911	C. G. Darwin (C) (1887–1962)	1915	J. Proudman (S) (1888–1966)
1913	R. H. Fowler (R) (1889–1944)	1918	K. A. Rau (S) (1893–1975)
1914	R. A. Frazer (R) (1891–1959)	1919	S. R. U. Savoor (S) (b. 1893)
1916	H. M. Garner (S) (1891–1977)	1914	B. P. Sen (S)
1913	A. H. S. Gillson (C) (1889–1954)	1916	W. M. Smart (R) (1889–1975)
1915	H. Glauert (R) (1892–1934)	1913	H. Spencer Jones (S) (1890–1967)
1913	A. R. Grieve (C) (1886–1952)	1914	C. A. Stewart (C) (1888–1959)
1918	E. L. Ince (S) (1891–1941)	1913	R. O. Street (R) (1890–1967)
1914	J. Jackson (S) (1887–1958)	1911	A. W. H. Thomson (C) (b. 1888)
1915	H. Jeffreys (S) (1891–1989)	1916	G. P. Thomson (S) (1892–1972)
1911	S. Lees (R) (1885–1940)	1917	H. Todd (S) (b. 1894)
1911	G. H. Livens (S) (1886–1950)	1918	H. W. Unthank (C) (1893–1979)
1919	C. N. H. Lock (S) (1894–1949)	1913	T. L. Wren (R) (1889–1972)

Table 2. Smith and Rayleigh Prize Students

1922	J. P. Duminy	(1897–1980)	Univ. 1923
1923	E. H.Saayman	(b. 1897)	New Coll.
1924	O. G. Sutton	(1903–1977)	Jesus 1927
1925	T. Vijayaraghavan	(1902–1955)	New Coll.
1925	G. K. Stanley	(1897–1974)	Home Stud. 1927
1925	N. L. Clapton	(b. 1903)	Hertford
1926	J. G. Nicholas	(b. 1908)	Jesus 1932

Table 3. Oxford B.Sc. Students

1919	W. R. Burwell	(1894–1971)	Merton
1921	F. V. Morley	(1899–1980)	New Coll. 1923
1922	E. C. Titchmarsh	(1899–1963)	Balliol
1924	G. L. Frewin	(b. 1902)	New Coll.
1925	P. L. Srivastava	(b. 1898)	New Coll. 1927
1926	U. S. Haslam-Jones	(1903–1962)	Queen's 1928
1926	E. H. Linfoot	(1905–1982)	Balliol 1928
1926	F. J. Brand	(1905–1995)	Jesus, Grad. B.Sc. 1929
1926	T. Vijayaraghavan	(1902–1955)	New Coll.
1927	L. S. Bosanquet	(1903–1984)	Balliol 1929
1928	M. L. Cartwright	(1900–1998)	St. Hugh's 1930
1929	E. M. Wright	(b. 1906)	Jesus 1932
1929	P. M. Owen*	(1906–1962)	Jesus 1933
1930	R. Profitt*	(1907–1974)	Exeter 1936
1930	E. G. Phillips*	(1909–1984)	Christ Ch. 1932
1930	A. C. Bassett*	(1908–1989)	Jesus 1936

Table 4. Oxford D.Phil. Students

	Name		University	B.A.	Ph.D.
1931	M. M. Ahmed	(b. 1908)	Edinburgh		
1932	M. Hall	(1910–1989)	Yale		
1933	J. M. Hyslop	(1908–1984)	CU	1932	1935
1933	R. Rado	(1906–1989)	Berlin		1935
1934	G. W. Morgan	(1911–1989)	CU	1932	1935
1934	A. C. Offord	(b. 1906)	London		1936
1934	F. Smithies	(b. 1912)	CU	1933	1937
1937	H. R. Pitt	(b. 1914)	CU	1935	1939
1938	F. M. C. Goodspeed	(b. 1914)	Winnipeg		1942
1939	Y. C. Chow	(b. 1918)	No degree		
1940	R. A. Rankin	(b. 1915)	CU	1937	1940
1942	S. M. Edmonds	(b. 1916)	CU	1938	1942

Table 5. Hardy's later Cambridge Research Students

I am indebted to Simon Bailey and Richard Hughes of Oxford University for the information contained in Tables 3 and 4.

F. V. Morley, a London author and publisher, wrote a book on Inversive Geometry together with his father Frank Morley, who was professor of mathematics at Johns Hopkins University.

5   Return to Cambridge

In Table 5 Hardy's research students are given as listed in the Annual Reports of the Board of Research Studies. The first column gives the year when the student's name first occurs with Hardy as supervisor. However, in nearly every case the date of commencement of research was earlier. There are various reasons for this, such as transferral from a former supervisor. For example, I transferred in 1938 to Hardy, but only appear in 1940 as having done so. For this reason I include, in the second last column, the year when students who had taken the Tripos examinations were awarded their B.A.

The third column lists the previous university CU, denoting Cambridge University. The first student listed is better known as M. Mursi. He was a student of E. T. Whittaker and had just obtained his Edinburgh Ph.D. for a dissertation on automorphic functions, and, in particular uniformization. It is very likely that both he and Marshall Hall intended, at the outset, to stay no more than one year in Cambridge. Y. C. Chow was an able young Chinese mathematician, with no University degree, and was an expert on inequalities on which he published two papers in 1939 in the Journal of the London Mathematical Society. He had to leave Cambridge when he got into financial difficulties and was, as I was informed, killed in an air crash. I knew him well and taught him how to ride a bicycle on Midsummer Common. From the tables it appears that Hardy had three female research students, Gertrude Stanley, Mary Cartwright and Sheila Edmonds.

Littlewood's Students

In view of the close connection between Hardy and Littlewood, I conclude by giving a list of the latter's research students. Apart from the years 1907–1910 when he was at Manchester University, Littlewood spent all his academic life in Cambridge. I have not been able to discover the names of any research students he may have supervised before 1920, but, with one exception that I mention later, Table 6 lists the names of his students who appear after that date in the reports of the Board of Research Studies. The second last column gives the date of obtaining the B.A. degree for those students who had been Cambridge undergraduates. As before, the last column gives the date of the award of the Ph.D. The exception mentioned above is Sir Peter Swinnerton-Dyer, who was a research student of Littlewood, but did not register with the Board of Research Studies. This was possible at that time, and I have discovered a number of additional Cambridge research students who did not enrol with the Board of Research Studies, but made private arrangements with their supervisors. One of those was Douglas C. Noerthcott, who was probably Hardy's last research student. There were others such as Norbert Wiener and Norman Levinson who, during their time in Cambridge, regarded Hardy as their teacher. Freeman Dyson was another who knew Hardy and Littlewood well and had a private arrangement with his supervisor A. S. Besicovitch.

The dates given for V. J. Levin are under the assumption that he was the Viktor Josifovich Levin who worked on complex function theory. His name only occurs once in 1933 in the Report of the Board of Research Studies. He was admitted 'conditional of being accepted by a college, or as a Non-Collegiate Student'. It is likely that he never came to Cambridge because no college wanted him or, more probably, he changed his mind.

Dr. Jeremy Bray was Member of Parliament for Motherwell South from 1983 until he retired in 1997.

1924	T. A. A. Broadbent	(1903–1973)	1924	
1924	R. Cooper	(1903–1979)	1924	1927
1925	F. W. Bradley	(1904–1953)		1931
1925	E. F. Collingwood	(1900–1970)	1921	1929
1925	G. A. A. H. Gyllensvard	(b. 1898)		
1926	F. G. Maunsell	(1898–1956)	1923	1928
1927	H. P. Mulholland	(1906–1977)	1926	1930
1929	S. Verblunsky	(1906–1996)	1927	1930
1930	S. D. S. Chowla	(1907–1995)		1931
1931	J. Cossar	(b. 1907)	1930	1933
1931	H. Davenport	(1907–1969)	1929 Sc.D.	1938
1932	S. Skewes	(1899–1988)	1925	1938
1933	A. E. Gwilliam	(1912–1984)		1935
1933	V. J. Levin	(1909–1986)		
1936	D. C. Spencer	(b. 1912)		1939
1947	A. O. L. Atkin	(b. 1925)	1946	1952
1947	T. M. Flett	(1923–1976)		1950
1948	A. C. Allan	(b. 1928)		
1948	N. DuPlessis	(1921–1983)	1951	
1948	G. R. Morris	(b. 1922)	1948	1952
1948	H. P. F. Swinnerton-Dyer	(b. 1927)	1948	
1949	S. R. Tims	(1926–1971)		1952
1949	E. J. Watson	(b. 1924)	1945	
1949	M. N. Ghabour	(b. 1915)		
1951	P. S. Bullen	(b. 1928)		1955
1951	F. R. Keogh	(1923–1991)		1954
1953	M. F. C. Woollett	(b. 1925)		
1954	J. W. Bray	(b. 1930)	1952	1957

Table 6. Littlewood's Research Students

I should be grateful for any information enabling me to supply dates missing from the above tables. I acknowledge with thanks the information on research students supplied by the archivists of Oxford and Cambridge Colleges. I am also grateful for help and advice received from Dr. Elizabeth Leedham-Green of Cambridge University Archives and Professor J. Milne Anderson of University College, London.

References

[1] *Cambridge Historical Register Supplement 1911–1920*, Cambridge, 1922.

[2] G. H. Hardy, *Ramanujan: Twelve Lectures on subjects suggested by his life and work*, University Press, Cambridge, 1940.

[3] R. Kanigel, *The man who knew infinity*, Charles Scribner's Sons, New York, 1991.

[4] R. A. Rankin, *A campanological problem in group theory*, Proc. Cambridge Phil. Soc. **44** (1948), 17–25.

[5] W. H. Thompson, *A note on Grandsire Triples*, Cambridge, 1886.

[6] D. H. J. Zebedee, *Lincoln Diocesan Training College 1862–1962*, Lincoln, 1962.

Discriminants of Some Painlevé Polynomials

David P. Roberts

1 Introduction

In this paper, we derive explicit formulas for the discriminants of the Yablonsky-Vorobiev polynomials $P_m(x)$, the bi-Hermite polynomials $H_{m,n}(x)$ and the Okamoto polynomials $Q_{m,n}(x)$, as well as some related resultants. In all three cases, the discriminant and related resultants factor into a product of small primes only. Our formula in the bi-Hermite case reduces when $n = 1$ to the nineteenth-century formula for the discriminant of a Hermite polynomial. Our other two discriminant formulas do not have direct antecedents. The introduction and final section provide some context for our results.

The Painlevé hierarchy was discovered by Painlevé and Gambier around 1900. It consists of six second-order non-linear ordinary differential equations I-VI, each depending on appropriate parameters. In brief, one can think of Painlevé VI as a non-linear analog of the usual linear hypergeometric equation. The others are limiting degenerations of Painlevé VI, thus analogs of the confluent hypergeometric equation:

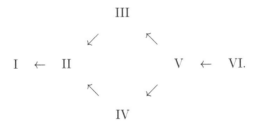

As one moves rightward in this diagram, one adds a parameter at each step, so that Painlevé I involves 0 parameters and Painlevé VI involves 4.

Most of the large and rapidly-expanding literature on the Painlevé equations is connected with mathematical physics. However, we are looking here at number-theoretic phenomena and familiarity with the literature is not necessary to understand this paper. For readers who nonetheless would like some general background, we recommend [3], [1], and [10] as places to start. The textbook [3] characterizes the six Painlevé equations as "the most important non-linear ordinary differential equations." The instructional conference proceedings [1] supports this characterization by describing many situations in mathematical physics which ultimately reduce to a Painlevé equation. The article [10] is an easy-to-read introduction aimed at pure mathematicians.

The bulk of this paper concerns Painlevé II and IV:

$$y'' = 2y^3 + xy + a \tag{1.1}$$

$$y'' = \frac{(y')^2}{2y} + \frac{3y^3}{2} + 4xy^2 + 2(x^2 - a)y + \frac{b}{y}. \tag{1.2}$$

Around 1960, Yablonsky [12] and Vorobiev [11] found rational solutions to Painlevé II. For m an integer, they defined polynomials $P_m \in \mathbf{Z}[x]$ and

$$f_m = \frac{d}{dx} \log\left(\frac{P_{m+1}}{P_m}\right).$$

Then $s_P f_m(s_P x)$ with $s_P = 2^{-2/3}$ is a solution to Painlevé II with $a = -m - 1$.

In the 1980's, Murata [6] found a family of solutions to Painlevé IV, indexed by $(m, n) \in \mathbf{Z}_{\geq 0}^2$. He defined polynomials $H_{m,n}$ and

$$h_{m,n} = \frac{d}{dx} \log\left(\frac{H_{m,n+1}}{H_{m,n}}\right) = \frac{H_{m+1,n}H_{m-1,n+1}}{H_{m,n+1}H_{m,n}}.$$

The function $s_H h_{m,n}(s_H x)$ with $s_H = \sqrt{-2}$ solves Painlevé IV for $a = m + 2n + 1$ and $b = -2m^2$. We call the polynomials $H_{m,n}$ bi-Hermite polynomials because they depend on the two indices m and n and $H_{m,1}(x) = i^{-m}H_{1,m}(ix)$ both coincide with the classical Hermite polynomial $H_m(x)$.

Also in the 1980's, Murata [6] found another family of solutions to Painlevé IV, this one indexed by $(m, n) \in \mathbf{Z}^2$. This family forms a generalization of two sequences of polynomials found several years earlier by Okamoto. Murata defined polynomials $Q_{m,n} \in \mathbf{Z}[x]$ and

$$g_{m,n} = \frac{d}{dx} \log\left(\frac{Q_{m,n+1}}{Q_{m,n}}\right) + x = \frac{Q_{m+1,n}Q_{m-1,n+1}}{Q_{m,n+1}Q_{m,n}}.$$

Then $s_Q g_{m,n}(s_Q x)$ with $s_Q = \sqrt{-2/3}$ solves Painlevé IV for $a = m + 2n$ and $b = (-2/9)(3m - 1)^2$.

It should be noted that notations and normalizations vary somewhat in the literature. We have opted to keep the most standard form of the Painlevé equations (1.1) and (1.2). Our versions of P_m, $H_{m,n}$, and $Q_{m,n}$ are monic and well-behaved at the primes 2 and 3. These good aspects of our normalizations force the presence of the unattractive irrational scale factors s_P, s_H, and s_Q. However, because of the symmetries (3.2), (4.3), (5.3), all rational solutions are in fact in $\mathbf{Q}(x)$.

In §2, we review the formalism of discriminants and resultants. Sections §3, §4, and §5 present our discriminant and resultant formulas for P_m, $H_{m,n}$, and $Q_{m,n}$ respectively. As our title suggests, our primary interest is discriminants. However our inductive method of proof requires us to compute related resultants as well. The induction proceeds by expressing a given discriminant or resultant as a product of earlier discriminants, earlier resultants, and small integers. Accordingly, all the final formulas have a common structure: discriminants and resultants are expressed as products of small integers to large powers. In carrying out the induction, we make fundamental use of quadratic contiguity relations among the polynomials in question.

The polynomials we consider here exhibit great regularity beyond what we capture in our discriminant and resultant formulas. Moreover, the polynomials P_m and $H_{m,n}$ are limiting cases of polynomials related to solutions to Painlevé III, V, and VI. In §6 we briefly indicate several ways in which our formulas here are fitting into an emerging larger picture.

We have implemented P_m, $H_{m,n}$ and $Q_{m,n}$ in a small *Mathematica* file. This file can be used as an aid in reading this paper. It is available at `http://cda.mrs.umn.edu/~roberts`.

2   Discriminants and Resultants

Let $f(x) = a_0 x^M + \cdots$ be a degree M polynomial in $\mathbf{Z}[x]$ with complex roots $\alpha_1, \ldots, \alpha_M$. Then the discriminant of f is defined to be

$$\mathrm{Disc}(f) = a_0^{2M-1} \prod_{1 \leq i < j \leq M} (\alpha_i - \alpha_j)^2 \in \mathbf{Z}. \tag{2.1}$$

Let $g(x) = b_0 x^N + \cdots$ be a degree N polynomial in $\mathbf{Z}[x]$ with complex roots $\beta_1, \ldots, \beta_N$. Then the resultant of f and g is

$$\mathrm{Res}(f, g) = a_M^N b_N^M \prod_{i=1}^{M} \prod_{j=1}^{N} (\alpha_i - \beta_j) \in \mathbf{Z}. \tag{2.2}$$

The connection between resultants and discriminants is

$$\mathrm{Disc}(f) = (-1)^{M(M-1)/2} \mathrm{Res}(f, f'). \tag{2.3}$$

While discriminants are just a special case of resultants, they are worthy of special attention.

It is clear from (2.2) that resultants behave homogeneously on constant polynomials, generalize evaluation, are multiplicative, and are symmetric up to sign:

$$\text{Res}(f, c) = c^{\deg(f)} \tag{2.4}$$

$$\text{Res}(f, x - c) = (-1)^{\deg(f)} f(c) \tag{2.5}$$

$$\text{Res}(f, g_1 g_2) = \text{Res}(f, g_1) \text{Res}(f, g_2) \tag{2.6}$$

$$\text{Res}(f, g) = (-1)^{\deg(f) \deg(g)} \text{Res}(g, f). \tag{2.7}$$

Another basic fact about resultants, not so immediate from (2.2), is that

$$\text{Res}(f, qf + r) = \text{Res}(f, r). \tag{2.8}$$

This fact allows one to compute resultants by the Euclidean algorithm. The initial step in all three of our proofs is several applications of (2.8). In these applications qf represents one or two terms in a quadratic relation. So these terms are not seen in our discriminant and resultant calculations.

The relations (2.3)–(2.8) completely characterize the functionals $\text{Disc}(\cdot)$ and $\text{Res}(\cdot, \cdot)$. We will not make further reference to (2.1) and (2.2) until §6. Thus we will be working purely over $\mathbf{Z}$, without any reference to complex roots.

Denote the absolute version of discriminants and resultants as follows:

$$D(f) = |\text{Disc}(f)|$$
$$R(f, g) = |\text{Res}(f, g)|.$$

In the absolute setting, the formalism of discriminants and resultants simplifies as the signs in (2.3), (2.5) and (2.7) disappear. To make our theorems and proofs more readable we will present them in the absolute setting; thus we will ignore minus signs and work with positive numbers only. Of course, it is easier to do the reverse, that is to ignore magnitudes and work with signs only. Doing this yields

$$\text{Disc}(P_m) < 0 \Leftrightarrow m \equiv 2 \quad (\text{mod } 4)$$

$$\text{Disc}(H_{m,n}) < 0 \Leftrightarrow (m, n) \equiv (1, 2), (1, 3), (3, 2), (3, 3) \quad (\text{mod } 4)$$

$$\text{Disc}(Q_{m,n}) < 0 \Leftrightarrow (m, n) \equiv \begin{array}{l} (2, 0), (2, 1), (2, 2), (2, 3), \\ (1, 2), (1, 3), (3, 0), (3, 1) \end{array} \quad (\text{mod } 4).$$

The signs of the resultants we treat are also determined by their indices modulo 4.

3 Yablonsky-Vorobiev Polynomials

For $m \in \mathbf{Z}$, the Yablonsky-Vorobiev polynomials are defined by

$$P_0 = 1$$
$$P_1 = x$$
$$P_{m+1}P_{m-1} = P'_m P'_m - P''_m P_m + x P_m P_m. \qquad (3.1)$$

From this definition it is apparent that the P_m are rational functions. It is not apparent, but it is true, that they are polynomials. The analogous remarks hold also in the bi-Hermite and Okamoto cases.

The polynomial P_m has degree

$$\Delta_m = \frac{m(m+1)}{2}$$

and satisfies the symmetry

$$P_m(\omega x) = \omega^{\Delta_m} P_m(x) \qquad (3.2)$$

for $\omega \in \mathbf{C}$ a third root of unity.

Besides the quadratic relation (3.1), we need the following quadratic relations.

$$P_{m+2}P_{m-2} = (4 - (2m+1)^2)P'_m P'_m \qquad (3.3)$$
$$+ (x^4 + (2m+1)^2 x)P_m P_m$$
$$- 4x^2 P_m P'_m + (2m+1)^2 P_m P''_m$$
$$P_{m+2}P_{m-1} = -(2m+3)P'_m P_{m+1} \qquad (3.4)$$
$$+ (2m+1)P'_{m+1}P_m + x^2 P_m P_{m+1}.$$

These can be derived from the determinant representation [4, Prop. 4.1] of P_m, following the model provided by the proof of [4, Prop. 2.2].

One has the symmetry $P_m = P_{1-m}$, so we will henceforth restrict attention to $m \in \mathbf{Z}_{\geq 0}$.

Theorem 3.1. *One has the discriminant formula*

$$D(P_m) = \prod_{j=3,5,\ldots}^{2m-1} j^{j(2m+1-j)^2/4}$$

and the resultant formulas

$$R(P_m, P_{m-1}) = \prod_{j=3,5,\ldots}^{2m-3} j^{j(2m+1-j)(2m-1-j)/4}$$

$$R(P_m, P_{m-2}) = (2m-1)^{\Delta_{m-1}} \prod_{j=3,5,\ldots}^{2m-5} j^{j(2m+1-j)(2m-3-j)/4}.$$

Here the products run over odd integers, as indicated.

Proof. Applying $R(P_m, \cdot)$ to (3.1), (3.3), and (3.4) yields

$$R(P_{m+1}, P_m)R(P_m, P_{m-1}) = D(P_m)^2 \tag{3.5}$$

$$R(P_{m+2}, P_m)R(P_m, P_{m-2}) = ((2m-1)(2m+3))^{\Delta_m} D(P_m)^2 \tag{3.6}$$

$$R(P_{m+2}, P_m)R(P_m, P_{m-1}) = (2m+3)^{\Delta_m} R(P_{m+1}, P_m)D(P_m). \tag{3.7}$$

As abbreviations, define

$$d_m = D(P_m)$$
$$r_m = R(P_m, P_{m+1})$$
$$(2m+3)^{\Delta_{m+2}} \rho_m = R(P_m, P_{m+2}),$$

so that

$$(2m-1)^{\Delta_m} \rho_{m-2} = R(P_{m-2}, P_m).$$

These abbreviations absorb most of the factors involving $(2m+3)$ and $(2m-1)$, and (3.5), (3.6), (3.7) can be written respectively as

$$d_m^{-2} r_m = r_{m-1}^{-1} \tag{3.8}$$

$$d_m^{-2} \rho_m = \rho_{m-2}^{-1}(2m+3)^{-2m-3} \tag{3.9}$$

$$d_m^{-1} r_m^{-1} \rho_m = r_{m-1}^{-1}(2m+3)^{-2m-3}. \tag{3.10}$$

The exponents of the three new variables in the three equations just displayed are as indicated in the following 3-by-3 matrix:

	d_m	r_m	ρ_m
(3.8)	-2	1	0
(3.9)	-2	0	1
(3.10)	-1	-1	1

The inverse of this 3-by-3 matrix is

$$
\begin{array}{cccc}
 & (3.8) & (3.9) & (3.10) \\
d_m & -1 & 1 & -1 \\
r_m & -1 & 2 & -2 \\
\rho_m & -2 & 3 & -2
\end{array}
$$

Accordingly, to isolate the new variable d_m we form the multiplicative combination $(3.8)^{-1}(3.9)^{1}(3.10)^{-1}$ and similarly for r_m and ρ_m. The result is

$$d_m = F_m^1 r_{m-1} \tag{3.11}$$

$$r_m = F_m^2 r_{m-1} \tag{3.12}$$

$$\rho_m = F_m^3 r_{m-1}(2m+3)^{-2m-3}, \tag{3.13}$$

where we have defined $F_m = r_{m-1}/\rho_{m-2}$. Using (3.12) twice and (3.13) once we deduce a two-step recursion:

$$
\begin{aligned}
F_m &= \frac{r_{m-1}}{\rho_{m-2}} \\
&= \frac{F_{m-1}^2 r_{m-2}(2m-1)^{2m-1}}{F_{m-2}^3 r_{m-3}} \\
&= \frac{F_{m-1}^2 F_{m-2}^2 r_{m-3}(2m-1)^{2m-1}}{F_{m-2}^3 r_{m-3}} \\
&= \frac{F_{m-1}^2 (2m-1)^{2m-1}}{F_{m-2}}.
\end{aligned}
$$

By direct calculation one has $F_2 = 3^3$ and $F_3 = 3^6 5^5$. The unique solution of the two-step recursion satisfying these initial conditions is

$$F_m = \prod_{j=3,5,\dots}^{2n-1} j^{j(2m+1-j)}. \tag{3.14}$$

Combining (3.14) with (3.11)–(3.13) and elementary steps gives the formulas in Theorem 3.1. $\qquad\square$

4 Bi-Hermite Polynomials

Our basic reference for bi-Hermite polynomials is [8], especially Theorem 4.2 which gives the quadratic relations we need. However we will renormalize the polynomials $H_{m,n}^{NY}$ there via

$$H_{m,n}(x) = \frac{H_{m,n}^{NY}(x/\sqrt{3})}{\sqrt{3}^{mn+m(m-1)+n(n-1)} \prod_{j=1}^{m-1} j^{m-j} \prod_{j=1}^{n-1} j^{n-j}}.$$

The polynomials $H_{m,n}$ are defined for $m, n \in \mathbf{Z}_{\geq 0}$ and characterized by

$$H_{m,0} = H_{0,n} = 1$$
$$H_{1,1} = x$$
$$nH_{m,n+1}H_{m+1,n-1} = H_{m,n}H'_{m+1,n} - H_{m+1,n}H'_{m,n} \qquad (4.1)$$
$$mH_{m+1,n}H_{m-1,n+1} = H_{m,n}H'_{m,n+1} - H_{m,n+1}H'_{m,n} \qquad (4.2)$$

One has the degree formula

$$\mathrm{degree}(H_{m,n}) = mn$$

and the symmetry

$$H_{m,n}(ix) = i^{mn}H_{n,m}(x). \qquad (4.3)$$

Finally Theorem 4.2 of [8] gives not only the quadratic relations (4.1) and (4.2), but also

$$H'_{m,n}H'_{m,n} - H_{m,n}H''_{m,n} = -m(H_{m+1,n}H_{m-1,n} - H^2_{m,n}) \qquad (4.4)$$
$$H'_{m,n}H'_{m,n} - H_{m,n}H''_{m,n} = n(H_{m,n+1}H_{m,n-1} - H^2_{m,n}). \qquad (4.5)$$

Define

$$e(j) = \begin{cases} j^2 - 2(m-j)(n-j) & \text{if } j \leq \min(m,n) \\ \min(m,n)^2 & \text{if } \min(m,n) \leq j \leq \max(m,n) \\ (m+n-j)^2 & \text{if } \max(m,n) \leq j. \end{cases}$$

For $n \in \mathbf{Z}_{\geq 0}$ define also

$$n^* = \prod_{j=1}^{n} j^j$$

$$n^{**} = \prod_{m=1}^{n} m^*.$$

We adopt the convention that $(-1)^* = 0^* = (-1)^{**} = 0^{**} = 1$.

Theorem 4.1. *The discriminant of $H_{m,n}$ is*

$$D(H_{m,n}) = \prod_{j=2}^{m+n-1} j^{je(j)}.$$

Related resultants are

$$R(H_{m,n}, H_{m+1,n}) = \frac{(m+n-1)^{**}}{(m-1)^{**}m^{*n}(n-1)^{**}} D(H_{m,n})$$

$$R(H_{m,n}, H_{m,n+1}) = \frac{(m+n-1)^{**}}{(m-1)^{**}n^{*m}(n-1)^{**}} D(H_{m,n})$$

$$R(H_{m,n}, H_{m+1,n-1}) = \frac{(n-1)^{*m}}{m^{*n}} D(H_{m,n}).$$

Proof. The theorem is true whenever $mn = 0$, as both sides of all four equations are 1. We totally order the remaining index set $\mathbf{Z}_{\geq 1}^2$ by saying

$$(\mu, \nu) < (m, n) \Leftrightarrow \begin{cases} \mu + \nu < m + n \\ \quad \text{or} \\ \mu + \nu = m + n \text{ and } \mu < m. \end{cases}$$

The proof proceeds inductively with respect to this order. So suppose the theorem is proved up through, but not including, the index $(m, n) \in \mathbf{Z}_{\geq 1}^2$.

Set

$$a_{m,n} = R(H_{m,n}, H_{m+1,n})$$
$$b_{m,n} = R(H_{m,n}, H_{m,n+1})$$
$$c_{m,n} = R(H_{m,n}, H_{m+1,n-1})$$
$$d_{m,n} = D(H_{m,n})$$

and

$$\begin{aligned} a &= a_{m,n} & A &= a_{m-1,n} \\ b &= b_{m,n} & B &= b_{m,n-1} \\ c &= c_{m,n} & C &= c_{m-1,n+1} \\ d &= d_{m,n}. \end{aligned}$$

The three index-pairs in the right column are all $< (m, n)$. So we can regard A, B, C as old variables and a, b, c, d as new variables.

Applying $R(H_{m,n}, \cdot)$ to (4.1), (4.2), (4.4), and (4.5), one gets

$$a^{-1}bcd^{-1} = n^{-mn}$$
$$ab^{-1}d^{-1} = C^{-1}m^{-mn}$$
$$a^{-1}d^2 = Am^{mn}$$
$$b^{-1}d^2 = Bn^{mn}.$$

We follow the basic pattern of (3.8)–(3.13). The exponents on the left side this time form a 4-by-4 matrix; this matrix is again invertible over the integers. Inverting this matrix, and taking corresponding multiplicative combinations, we get

$$a = A^{-3}B^2C^2n^{2mn}m^{-mn}$$
$$b = A^{-2}BC^2n^{mn}$$
$$c = A^{-2}B^2Cn^{mn}m^{-mn}$$
$$d = A^{-1}BCn^{mn}.$$

Plugging in the known formulas on the right and carrying out a number of elementary steps yields the formulas to be proved for the quantities on the left. □

5   Okamoto Polynomials

Our basic reference for Okamoto polynomials is again [8], especially Theorem 4.2, which again gives the quadratic relations we need. However we will reindex the polynomials $Q_{m,n}^{NY}$ there by

$$Q_{m,n,r} = Q_{m,n} = Q_{m+n,n}^{NY}.$$

Here r is a redundant parameter always satisfying

$$m + n + r = 1.$$

We include r sometimes to see an S_3 symmetry more clearly. For example,

$$d_{m,n} = \frac{1}{2}(m^2 + n^2 + r^2 - 1) \tag{5.1}$$

is the degree of $Q_{m,n}$.

The polynomials $Q_{m,n} \in \mathbf{Z}[x]$ are defined for $m, n \in \mathbf{Z}$ and characterized by the following equations:

$$Q_{0,0} = Q_{1,0} = Q_{0,1} = 1$$
$$Q_{1,1} = x$$
$$Q_{m+1,n}Q_{m-1,n} = (x^2 + 2m + n - 1)Q_{m,n}^2 \tag{5.2}$$
$$+ Q''_{m,n}Q_{m,n} - Q'_{m,n}Q'_{m,n}$$
$$Q_{m,n}(ix) = i^{d_{m,n}}Q_{n,m}(x). \tag{5.3}$$

One has the further symmetry

$$Q_{m,n,r} = Q_{r,m,n} = Q_{n,r,m}. \tag{5.4}$$

This symmetry means that every Okamoto polynomial can be written $Q_{m,n}$ with $mn \geq 0$.

From Theorem 4.2 of [8] one gets not only the quadratic relation (5.2), but also

$$Q'_{m,n}Q_{m+1,n} = Q_{m,n}Q'_{m+1,n} - xQ_{m,n}Q_{m+1,n} + Q_{m+1,n-1}Q_{m,n+1} \tag{5.5}$$

$$0 = (3r - 1)Q_{m+1,n-1,r}Q_{m-1,n+1,r} \tag{5.6}$$
$$+ (3m - 1)Q_{m,n+1,r-1}Q_{m,n-1,r+1}$$
$$+ (3n - 1)Q_{m-1,n,r+1}Q_{m+1,n,r-1}.$$

For $x, x' \in \mathbf{Z}$ with $|x - x'| \leq 1$ define

$$f_{x,x'} = \begin{cases} \displaystyle\prod_{j=2,5,\ldots}^{3\min(x,x')-4} j^{j(x-(j+1)/3)(x'-(j+1)/3)} & \text{if } x, x' \geq 2 \\ \displaystyle\prod_{j=4,7,\ldots}^{3\min(|x|,|x'|)-2} j^{j(|x|-(j-1)/3)(|x'|-(j-1)/3)} & \text{if } x, x' \leq -2 \\ 1 & \text{else.} \end{cases}$$

Clearly $f_{x,x'} = f_{x,x'}$. Also, straightforward calculation yields

$$f_{x-1,x}f_{x+1,x} = f_{x,x}^2 \tag{5.7}$$
$$f_{x,x-1}^2 f_{x+1,x+1} = f_{x,x+1}^2 f_{x-1,x-1}(3x - 1)^{3x-1}. \tag{5.8}$$

Theorem 5.1. *The discriminant of $Q_{m,n,r}$ depends separately on m, n, and r via*

$$D(Q_{m,n,r}) = f_{m,m}f_{n,n}f_{r,r}. \tag{5.9}$$

If $\{m - m', n - n', r - r'\} = \{-1, 0, 1\}$ then the corresponding resultant factors as

$$R(Q_{m,n,r}, Q_{m',n',r'}) = f_{m,m'} f_{n,n'} f_{r,r'}. \qquad (5.10)$$

Proof. A difference between the present Okamoto case and the bi-Hermite case of §4 is that here we do not have a fact analogous to $H_{m,0} = H_{0,n} = 1$. Accordingly, we have to replace the induction there by a different inductive argument here.

Abbreviate

$$D_{m,n,r} = D(Q_{m,n,r})$$
$$R_{m,n,r}^{m',n',r'} = R(Q_{m,n,r}, Q_{m',n',r'}).$$

Applying $R(Q_{m,n,r}, \cdot)$ to (5.2) and (5.5) respectively gives

$$D_{m,n,r}^2 = R_{m,n,r}^{m+1,n,r-1} R_{m,n,r}^{m-1,n,r+1} \qquad (5.11)$$

$$D_{m,n,r} R_{m,n,r}^{m+1,n,r-1} = R_{m,n,r}^{m+1,n-1,r} R_{m,n,r}^{m,n+1,r-1}. \qquad (5.12)$$

In comparison with the bi-Hermite case, we need to use a new relation to substitute for the lack of infinitely many initial conditions. We get this relation by applying $R(Q_{m+1,n,r-1}, \cdot)$, $R(Q_{m-1,n+1,r}, \cdot)$ and $R(Q_{m,n-1,r+1})$ in turn to (5.6), and taking the product of these three relations so that unwanted factors drop out. The result, after the substantial straightforward cancellation, is

$$R_{m,n-1,r+1}^{m-1,n,r+1} R_{m-1,n+1,r}^{m,n+1,r-1} R_{m+1,n,r-1}^{m+1,n-1,r}$$
$$= R_{m+1,n-1,r}^{m,n-1,r+1} R_{m-1,n,r+1}^{m-1,n+1,r} R_{m,n+1,r-1}^{m+1,n,r-1} \qquad (5.13)$$
$$\cdot (3m - 1)^{3m-1} (3n - 1)^{3n-1} (3r - 1)^{3r-1}.$$

If one replaces the D's and R's in (5.11), (5.12), and (5.13) by f's using the not-yet-known formulas (5.9) and (5.10) one gets relations among the f's which need to be checked. For example, the f-version of (5.11) is

$$f_m^m f_m^m f_n^n f_n^n f_r^r f_m^m = f_{m+1}^m f_n^n f_r^{r-1} f_m^{m-1} f_n^n f_r^{r+1}.$$

This equation is true because of (5.7) applied twice, once with $x = m$ and once with $x = r$. Similarly, the f-version of (5.12) is true because of (5.7) applied with $x = n$. Finally, the f-version of (5.13) is true because of three applications of (5.8), one each to $x = m$, n, and r.

It is best to present the inductive aspects of our argument geometrically. We will illustrate them by Figure 5.1. View m, n, and r as coordinate functions on the plane of Figure 5.1, satisfying $m + n + r = 1$.

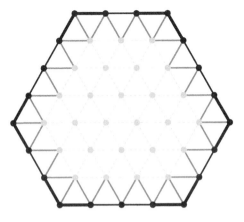

Figure 5.1. Indices for the inductive step $G_3 \to G_4$. Vertices index discriminants and edges index resultants.

The central point in this figure is the non-lattice point $(1/3, 1/3, 1/3)$. If (m, n, r) is a lattice point then we call the sum of the positive numbers among m, n, and r its height. So there are three points of height one, those with $\{m, n, r\} = \{1, 0, 0\}$. There are nine points of height two, three with $\{m, n, r\} = \{1, 1, -1\}$ and six with $\{m, n, r\} = \{2, 0, -1\}$. In general, there are $6h - 3$ lattice points of height h. Because of the S_3 symmetry, we do not need to indicate which collections of parallel lines correspond to constant integral values of which variable.

Let G_h be the graph with vertices the lattice points of height $\leq h$ and edges connecting adjacent vertices. So Figure 5.1 shows all of G_3 in gray, with edges dashed. It shows the rest of G_4 in gray and black, edges not being dashed. The union of the G_h is an infinite planar graph we call G.

To the vertex $(m, n, r) \in G$ there corresponds the real number $D_{m,n,r}$. To the edge in G connecting (m, n, r) with the adjacent vertex (m', n', r') corresponds the real number $R_{m,n,r}^{m',n',r'}$. Note that the two functions

$$D: \text{Vertices} \to \mathbf{R}_{\geq 0}$$
$$R: \text{Edges} \to \mathbf{R}_{\geq 0}$$

are S_3-invariant. So are the formulas on the right side of (5.9) and (5.10). The diagrams in Figure 5.2 correspond to relations (5.11), (5.12), and (5.13) respectively. Consider the edge graph rigidly embedded somehow into G; "rigidly" means here "no bending in the middle"; however different orientations are allowed, according to the S_3 symmetry. One has a discriminant

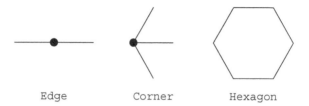

Figure 5.2. Diagrams indicating factors present in Equations (5.11), (5.12), (5.13), respectively.

corresponding to the middle vertex and two resultants corresponding to the two edges. If our formula is true for two of these quantities then it is true for the third. Similarly, consider the corner graph rigidly embedded into G. Here one has a discriminant and three resultants. If our formula is true for three of these quantities it is true for the fourth. Finally, consider the hexagon graph rigidly embedded. Here one has six resultants corresponding to the six edges. If our formula is true for five of these six resultants it is true for the last.

We begin the induction at G_3, for which the truth of the formulas can be checked by direct computation. Next we use the edge relation to extend the domain on which the formula is known to the edges which have a vertex in G_3; these 36 edges are drawn in gray. Next we use the hexagon relation to extend this checked domain to the remaining edges except the twelve remaining edges which are incident upon one of the six corners; these 9 edges are drawn as thin black lines. Third, all vertices except the six corners are obtained by using corner relations. Fourth, the twelve remaining edges are obtained by edge relations. Fifth, the six vertices are obtained by corner relations. These last three parts of G_4 are all shown in black, with thick edges. The remaining inductive steps $G_n \to G_{n+1}$ are done in the same way. □

6   Larger Picture

It is known that Painlevé I has no rational solutions. It is also known that the rational functions discussed in §1 are the only rational solutions to Painlevé II and Painlevé IV. Finally, there are complete lists of rational solutions to Painlevé III, V, VI, coming from polynomials analogous to P_m, $H_{m,n}$, and $Q_{m,n}$. The extreme case of Painlevé VI is particularly interesting; see [7] for explicit polynomials and [5] for the completeness result.

The Painlevé polynomials relating to Painlevé III, V, and VI fit into continuous families. They can be divided into two types:

- Deformations of Yablonsky-Voboriev polynomials. We call these *triangular* Painlevé polynomials.

- Deformations of bi-Hermite polynomials. We call these *bi-classical* polynomials because the case $n = 1$ gives the classical Hermite-Laguerre-Jacobi hierarchy.

We have computationally investigated all these Painlevé polynomials in considerable detail. For example, from our computations we have extracted conjectural discriminant formulas for the remaining Painlevé polynomials. Presumably these discriminant formulas have proofs similar to the three presented here. The discriminant formulas in the bi-classical case reduce for $n = 1$ to classical formulas dating back to Hilbert [2] and Stieltjes [9].

We are quite convinced that all these discriminant formulas form just part of a larger picture. Here are very brief sketches of topics we plan to treat more fully in the future.

Complex roots of P_m, $H_{m,n}$, and $Q_{m,n}$. In Figure 6.1, each of the four squares represents the region $|\mathrm{Re}(x)|, |\mathrm{Im}(x)| \leq 5$ in the complex x-plane. In each square, we have plotted the roots of the indicated Painlevé polynomial. In general, one has the following experimental observations. The roots of P_m approximately form a triangle with m roots on a side. The roots of $H_{m,n}$ approximately form an $m \times n$ rectangle for $m/n \sim 1$; if m/n is far from 1 then the rectangle becomes quite distorted but the rectangular structure remains clearly visible. For $m, n \geq 0$, the roots of $Q_{m,n}$ and $Q_{-n,-m}$ approximately form an $m \times n$ rectangle appropriately framed by triangles in accordance with the degree formulas

$$\mathrm{degree}(Q_{m,n}) = mn + 2\Delta_{m-1} + 2\Delta_{n-1} \tag{6.1}$$

$$\mathrm{degree}(Q_{-n,-m}) = mn + 2\Delta_m + 2\Delta_n. \tag{6.2}$$

Here both (6.1) and (6.2) are rewritten versions of the symmetric degree formula (5.1). In all three cases the roots behave so regularly that, after separating out polynomials according to their indices modulo 3, 2, and 2, respectively, one can interpolate roots with m and $n \in \mathbf{R}$.

p-adic roots of P_m, $H_{m,n}$, and $Q_{m,n}$. Our discriminant formulas say that 2 doesn't divide $D(P_m)$ and 3 doesn't divide $D(Q_{m,n})$. The separable polynomials $P_m \in \mathbf{F}_2[x]$ and $Q_{m,n} \in \mathbf{F}_3[x]$ factor in an unusually regular way.

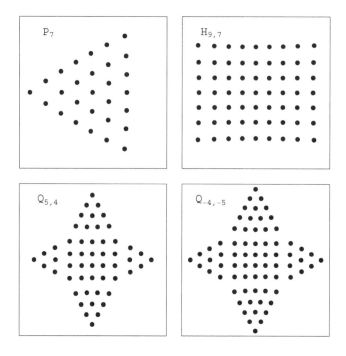

Figure 6.1. Complex roots of four Painlevé polynomials

In all other cases, one can use Newton polygons to investigate the ramification at p. These polygons behave very regularly. Combining Newton polygon information with our discriminant formulas gives prime-by-prime bounds on the algebra discriminants of $\mathbf{Q}[x]/P_m(x)$, $\mathbf{Q}[x]/H_{m,n}(x)$, and $\mathbf{Q}[x]/Q_{m,n}(x)$. The simplest behavior is for large primes dividing the polynomial discriminant, namely the primes which contribute to only one factor in the polynomial discriminant formula. The p-adic roots exhibit sufficient periodicity that it makes sense to interpolate them with m and n in $\mathbf{Z}_p$.

Complex monodromy of Painlevé III, V, and VI polynomials.
As mentioned above, these remaining Painlevé polynomials depend on free parameters. Geometrically, the new parameters let complex roots as in the top two frames of Figure 6.1 move. Our conjectural discriminant formulas identify the locus D in the complex parameter space P where complex roots come together. Computation suggests extremely regular behavior for how roots are interchanged as one moves the complex parameters through $P-D$.

References

[1]   R. Conte (ed.), *The Painlevé property: One century later*, Springer-Verlag, 1999.

[2]   D. Hilbert, *Über die Diskriminante der im Endlichen abbrechenden hypergeometrischen Reihe*, J. Reine Angew. Math. **103** (1888), 337–345.

[3]   L. Iwasaki, H. Kimura, S. Shimomura, and M. Yoshida, *From Gauss to Painlevé: A modern theory of special functions*, Vieweg, 1991.

[4]   K. Kajiwara, K. Yamamoto, and Y. Ohta, *Rational solutions for the discrete Painlevé II equation*, Physics Letters A **232** (1997), 189–199.

[5]   M. Mazzocco, *Rational solutions of the Painlevé VI equation*, J. Phys. A: Math. Gen. **34** (2001), 2281–2294.

[6]   Y. Murata, *Rational solutions of the second and fourth Painlevé equations*, Funkcial. Ekvac. **28** (1985), 1–32.

[7]   M. Noumi, S. Okada, K. Okamoto, and H. Umemura, *Special polynomials associated with the Painlevé equations II*, Integrable systems and algebraic geometry, Kobe/Kyoto, World Sci. Publishing, 1997, pp. 349–372.

[8]   M. Noumi and Y. Yamada, *Symmetries in the fourth Painlevé equation and Okamoto polynomials*, Nagoya Math J. **153** (1999), 53–86.

[9]   T. Stieltjes, *Sur les polynômes de Jacobi*, C.-R. Acad. Sci. **100** (1885), 620–622.

[10] H. Umemura, *The Painlevé equation and classical functions*, Sugaku Expositions **11** (1998), 77–100.

[11] A. P. Vorobiev, *On the rational solutions of the second Painlevé equation*, Differential Equations **I** (1965), 58–59.

[12] A. I. Yablonsky, Vestsi Akad. Navuk BSSR Ser. Fiz.-Tekh. Navuk **3** (1959), 30–35.

Identities Between Mahler Measures

Fernando Rodriguez-Villegas[1]

1 Introduction

The purpose of this short note is to give a proof of the following identity between (logarithmic) Mahler measures

$$m(y^2 + 2xy + y - x^3 - 2x^2 - x) = \frac{5}{7} m(y^2 + 4xy + y - x^3 + x^2), \quad (1.1)$$

which is one of many examples that arise from the comparison of Mahler measures and special values of L-functions [3], [4], [5]. Let us recall that the *logarithmic Mahler measure* of a Laurent polynomial $P \in \mathbb{C}[x_1^{\pm 1}, \ldots, x_n^{\pm 1}]$ is defined as

$$m(P) = \int_0^1 \cdots \int_0^1 \log \left| P(e^{2\pi i \theta_1}, \ldots, e^{2\pi i \theta_n}) \right| d\theta_1 \cdots d\theta_n .$$

The conjecture of Bloch–Beilinson [1], [2] for elliptic curves predicts that both sides of (1.1) are rationally related to $L'(E,0)$ (and hence to each other), where E is the elliptic curve of conductor 37

$$E : \quad y^2 + y = x^3 - x ,$$

and $L(E,s)$ is its L-function. More precisely, we expect that the two numbers a and b defined by

$$m(y^2 + 2xy + y - x^3 - 2x^2 - x) = a\, L'(E,0),$$
$$m(y^2 + 4xy + y - x^3 + x^2) = b\, L'(E,0),$$

are rational. A proof of this fact is not without reach but will not be attempted here; we will prove instead that $a/b = 5/7$.

[1] Support for this work was provided in part by a grant of the NSF and by a Sloan Research Fellowship.

2 Computing in $K_2(E)$

We first recall the definition of the group $K_2(A)$ of an elliptic curve A. Given a field F the group $K_2(F)$ can be defined as $F^* \otimes F^*$ modulo the Steinberg relations $x \otimes (1 - x)$ for $x \neq 0, 1$ in F.

Given a discrete valuation v on F with maximal ideal $\mathcal{M}$ and residue field k we have the *tame symbol* at v defined by

$$(x, y)_v \equiv (-1)^{v(x)v(y)} \frac{x^{v(y)}}{y^{v(x)}} \quad \text{mod } \mathcal{M},$$

which determines a homomorphism

$$\lambda_v : K_2(F) \longrightarrow k^*.$$

For an elliptic curve A defined over $\mathbb{Q}$ we let $K_2(A)$ be the elements of $K_2(\mathbb{Q}(A))$ annihilated by all λ_v with v the valuations associated to $\overline{\mathbb{Q}}$ points of A.

Our E appears as a fiber in several of Boyd's families of elliptic curves (see [3], [5] for a discussion of these families). For example, in its original form $y^2 + y = x^3 - x$, but also as the two Weierstrass equations

$$E_1 : \quad y_1^2 + 4x_1 y_1 + y_1 = x_1^3 - x_1^2$$

and

$$E_2 : \quad y_2^2 + 2x_2 y_2 + y_2 = x_2^3 + 2x_2^2 + x_2 \,.$$

It is easy to check that

$$x_1 = x - 1$$
$$y_1 = y - 2x + 2$$

and

$$x_2 = x - 1$$
$$y_2 = -x + y + 1$$

give isomorphisms

$$E \simeq E_1, \qquad E \simeq E_2 \,.$$

It follows from [5] therefore that some integer multiple of each of

$$\xi = \{x, y\}, \quad \xi_1 = \{x_1, y_1\}, \quad \xi_2 = \{x_2, y_2\}$$

is in $K_2(E)$.

The divisors of the six functions x, y, x_1, y_1, x_2, y_2 are supported on $E(\mathbb{Q})$, which is generated by the point P with $x = 0, y = 0$. More precisely, we have

$$(x) = [P] + [-P] - 2[O]$$
$$(y) = [P] + [2P] + [-3P] - 3[O]$$

$$(x_1) = [2P] + [-2P] - 2[O]$$
$$(y_1) = 2[2P] + [-4P] - 3[O]$$

$$(x_2) = [-2P] + [2P] - 2[O]$$
$$(y_2) = [2P] + 2[-P] - 3[O]$$

where $[O]$ denotes the point at infinity on E.

Given a pair of functions f and g on E with divisors supported on $E(\mathbb{Q})$,

$$(f) = \sum_{n \in \mathbb{Z}} a_n[nP], \qquad (g) = \sum_{n \in \mathbb{Z}} b_n[nP]$$

we define

$$(f) \diamond (g) = \sum_{m,n} a_n b_m [(n-m)P],$$

which we will view as an element of

$$\mathbb{Z}[E(\mathbb{Q})]^- = \mathbb{Z}[E(\mathbb{Q})]/\sim,$$

where $\sim$ is the equivalence relation determined by

$$[-nP] \sim -[nP], \qquad n \in \mathbb{Z}.$$

We may and will represent elements of $\mathbb{Z}[E(\mathbb{Q})]^-$ as vectors $[a_1, a_2, \ldots]$ with $a_i \in \mathbb{Z}$ almost all zero where

$$[a_1, a_2, \ldots] \qquad \longleftrightarrow \qquad \sum_{n=1}^{\infty} a_n[nP]$$

In fact, we will only consider elements where $a_n = 0$ for $n > 6$ and hence simply write $[a_1, \ldots, a_6]$.

We now compute

$$(x) \diamond (y) = [1, 2, -3, 1, 0, 0]$$
$$(x_1) \diamond (y_1) = [0, 5, 0, -4, 0, 1]$$
$$(x_2) \diamond (y_2) = [-6, 2, 2, -1, 0, 0].$$

On the other hand, we also find

$$(-y) \diamond (1+y) = [-8, -7, 8, 1, 0, -1]$$
$$(x - y) \diamond (1 - x + y) = [-9, 5, -5, 5, 0, -1]$$

and verify easily that

$$
\begin{aligned}
7(x) \diamond (y) + (x_1) \diamond (y_1) &= -2(-y) \diamond (1+y) + (x-y) \diamond (1-x+y) \\
5(x) \diamond (y) + (x_2) \diamond (y_2) &= -(-y) \diamond (1+y) + (x-y) \diamond (1-x+y).
\end{aligned}
\tag{2.1}
$$

3 The Regulator

Let

$$r : \quad K_2(E) \longrightarrow \mathbb{R}$$

be the regulator map. It can be defined as follows. If f, g are two non-constant functions on E with $\{f, g\} \in K_2(E)$ then

$$r(\{f, g\}) = \int_\gamma \eta(f, g), \tag{3.1}$$

where

$$\eta(f, g) = \log |f| \, d\arg g - \log |g| \, d\arg f$$

and γ is a closed path not going through poles or zeroes of f or g which generates the subgroup $H_1(E, \mathbb{Z})^-$ of $H_1(E, \mathbb{Z})$ where complex conjugation acts by -1, properly oriented. The fact that the integral only depends on the homology class of γ is a consequence of $\{f, g\} \in K_2(E)$; see [5] for details. (However, note that in [5] we inaccurately said γ should generate the cycles fixed by complex conjugation; we take the opportunity to correct this.)

The regulator may also be expressed in terms of the elliptic dilogarithm [2], [6]

$$\mathcal{L} : \quad E(\mathbb{C}) \longrightarrow \mathbb{R}.$$

In our context, this works as follows. We extend it by linearity to $\mathbb{Z}[E(\mathbb{Q})]$ and since $\mathcal{L}$ is odd it actually gives a map

$$\mathcal{L} : \quad \mathbb{Z}[E(\mathbb{Q})]^- \longrightarrow \mathbb{R}.$$

If f, g are two non-constant functions on E with divisors supported on $E(\mathbb{Q})$ and such that $\{f, g\} \in K_2(E)$ then

$$r(\{f, g\}) = c \, \mathcal{L} \left((f) \diamond (g) \right),$$

for some explicit non-zero constant c, which is not relevant for our purposes. In particular, in the case that $g = 1 - f$

$$\mathcal{L}((f) \diamond (1 - f)) = 0.$$

The above discussion extends naturally to $K_2(E) \otimes \mathbb{Q}$, which contains ξ, ξ_1 and ξ_2.

It follows from (2.1) therefore, that

$$r(\xi_1) = -7r(\xi)$$
$$r(\xi_2) = -5r(\xi). \tag{3.2}$$

4 The Regulator and Mahler's Measure

In [5] we showed that if $P_k(x, y) = 0$ is one of Boyd's families of elliptic curves and k is such that P_k does not vanish on the torus $|x| = |y| = 1$ then

$$r(\{x, y\}) = c_k \pi\, m(P_k)$$

for some nonzero integer c_k. We will now make this precise for

$$P_k(x, y) = y^2 - kxy + y - x^3 + x^2.$$

We consider the region $\mathcal{K}$ of $k \in \mathbb{C}$ such that P_k vanishes somewhere on the torus. It is the image of the torus under the rational map

$$R: \qquad (x, y) \mapsto \frac{y^2 + y - x^3 + x^2}{xy}.$$

We can get a pretty good idea of what $\mathcal{K}$ looks like by graphing the image of a grid under $(\theta_1, \theta_2) \mapsto R(e^{2\pi i \theta_1}, e^{2\pi i \theta_2})$. Dividing the square $0 \le \theta_1 < 1$, $0 \le \theta_2 < 1$ in 40 equal parts we obtain Figure 1 below.

It is not hard to verify directly that the boundary of $\mathcal{K}$ meets the real axis at $k = -4$ and $k = 2$.

If $k \notin \mathcal{K}$ then as x moves counterclockwise on the circle $|x| = 1$ one root $y_1(x)$ of $P_k(x, y) = 0$ satisfies $|y_1(x)| < 1$ and the other $y_2(x)$ satisfies $|y_2(x)| > 1$ and in particular $y_1(x)$ and $y_2(x)$ do not meet. To see this, note that when $x = 1$ the roots are 0 and $k - 1$. Hence, for $|k|$ large these roots are one inside and the other outside the unit circle. The claim follows since the roots depend continuously on k. We let σ_k be the resulting smooth closed path $(x, y_1(x))$ on the elliptic curve E_k determined by $P_k(x, y) = 0$.

Using Jensen's formula we find that

$$m(P_k) = \frac{1}{2\pi i} \int_{\sigma_k} \log |y| \, \frac{dx}{x}$$

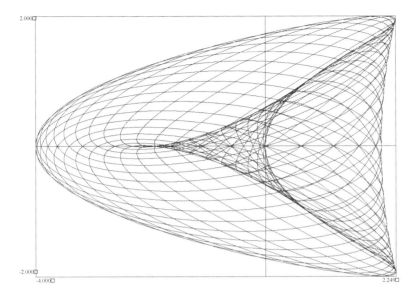

Figure 1. Region $\mathcal{K}$

and note that since $|x| = 1$ on σ_k we can write this identity as

$$m(P_k) = \frac{1}{2\pi} \int_{\sigma_k} \eta(x, y). \tag{4.1}$$

We now show that for k real, $k \notin \mathcal{K}$, the homology class of σ_k generates $H_1(E_k, \mathbb{Z})^-$. We complete the square and write $P_k = (2y - kx + 1)^2 - f(x)$, where $f(x) = 4x^3 + (k^2 - 4)x^2 - 2kx + 1$. The discriminant $\Delta(k) = k^4 - k^3 - 8k^2 + 36k - 1$ of f has two real roots $\alpha = -3.7996\ldots$ and $\beta = .3305\ldots$. Hence, for $k < \alpha$ or $k > \beta$, $\Delta(k) > 0$ and f has three real roots $e_1 < e_2 < e_3$. As $|k|$ increases the roots of f tend to $e_1 = -\infty$ and $e_2 = e_3 = 0$ and by continuity the circle $|x| = 1$ encircles e_2 and e_3 once. Since f is negative in the interval $e_2 < x < e_3$ the period

$$\int_{\sigma_k} \frac{dx}{2y - kx + 1}$$

is purely imaginary and our claim follows.

Combined with (3.1) and (4.1) this proves that in fact

$$r(\{x, y\}) = \pm 2\pi \, m(P_k), \qquad k \in \mathbb{R}, \quad k \notin \mathcal{K}. \tag{4 2}$$

By continuity (4.2) also holds for $k = -4$ and $k = 2$, which are on the boundary of $\mathcal{K}$. In particular, in the notation of Section 3, we obtain the

identity
$$r(\xi_1) = \pm 2\pi \, m(y^2 + 4xy + y - x^3 + x^2) \,. \tag{4.3}$$

A completely analogous analysis yields
$$r(\xi_2) = \pm 2\pi \, m(y^2 + 2xy + y - x^3 - 2x^2 - x) \tag{4.4}$$

(and again $k = -2$ is on the boundary of the corresponding set $\mathcal{K}$). Putting together (3.2), (4.3) and (4.4) (and a simple check for the right sign) we obtain (1.1).

Remarks. 1. We should point out that we do not expect $m(y^2 + y - x^3 + x)$ to be rationally related to either side of (1.1) (and numerically it indeed does not appear to be). The reason is that $y^2 + y - x^3 + x$ vanishes on the torus and in fact $k = 0$ is in the interior of the region $\mathcal{K}$ corresponding to the Boyd family $y^2 - kxy + y - x^3 + x$. Hence the analogue of (4.1) gives the integral of $\eta(x, y)$ on a non-closed cycle.

2. One can prove in a similar way an identity relating either side of (1.1) with $m(y^2 + 2xy + y - x^3 + x^2)$.

References

[1] A. Beilinson, *Higher regulators of modular curves*, Applications of algebraic K-theory to algebraic geometry and number theory, Part I, II (Boulder, Colo., 1983) (Providence, RI), Contemp. Math., vol. 55, Amer. Math. Soc., 1986, pp. 1–34.

[2] S. Bloch and D. Grayson, *K_2 and $L-$functions of elliptic curves: Computer Calculations*, Applications of algebraic K-theory to algebraic geometry and number theory, Part I, II (Boulder, Colo., 1983) (Providence, RI), Contemp. Math., vol. 55, Amer. Math. Soc., 1986, pp. 79–88.

[3] D. W. Boyd, *Mahler's measure and special values of L-functions*, Experiment. Math. **7** (1998), 37–82.

[4] C. Deninger, *Deligne periods of mixed motives, K-theory and the entropy of certain $\mathbf{z}^n$-actions*, J. Amer. Math. Soc. **10** (1997), 259–281.

[5] F. Rodriguez Villegas, *Modular Mahler measures I*, Topics in number theory (University Park, PA, 1997), Math. Appl., vol. 467, Kluwer Acad. Pub., 1999, pp. 17–48.

[6] D. Zagier, *The Bloch–Wigner–Ramakrishnan polylogarithm function*, Math. Ann. **286** (1990), 613–624.

Normal Integral Bases, Swan Modules and p-Adic L-Functions

Anupam Srivastav

1 Introduction

A central theme in Galois module theory is the dominance of analytic functions over class invariants. Let N/K be a finite extension of number fields with the Galois group isomorphic to a finite group Γ. We use the standard notation and denote by $\mathcal{O}_F$ the ring of algebraic integers of a number field F. We are concerned with the problem of obtaining an explicit description of $\mathcal{O}_N$ as a module over either $\mathcal{O}_K[\Gamma]$ or $\mathbf{Z}[\Gamma]$.

It is well known that the extension N/K is tame, i.e., at most tamely ramified, if and only if $\mathcal{O}_N$ is $\mathcal{O}_K[\Gamma]$-projective (if and only if $\mathcal{O}_N$ is a locally free $\mathcal{O}_K[\Gamma]$-module). Therefore, if the extension N/K is also assumed to be tame, then $\mathcal{O}_N$ determines a class $[\mathcal{O}_N]$ in $\mathrm{Cl}(\mathcal{O}_K[\Gamma])$, the class group of locally free $\mathcal{O}_K[\Gamma]$-modules, and by restricting scalars a class $[\mathcal{O}_N]_{\mathbf{Z}[\Gamma]}$ in $\mathrm{Cl}(\mathbf{Z}[\Gamma])$, the class group of locally free $\mathbf{Z}[\Gamma]$-modules.

The main result of classical Galois module theory is the dominance of Artin L-functions associated to the extension N/K over the class $[\mathcal{O}_N]_{\mathbf{Z}[\Gamma]}$. This is the famous Fröhlich conjecture proved by M. J. Taylor in [19]. At present there is no analogous result for the class $[\mathcal{O}_N]$ in $\mathrm{Cl}(\mathcal{O}_K[\Gamma])$. However, there are some other instances of analytic functions governing the Galois module structure. We now list them as follows.

1. The extension N/K is an unramified extension of prime degree p, where $K = \mathbf{Q}(\mu_p)$ is such that p divides the class number of K. Here the analytic functions are p-adic L-functions. This is due to Taylor [21].

2. The extension N/K is an unramified quadratic extension of certain real quadratic fields. This is a 2-adic analog of Taylor's work mentioned above (cf. [17]).

3. There is a Galois module theory for certain Kummer extensions associated with elliptic curves (cf. [20]). There are some partial analogs of Taylor's classical theorem in this setting; see [16], [21], [4], [1].

Let us now restrict our attention to finite tame abelian extensions N/K. The classical Hilbert-Speiser theorem proves that each finite tame abelian extension of the field of rational numbers admits a normal integral basis, i.e., if $K = \mathbf{Q}$ and Γ is abelian, then $\mathcal{O}_N$ is free as a module over $\mathcal{O}_K[\Gamma]$. Greither, Replogle, Rubin and Srivastav [8] recently proved that if K is a number field such that for each finite tame abelian extension N/K, $\mathcal{O}_N$ admits a normal integral basis over $\mathcal{O}_K$, then $K = \mathbf{Q}$. This converse of the Hilbert-Speiser theorem was proved using Swan modules.

In this article we consider the following situation: Let $M = \mathbf{Q}(\mu_p)$, the cyclotomic field of conductor p where p is an odd prime. Let K be a subfield of M^+, the maximal real subfield of M. We let Γ be a cyclic group of order p.

We consider the set R of all realisable classes in the locally free class group $\mathrm{Cl}(\mathcal{O}_K[\Gamma])$, i.e., the set of classes $[\mathcal{O}_N]$, where N/K is a tame cyclic extension of degree p. McCulloh showed in [11] that R is a subgroup of $\mathrm{Cl}(\mathcal{O}_K[\Gamma])$. There is also the kernel group D, the subgroup of $\mathrm{Cl}(\mathcal{O}_K[\Gamma])$ consisting of those classes that become trivial under extension of scalars to the maximal order. We now consider a canonical subgroup of D. This is the Swan subgroup T formed by classes of Swan modules $(s, \Sigma) = s\mathcal{O}_K[\Gamma] + \Sigma\mathcal{O}_K[\Gamma]$ where $s \in \mathcal{O}_K$ is relatively prime to p and Σ is the sum of the elements of the group Γ. The idea of Swan classes is due to R. G. Swan [18], and Ullom [22] studies them from the point of view of Hom-description. Replogle considered the relationship between R and T for any base field containing M in his Ph.D. thesis [13]. It follows essentially from his thesis (cf. [14]) that $T \subseteq R \cap D$ for number fields K considered in this article. Thus under the above setting each Swan class is a realisable class in the kernel group.

We shall describe the Swan subgroup T in terms of p-adic L-functions. Let $G = \mathrm{Gal}(K/\mathbf{Q})$ and let Ψ denote a character of G taking values in $\mathbf{Z}_p$, the ring of p-adic integers. Let $\mathrm{char}(G)$ denote the group of characters of G. For each character Ψ of G, the Ψ-idempotent is defined as

$$e(\Psi) = (1/ \mid G \mid) \sum_{g \in G} \Psi^{-1}(g)g.$$

For a $\mathbf{Z}_p[G]$-module X, we write

$$X(\Psi) = e(\Psi)X,$$

and note that X has a canonical direct sum decomposition

$$X = \sum_{\Psi \in \mathrm{char}(G)} X(\Psi).$$

Let C denote the p-Sylow subgroup of the ideal class group of K. We know that C is a $\mathbf{Z}_p[G]$-module. Let h_p denote the order of C. For each character

Ψ of G let $h_p(\Psi)$ denote the order of $C(\Psi)$ and let $L_p(s, \Psi)$ denote the p-adic L-function of Kubota-Leopoldt. We now state the main result of this article.

Theorem 1. *Let K be a subfield of $\mathbf{Q}(\mu_p)^+$ and let Γ be a cyclic group of prime order $p > 2$. The Swan subgroup T is a $\mathbf{Z}_p[G]$-module. Moreover, for each character Ψ of G we have:*
  (1) $T(\Psi) \cong 0$ or $\mathbf{Z}_p/p\mathbf{Z}_p$; and
  (2) $T(\Psi) = 0$ if and only if either $\Psi = 1$ or $L_p(1, \Psi)h_p(\Psi)^{-1} \not\equiv 0$
(mod p).

We have remarked earlier that T is a subgroup of R in this setting. According to Fröhlich's conjecture the structure of classes in R as $\mathbf{Z}[\Gamma]$-module is governed by L-functions. Therefore, this result is a partial extension of Fröhlich's conjecture in the above setting. We now pose two problems.

Problems. *Let K be a subfield of $\mathbf{Q}(\mu_p)^+$ and let Γ be a cyclic group of prime order $p > 2$.*
  1. Is the p-Sylow subgroup of $R \cap D$ governed by p-adic L-functions?
  2. What is the index $[R \cap D : R \cap T]$?

Remarks. 1. If p is a regular prime, then for $K \subseteq \mathbf{Q}(\mu_p)^+$ and Γ a cyclic group of order p, $T = 0$ (cf. Prop. 5.23 in [23]).

2. If $K = \mathbf{Q}$, then $R = 0$ by the Hilbert-Speiser theorem.

3. If $K = \mathbf{Q}(\sqrt{p})$ $(p \equiv 1 \pmod 4)$, then $h_p = 1$, since the class number of K can not exceed $\sqrt{p}$ (cf. Proposition 8.7 in [12]). Moreover, the Kummer-Vandiver conjecture implies that $h_p = 1$ for all K as in Theorem 1.

4. Under the hypothesis that $h_p = 1$ for the field $K = \mathbf{Q}(\mu_p)^+$, Theorem 1 for this field follows essentially from the thesis of Replogle [13]. See also [15].

5. If $h_p \neq 1$ for some subfield K of $\mathbf{Q}(\mu_p)^+$ and N/K is an unramified cyclic extension of degree p, then from a result of Brinkhuis [3] it follows that $[\mathcal{O}_N] \notin D$.

In Section 2 we describe results on Swan modules and realisable classes. We prove Theorem 1 for $K = \mathbf{Q}(\sqrt{p})$ $(p \equiv 1 \pmod 4)$ in Section 3 and more generally for any $K \subseteq \mathbf{Q}(\mu_p)^+$ in Section 4. The author would like to thank the referee for critical comments and suggestions that greatly improved the presentation of this paper.

2   Swan Modules and Realisable Classes

We fix some notation. Let $M = \mathbf{Q}(\mu_p)$, the cyclotomic field of conductor p, where p is an odd prime. We assume that K is a subfield of M^+, the maximal real subfield of M. We let Γ be a cyclic group of order p. For any ring H, let H^* denote its group of units.

The Swan subgroup T of $\mathrm{Cl}(\mathcal{O}_K[\Gamma])$ is formed by classes of Swan modules $(s, \Sigma) = s\mathcal{O}_K[\Gamma] + \Sigma\mathcal{O}_K[\Gamma]$ where $s \in \mathcal{O}_K$ is relatively prime to p and Σ is the sum of the elements of the group Γ. Consider the Milnor square

$$\begin{array}{ccc}
\mathcal{O}_K[\Gamma] & \xrightarrow{\phi} & \mathcal{O}_K[\Gamma]/\mathcal{O}_K\Sigma = B \\
\epsilon \downarrow & & \downarrow \overline{\epsilon} \\
\mathcal{O}_K & \xrightarrow{\overline{\phi}} & \overline{\mathcal{O}_K} = \mathcal{O}_K/p\mathcal{O}_K,
\end{array}$$

where ϕ and $\overline{\phi}$ are the canonical quotient maps, ϵ is the augmentation map, and $\overline{\epsilon}$ is induced by the augmentation map.

There is an exact sequence

$$\mathcal{O}_K^* \times B^* \xrightarrow{h} \overline{\mathcal{O}_K}^* \xrightarrow{\delta} K_0(\mathcal{O}_K[\Gamma]) \longrightarrow K_0(\mathcal{O}_K) \oplus K_0(B) \longrightarrow 0.$$

This sequence is a special case of the Mayer-Vietoris sequence from K-theory for any Milnor square of rings (cf. Theorem 49.27 in [5]). Note that the groups on the left are K_1 of the appropriate rings. A key idea in Ullom [22] is that the Swan subgroup can be defined as the image of the boundary homomorphism δ. In fact, for $\overline{s} \in \overline{\mathcal{O}_K}^*$, $\delta(\overline{s}) = [s, \Sigma]$, the class of the Swan module (s, Σ) (cf. (53.3) in [5]).

Now let A denote the group of automorphisms of Γ. We let A act trivially on $\mathcal{O}_K$, fixing each element, and then by linearity as a group of ring automorphisms on $\mathcal{O}_K[\Gamma]$. Since Σ is fixed by A, we obtain an induced action of A on B. These A-actions also allow us to view K-groups of appropriate rings as $\mathbf{Z}[A]$-modules. It is easy to see now that the above Mayer-Vietoris sequence is, in fact, an exact sequence of $\mathbf{Z}[A]$-modules.

Lemma 1. *The Swan subgroup T is isomorphic to $\overline{\mathcal{O}_K}^*/\mathrm{Im}(B^*)$ as a $\mathbf{Z}[A]$-module.*

Proof. The map h in the above Mayer-Vietoris sequence is given by $h([u, v]) = \overline{u}\overline{\epsilon}(v)^{-1}$, where $\overline{u}$ denotes the image of $u \in \mathcal{O}_K^*$ under the canonical map $\overline{\phi}$ and $v \in B^*$. Note that the image of $\mathcal{O}_K^*$ under h is contained in the image of B^* under h.

Note that $\mathbf{Z}[A]$ acts on T via augmentation. Let $S = \mathcal{O}_K$ and $\overline{S} = S/pS$. Moreover, there is a commutative diagram

$$\begin{array}{ccc}
B^* & \xrightarrow{N} & (B^A)^* \cong S^* \\
\downarrow \overline{\epsilon} & & \downarrow \overline{\epsilon} \quad \downarrow can \\
\overline{S}^* & \xrightarrow{()^{p-1}} & \overline{S}^* \quad = \overline{S}^*
\end{array}$$

where the "norm" map N is defined by $N(b) = \prod_{\alpha \in A} b^\alpha$. To see this, note that $\bar{\epsilon}(v^\alpha) = \bar{\epsilon}(v)$ for every $\alpha \in A, v \in B^*$. Thus $\bar{\epsilon}(v)^{p-1} = \bar{\epsilon}(N(v))$. But $B = \mathbf{Z}[G]/\mathbf{Z}\Sigma \otimes S$ so $B^A = (\mathbf{Z}[G]/\mathbf{Z}\Sigma)^A \otimes S = S$ (note that $\mathbf{Z}[G]/\mathbf{Z}\Sigma$ is the ring of integers in $\mathbf{Q}(\mu_p)$ and A is the group of automorphisms of this field). Thus $N(v) \in S^*$ and therefore $\bar{\epsilon}(v)^{p-1} = \bar{\epsilon}(N(v))$ is in the image of S^*.

This commutative diagram shows that in the Mayer-Vietoris sequence,

$$\mathrm{Im}(h) \subseteq \{x \in \overline{S}^* : x^{p-1} \in \mathrm{Im}(S^*)\}. \qquad \Box$$

Proposition 1. *Assume that K is a subfield of $\mathbf{Q}(\mu_p)^+$ and let Γ be a cyclic group of prime order p. Then the connecting homomorphism δ of the exact Mayer-Vietoris sequence induces an isomorphism of $\mathbf{Z}[G]$-modules*

$$(\mathcal{O}_K/p\mathcal{O}_K)^* / \mathrm{Im}((\mathbf{Z}/p\mathbf{Z})^*) \, \mathrm{Im}(\mathcal{O}_K^*) \cong T.$$

Proof. Since p ramifies totally in $K/\mathbf{Q}$ and $p > [K : \mathbf{Q}]$, we see that $\overline{S}^*$ is an abelian group of exponent $p(p-1)$. In addition, $\mathrm{Im}((\mathbf{Z}/p\mathbf{Z})^*)$ is the $(p-1)$-torsion subgroup of $\overline{S}^*$. Hence

$$\mathrm{Im}(h) \subseteq \mathrm{Im}((\mathbf{Z}/p\mathbf{Z})^*) \, \mathrm{Im}(S^*).$$

It is well known that the image of h contains $\mathrm{Im}((\mathbf{Z}/p\mathbf{Z})^*$ since Γ is a cyclic group: $1 + g + \ldots + g^i + \mathcal{O}_K\Sigma$ are units of B (they correspond to the cyclotomic units of $\mathbf{Z}[\mu_p]$). Here g is a generator of Γ and $0 < i < p$. Of course, $\mathrm{Im}(h) \supseteq \mathrm{Im}(S^*)$. Thus

$$\mathrm{Im}(h) = \ker(\delta) \supseteq \mathrm{Im}((\mathbf{Z}/p\mathbf{Z})^*) \, \mathrm{Im}(S^*).$$

We let $G = \mathrm{Gal}(K/\mathbf{Q})$ act on T by $[s, \Sigma]^g = [s^g, \Sigma]$, for $s \in S$ relatively prime to p and $g \in G$, so that $\delta(\overline{s}^g) = \delta(\overline{s})^g$. It is easy see that the isomorphism proved is in fact a G-isomorphism. $\qquad \Box$

Note that, since p ramifies totally in $K/\mathbf{Q}$ and $p > [K : \mathbf{Q}]$, we see that $(\mathcal{O}_K/p\mathcal{O}_K)^* / \mathrm{Im}((\mathbf{Z}/p\mathbf{Z})^*)$ is an abelian group of exponent p and of p-rank $[K : \mathbf{Q}] - 1$. Hence it follows from Proposition 1 that T is an abelian group of exponent p. This allows us to consider T as a $\mathbf{Z}_p$-module.

Corollary 1. *Let K be a subfield of $\mathbf{Q}(\mu_p)^+$ and let Γ be a cyclic group of prime order $p > 2$. Then $T \neq 0$ if and only if there exists a unit ϵ of K congruent to a rational integer mod p such that ϵ is not the pth power of a unit of K.*

Proof. Note that here $(\mathcal{O}_K/p\mathcal{O}_K)^*$ is a direct product of $\mathrm{Im}((\mathbf{Z}/p\mathbf{Z})^*)$ and $[K : \mathbf{Q}] - 1$ cyclic groups each of order p. By the Dirichlet unit theorem $\mathcal{O}_K^*$ is a direct product of $\mathbf{Z}^*$ and $[K : \mathbf{Q}] - 1$ infinite cyclic groups. Since $\mathrm{Im}((\mathcal{O}_K^*)^p) \subset \mathrm{Im}((\mathbf{Z}/p\mathbf{Z})^*)$ in $(\mathcal{O}_K/p\mathcal{O}_K)^*$, we deduce from the isomorphism in Proposition 1 that $T = 0$ if and only if the canonical surjection from $\mathcal{O}_K^*/(\mathcal{O}_K^*)^p$ to $(\mathrm{Im}((\mathbf{Z}/p\mathbf{Z})^*)\,\mathrm{Im}(\mathcal{O}_K^*))/\mathrm{Im}((\mathbf{Z}/p\mathbf{Z})^*)$ is injective. Hence $T \neq 0$ if and only if the above surjective map is not injective. Now the kernel of this map is non-trivial if and only if there exists a unit ϵ of K congruent to a rational integer mod p such that ϵ is not the pth power of a unit of K. $\qquad\square$

Corollary 2. *If p is a regular prime, then for $K \subseteq \mathbf{Q}(\mu_p)^+$ and Γ a cyclic group of order p, $T = 0$ and Theorem 1 holds.*

Proof. If p is a regular prime, then $h_p = 1$, and for any character $\Psi \neq 1$ of $G = \mathrm{Gal}(K/\mathbf{Q})$, $L_p(1, \Psi) \not\equiv 0 \pmod{p}$ (cf. Proposition 5.23 in [23]). In view of Corollary 1, let ϵ be a unit of K which is congruent to a rational integer mod p. From Theorem 5.36 of [23] we conclude that ϵ is the pth power of a unit of $M = \mathbf{Q}(\mu_p)$. Since $[M : K]$ is prime to p we deduce that ϵ is in fact the pth power of a unit of K. $\qquad\square$

3 $\mathbf{Q}(\sqrt{p})$

Let p be a prime number such that $p \equiv 1 \pmod{4}$. In this section we set $K = \mathbf{Q}(\sqrt{p})$ and prove Theorem 1 for K in a classical way. We shall prove Theorem 1 in Section 4 for any subfield of $\mathbf{Q}(\mu_p)^+$ (including $\mathbf{Q}(\sqrt{p})$). We know that here $h_p = 1$, as pointed out in the third remark at the end of Section 1. Let Ψ denote the nontrivial character of G. Let $\epsilon = (t + u\sqrt{p})/2 > 1$ be the fundamental unit for K. Of course, t and u are rational integers, and $t \not\equiv 0 \pmod{p}$.

It is clear from Proposition 1 that the Swan subgroup T is either trivial or cyclic of order p. Now $(\mathcal{O}_K/p\mathcal{O}_K)^*$ is a direct product of $\mathrm{Im}((\mathbf{Z}/p\mathbf{Z})^*)$ and a cyclic group C of order p generated by the image of $1 + \sqrt{p}$. The action of G on C is clearly given by the non-trivial character of G, so $T(1) = 0$ and $T = T(\Psi)$.

Lemma 2. $T(\Psi) \neq 0$ *if and only if $u \equiv 0 \pmod{p}$.*

Proof. It follows from Proposition 1 and Dirichlet's unit theorem that $T(\Psi) \neq 0$ if and only if ϵ is congruent to a rational integer $\pmod{p}$, which in turn holds if and only if $u\sqrt{p} = 2\epsilon - t$ is congruent to a rational integer $\pmod{p}$ (since $p > 2$). The latter condition is equivalent to $u \equiv 0 \pmod{\sqrt{p}}$, i.e., $u \equiv 0 \pmod{p}$ (since u is a rational integer). $\qquad\square$

It has been conjectured by Ankeny-Artin-Chowla in [2] that $u \not\equiv 0$ (mod p). Since $h_p(\Psi) = 1$, Theorem 1 follows for $K = \mathbf{Q}(\sqrt{p})$ immediately from Lemma 2 and the following result.

Theorem (Ankeny-Artin-Chowla). *We have $u \equiv 0$ (mod p) if and only if $L_p(1, \Psi) \equiv 0$ (mod p).*

This follows at once from Theorem 5.37 of [23]. We also refer the reader to [2].

4   Subfields of $\mathbf{Q}(\mu_p)^+$

In this section we assume that K is a subfield of $\mathbf{Q}(\mu_p)^+$. We also introduce some notation which is mostly standard. We fix a primitive p-th root of unity ζ. We write $\kappa = N(1-\zeta)$, where N is the norm from $\mathbf{Q}(\mu_p)$ to K, and $d = [K : \mathbf{Q}]$. The prime ideal of $\mathcal{O}_K$ that lies above the rational prime p is generated by κ as $p\mathcal{O}_K = (\kappa^d)\mathcal{O}_K$. The field $K \otimes_\mathbf{Q} \mathbf{Q}_p$ is the completion of K under the embedding of K in $K \otimes_\mathbf{Q} \mathbf{Q}_p$. Any completion of K above p is isomorphic to this field. Let U denote the group of local units of the completion of K above p (from now on identified with the field $K \otimes_\mathbf{Q} \mathbf{Q}_p$) which are congruent to 1 modulo the maximal ideal (generated by κ). We denote this maximal ideal by $\mathcal{P}$. Thus $\mathcal{P}$ is identified with $\kappa\mathcal{O}_K \otimes_\mathbf{Z} \mathbf{Z}_p$ and U with $(1 + \kappa\mathcal{O}_K) \otimes_\mathbf{Z} \mathbf{Z}_p$. We denote by E the group of global units, i.e., $E = \mathcal{O}_K^*$, and denote the group of cyclotomic units of $\mathcal{O}_K$ by $\mathcal{E}$.

For $i \geq 1$, we define

$$U^{(i)} = \{x \in U : x \equiv 1 \pmod{\mathcal{P}^i}\},$$

$$E^{(i)} = E \cap U^{(i)}, \quad \mathcal{E}^{(i)} = \mathcal{E} \cap U^{(i)}.$$

Let $\overline{E^{(i)}}$ and $\overline{\mathcal{E}^{(i)}}$ denote the closures of $E^{(i)}$ and $\mathcal{E}^{(i)}$, respectively, in U. Notice that $U^{(1)} = U$, and $U^p \subseteq U^{(d)}$ (by the binomial theorem).

Now U is identified with $(1 + \kappa\mathcal{O}_K) \otimes_\mathbf{Z} \mathbf{Z}_p$. We know that Leopoldt's conjecture on the non vanishing of the p-adic regulator holds for all finite abelian extensions of $\mathbf{Q}$ (see for instance [23, Chapter 5]). In particular, for K we have the following result.

Theorem. *The $\mathbf{Z}_p$-rank of $\overline{E^{(1)}}$ is $d - 1$.*

Of course, by Dirichlet's unit theorem, the $\mathbf{Z}$-rank of $E^{(1)}$ is also $d - 1$. As a consequence of these ranks being equal, it follows that $\overline{E^{(i)}} = E^{(i)} \otimes \mathbf{Z}_p$ for all $i \geq 1$.

Proposition 2. *There is an isomorphism of $\mathbf{Z}_p[G]$-modules*

$$T \cong (U/U^{(d)})/(\overline{E^{(1)}}/\overline{E^{(d)}}).$$

Proof. Since p ramifies totally in $K/\mathbf{Q}$, there is an isomorphism of $\mathbf{Z}[G]$-modules

$$(1 + \kappa \mathcal{O}_K)/(1 + p\mathcal{O}_K) \overset{f}{\cong} (\mathcal{O}_K/p\mathcal{O}_K)^*/\operatorname{Im}((\mathbf{Z}/p\mathbf{Z})^*),$$

where f is the canonical projection (note that f is injective and both modules have the same p-rank of $d - 1$).

Note that $(1 + \kappa \mathcal{O}_K)/(1 + p\mathcal{O}_K)$ is also a $\mathbf{Z}_p$-module (being an abelian group of exponent p). So f is an isomorphism of $\mathbf{Z}_p[G]$-modules. We have identified U with $(1 + \kappa \mathcal{O}_K) \otimes \mathbf{Z}_p$. Tensoring with $\mathbf{Z}_p$, we find that under this identification

$$(1 + \kappa \mathcal{O}_K)/(1 + p\mathcal{O}_K) \overset{\otimes \mathbf{Z}_p}{\cong} (1 + \mathcal{P})/(1 + \mathcal{P}^d) = U/(U^{(d)}).$$

We also find that $f^{-1}(\operatorname{Im}(\mathcal{O}_K^*))$ is the image of $(E^{(1)} \otimes \mathbf{Z}_p)$ in $U/(U^{(d)})$ under the above identification (recall that $\mathcal{O}_K^* = E$, $U^{(1)} = U$ and $E \cap U^{(1)} = E^{(1)}$). Since $\overline{E^{(1)}}$, the closure of $E^{(1)}$ in U, is $E^{(1)} \otimes \mathbf{Z}_p$ as noted just before stating this proposition, we see that the above image of $(E^{(1)} \otimes \mathbf{Z}_p)$ is the image of $\overline{E^{(1)}}$ in $U/U^{(d)} = (\overline{E^{(1)}})/(\overline{E^{(1)}} \cap U^{(d)}) = (\overline{E^{(1)}})/(\overline{E^{(d)}})$, since $U^{(d)}$ is closed in U. Now the result follows from Proposition 1. $\qquad\square$

Let Ψ be a character of $G = \operatorname{Gal}(K/\mathbf{Q})$. Then we can also view Ψ as a character of $\operatorname{Gal}(\mathbf{Q}(\mu_p)^+/\mathbf{Q})$. We do this throughout the article.

Lemma 3. *For each character Ψ of G, we have $T(\Psi) \cong 0$ or $\mathbf{Z}_p/p\mathbf{Z}_p$.*

Proof. Since T is an abelian group of exponent p, this follows at once from Proposition 2 upon noting that

(i) $U(\Psi)$ is $\mathbf{Z}_p$-cyclic for $\Psi \neq 1$ (cf. Lemma 13.36 of [23]; this also holds for K since the units of K form a submodule of the units of the full cyclotomic field); and

(ii) $U(1) = U^{(d)}(1) = 1 + p\mathbf{Z}_p$, so that $T(1) = 0$. $\qquad\square$

Lemma 4. *For each character $\Psi \neq 1$ of G, $T(\Psi) = 0$ if and only if $U(\Psi) = \overline{E^{(1)}}(\Psi)$.*

Proof. From Proposition 2 it follows that, for any character Ψ of G, $T(\Psi) \cong (U(\Psi)/U^{(d)}(\Psi))/(\overline{E^{(1)}}(\Psi)/\overline{E^{(d)}}(\Psi))$ as $\mathbf{Z}_p$-modules. It is clear that, if $U(\Psi) = \overline{E^{(1)}}(\Psi)$, then $T(\Psi) = 0$.

Now suppose that $U(\Psi) \neq \overline{E^{(1)}}(\Psi)$, and $\Psi \neq 1$. Since $U(\Psi)$ is $\mathbf{Z}_p$-cyclic, we have $\overline{E^{(1)}}(\Psi) \subset (U(\Psi))^p \subset U^{(d)}(\Psi) \neq U(\Psi)$, and $U(\Psi)/U^{(d)}(\Psi) \cong \mathbf{Z}_p/p\mathbf{Z}_p$. Therefore, $\overline{E^{(1)}}(\Psi) = \overline{E^{(d)}}(\Psi)$, and $T(\Psi) \cong \mathbf{Z}_p/p\mathbf{Z}_p \neq 0$. $\qquad\square$

We compute the index $[U(\Psi) : \overline{E^1}(\Psi)]$, following the strategy of Taylor in [21]. As a consequence of their celebrated proof of the "main conjecture" of the Iwasawa theory of cyclotomic fields, Mazur-Wiles proved a conjecture of Gras [10]:

Theorem (Mazur-Wiles). *For each character Ψ of G, we have $[\overline{E^{(1)}}(\Psi) : \overline{\mathcal{E}^{(1)}}(\Psi)] = h_p(\Psi)$.*

Another proof of this result has been given by Kolyvagin [9]. As for the index of the closure of the cyclotomic units in the group of local units, this is proved in [21] (or Theorem 8.25 of [23]):

Proposition 3. *For each character $\Psi \neq 1$ of G, $[U(\Psi) : \overline{\mathcal{E}^{(1)}}(\Psi)] \sim L_p(1, \Psi)$, where $a \sim b$ if and only if $a = ub$ for some p-adic unit u.*

Remark. Both of these results were first proved for $\mathbf{Q}(\mu_p)^+$. They follow for any subfield K of $\mathbf{Q}(\mu_p)^+$, given that $[\mathbf{Q}(\mu_p)^+ : K]$ is relatively prime to p; we refer the reader to the proofs of Theorem 4.11, Theorem 4.14 and Corollary 4.15 in Greither [7]. We also refer the reader to [6] for a related result of Greenberg.

Combining these results we find that for $\Psi \neq 1$, $[U(\Psi) : \overline{E^1}(\Psi)] \sim L_p(1, \Psi)h_p(\Psi)^{-1}$. Thus, $T(\Psi) = 0$, if and only if $U(\Psi) = \overline{E^1}(\Psi)$, which in turn is equivalent to $[U(\Psi) : \overline{E^1}(\Psi)] \sim 1$, or $L_p(1, \Psi)h_p(\Psi)^{-1} \not\equiv 0 \pmod{p}$.

This completes the proof of Theorem 1.

Remark. The Swan subgroup for subfields of $\mathbf{Q}(\mu_p)$ that are not totally real can be described as follows. Let K be a subfield of $\mathbf{Q}(\mu_p)$ that is not totally real and let Γ be a cyclic group of prime order $p > 2$. We set $K^+ = K \cap \mathbf{Q}(\mu_p)^+$. Then $[K : K^+] = 2$ and K^+ is the maximal real subfield of K. We can prove (following the proof of Proposition 1) that the Swan subgroup T has a direct sum decomposition: $T = T^+ \oplus T^-$, where, as usual T^+ is the invariant subgroup of T under the action of complex conjugation on $\mathcal{O}_K$ (and extended to $\mathcal{O}_K[\Gamma]$ by Γ-linearity). Moreover, $T^+ \cong T(\mathcal{O}_{K^+}[\Gamma])$ (which is considered in this article), and T^- is an elementary abelian p-group of rank $(p-3)/2$ if $K = M = \mathbf{Q}(\mu_p)$, and of rank $[K^+ : \mathbf{Q}]$ otherwise. Under the Kummer-Vandiver hypothesis, $T(\mathcal{O}_M[\Gamma])$ has been computed in Replogle's thesis [13] (cf. [15]).

References

[1] A. Agboola, *Torsion points on elliptic curves and Galois module structure*, Invent. Math. **123** (1996), 105–122.

[2]   N. Ankeny, E. Artin, and S. Chowla, *The class number of real quadratic number fields*, Ann. of Math. **56** (1952), 479–493.

[3]   J. Brinkhuis, *On the Galois module structure over CM-fields*, Manuscripta Math. **75** (1992), 333–347.

[4]   Ph. Cassou-Nogues and M. J. Taylor, *Espaces homogenes principaux, unites elliptiques et fonctions L*, Ann. Inst. Fourier **44** (1994), 631–661.

[5]   C. W. Curtis and I. Reiner, *Methods of representation theory*, Wiley-Interscience, New York, 1987.

[6]   R. Greenberg, *On p-adic L-functions and cyclotomic fields II*, Nagoya Math. J. **67** (1977), 139–158.

[7]   C. Greither, *Class groups of abelian fields and the main conjecture*, Ann. Inst. Fourier **42** (1992), 449–499.

[8]   C. Greither, D. R. Replogle, K. Rubin, and A. Srivastav, *Swan modules and Hilbert-Speiser number fields*, J. Number Theory **79** (1999), 164–173.

[9]   V. A. Kolyvagin, *Euler systems*, The Grothendieck Festschrift, Vol.II, Birkhäuser, Boston, 1990, Progr. Math., vol. 87, pp. 435–483.

[10] B. Mazur and A. Wiles, *Class fields of abelian extensions of* **Q**, Invent. Math. **76** (1984), 179–330.

[11] L. R. McCulloh, *Galois module structure of elementary abelian extensions*, J. Algebra **82** (1983), 102–134.

[12] W. Narkiewicz, *Elementary and analytic theory of algebraic numbers*, Springer-Verlag, Berlin, 1989.

[13] D. R. Replogle, *Swan classes and realisable classes for integral group rings over groups of prime order*, Ph.D. thesis, State University of New York at Albany, 1997.

[14] _____, *Swan modules and realisable classes for Kummer extensions of prime degree*, J. Algebra **212** (1999), 482–494.

[15] _____, *Cyclotomic Swan subgroups and irregular indices*, Rocky Mountain J. Math. **31** (2001), 611–618.

[16] A. Srivastav and M. J. Taylor, *Elliptic curves with complex multiplication and Galois module structure*, Invent. Math. **99** (1990), 165–184.

[17] A. Srivastav and S. Venkataraman, *Unramified quadratic extensions of real quadratic fields, normal integral bases and 2-adic L-functions*, J. Number Theory **67** (1997), 139–145.

[18] R.G. Swan, *Periodic resolutions for finite groups*, Ann. of Math. . **72** (1960), 267–291.

[19] M. J. Taylor, *On Fröhlich's conjecture for ring of integers of finite extensions*, Invent. Math. **63** (1981), 41–79.

[20] _____ , *Mordell-Weil groups and the Galois module structure of rings of integers*, Illinois J. Math. **32** (1988), 428–452.

[21] _____ , *The Galois module structure of certain arithmetic principal homogeneous spaces*, J. Algebra **153** (1992), 203–214.

[22] S. V. Ullom, *Nontrivial lower bounds for class groups of integral group rings*, Illinois J. Math. **20** (1976), 361–371.

[23] L. C. Washington, *Cyclotomic fields*, Springer-Verlag, New York-Berlin, 1982, Graduate Texts in Mathematics, Vol. 83.

Discriminants and Divisibility for Chebyshev-Like Polynomials

Kenneth B. Stolarsky[1]

1 Introduction

There's something special about

$$D(t,x) = (1-t)^8 - xt(1-t)^2(1+14t+34t^2+14t^3+t^4) + x^2t^4.$$

Its discriminant is simpler than expected:

$$\Delta_t D(t,x) = 3^6 x^{10}(x+64)^4(x+16)^2(x-64)^4.$$

Another special polynomial is

$$N(t,x) = (1-t)^7 - xt^2(1-t)(t+3)(1+3t)$$

for which

$$\Delta_t N(t,x) = 2^{18}3^6 x^7(x-64)^2(x+16).$$

The repeated polynomial factors for N and D when $x = -16$ and $x = 64$ turn out to be identical, namely $(1+t)^2$ for $x = -16$ and $(1+6t+t^2)^2$ for $x = 64$. This suggests forming their ratio, namely

$$R(x,t) = \frac{N(t,x)}{D(t,x)} = 1 + \sum_{m=1}^{\infty} P_m(x)t^m.$$

Computer algebra now suggests that all the zeros of the $P_m(x)$ are real and in fact lie in the interval $[-64,0]$. Observe that -64 is the one zero of $\Delta_t D(t,x)$ that did not occur in $\Delta_t N(t,x)$. Moreover, it seems that

$$P_m(x) \mid P_{3m+1}(x)$$

for all positive integers m.

There are more mysteries here than the author can explain. Our present goal is to convince the reader that from one point of view the $P_m(x)$ are

[1]Partially supported by an NSF SCREMS grant DMS 9871951.

Chebyshev-like polynomials in disguise, and that from another point of view their properties arise from a certain "discriminant transform." As to the first point of view, we'll note for now that the above $R(x,t)$ is in fact the third element in a (conjectural) infinite sequence of rational functions, the first two of which are

$$R_1(x,t) = \frac{1-t}{(1-t)^2 - xt} \text{ and } R_2(x,t) = \frac{(1-t)^3}{(1-t)^4 - xt(t+1)^2}.$$

Here R_1 resembles the generating function for Chebyshev T-polynomials.

In §2 we introduce the general concept of a discriminant transform, and in §3 we examine some specific such transforms that lead to our rational functions $R_n(x,t)$. The conjecture of §4 states how the x-polynomials produced by any $R_m(x,t)$ can be written rather simply in terms of the "Chebyshev-like" polynomials produced by $R_1(x,t)$. In §5 we show how to produce some further two variable polynomials with nice discriminants. We end in §6 with a brief remark on the genesis of this investigation.

Remark. By $\triangle_x P$ we mean the resultant of $P(x)$ and its derivative with respect to x. For example,

$$\triangle_x(ax^2 + bx + c) = a(-b^2 + 4ac).$$

Other normalizations are common.

2 Discriminant Transforms

A considerable part of mathematics is concerned with linear transformations of the form

$$U: f(x) \to g(q) = \int K(x,q)\, f(x)\, dx$$

and their discrete analogues. We shall restrict both $f(x)$ and the kernel $K(x,q)$ to be polynomials over $\mathbb{Q}$, and introduce

$$\Delta: f(x) \to g(q) = \Delta_x(K(x,q)\, f(x))$$

where Δ_x represents the discriminant with respect to x. This is of course a non-linear transformation that lacks the plethora of useful properties that U has, but we believe it has a role nonetheless. The first proposition of this section, which explains why this transform creates families of polynomials with all zeros real, is relatively transparent. The problem is to find interesting cases in which the hypotheses are valid. For this the second proposition is useful.

Proposition 1. *Let $a < b$ be real. Say $K(x,q)$ is a polynomial in x and q such that*

(i) *$K(x,q)$ is never zero for $|x| = 1$ unless*

$$a \leq q \leq b.$$

(ii) *$K(x,q)$ has no multiple zeros unless $a \leq q \leq b$.*

Say $f_m(x)$ is a sequence of polynomials with no multiple zeros, and the zeros of $f_m(x)$ all lie on $|x| = 1$. Then

$$g_m(q) = \Delta f_m(x) := \Delta_x(K(x,q) f_m(x))$$

is a sequence of polynomials in q whose roots are all in $[a,b]$.

Proof. If $g_m(q) = 0$ then $K f_m$ has a multiple root. If this is because K has a multiple root, we are done by (ii). Otherwise, K and f_m must have a root in common. Since this root must be on $|x| = 1$ we have $a \leq q \leq b$ by (i). This completes the proof.

We now show that non-trivial K and f_m exist. For f_m we may of course choose

$$f_m = f_m(x) = (x^m - 1)/(x - 1).$$

To get a suitable K we simply remove the middle term from the expansion of $(1 + x)^{2n}$ and replace it by qx^n. $\qquad\square$

Proposition 2. *The polynomial*

$$K = K_n(x,q) = (1 + x)^{2n} - \binom{2n}{n} x^n + qx^n$$

cannot have roots on $|x| = 1$ unless

$$\binom{2n}{n} - 2^{2n} \leq q \leq \binom{2n}{n}.$$

Proof. Assume $K = 0$ with $|x| = 1$. Set

$$\lambda = q - \binom{2n}{n}.$$

Then $K = 0$ becomes

$$(1 + x)^{2n} + \lambda x^n = 0$$

and with $y = x^{1/2}$ this becomes

$$y^2 + by + 1 = 0$$

for some complex w with $b = -w$ and $w^{2n} = -\lambda$. Now for any complex b, if the above quadratic equation has a root of modulus 1 then b must be real. This tells us that w is real and hence that λ is negative, i.e., the right side of the inequality holds. Next, rewrite $K = 0$ in the form

$$\left(\frac{1+x}{2}\right)^{2n} = \left(\frac{-\lambda}{2^{2n}}\right) x^n = c x^n$$

with $c \geq 0$. If $c > 1$ and $|x| = 1$, the triangle inequality provides an immediate contradiction. Hence $-\lambda \leq 2^{2n}$ and the left side of the inequality holds. This completes the proof. $\square$

There is a bit more to be said about $K_n(x, q)$. When $q = \binom{2n}{n}$ it has a high-order zero, namely a zero of order $2n$ at $x = -1$. If $q = \binom{2n}{n} - 2^{2n}$, it does have a multiple zero at $x = 1$, but only of multiplicity 2. (In fact, consider $p(x) = (1 + x)^{2n} - 2^{2n} x^n$. Then $p(1) = 0$ and $p'(x) = 2n(1 + x)^{2n-1} - n2^{2n} x^{n-1}$ so $p'(1) = 0$. However,

$$p''(x) = 2n(2n - 1)(1 + x)^{2n-2} - n(n - 1)2^{2n} x^{n-2}$$

so $p''(1) = 2^{2n-1} n \neq 0$.) This explains why in the formula

$$\Delta_x K_n(x, q) = C_0(n) \left(q - \binom{2n}{n}\right)^{2n-1} \left(q + 2^{2n} - \binom{2n}{n}\right),$$

where $C_0(n)$ is some scalar from $\mathbb{Q}$, only the $q - \binom{2n}{n}$ factor has high multiplicity.

3 Specific Cases

We now take
$$f_m(x) = (x^{2m+1} - 1)/(x - 1)$$
and use the above kernel with $n = 1$:

$$K_1(x, q) = 1 + qx + x^2.$$

Computer algebra suggests that

$$\Delta_x(K_1 f_m) = C_0(m)(q - 2)(q + 2) E_m(q)^4,$$

where $C_0(m)$ is a scalar from $\mathbb{Q}$ and $E_m(q)$ is a monic polynomial. The factors $(q - 2)$ and $(q + 2)$ arise solely from the K_1 polynomial, so $E_m(q)$ represents all of the "interaction" between K_1 and $f_m(x)$.

Since the variable x has been eliminated, we are free to use it again. Make the change of variable $q = x + 2$ in $E_m(q)$. The resulting polynomials $C_m(x) = E_m(x + 2)$ have all their zeros in $[-4, 0]$ and computer algebra supports the following conjecture.

Conjecture 1. *The polynomials $C_m(x)$ satisfy*

$$R_1(x, t) = \frac{1 - t}{(1 - t)^2 - xt} = 1 + \sum_{m=1}^{\infty} C_m(x)t^m.$$

We note that

$$\Delta_t((1 - t)^2 - xt) = -x(4 + x).$$

Thus the denominator of $R_1(x, t)$ has no multiple roots as a polynomial in t except when x is at one of the endpoints of the critical interval $[-4, 0]$. Of course, the $C_m(x)$ are closely related to Chebyshev T-polynomials. (We add that $C_m(1) = F_{2m+1}$, the $(2m + 1)$st Fibonacci number.) Recall the generating function definition of $T_n = T_n(x)$:

$$\frac{1 - xt}{1 + t^2 - 2tx} = 1 + \sum_{n=1}^{\infty} T_n(x)t^n.$$

Here $T_n(x) = \cos n\theta$ for $x = \cos\theta$, so the zeros of T_n lie in $[-1, 1]$. Now replace x by $(x + 2)/2$, multiply both sides by 2, and subtract 1 from each side. The result is

$$\frac{1 - t}{(1 - t)^2 - xt} = \frac{1}{1 + t}\left(1 + \sum_{n=1}^{\infty} 2T_n\left(\frac{x + 2}{2}\right)t^n\right).$$

Upon expanding $(1 + t)^{-1}$ into a geometric series we discover that $C_m(x)$ is twice an alternating sum of Chebyshev polynomials plus or minus 1 depending on the parity of m.

Proposition 3. $C_n(x) \mid C_{3n+1}(x)$.

Proof. Write $2\cos m\theta = z^m + z^{-m}$. Then

$$S_n(z) := 2\cos n\theta - 2\cos(n-1)\theta + 2\cos(n-2)\theta - \cdots \mp 2\cos\theta \pm 1 = \frac{z^{2n+1} + 1}{z^n(z + 1)},$$

so

$$\frac{S_{3n+1}(z)}{S_n(z)} = \frac{z^{6n+3} + 1}{z^{2n+1}(z^{2n+1} + 1)} = z^{2n+1} + z^{-(2n+1)} - 1$$

$$= 2\cos(2n + 1)\theta - 1.$$

Now for $-4 \leq x \leq 0$ we have $-1 \leq (x+2)/2 \leq 1$ and by the appropriate choice of θ (to pass back to the polynomial forms from the trigonometric forms) we conclude that

$$C_{3n+1}(x) = C_n(x) \left(2T_{2n+1} \left(\frac{x+2}{2} \right) - 1 \right).$$

We now turn to the case $n = 2$. Here

$$K_2(x, q) = 1 + 4x + qx^2 + 4x^3 + x^4.$$

Computer algebra suggests that

$$\Delta_x(K_2 f_m) = C_0(m)(q-6)^3(q+10) G_m(q)^4$$

where $C_0(m)$ is a scalar from $\mathbb{Q}$ and $G_m(q)$ is a monic polynomial. The factors $(q-6)^3$ and $(q+10)$ arise solely from the K_2 polynomial, so $G_m(q)$ represents all of the "interaction" between K_2 and $f_m(x)$.

Since the variable x has been eliminated we are free to use it again. Let $q = x+6$. The resulting polynomials $H_m(x) = G_m(x+6)$ have all their zeros in $[-16, 0]$, and computer algebra suggests the following conjecture.   □

Conjecture 2. *The polynomials $H_m(x)$ satisfy*

$$R_2(x, t) = \frac{(1-t)^3}{(1-t)^4 - xt(t+1)^2} = 1 + \sum_{m=0}^{\infty} H_m(x)t^m.$$

It is a consequence of this conjecture that $H_m(x)$ has non-negative integer coefficients with lead and constant coefficients equal to 1. In particular, this means that all of its roots are units. In fact, the above statement is true for the x-polynomials generated by the expansion of any

$$L(x, t) = \frac{1}{1-t} \left(1 - xt \left(\frac{1 + tp_2(t)}{(1 - tp_1(t))^n} \right) \right)^{-1},$$

where p_1, p_2 are polynomials in t with non-negative integer coefficients and n is a non-negative integer. To see this one expands the factors into geometric series, and then similarly expands the denominators of the coefficients of the powers of x. Tables of $H_m(x)$ indicate that $H_m(x) \mid H_{3m+1}(x)$.

We now examine the denominator of $R_2(x, t)$, namely

$$D(t, x) = (1-t)^4 - xt(t+1)^2.$$

Here

$$\Delta_t(D(t, x)) = -64x^3(16 + x)^2$$

so $D(t, x)$ has no multiple roots as a polynomial in t except when x is at one of the endpoints of the critical interval $[-16, 0]$. We observe that

$$D(t, 0) = (t - 1)^4 \text{ and } D(t, -16) = (1 + 6t + t^2)^2.$$

We also note that the conjecture implies that $H_m(x)$ satisfies the irreducible 4th order recurrence

$$H_{n+4} = (4 + x)H_{n+3} + (2x - 6)H_{n+2} + (4 + x)H_{n+1} - H_n$$

and hence the $H_m(x)$ cannot constitute a family of orthogonal polynomials, in spite of the nature of their (conjectural) zero distribution.

We now sketch a similar investigation for $n = 3$ where the kernel is

$$K_3(x, q) = 1 + 6x + 15x^2 + qx^3 + 15x^4 + 6x^5 + x^6.$$

Computer algebra suggests

$$A_x(K_3 f_m) = C_0(m)(q - 20)^5(q + 44)A_m(q)^4.$$

We set $J_m(x) = A_m(q+20)$, to obtain a sequence of polynomials $J_m(x)$ that have all their zeros in $[-64, 0]$. The conjecture here, of course, is that the $J_m(x)$ are generated by $R_3 = R_3(x, t)$ where R_3 is the $R(x, t)$ mentioned in §1.

4 A Unifying Conjecture

Although R_1, R_2 and $R = R_3$ look rather different from one another, the polynomials occurring in their expansions have similar properties. Here's a conjectural explanation. Let

$$P_{m,n}(x) = \Delta_x(K_n f_m)$$

where K_n and f_m are as in §3.

Conjecture 3. *The polynomial $P_{m,n}$ has the factorization*

$$P_{m,n}(x) = \prod_{k=1}^{m} (x + x_i^n)$$

where

$$P_{m,1}(x) = \Delta_x(K_1 f_m) = \prod_{k=1}^{m} (x + x_i).$$

Thus, replacing the kernel K_1 by K_n has the effect of raising the negatives of the zeros of the polynomial coefficients (of t in the $R_1(x, t)$ expansion) to the nth power.

This conjecture would of course explain why the polynomials generated by different K_n are so similar, and share similar divisibility properties.

5   Further Nice Factorizations

Since the $P_{3m+1,n}$ seem special, it is natural to consider not only the various $R(x, t)$ generating functions, but also the functions

$$R^*(x, t) = P_1(x)t^3 + P_4(x)t^6 + P_7(x)t^9 + \cdots.$$

We shall see that they produce further examples of polynomials with unusually nice discriminant factorizations. Introduce the operator θ given by

$$\theta g(t) = \big(g(t) + g(wt) + g\left(w^2 t\right)\big)/3$$

where w is a primitive cube root of unity. Then

$$R^*(x, t) = \theta\left(t^2 R(x, t)\right).$$

By direct calculation (preferably with computer algebra) we find that

$$R_1^*(x, t) = \frac{t^3(1 - t^3)(1 + x)}{((1 - t)^2 - xt)\left(1 + 2t + 3t^2 + 2t^3 + t^4 + xt\left(1 + 4t + t^2\right) + x^2 t^2\right)}$$

and

$$R_2^*(x, t) = \frac{t^3(1 - t^3)(1 + x)((1 - t^3)^2 + 12xt^3)}{((1 - t)^4 - tx(1 + t)^2)\, E(t, x)}$$

where

$$\begin{aligned}
E(t, x) = {}& 1 + 4t + 10t^2 + 16t^3 + 19t^4 + 16t^5 + 10t^6 + 4t^3 + t^8 \\
& + xt\left(1 + 10t - 4t^2 - 32t^3 - 4t^4 + 10t^5 + t^6\right) \\
& + \left(x^2 t^2\left(1 - 2t + 3t^2 - 2t^3 + t^4\right)\right).
\end{aligned}$$

The corresponding expression for $R_3^*(x, t)$ is rather lengthy, but we note that it too has a $t^3(1 - t^3)(1 + x)$ factor in the numerator. Thus we have proved the cases $1 \le n \le 3$ of the following conjecture.

Conjecture 4. *For any m, the $P_{3m+1,n}(x)$ polynomials are always divisible by $(x + 1)$.*

Also, we now have two more variable polynomials whose discriminants are nice. For example,

$$\Delta_t(E(t, x)) = 2^{12} 3^4 (x - 9)^2 x^6 (x + 1)^4 (x + 9)^4 (x + 16)^4 (3x - 1)^2.$$

Next, what would happen if we chose our first rational function to be

$$S_1(t, x) = \frac{(1 - t)^2}{(1 - t)^3 - xt(1 + t)} = 1 + \sum_{m=1}^{\infty} W_m(x)t^m,$$

wrote $W_m = \prod (x + x_i)$, $W_m^* = \prod (x + x_i^2)$, and in the spirit of the conjecture in §4 calculated

$$S_2(t, x) = 1 + \sum_{m=1}^{\infty} W_m^*(x)t^m \ ?$$

Our guess (after much computer algebra) is that

$$S_2(t, x) = \frac{(1 - t)^5 - xt^2(1 + 3t)(7 + 8t + 12t^2)}{(1 - t)^6 - xt(1 + 23t + 60t^2 + 23t^3 + t^4) - x^2t^3}$$

For the denominator of $S_1(t, x)$ we have

$$\Delta_t = x^2(x^2 - 108)$$

and for the denominator of $S_2(t, x)$ we have

$$\Delta_t = -16(-27 + x)^4 x^5 (x + 8)^2 (108 + x)^3,$$

again a nice factorization. And indeed, the most negative zeros of the W_m^* polynomials seem to be approaching -108. However, the t-discriminant of the numerator of S_2 does not seem at all nice, and aside from an x^4 factor has no factors in common with the t-discriminant of the denominator.

A further example of a completely factoring discriminant of a "nice" polynomial resembling the denominator of $S_2(t, x)$ is

$$\Delta_t \left((1 - t)^6 - xt(1 + 10t + 2t^2 + 10t^3 + t^4) - x^2 t^3\right)$$
$$= -(x - 8)^6 x^5 (24 + x)(243 + 4x)^2.$$

We end this section with a lengthier example more intimately connected with $S_2(t, x)$:

$$\Delta_t \left((1 + t^2)^6 - 2t^2 x(29 - 748t^2 + 1902t^4 - 748t^6 + 29t^8)\right.$$
$$+ t^2 x^2 (1 + 397t^2 + 2584t^4 + 397t^6 + t^8)$$
$$+ 2t^4 x^3 (1 - 60t^2 + t^4) + x^4 t^6)$$
$$= 2^{44}(x - 108)^6 (-8 + x)^{12} x^{10} (27 + x)^8 (1 + 3x)^8 (8 - 58x + x^2)^4.$$

The quadratic factor at the end is easily split into real linear factors at the expense of introducing quadratic surds. We mention (but omit the proof) that if the power series expansion of the reciprocal of the denominator of $S_2(t, x)$ is used to define a sequence $V_n(x)$ of polynomials in x, then V_{4m+3} is always divisible by $8 - 58x + x^2$.

6 Remarks

In [2] the authors investigate special sequences of polynomials, which we shall denote here by $\alpha_n(x)$ and $\beta_n(x)$, such that their zero distributions, and also the zero distributions of all $\alpha_n(x) + \beta_n(x)$, can be determined in detail. The actual polynomials involved are related to Chebyshev polynomials. This is a sort of ABC problem for polynomials rather than integers. Further general investigations of this nature can be found in the references in [2] and also in the thesis of S.-H. Kim [6]. After speaking on these matters at the Millennial Conference on Number Theory the present author wondered if [2] could be strengthened by not only determining all the zero distributions, but also all the discriminants of all the polynomials involved, and certain generalizations thereof. (Certain obvious repeated factors are to be removed to avoid trivialities.) A preliminary attempt at this led to the various "discoveries" and conjectures of this paper.

For a comprehensive treatment of discriminants see [3]. A great deal of information about them is also present in [1] and [4]. For connections with the theory of transfinite diameters see [5].

References

[1] D. Cox, J. Little, , and D. O'Shea, *Ideals, Varieties, and Algorithms*, 2nd ed., Springer-Verlag, New York, 1996.

[2] K. Dilcher and K. Stolarsky, *Sequences of polynomials whose zeros lie on fixed lemniscates*, Periodica Math. Hung. **25** (1992), 179–190.

[3] I. M. Gelfand, M. M. Kapranov, and A. V. Zelevinsky, *Discriminants, Resultants and Multidimensional Determinants*, Birkhäuser, Boston, 1994.

[4] ———, *Using Algebraic Geometry*, Springer-Verlag, New York, 1998.

[5] E. Hille, *Analytic Function Theory*, vol. 2, Ginn and Company, Boston, 1962.

[6] S.-H. Kim, *Sums of Polynomials, Minmax Problems and Number Theory*, Ph.D. thesis, University of Illinois at Urbana-Champaign, 2000.

Remarks on Adelic Geometry of Numbers

Jeffrey Lin Thunder

A major result in the geometry of numbers is Minkowski's second convex bodies theorem. This result gives bounds for the product of the successive minima of a convex body with respect to a lattice in terms of the volume of the body and the determinant (covolume) of the lattice. Here by convex body we mean a non-empty compact and convex subset of $\mathbb{R}^n$ such that $r\mathbf{x}$ is an interior point whenever $\mathbf{x}$ is in the subset and $|r| < 1$, and by lattice we mean a discrete $\mathbb{Z}$-module spanning $\mathbb{R}^n$. If C is a convex body and Λ is a lattice, the successive minima $\lambda_i(C, \Lambda)$ for $i = 1, \dots, n$ are defined by

$$\lambda_i(C, \Lambda) = \inf_{r>0} \{r : rC \cap \Lambda \text{ contains } i \text{ linearly independent vectors}\}.$$

Since C is closed and Λ is discrete, these infima are attained (i.e., they are minima). Minkowski's theorem states that

$$\frac{2^n \det(\Lambda)}{n! \mathrm{Vol}(C)} \leq \lambda_1(C, \Lambda) \dots \lambda_n(C, \Lambda) \leq \frac{2^n \det(\Lambda)}{\mathrm{Vol}(C)}. \tag{1}$$

Here $\mathrm{Vol}(C)$ denotes the Jordan volume of C (which can be shown to exist) and $\det(\Lambda)$ denotes the determinant of Λ. Both the upper and the lower bounds are sharp. The lower bound is rather simple to prove, whereas there are no truly simple proofs of the upper bound.

This result can be restated using the language of adeles of $\mathbb{Q}$. Moreover, one can give such an adelic statement for arbitrary number fields. This was done by MacFeat [2] and later independently by Bombieri and Vaaler [1]. It is the purpose of this note to show that, contrary to what one might initially think (see, for example, the first remark following the statement of Theorem 4 of [1]), Minkowski's original result implies these generalizations. In fact, we'll show that (1) implies something *stronger* than the results in [1] and [2] when $K \neq \mathbb{Q}$ (see Theorem 1 below).

In [1] the authors apply their number field version of Minkowski's theorem and a generalization of Vaaler's cube-slicing inequality to local fields to prove a formulation of Siegel's lemma using heights. Here we'll show how (1) along with Vaaler's original inequality and an earlier result due to W. M. Schmidt suffice to prove a stronger formulation of Siegel's lemma (Theorem 2 below).

The author gratefully acknowledges many helpful conversations with D. Roy concerning the writing of this paper.

We now give some definitions. For K a number field we let $M(K)$ denote the set of places of K. For each $v \in M(K)$ we let $|\cdot|_v$ be the v-adic absolute value normalized so that it extends the usual w-adic absolute value on $\mathbb{Q}$, where w is the place of $\mathbb{Q}$ lying below v. For each $v \in M(K)$ we let K_v denote the completion of K at v. Let $\mathfrak{O}_K$ denote the ring of integers of K and let $\mathfrak{O}_v \subset K_v$ denote the ring of $x \in K_v$ with $|x|_v \leq 1$ when $v \nmid \infty$. We let α_v be the Haar measure on K_v normalized so that $\alpha_v(\mathfrak{O}_v) = 1$ whenever $v \nmid \infty$, α_v is the Lebesgue measure when v is real, and α_v is twice the usual measure on $\mathbb{C}$ when v is complex. We let α be the Haar measure on the adele ring $K_{\mathbb{A}}$ given by

$$\alpha = |D(K)|^{-1/2} \prod_{v \in M(K)} \alpha_v,$$

where $D(K)$ denotes the discriminant of K. We abuse notation somewhat and also denote by α the measure on $(K_{\mathbb{A}})^n$ given by the n-fold product measure.

Fix a number field K and let $d = [K:\mathbb{Q}]$. Write $d = r_1 + 2r_2$, where r_1 denotes the number of real embeddings and r_2 denotes the number of pairs of complex conjugate embeddings of K into $\mathbb{C}$. Order the embeddings $a \mapsto a^{(j)}$ of K into $\mathbb{C}$ so that the first r_1 are real and $a^{(j+r_2)}$ is the complex conjugate of $a^{(j)}$ for $r_1 < j \leq r_1 + r_2$. Letting $(K^n)_\infty = \prod_{v|\infty}(K_v)^n$, we get the embedding $K^n \to (K^n)_\infty$ defined by $\mathbf{a} \mapsto (\mathbf{a}^{(1)}, \ldots, \mathbf{a}^{(r_1+r_2)})$. For $\mathbf{A} = (\mathbf{a}_1, \ldots, \mathbf{a}_{r_1+r_2}) \in (K^n)_\infty$ we define

$$\rho(\mathbf{A}) = \big(\mathbf{a}_1, \ldots, \mathbf{a}_{r_1}, \mathfrak{R}(\mathbf{a}_{r_1+1}), \mathfrak{I}(\mathbf{a}_{r_1+1}), \ldots, \mathfrak{R}(\mathbf{a}_{r_1+r_2}), \mathfrak{I}(\mathbf{a}_{r_1+r_2})\big) \in \mathbb{R}^{nd},$$

where $\mathfrak{R}$ and $\mathfrak{I}$ denote the real and imaginary part, respectively. We will also view ρ as a map on K^n via the embedding defined above.

Our generalization of (1) to number fields will use only (1) and the following elementary lemma.

Lemma. *For each $v \nmid \infty$ let $S_v \subset (K_v)^n$ be an $\mathfrak{O}_v$-module of rank n spanning $(K_v)^n$, with $S_v = (\mathfrak{O}_v)^n$ for all but finitely many v. Then*

$$\mathfrak{M} = \bigcap_{v \nmid \infty} K^n \cap S_v$$

is an $\mathfrak{O}_K$-module spanning K^n and $\rho(\mathfrak{M}) \subset \mathbb{R}^{nd}$ is a lattice with

$$\det\big(\rho(\mathfrak{M})\big) \times \prod_{v \nmid \infty} \alpha_v(S_v) = \left(2^{-r_2}\sqrt{|D(K)|}\right)^n.$$

Proof. The first statement is a standard fact [4, Chap. 5, Theorem 2] and the second follows from the first since ρ is $\mathbb{Q}$-linear with trivial kernel. Suppose we have T_v's that also satisfy the hypotheses of the lemma and further that $S_v \subseteq T_v$ for all v. Get $\mathfrak{M}'$ as above using T_v in place of S_v. We claim that

$$\frac{\det\left(\rho(\mathfrak{M})\right)}{\det\left(\rho(\mathfrak{M}')\right)} = \frac{\prod_{v\nmid\infty} \alpha_v(T_v)}{\prod_{v\nmid\infty} \alpha_v(S_v)}. \tag{2}$$

Indeed, by [4, Chap. 5, Theorem 4],

$$[\mathfrak{M}' : \mathfrak{M}] = \prod_{v\nmid\infty} [T_v : S_v] = \frac{\prod_{v\nmid\infty} \alpha_v(T_v)}{\prod_{v\nmid\infty} \alpha_v(S_v)},$$

and since ρ is $\mathbb{Q}$-linear with trivial kernel,

$$[\mathfrak{M}' : \mathfrak{M}] = \frac{\det(\rho(\mathfrak{M}))}{\det(\rho(\mathfrak{M}'))}.$$

This shows that (2) is true.

Now suppose that the T_v's also satisfy $T_v \supseteq (\mathfrak{O}_v)^n$ for all finite v. Then by two applications of (2) we have

$$\det\left(\rho(\mathfrak{M})\right) \times \prod_{v\nmid\infty} \alpha_v(S_v) = \det\left(\rho(\mathfrak{M}')\right) \times \prod_{v\nmid\infty} \alpha_v(T_v)$$

$$= \det\left(\rho((\mathfrak{O}_K)^n)\right) \times \prod_{v\nmid\infty} \alpha_v\left((\mathfrak{O}_v)^n\right)$$

$$= \left(2^{-r_2}\sqrt{|D(K)|}\right)^n,$$

proving the lemma. □

Now suppose $S_\infty \subset (K^n)_\infty$ is such that $\rho(S_\infty)$ is a convex body in $\mathbb{R}^{nd}$. Let $S_v \subset K_v^n$ for $v \nmid \infty$ and $\mathfrak{M}$ be as in the statement of the lemma. Via the definitions and the lemma, one easily verifies that

$$\frac{\operatorname{Vol}(\rho(S_\infty))}{\det\left(\rho(\mathfrak{M})\right)} = \alpha\Big(S_\infty \times \prod_{v\nmid\infty} S_v\Big).$$

Thus, by (1) we have an "adelic" version of Minkowski's convex bodies theorem for number fields.

Theorem 1. *Let K be a number field of degree d over $\mathbb{Q}$ and let $S \subset (K_\mathbb{A})^n$ be of the form $S = S_\infty \times \prod_{v\nmid\infty} S_v$, where $S_\infty \subset (K^n)_\infty$ and $S_v \subset (K_v)^n$ are as above. Let $\mathfrak{M}$ be as in the lemma. Then*

$$\frac{2^{nd}}{(nd)!} \le \alpha(S)\lambda_1\left(\rho(S_\infty), \rho(\mathfrak{M})\right) \ldots \lambda_{nd}\left(\rho(S_\infty), \rho(\mathfrak{M})\right) \le 2^{nd}.$$

Now suppose S_∞ has the form $S_\infty = \prod_{v|\infty} S_v$, where each $S_v \subset (K_v)^n$ is a convex body (in the sense defined above). Clearly $\rho(S_\infty)$ is a convex body in $\mathbb{R}^{nd}$. Following [1], we define the successive minima $\mu_1(S) \leq \cdots \leq \mu_n(S)$, by letting $\mu_i(S)$ be the infimum of the set of all $r > 0$ such that $rS_\infty \prod_{v\nmid\infty} S_v$ contains at least i elements of K^n which are linearly independent over K. (As usual, view K^n as a subset of $(K_{\mathbb{A}})^n$ via the diagonal embedding.) Choose $\mathbf{x}_1, \ldots, \mathbf{x}_{nd} \in \mathfrak{M}$, linearly independent over $\mathbb{Q}$, satisfying $\rho(\mathbf{x}_i) \in \lambda_i \rho(S_\infty)$ for each i. Let $i_1 = 1$ and for $1 < j \leq n$ let i_j be least such that $\mathbf{x}_{i_1}, \ldots, \mathbf{x}_{i_j}$ are linearly independent over K. Since $[K : \mathbb{Q}] = d$, we have $i_j \leq (j-1)d + 1$, and in this way we see that

$$(\mu_1(S) \ldots \mu_n(S))^d \leq \left(\lambda_{i_1}\big(\rho(S_\infty), \rho(\mathfrak{M})\big) \ldots \lambda_{i_n}\big(\rho(S_\infty), \rho(\mathfrak{M})\big)\right)^d$$
$$\leq \prod_{i=1}^{nd} \lambda_i\big(\rho(S_\infty), \rho(\mathfrak{M})\big).$$

Theorem 1 thus implies (but is not implied by) the following "adelic" version of Minkowski's second convex bodies theorem.

Corollary ([1, Theorem 3], [2, Theorem 5]). *Let K be a number field of degree d over $\mathbb{Q}$ and let $S \subset (K_{\mathbb{A}})^n$ be of the form above. Then*

$$\alpha(S)\big(\mu_1(S) \ldots \mu_n(S)\big)^d \leq 2^{nd}.$$

We now consider Siegel's lemma. We first give the definition of the heights used in [1]. Let K be a number field as above. For $v \mid \infty$ we let $\|\cdot\|_v'$ be the usual L^2 norm on $(K_v)^n$:

$$\|\mathbf{x}\|_v' = \left(|x_1|_v^2 + \cdots + |x_n|_v^2\right)^{1/2}.$$

For all v we let $\|\cdot\|_v$ be the maximum norm on $(K_v)^n$:

$$\|\mathbf{x}\|_v = \max_{1 \leq i \leq n} \{|x_i|_v\}.$$

We then define the height $h(\mathbf{x})$ for $\mathbf{x} \in K^n$ by

$$h(\mathbf{x}) = \prod_{v \in M(K)} \|\mathbf{x}\|_v^{n_v/d},$$

where n_v is the local degree; $n_v = [K_v : \mathbb{Q}_w]$ where w is the place of $\mathbb{Q}$ lying below v. For $W \subset K^n$ a subspace of dimension m we define

$$H(W) = \prod_{v|\infty} \left(\|(\mathbf{x}_1 \wedge \cdots \wedge \mathbf{x}_m)\|_v'\right)^{n_v/d} \times \prod_{v\nmid\infty} \|(\mathbf{x}_1 \wedge \cdots \wedge \mathbf{x}_m)\|_v^{n_v/d},$$

where $\{\mathbf{x}_1, \ldots, \mathbf{x}_m\}$ is any basis for W over K. This definition is independent of the choice of basis (see [1] or [3]). For A an $M \times N$ matrix over K we define $H(A)$ to be the height of the nullspace of A. This is also the height of the rowspace of A (see [1] or [3]).

Theorem 2. *Let K be a number field of degree d over $\mathbb{Q}$ and let A be an $M \times N$ matrix over K with rank $M < N$. Then there exist $\mathbf{x}_1, \ldots, \mathbf{x}_{d(N-M)}$ $\in (\mathfrak{O}_K)^N$ in the nullspace of A which are linearly independent over $\mathbb{Q}$ and satisfy*

$$\prod_{i=1}^{d(N-M)} h(\mathbf{x}_i) \le \left(\frac{2^{r_2} \sqrt{|D(K)|}}{\pi^{r_2}} \right)^{N-M} H(A)^d.$$

An immediate consequence is the following formulation of Siegel's lemma in [1].

Corollary ([1, Theorem 9]). *With the notation above, there exist $\mathbf{y}_1, \ldots,$ $\mathbf{y}_{N-M} \in (\mathfrak{O}_K)^N$ in the nullspace of A which are linearly independent over K and satisfy*

$$\prod_{j=1}^{(N-M)} h(\mathbf{y}_j) \le \left(\frac{2^{r_2} \sqrt{|D(K)|}}{\pi^{r_2}} \right)^{(N-M)/d} H(A).$$

Proof of Theorem 2. Let

$$C_1(N) = \{\mathbf{x} \in \mathbb{R}^N : |x_i| \le 1 \text{ for all } i\},$$
$$C_2(N) = \{(\mathbf{x}, \mathbf{y}) \in \mathbb{R}^{2N} : \mathbf{x}, \mathbf{y} \in \mathbb{R}^N \text{ and } x_i^2 + y_i^2 \le 1 \text{ for all } i\},$$

and let $C(N) \subset \mathbb{R}^{Nd}$ be the cartesian product of r_1 copies of $C_1(N)$ and r_2 copies of $C_2(N)$. Suppose V is an $(N - M)$-dimensional subspace of $\mathbb{R}^N$. Then Vaaler's cube-slicing inequality says that the $(N - M)$-dimensional volume of $C_1(N) \cap V$ is at least the volume of $C_1(N - M)$, which is 2^{N-M}. Similarly, if V is an $(N - M)$-dimensional subspace of $\mathbb{C}^N$, then when we view $\mathbb{C}^N$ as $\mathbb{R}^{2N}$, the $(2N - 2M)$-dimensional volume of $C_2(N) \cap V$ is at least the volume of $C_2(N - M)$, which is π^{N-M}.

Now let W be the nullspace of A (as a vector space over K). Let $W_\infty = W \otimes_{\mathbb{Q}} \mathbb{R}$, which we view as a subset of $(K^N)_\infty$ in the usual way (see [4, Chap. 5, §2]). The previous discussion shows that the $d(N - M)$-dimensional volume of $\rho(W_\infty) \cap C(N)$ is at least $(2^{r_1} \pi^{r_2})^{(N-M)}$. In other words, if U is an orthonormal transformation of $\mathbb{R}^{dN}$ taking $\rho(W_\infty)$ to $\mathbb{R}^{d(N-M)}$, then

$$\mathrm{Vol}\left(U\left(\rho(W_\infty) \cap C(N) \right) \right) \ge \left(2^{r_1} \pi^{r_2} \right)^{N-M}. \tag{3}$$

Next, let $I(W) = W \cap (\mathfrak{O}_K)^N$. Then, by [3, Theorem 1], $\Lambda(W) = U(\rho(I(W)))$ is a lattice in $\mathbb{R}^{d(N-M)}$ with

$$\det(\Lambda(W)) = \left(2^{-r_2}\sqrt{|D(K)|}\right)^{N-M} H(W)^d. \tag{4}$$

We have

$$h(\mathbf{x}) \le \prod_{v|\infty} \|\mathbf{x}\|_v^{n_v/d}$$

for $\mathbf{x} \in (\mathfrak{O}_K)^N$. By the definition of $\|\cdot\|_v$, we see that

$$\prod_{v|\infty} \|\mathbf{x}\|_v^{n_v/d} \le \lambda \tag{5}$$

whenever $\rho(\mathbf{x}) \in \lambda C(N)$. Thus, if $\lambda_1 \le \cdots \le \lambda_{d(N-M)}$ are the successive minima of $U(\rho(W_\infty) \cap C(N))$ with respect to $\Lambda(W)$, then there are $\mathbf{x}_1, \ldots, \mathbf{x}_{d(N-M)} \in I(W)$ which are linearly independent over $\mathbb{Q}$ with

$$h(\mathbf{x}_1)\ldots h(\mathbf{x}_{d(N-M)}) \le \lambda_1 \ldots \lambda_{d(N-M)}.$$

Theorem 2 now follows from (1), (3), and (4).   □

We make several remarks. First, note that our proof of Siegel's lemma didn't use an adelic version of Minkowski's convex bodies theorem at all. Also, if one were to use the L_2 norms $\|\cdot\|_v'$ in the definition of $h(\mathbf{x})$, then one wouldn't need Vaaler's cube slicing inequality and the corresponding result (one might call it "Siegel's lemma with L_2 norms") would be quicker to prove.

Finally, what we have proven is actually stronger than the statements of either Theorem 2 or its corollary. To see what we mean here, consider the inhomogeneous height h_1 defined by

$$h_1(\mathbf{x}) = h((1, \mathbf{x})).$$

Suppose $\mathbf{x} = (x_1, \ldots, x_N) \in I(W)$ is non-zero and satisfies $\rho(\mathbf{x}) \in \lambda C(N)$. Note that

$$|\mathcal{N}(x_i)| \le \prod_{v|\infty} \|\mathbf{x}\|_v^{n_v}$$

for each i, where $\mathcal{N}$ denotes the norm. Since the x_i are in $\mathfrak{O}_K$ and not all are zero, we see by (5) that $\lambda \ge 1$. This shows that the successive minima of $U(\rho(W_\infty) \cap C(N))$ with respect to $\Lambda(W)$ are all at least 1. But by the same reasoning as above, we have

$$h_1(\mathbf{x}) - \prod_{v|\infty} \max\{1, \|\mathbf{x}\|_v^{n_v/d}\} \le \max\{1, \lambda\}$$

whenever $\mathbf{x} \in (\mathfrak{O}_K)^N$ satisfies $\rho(\mathbf{x}) \in \lambda C(N)$. This gives the following.

Theorem 3. *In the statement of Theorem 2, the heights* $h(\mathbf{x}_1), \ldots,$ $h(\mathbf{x}_{d(N-M)})$ *may all be replaced by the inhomogeneous heights* $h_1(\mathbf{x}_1), \ldots,$ $h_1(\mathbf{x}_{d(N-M)})$. *Similarly, the heights* $h(\mathbf{y}_1), \ldots, h(\mathbf{y}_{N-M})$ *in the statement of the corollary may all be replaced by the inhomogeneous heights* $h_1(\mathbf{y}_1),$ $\ldots, h_1(\mathbf{y}_{N-M})$.

Theorem 3 improves on [1, Corollary 10] where, with the notation in the corollary to Theorem 2 above, the upper bound

$$h_1(\mathbf{y}_1) \ldots h_1(\mathbf{y}_{N-M}) \leq \sqrt{|D(K)|}^{(N-M)/d} H(A)$$

is given for the product of the inhomogeneous heights.

References

[1] E. Bombieri and J. Vaaler, *On Siegel's Lemma*, Invent. Math. **73** (1983), 11–32.

[2] R. B. MacFeat, *Geometry of numbers in adele spaces*, Dissertationes Mathematicae **88** (1971).

[3] W. M. Schmidt, *On heights of algebraic subspaces and Diophantine approximations*, Annals of Math. **85** (1967), 430–472.

[4] A. Weil, *Basic Number Theory*, Springer-Verlag, New York, 1974.

Some Applications of Diophantine Approximation

R. Tijdeman

0   Introduction

The paper gives a survey of some results on diophantine approximation (Sections 1 and 2) and their applications (Sections 3, 4 and 5). Section 1 contains an introduction to the theory of linear forms in logarithms of algebraic numbers, and Section 2 describes some results following from the Subspace Theorem. Section 3 gives applications to the local behaviour of sequences of numbers composed of small primes and of sums of two such numbers. Section 4 deals with the transcendence of infinite sums of values of a rational function and related sums, and in Section 5 some recent applications to diophantine equations and recurrence sequences are described. The Appendix contains some elaborations of Section 4. Section 3 and the Appendix contain some new results.

1   Linear Forms in Logarithms

In 1966–1968 Baker extended the Gelfond-Schneider theorem by proving the following result.

Theorem 1 ([2], p. 10). *If $\alpha_1, \alpha_2, \ldots, \alpha_n$ are non-zero algebraic numbers such that $\log \alpha_1, \ldots, \log \alpha_n$ are linearly independent over the rationals, then $1, \log \alpha_1, \ldots, \log \alpha_n$ are linearly independent over the field of all algebraic numbers.*

For our convenience we shall assume in the sequel that log denotes the principal value of the logarithm. I record some consequences of Theorem 1.

Corollary 1.1. $e^{\beta_0} \alpha_1^{\beta_1} \ldots \alpha_n^{\beta_n}$ *is transcendental for any non-zero algebraic numbers* $\alpha_1, \ldots, \alpha_n, \beta_0, \beta_1, \ldots, \beta_n$.

The fact that $\beta_0 \neq 0$ says that we are dealing with the inhomogeneous case. The corollary implies that numbers like $e \cdot 2^{\sqrt{2}}$ and $\pi + \log 2$ are transcendental, but it is less useful for applications than its homogeneous counterpart which reads as follows.

Corollary 1.2. $\alpha_1^{\beta_1} \ldots \alpha_n^{\beta_n}$ *is transcendental for any algebraic numbers* $\alpha_1, \ldots, \alpha_n$, *other than* 0 *or* 1, *and any algebraic numbers* $\beta_1, \ldots, \beta_n$ *with* $1, \beta_1, \ldots, \beta_n$ *linearly independent over the rationals.*

An alternative formulation is as follows.

Corollary 1.3. *If* $\alpha_1, \ldots, \alpha_n$, $\beta_1, \ldots, \beta_n$ *are non-vanishing algebraic numbers and* $\Lambda = \beta_1 \log \alpha_1 + \cdots + \beta_n \log \alpha_n$, *then*

$$\Lambda = 0 \quad or \quad \Lambda \text{ is transcendental.}$$

Baker used the effectiveness of his method to give lower bounds for $|\Lambda|$ in the case $\Lambda \neq 0$. Of course, these bounds depend on $\alpha_1, \ldots, \alpha_n, \beta_1, \ldots, \beta_n$. They are functions of the degrees and the heights of these numbers. The earlier bounds were expressed in terms of the classical height, which is defined as the maximal absolute value of the coefficients of the minimal defining polynomial of the number. These bounds have been successively improved, and they are now stated in terms of the Mahler height or its variants. This is more convenient for the proofs and leads to better constants. Since the different heights used are more or less equivalent, I do not want to go into this rather technical aspect here.

Let $\alpha_1, \ldots, \alpha_n$ be algebraic numbers $\neq 0, 1$. Let $\mathbb{Q}(\alpha_1, \ldots, \alpha_n)$ have degree at most d over $\mathbb{Q}$. Let the classical height of α_j be at most $A_j \geq 4$ for $j = 1, \ldots, n$. Put $A = \max A_j$, $\Omega = (\log A_1) \ldots (\log A_n)$, $\Omega' = (\log A_1) \ldots (\log A_{n-1})$. Let $\beta_1, \ldots, \beta_n$ be rational integers of absolute values at most $e^B \geq 4$. In 1977 Baker [3] proved that

$$\Lambda = 0 \text{ or } \log|\Lambda| \geq -(16nd)^{200n} B\Omega \log \Omega'. \tag{1.1}$$

This bound has three weaknesses: The constants are rather large; it is expected that the factor $(16nd)^{200n}$ can be replaced by a polynomial expression in n and d; and it is conjectured that the product of the logarithms in Ω may be replaced by the sum of the logarithms. The first drawback has been largely overcome. After several improvements by Waldschmidt and others, Baker and Wüstholz [6] published in 1993 a result which implies that

$$\Lambda = 0 \text{ or } \log|\Lambda| \geq -(16nd)^{2(n+2)} B\Omega.$$

In addition to having a smaller constant in the exponent, we see that the factor $\log \Omega'$ in Baker's bound has been removed. Baker and Wüstholz state that a similar improvement in the constant can be obtained in the case of arbitrary algebraic coefficients β_j, but they have not worked out the details. For bounds in the general case, see [62]. In many applications only two or three logarithms occur. In these cases bounds with better constants are available. For the case of two logarithms, see Laurent, Mignotte and Nesterenko [33] [39]; for linear forms in three logarithms see [7].

In some applications the dependence on the number n of logarithms is crucial. In this direction Matveev [36] proved the following result:

Let $\alpha_1, \ldots, \alpha_n$ be algebraic numbers distinct from 0, let $\log \alpha_j$ be fixed values of the logarithms distinct from 0, and let $h(\alpha_j)$ denote the absolute logarithmic heights. Consider the linear form

$$\Lambda := b_1 \log \alpha_1 + \cdots + b_n \log \alpha_n \neq 0, \quad b_j \in \mathbf{Z}.$$

Let K be an algebraic number field containing all numbers α_j and having degree at most D. Define $\kappa = 1$ if K is contained in the field of real numbers and $\kappa = 2$ otherwise. Put

$$B = \max(|b_1|, \ldots, |b_n|).$$

Let A_j be real numbers such that

$$A_j \geq \max(Dh(\alpha_j), |\log \alpha_j|, 0.16), \quad j = 1, \ldots, n.$$

Then

$$\log |\Lambda| \geq -C_1 D^2 A_1 \ldots A_n \log(eD) \log(eB),$$

where

$$C_1 = \min\left(\kappa^{-1}\left(\frac{1}{2}en\right)^\kappa 30^{n+3} n^{3.5}, \ 2^{6n+20}\right).$$

The dependence on n in this result is reduced to c^n; furthermore, the constants are remarkably small.

There exist p-adic analogues of the complex linear forms estimates which are important for many applications. Generally speaking, the estimates are similar and the constants are slightly better. Kunrui Yu obtained the p-adic analogues of the Baker-Wüstholz estimate [64] and of a preliminary version of Matveev's result [65]. A p-adic analogue of the result of Laurent, Mignotte and Nesterenko on linear forms in two logarithms was given by Bugeaud and Laurent [16]. In general the upper bounds for

$$\text{ord}_\mathfrak{p}(\alpha_1{}^{\beta_1} \ldots \alpha_n{}^{\beta_n} - 1),$$

where $\text{ord}_\mathfrak{p}(\alpha)$ denotes the exponent to which the prime ideal $\mathfrak{p}$ divides the principal fractional ideal generated by $\alpha \in \mathbb{Q}(\alpha_1, \ldots, \alpha_n)$, $\alpha \neq 0$, are similar to the estimates for $-\log |\Lambda|$ with respect to $n, \alpha_1, \ldots, \alpha_n$ and $\beta_1, \ldots, \beta_n$. There is, however, an additional factor $p^{f_\mathfrak{p}}$ where $f_\mathfrak{p}$ denotes the residue class degree of $\mathfrak{p}$. For some applications this dependence on p is an impediment.

2   The Subspace Theorem

In this section we state some consequences of the Subspace Theorem which
are quite useful in arithmetical applications. The Subspace Theorem is a
generalization of Roth's theorem proved by W. M. Schmidt [45] in 1972.
A p-adic analogue was obtained by Schlickewei [42] in 1977. Both van der
Poorten and Schlickewei [59] and Evertse [27] applied the p-adic Subspace
Theorem to S-unit equations. For the general statements I refer to the
original papers. Here I state the consequence for rational integers of a
result of Evertse.

Theorem 2 ([27], p. 227). *Let c, d be constants with $c > 0$, $0 < d < 1$.
Let S be a finite set of prime numbers and let n be a positive integer. Then
there are only finitely many tuples $\boldsymbol{x} = (x_0, x_1, \ldots, x_n)$ of rational integers
such that*

$$x_0 + x_1 + \cdots + x_n = 0,$$

$$\sum_{i \in I} x_i \neq 0$$

for each proper, non-empty subset I of $\{0, 1, \ldots, n\}$,

$$\gcd(x_0, x_1, \ldots, x_n) = 1,$$

and

$$\prod_{k=0}^{n} \left(|x_k| \prod_{p \in S} |x_k|_p \right) \leq c \cdot |\boldsymbol{x}|^d,$$

where $|a|_p = p^{-\operatorname{ord}_p(a)}$.

In 1989 Schmidt [46] made another breakthrough by proving a quantitative version of his Subspace Theorem. Soon afterwards Schlickewei [43]
proved the p-adic analogue of this result and extended it to number fields.
In 1995 Evertse [28] improved upon earlier upper bounds for the number of
equivalence classes of solutions of S-unit equations. For the general formulation and related results I again refer to the original papers. Here I state
the result for S-unit equations in the case of rational integers.

Theorem 3 ([28], p. 564). *Let $a_1, \ldots, a_n$ be non-zero rational numbers.
Let S be a finite set of s prime numbers. Then the equation*

$$a_1 u_1 + \cdots + a_n u_n = 1$$

with

$$\sum_{i \in I} a_i u_i \neq 0$$

for each non-empty subset I of $\{0, 1, \ldots, n\}$, has at most $(2^{35}n^2)^{n^3(s+1)}$ solutions in rational numbers $u_1, \ldots, u_n$ which are entirely composed of primes from S.

It is remarkable that the bound depends only on n and s, and not on the primes in S or the coefficients a_i. The upper bound is large, but the dependence on n and s cannot be polynomial [26], [30]. For a corresponding estimate for linear equations in variables which lie in an arbitrary finitely generated multiplicative group, see [29]. The existing proofs of the Subspace Theorem do not enable us to give upper bounds for the solutions themselves. We call them therefore *ineffective*.

3   Numbers Composed of Small Primes

The results mentioned in Sections 1 and 2 have immediate consequences for the local distribution of integers composed of small primes. In this section I give a survey of such results.

Let $p_1, \ldots, p_k$ be fixed primes, each at most P. Consider all integers $p_1^{r_1} \ldots p_k^{r_k}$, where $r_1, \ldots, r_k$ are non-negative integers and order them in increasing order of magnitude as $1 = n_0 < n_1 < n_2 < \ldots$. For example, for $k = 2$, $p_1 = 2$, $p_2 = 3$ we get

$$(n_i)_{i=0}^{\infty} = 1, 2, 3, 4, 6, 8, 9, 12, 16, 18, 24, 27, \ldots.$$

Improving upon earlier lower bounds by Störmer, Thue and Erdős, I noticed in 1973 [55] that there is a computable number C_1 depending only on P such that

$$n_{i+1} - n_i > \frac{n_i}{(\log n_i)^{C_1}}$$

for $n_i \geq 3$. The result is an immediate consequence of linear forms estimates. From

$$\frac{n_{i+1}}{n_i} - 1 > \log \frac{n_{i+1}}{n_i} = t_1 \log p_1 + \cdots + t_k \log p_k > 0$$

we obtain, by (1.1),

$$n_{i+1} - n_i > n_i (\max_j t_j)^{-C_2} > \frac{n_i}{(\log n_i)^{C_1}}.$$

Apart from the value of C_1 the result is the best possible. In another paper [56] I proved that there are computable numbers C_3 and N depending only on P such that

$$n_{i+1} - n_i < \frac{n_i}{(\log n_i)^{C_3}}$$

for $n_i \geq N$. The average order of the difference is

$$\frac{n_i}{(\log n_i)^{k-1}}$$

so that $C_3 \leq k - 1 \leq C_1$. Actually we need not assume that $p_1, \ldots, p_k$ are fixed. Using the Baker-Wüstholz estimate and the inequality $t_i \leq 2 \log n_i$ we see that we can take $C_1 = (ck^2 \log P)^k$ where c is some absolute constant.

Let $1 = m_0 < m_1 < m_2 < \ldots$ be a sequence of positive integers each of which is the sum of at most two terms from the sequence $(n_i)_{i=1}^\infty$ such that m_i and m_{i+1} do not use the same n_i. For example, for $k = 2$, $p_1 = 2$, $p_2 = 3$ we may take

$$(m_i)_{i=0}^\infty = 1, 2, 3, 4, 5, 6, 7, 8, 9, 10, 11, 12, 13, 14, 15, 16, \ldots,$$

but 23 will not occur. What can be said about the differences $m_{i+1} - m_i$? By our restriction we avoid large neighbouring numbers like

$$2^k 3^l + 2^3, \quad 2^k 3^l + 3^2.$$

Baker-type estimates are of no use here. I do not know of any method giving explicit lower bounds. However, the ineffective methods from Section 2 show that $m_{i+1} - m_i \to \infty$. In fact, it follows from Theorem 2 that for every $\epsilon > 0$ there are only finitely many i such that

$$m_{i+1} - m_i < m_i^{1-\epsilon}. \tag{3.1}$$

By applying Theorem 3 to the equation

$$\frac{n_{i_1}}{n_{i_4}} + \frac{n_{i_2}}{n_{i_4}} - \frac{n_{i_3}}{n_{i_4}} = 1$$

we see that there are at most

$$\left(2^{35} \cdot 9\right)^{27(k+1)}$$

integers m which can be written as

$$m = n_{i_1} + n_{i_2} = n_{i_3} + n_{i_4}$$

with $\gcd(n_{i_1}, n_{i_2}, n_{i_3}, n_{i_4}) = 1$ and $n_{i_1} \neq n_{i_3}, n_{i_4}$. Similar results can be derived for numbers which can be written in more than one way as the sum of at most t numbers n_i, but if $t > 2$ the vanishing subsums cause complications.

It is not necessary that the n_i be entirely composed of $p_1, \ldots, p_k$. For example, let $0 \le \delta < 1/4$. Let $1 = l_0 \le l_1 \le l_2 \le \ldots$ be a sequence of positive integers which are the sum of two numbers of the form $p_1^{r_1} \ldots p_k^{r_k} q$ where q is some positive integer with $q \le (p_1^{r_1} \ldots p_k^{r_k})^\delta$ such that l_i and l_{i+1} do not use the same summand for any i. Then, by Theorem 2, for every positive ϵ there are only finitely many i such that

$$l_{i+1} - l_i < l_i^{1-4\delta-\epsilon}.$$

The inequality reduces to (3.1) if $\delta = 0$.

4   On the Transcendence of Infinite Sums

It is well known that

$$\sum_{n=1}^{\infty} \frac{1}{n(n+1)} = 1 - \frac{1}{2} + \frac{1}{2} - \frac{1}{3} + \frac{1}{3} - \frac{1}{4} + \frac{1}{4} - \frac{1}{5} + \cdots = 1$$

is an integer, but

$$\sum_{n=0}^{\infty} \frac{1}{(2n+1)(2n+2)} = 1 - \frac{1}{2} + \frac{1}{3} - \frac{1}{4} + \frac{1}{5} - \frac{1}{6} + \cdots = \log 2$$

is transcendental. Remarkably, little attention has been given in the literature to the transcendence of sums $\sum_{n=0}^{\infty} f(n)$ where $f \in \mathbb{Z}(x)$, whereas a number of exotic series such as

$$\sum_{n=1}^{\infty} \frac{1}{2^{n!}}, \qquad \sum_{n=0}^{\infty} \frac{1}{2^{2^n}}, \qquad \sum_{h=0}^{\infty} \frac{1}{F_{2^h+1}}, \qquad \sum_{n=1}^{\infty} \frac{(-1)^n}{F_n^2},$$

where $(F_h)_{h=0}^{\infty}$ is the Fibonacci sequence, have been studied in the literature. As to sums $\sum_{n=0}^{\infty} f(n)$ with $f \in \mathbb{Z}(x)$, Lehmer [34] gives the following examples:

$$\sum_{n=0}^{\infty} \frac{1}{(n+1)(2n+1)(4n+1)} = \frac{\pi}{3},$$

$$\sum_{n=0}^{\infty} \frac{1}{(6n+1)(6n+2)(6n+3)(6n+4)(6n+5)(6n+6)}$$
$$= \frac{1}{4320}(192 \log 2 - 81 \log 3 - 7\pi\sqrt{3}).$$

We see that both sums are transcendental. The latter example shows that Baker's theory is relevant here. If one replaces π by $\log(-1)$, the sum becomes a linear form in logarithms of algebraic numbers with algebraic coefficients. Corollary 1.3 tells us that such a sum is 0 or transcendental, and a sum of positive numbers cannot be 0. Of course, there are also sums of this form, such as $\sum_{n=1}^{\infty} n^{-3}$ and $\sum_{n=1}^{\infty} n^{-5}$, for which we do not know whether they are transcendental.

We assume from now on that $f(x) = P(x)/Q(x)$ for $P(x), Q(x) \in \mathbb{Q}[x]$, where the zeros of Q are simple and rational and $\deg P < \deg Q$. Hence we can split $f(x)$ into partial fractions and write

$$f(x) = \frac{P(x)}{Q(x)} = \sum_{j=1}^{m} \frac{c_j}{k_j x + r_j} \quad \text{where } k_j, r_j \in \mathbb{Z}, c_j \in \mathbb{Q}.$$

An obvious simplification is possible if $k_i = k_j$ and $k_i \mid r_i - r_j$ for some i, j. Changing the summation index we get a rational initial term and we can combine the remaining terms. In this way we can arrange that $0 < r_j \le k_j$ for all j (starting the summation from $n = 0$). For example,

$$\sum_{n=0}^{\infty} \left(\frac{4}{5n+3} - \frac{3}{5n+7} - \frac{1}{5n+8} \right)$$

$$= \frac{3}{2} + \frac{1}{3} + \sum_{n=0}^{\infty} \left(\frac{4}{5n+3} - \frac{3}{5n+2} - \frac{1}{5n+3} \right)$$

$$= \frac{11}{6} + \sum_{n=0}^{\infty} \left(\frac{3}{5n+3} - \frac{3}{5n+2} \right).$$

We call an infinite sum *reduced* if it can be written as

$$\sum_{n=0}^{\infty} \sum_{j=1}^{m} \frac{c_j}{k_j n + r_j} \quad \text{with } k_j, r_j \in \mathbb{Z}_{>0}, r_j \le k_j, c_j \in \mathbb{Q},$$

so that all zeros of Q are in the interval $[-1, 0)$. It may happen that the infinite sum vanishes, in which case the sum is rational; for example,

$$\sum_{n=1}^{\infty} \frac{1}{n(n+1)} = 1 + \sum_{n=0}^{\infty} \left(\frac{1}{n+1} - \frac{1}{n+1} \right) = 1$$

and

$$\sum_{n=0}^{\infty} \left(\frac{1}{5n+2} - \frac{3}{5n+7} + \frac{2}{5n-3} \right)$$

$$= \frac{3}{2} - \frac{2}{3} + \sum_{n=0}^{\infty} \left(\frac{1}{5n+2} - \frac{3}{5n+2} + \frac{2}{5n+2} \right) = \frac{5}{6}.$$

If this is not the case, it is useful to apply formulas for $\sum_{n=0}^{N} 1/(kn+r)$ which D. H. Lehmer [34] gave in 1975. The formulas are of the form

$$\sum_{n=0}^{N} \frac{1}{kn+r} = \frac{1}{k} \log N + \gamma_{k,r} + o(1),$$

where the numbers $\gamma_{k,r}$ are generalized Euler constants. The arithmetic nature of these numbers is unknown, but fortunately they cancel in cases where $\sum_{n=0}^{\infty} P(n)/Q(n)$ converges. By using Lehmer's formulas, Adhikari, Saradha, Shorey and I proved that if the reduced sum $\sum_{n=0}^{\infty} \sum_{j=1}^{m} c_j/(k_j n + r_j)$ converges, then it can be written as a finite sum of the form

$$\sum_{j=1}^{m} \sum_{t=0}^{k_j-1} \frac{c_j}{k_j}(1 - \zeta^{-r_j t}) \log(1 - \zeta_j^t),$$

where ζ_j is a primitive k_jth root of unity. Here we recognize the linear form in logarithms of algebraic numbers with algebraic coefficients. According to Corollary 1.3 the double sum is either 0 or transcendental. Hence we obtain the following result.

Theorem 4 ([1]). *Let $P(x) \in \mathbb{Q}[x]$. Let $Q(x) \in \mathbb{Q}[x]$ have only simple rational zeros. If $S := \sum_{n=0}^{\infty} P(n)/Q(n)$ converges, then either S equals a computable rational number or S is transcendental.*

I call a polynomial $Q \in \mathbb{Q}[x]$ *reduced* if it has only simple rational zeros which are all in the interval $[-1, 0)$. If the polynomial Q in Theorem 4 is reduced, then the computable number equals 0. Hence, for example, it is obvious that the sum

$$\sum_{n=0}^{\infty} \frac{1}{(5n+1)(5n+3)(5n+5)}$$

is transcendental.

When applying Theorem 4 it may be important to exclude the case where the infinite sum S has the exceptional rational value, say q. If $S \neq q$, then this can be checked by numerical methods. If $S = q$, then the logarithms are linearly dependent over $\mathbb{Q}$ and the equality may be proved by simplifying the linear forms in logarithms to 0. For example, consider

$$\sum_{n=0}^{\infty} \left(\frac{1}{4n+1} - \frac{3}{4n+2} + \frac{1}{4n+3} + \frac{1}{4n+4} \right). \tag{4.1}$$

The corresponding linear form is

$$\log 2 - \log(1 - i) - \log(1 + i),$$

which happens to be 0. This example shows that a sum may vanish without an obvious reason.

The same method can be applied to other infinite sums. In this way we derived the following results. We denote here the field of algebraic numbers by $\overline{\mathbb{Q}}$.

Theorem 5 ([1]). *Let $f\colon \mathbb{Z}_{\geq 0} \to \overline{\mathbb{Q}}$ be periodic mod q. Let $Q(x) \in \mathbb{Q}[x]$ have only simple rational zeros. If $S = \sum_{n=0}^{\infty} f(n)/Q(n)$ converges, then either S equals a computable algebraic number or S is transcendental.*

Again the computable algebraic number equals 0 if Q is reduced. On combining Theorem 5 with Dirichlet's result that $L(1, \chi) \neq 0$ for an arbitrary non-principal Dirichlet character χ, we immediately obtain the following application.

Corollary 5.1 ([1]). *Let $q \geq 2$ be an integer and χ a non-principal Dirichlet character mod q. Then $L(1, \chi) = \sum_{n=1}^{\infty} \chi(n)/n$ is transcendental.*

We call $f\colon \mathbb{Z}_{\geq 0} \to \mathbb{Q}$ *completely multiplicative* if $f(mn) = f(m) \cdot f(n)$ for all integers m, n. Dirichlet characters are completely multiplicative. In Theorem 9, proved in the Appendix, it is shown that $\sum_{n=1}^{\infty} f(n)/n \neq 0$ if $f\colon \mathbb{Z} \to \mathbb{Q}$ is periodic and completely multiplicative. Using this we obtain the following variant of Corollary 5.1.

Corollary 5.2. *Let $f\colon \mathbb{Z} \to \mathbb{Q}$ be completely multiplicative and periodic. Then $\sum_{n=1}^{\infty} f(n)/n$ is transcendental.*

Another result obtained by the method above is as follows.

Theorem 6 ([1]). *Let $P_1(x), \ldots, P_l(x) \in \overline{\mathbb{Q}}[x]$, $\alpha_1, \ldots, \alpha_l \in \overline{\mathbb{Q}}$. Put $g(x) = \sum_{\lambda=1}^{l} P_\lambda(x)\alpha_\lambda^x$. Let $Q(x)$ have only simple rational zeros. If $S = \sum_{n=0}^{\infty} g(n)/Q(n)$ converges, then either S equals a computable algebraic number or S is transcendental.*

Again the special value of S equals 0 if Q is reduced. It follows immediately from this observation and Theorem 6 that $\sum_{n=1}^{\infty} F_n/(n \cdot 2^n)$ is transcendental, in view of

$$F_n = \frac{1}{\sqrt{5}} \left(\left(\frac{1 + \sqrt{5}}{2} \right)^n - \left(\frac{1 - \sqrt{5}}{2} \right)^n \right).$$

On the other hand, we have $\sum_{n=1}^{\infty} F_n/2^n = 2$. The example (4.1) also shows that in Theorems 5 and 6 the exceptional value can be attained for a non-trivial reason.

The question when $S = 0$ can occur in the special case that $Q(x) = x$ has been the subject of some conjectures of Chowla and Erdős. In 1952 Chowla [20, p. 300] incorrectly attributed the following problem to Erdős:

Let $f(x)$ be a number-theoretic (integer-valued) function which is periodic mod q. Suppose that not all $f(n)$ are 0. Then $\sum_{n=1}^{\infty} f(n)/n \neq 0$.

Already in 1949 Siegel had shown that this is true if q is prime (cf. [21]). Example (4.1) shows that Chowla's assertion is not true in general. Another example was given in 1973 by Baker, Birch and Wirsing ([4, p. 225]). According to Livingston [35] Erdős made the conjecture:

If q is a positive integer and $f(x)$ is a number-theoretic function mod q for which $f(n) \in \{-1, 1\}$ when $n = 1, 2, \ldots, q - 1$ and $f(q) = 0$, then $\sum_{n=1}^{\infty} f(n)/n \neq 0$ whenever the series is convergent.

In a lecture at the Stony Brook conference in 1969 Chowla raised the question whether there exists a rational-valued function $f(n)$ that is periodic with prime period p, such that $\sum_{n=1}^{\infty} f(n)/n = 0$. Chowla proved that, under some additional conditions, this is not possible. The general question, and more, was answered by Baker, Birch and Wirsing [4] in 1973, using Baker's theory.

Theorem 7 ([4]). *Suppose f is a non-vanishing function defined on the integers with algebraic values and period q such that*
  (i) $f(r) = 0$ if $1 < \gcd(r, q) < q$,
  (ii) *the cyclotomic polynomial Φ_q is irreducible over $\mathbb{Q}(f(1), \ldots, f(q))$.*
Then

$$\sum_{n=1}^{\infty} \frac{f(n)}{n} \neq 0.$$

If q is prime, then condition (i) is vacuous, and if f is rational-valued then (ii) holds trivially. Baker, Birch and Wirsing showed further that their theorem becomes false if (i) or (ii) is dropped. On the other hand, they showed that all functions f with algebraic values that are periodic mod q and for which (i) and $\sum_{n=1}^{\infty} f(n)/n = 0$ hold, are odd.

In 1982 Okada [41] gave a description of the functions f which satisfy condition (ii) of Theorem 7 and do not satisfy the conclusion of Theorem 7. Denoting the number of distinct prime factors of an integer m by $\omega(m)$ and Euler's function by $\phi(m)$, he gave a system of $\phi(m) + \omega(m)$ homogeneous linear equations with rational coefficients which is satisfied if and only if (ii) implies $\sum_{n=1}^{\infty} f(n)/n = 0$. The precise statement can be found in the Appendix. The non-vanishing result used in the proof of Corollary 5.2 is derived from Okada's theorem. Okada used his characterization to show that if $2\phi(q) + 1 > q$ and $f(n) \in \{1, -1\}$ when $n = 1, \ldots, q - 1$ and

$f(q) = 0$, then

$$\sum_{n=1}^{\infty} \frac{f(n)}{n} \neq 0$$

whenever the series is convergent. This gives a partial answer to Erdős' conjecture mentioned above. Okada's result covers the cases of Erdős' conjecture in which q is either a prime, a prime power or the product of two odd primes. As shown in the Appendix, it follows from Okada's criterion that Erdős' conjecture is also true if q is even. I am not able to establish Erdős' conjecture in full, let alone answer the question on the vanishing of S in general.

5   Diophantine Equations and Related Questions

There has been a tremendous stream of important results on diophantine equations in the past decades, culminating in the proof of Fermat's Last Theorem by Wiles and Taylor. In the shadow of the heavy arithmetic algebraic geometry machinery of Faltings and Wiles, which may be the subjects of other speakers, there are some remarkable new results obtained by diophantine approximation methods or combinations of these methods with other methods. Here I present a personal selection of some results which may be of general interest, but may not be well known.

5.1   Modular Curves

In 1929 Siegel proved that if a curve over a number field K has genus at least 1, then the number of integral points on the curve is finite. (In 1982 Faltings proved that if the genus is at least 2, then the number of K-points on the curve is finite.) In 1970 Baker and Coates [5] made Siegel's theorem effective for curves of genus 1 by using linear forms estimates. At the Oberwolfach meeting in April 2000 Bilu announced that using linear forms estimates he had made Siegel's theorem effective for congruence subgroups of finite index of modular curves.

5.2   Catalan's Equation

One of the first spectacular applications of Baker's theory on linear forms was the (effective) proof that the equation

$$x^p - y^q = 1 \text{ in integers } p > 1, q > 1, x > 1, y > 1$$

admits only finitely many solutions [57]. The initial upper bounds for the unknowns were huge. Mignotte and others have improved the upper and lower bound for unknown solutions. Last year Mihailescu [40] made a breakthrough by proving that, if there is a solution different from $3^2 - 2^3 = 1$ with p, q prime, then

$$p^{q-1} \equiv 1 \pmod{q^2} \quad \text{and} \quad q^{p-1} \equiv 1 \pmod{p^2}.$$

Using this result Mignotte obtained the following improvement (announced during this conference) on previous bounds (cf. [38]):

$$15 \cdot 10^6 < \min(p, q) < 8 \cdot 10^{11}.$$

5.3 The Equation $ax^n - by^n = c$

Another early application of Baker's theory was that, given a, b and $c \neq 0$, the equation $ax^n - by^n = c$ admits only finitely many solutions in integers $n \geq 3$, x, y; see [49, Ch. 5]. Bennett [8] has found a useful upper bound for the number of solutions if $c = \pm 1$ and n is fixed:

Let a, b and n be integers with $a > b \geq 1$ and $n \geq 3$. Then $|ax^n - by^n| = 1$ admits at most one solution in positive integers x, y.

5.4 Perfect Powers with Identical Decimal Digits

An old question is to determine all perfect powers which have identical digits in their decimal representation. This yields the diophantine equation

$$\frac{10^n - 1}{10 - 1} = ay^q \quad (1 \leq a \leq 9).$$

Obláth proved in 1956 that there are no solutions with $1 < a \leq 9$. In 1976 Shorey and Tijdeman showed that there are only finitely many exceptions (cf. [49, Ch. 12]). Recently Bugeaud [17] solved the problem completely by showing that there are no perfect powers with identical digits whatsoever.

More generally, Shorey and Tijdeman proved that the more general equation

$$\frac{x^n - 1}{x - 1} = y^q \text{ in integers } x > 1, y > 1, n > 2, q > 1 \qquad (5.1)$$

admits only finitely many solutions if x or n is fixed, or if y has a fixed prime divisor. Bugeaud, Mignotte and Roy [18] have extended Bugeaud's result as follows:

If (5.1) admits a solution x, y, n, q with $n \geq 5$, then there exists a prime p such that p divides x and q divides $p - 1$. In particular, $x \geq 2q + 1$.

The proof uses Skolem's method and a slight refinement of the estimate for p-adic linear forms in two logarithms due to Bugeaud and Laurent. For more results on (5.1) and related equations I refer to [48].

5.5   The Equation of Goormaghtigh

One step further is the equation

$$\frac{x^m - 1}{x - 1} = \frac{y^n - 1}{y - 1} \quad \text{in integers } x > 1, \, y > 1, \, m > n > 1.$$

The only known solutions are due to Goormaghtigh:

$$\frac{2^5 - 1}{2 - 1} = \frac{5^3 - 1}{5 - 1}, \quad \frac{2^{13} - 1}{2 - 1} = \frac{90^3 - 1}{90 - 1}.$$

This says that the numbers 31 and 8191 both have identical digits in their expansions with respect to two different bases. Shorey and Tijdeman (cf. [49, Ch. 12]) showed that if x and y are fixed, then m and n are bounded in terms of x and y. A recent result of Bugeaud and Shorey [19] says that for given x and y there are at most two solutions (m, n).

5.6   The abc-Conjecture

In 1985 Masser stated the following conjecture which refines a conjecture of Oesterlé and is now known as the abc-conjecture.

Let $a, b, c \in \mathbb{Z}_{>0}$. For every $\epsilon > 0$ there exists a number $C(\epsilon) > 0$ such that if $a + b = c$ and $\gcd(a, b, c) = 1$, then

$$c < C(\epsilon) N^{1+\epsilon} \text{ where } N = \prod_{p \mid abc} p.$$

By using p-adic linear forms in logarithms, Stewart and I [51] proved that for all such a, b, c,

$$\log c < c_1 N^{15} \qquad (c_1 \text{ some constant}).$$

In the other direction, we showed that the conjecture is not true with

$$N \cdot \exp\left((4 - \epsilon) \frac{\sqrt{\log N}}{\log\log N} \right) \tag{5.2}$$

as upper bound for c, for any $\epsilon > 0$. The constants in both of these results have recently been improved. By using the p-adic version of Matveev's refinement Stewart and Yu [52], [53] derived the inequality

$$\log c < c_2 N^{1/3} (\log N)^3 \qquad (c_2 \text{ constant}).$$

Using an idea of H. W. Lenstra, van Frankenhuysen [60] replaced the constant $(4 - \epsilon)$ in the exponent of (5.2) by 6.068.

5.7 Sums of Two Powers Being a Power

Darmon and Granville [23] proved that, for given positive integers k, l, m with $(1/k) + (1/l) + (1/m) < 1$, the equation

$$x^k + y^l = z^m \text{ in coprime integers } x, y, z \tag{5.3}$$

admits only finitely many solutions. If the *abc*-conjecture is true, then the total number of solutions k, l, m, x, y, z is finite ([58, p. 234]). The proof of Fermat's Last Theorem by Wiles and Taylor [63], [54] shows that (5.2) has no solutions if $k = l = m$. While preparing for a "Fermat day" for a broad audience in Utrecht in November 1993 on this celebrated result, Beukers and Zagier [9] found five new large solutions of (5.3). At present, the following solutions are known:

$$1^m + 2^3 = 3^2,$$
$$13^2 + 7^3 = 2^9,$$
$$2^7 + 17^3 = 71^2,$$
$$2^5 + 7^2 = 3^4,$$
$$3^5 + 11^4 = 122^2,$$
$$17^7 + 76721^3 = 21063928^2,$$
$$1414^3 + 2213459^2 = 65^7,$$
$$33^8 + 1549034^2 = 15613^3,$$
$$43^8 + 96222^3 = 30042907^2,$$
$$9262^3 + 15312283^2 = 133^7$$

When I presented these examples during this Fermat day, I noticed that in all examples a square occurs, and I formulated the following conjecture. Later, Beal attached a reward to the resolution of this problem (cf. [37]), and the conjecture has become known as the Beal Prize Problem.

Let x, y, z, k, l, m be positive integers with $k > 2$, $l > 2$, $m > 2$. If $x^k + y^l = z^m$, then x, y, z have a factor in common.

For certain triples (k, l, m) it has recently been established that there are no solutions of (5.3) other than those listed above. All of these results were obtained by geometric methods, and some of them were subject to the Taniyama-Shimura-Weil conjecture, which has recently been proved; see [12]. Darmon [22] proved that (5.3) admits no non-trivial solutions when $k = l = p > 13$ is a prime $\equiv 1 \pmod 4$ and $m = 2$. Kraus [32] proved that there are no non-trivial solutions of (5.3) if $k = l = 3$ and m is a prime with $16 < m < 10000$. Bruin [15] used Chabauty's method to extend Kraus' result to $m = 4$ and $m = 5$. In addition, Bruin [13], [14] showed that the above list contains all the solutions (modulo $\pm$ signs) for the exponent triples $(2, 3, 8)$, $(2, 4, 6)$ and $(2, 4, 5)$.

5.8 Arithmetic Progressions of Powers and Binomials Which Are Powers

Fermat claimed, and Euler proved, that four positive integers in arithmetical progression cannot be all squares. There are many triples of squares in arithmetic progression; examples are $1, 25, 49$ and $1, 841, 1641$. Darmon and Merel [24] used an extension of Wiles' method to prove that there exists no integer $n > 2$ for which there are triples of n-th powers in arithmetic progression. This beautiful result was used by Győry [31] to complete the solution of another classical problem. Erdős [25] had shown that

$$\binom{n}{k} = y^l$$

has no solution in positive integers $n \geq 2k \geq 8$, $y > 1, l > 1$. Győry treated the remaining cases $k = 2, 3$ by a combination of estimates for p-adic linear forms in logarithms, Eisenstein's reciprocity theorem and the above-mentioned result of Darmon and Merel. The only solution with $kl > 4$ turns out to be

$$\binom{50}{3} = 140^2.$$

There are many related results in the literature.

5.9 Primitive Divisors of Lucas and Lehmer Sequences

A *primitive divisor* of the n-th term of the sequence $(u_n)_{n=1}^{\infty}$ is a prime which divides u_n, but not u_m for any $m < n$. It follows from old results of Bang, Zsigmondy, and Birkhoff and Vandiver that $a^n - b^n$ has a primitive divisor for $n > 6$ when $a > b > 0$ are integers. For example, for $a = 2$, $b = 1$ we obtain the sequence $1, 3, 7, 15, 31, 63, 127, \ldots$, and according to this result 63 is the last term which does not have new a prime factor in the sequence. *Lucas numbers* are defined as follows: Let P and Q be coprime integers and let α and β be distinct roots of the equation $x^2 - Px - Q = 0$. Then $u_n := (\alpha^n - \beta^n)/(\alpha - \beta)$ and $v_n := \alpha^n + \beta^n$. Lehmer sequences are defined in a similar way. It follows from work of Carmichael, Ward and Durst that if α and β are reals, then there are primitive divisors in Lucas and Lehmer numbers for any $n > 12$. The Fibonacci sequence shows that this result cannot be improved in general:

$1, 1, 2, 3, 5, 8, 13, 21, 34, 55, 89, 144, 233, \ldots$ and the twelfth term 144 has no primitive divisor as its prime factors 2 and 3 both occur in previous terms of the sequence. For non-real α and β upper bounds for the integers n for which the n-th term of the sequence has not a primitive divisor were given by Schinzel and Stewart in the 1970s (see e.g. [50]). Voutier [61] found the much smaller upper bound 30030. Bilu, Hanrot and Voutier [11] have now reached the best possible upper bound 30. This result has been used in quite a few applications.

5.10 Linear Recurrence Sequences

Lucas and Lehmer sequences are special cases of linear recurrence sequences. A *(homogeneous linear) recurrence sequence* $(u_n)_{n=1}^{\infty}$ *of order* r satisfies a relation

$$u_n = a_1 u_{n-1} + \cdots + a_r u_{n-r} \text{ for } n > r,$$

where the numbers $a_1, \ldots, a_r$ and the initial values $u_1, \ldots, u_r$ are given. A strong feature of the p-adic Subspace Theorem is that it gives upper bounds for the numbers of solutions which depend on surprisingly few parameters. A major result in this direction, due to Schlickewei [44], states that if the zeros of the companion polynomial $x^r - a_1 x^{r-1} \cdots - a_r$ lie in a number field of degree d and no two roots have a root of unity as their ratio, then a linear recurrence sequence of order r has at most

$$d^{6r^2} \cdot 2^{28r!}$$

terms equal to 0. Beukers and Schlickewei [10] found a much better upper bound for a recurrence sequence of complex numbers of order 3 with a companion polynomial having simple zeros, namely 61. Schmidt [47] proved that for any recurrence sequence of complex numbers of order r for which the companion polynomial does not have two distinct roots with a root of unity as ratio, the number of terms 0 is bounded by a bound which depends only on r. Schmidt's bound is triply exponential in r. I refer to [29] for bounds which are "only" doubly exponential in r.

Appendix: On Erdős' Conjecture on the Vanishing of Infinite Sums

As mentioned in Section 4, the following conjecture is attributed to Erdős:

Suppose q is a positive integer and $f(x)$ is a number-theoretic function that is periodic mod q with values $f(n) \in \{-1, 1\}$ when $n = 1, 2, \ldots, q-1$ and $f(q) = 0$. Then $\sum_{n=1}^{\infty} f(n)/n \neq 0$ whenever the series is convergent.

It follows from Theorem 7 that this holds if q is prime. Furthermore, Okada [41] proved that the assertion is true if $2\phi(q) + 1 > q$. He derived this result from the following criterion (cf. [41], Theorem 10 and the proof of Corollary 15).

Theorem 8 ([41]). *If f satisfies condition* (ii) *of Theorem 7, then we have* $\sum f(n)/n = 0$ *if and only if* $(f(1), \ldots, f(q))$ *is a solution of the following system of* $\phi(q) + \omega(q)$ *homogeneous linear equations with rational coefficients*

$$f(a) + \sum_{d|q, 1 < d < q} \prod_{p \in P(d)} \left(1 - \frac{1}{p^{\phi(q)}}\right)^{-1} \sum_{n \in S(d)} \frac{f(adn)}{dn}$$

$$+ \frac{f(q)}{\phi(q)} = 0 \ (a \in J), \tag{5.4}$$

$$\sum_{r \in L} f(r)\epsilon(r, p) = 0 \text{ for all prime divisors } p \text{ of } q, \tag{5.5}$$

where

$$J = \{a \mid 1 \le a < q, \gcd(a, q) = 1\},$$
$$L = \{r \mid 1 \le r \le q, \gcd(r, q) > 1\},$$
$$P(d) = \left\{p \text{ prime} \mid p \mid q, \operatorname{ord}_p(d) \ge \operatorname{ord}_p(q)\right\},$$
$$\epsilon(r, p) = \begin{cases} \operatorname{ord}_p(q) + \dfrac{1}{p-1} & \text{if } p \in P(r), \\ \operatorname{ord}_p(r) & \text{otherwise} \end{cases}$$

and

$$S(d) = \left\{\prod_{p \in P(d)} p^{\alpha(p)} \mid 0 \le \alpha(p) < \phi(q)\right\}.$$

Okada derived from (5.4) that $2\phi(q) + 1 \le q$. Since (ii) is satisfied, this implies that $\sum_{n=1}^{\infty} f(n)/n \neq 0$, if q is either a prime, a prime power, or the product of two odd primes. Note that it follows from the convergence of $\sum f(n)/n$ that $\sum_{n=1}^{q} f(n) = 0$. Hence Erdős' conjecture is true if q is even.

Let $f : \mathbb{Z} \to \mathbb{Q}$ be periodic mod q such that $\sum_{n=0}^{\infty} f(n)/n = 0$. From Theorem 8 we immediately obtain the result of Baker, Birch and Wirsing [4] that q is composite. Okada's criterion gives a good way to construct such functions f: the values $f(n)$ with $\gcd(n, q) > 1$ and n composite can be chosen arbitrarily; then the values $f(n)$ with n a prime divisor of q are fixed by (5.5); finally, values $f(n)$ with $\gcd(n, d) = 1$ are determined by (5.4).

The structure of (5.4) and (5.5) becomes much more transparent if f is multiplicative. Let $f\colon \mathbb{Z} \to \mathbb{Q}$ be a multiplicative function that is periodic mod q. Observe that it follows from $f(a)f(q) = f(aq) = f(q)$ for every $a \in J$ that

$$f(q) = 0 \text{ or } f(a) = 1 \text{ for every } a \in J. \tag{5.6}$$

In (5.4) we have that $\gcd(a, dn) = 1$. Hence $f(adn) = f(a) \cdot f(dn)$. Since $f(1) = 1$, we see that (5.4) is equivalent to the single equation

$$1 + \sum_{d|q, 1<d<q} \prod_{p \in P(d)} \left(1 - \frac{1}{p^{\phi(q)}}\right)^{-1} \sum_{n \in S(d)} \frac{f(dn)}{dn} + \frac{f(q)}{\phi(q)} = 0.$$

Since

$$\frac{f(q)}{\phi(q)} = \sum_n^* \frac{f(qn)}{qn},$$

where the summation is extended over all positive integers n which are composed of prime divisors of q, the left-hand side equals

$$\sum_n^* \frac{f(n)}{n} = \prod_{p|q} \left(1 + \frac{f(p)}{p} + \frac{f(p^2)}{p^2} + \dots\right).$$

Hence, for multiplicative functions f with period q, condition (5.4) is equivalent to

$$1 + \frac{f(p)}{p} + \frac{f(p^2)}{p^2} + \dots = 0 \text{ for some prime divisor } p \text{ of } q. \tag{5.7}$$

It follows immediately from (5.7) and Theorem 8 with (5.7) in place of (5.4) that the only periodic multiplicative functions $f\colon \mathbb{Z} \to \mathbb{Q}$ with period 4 are given by (4.1); those with period 6 are given by $f(n) = 0, 1, -3, 0, 3, -1$ and $f(n) = 2, 1, -1, -2, -1, 1$ for $n \equiv 0, 1, 2, 3, 4, 5 \bmod 6$, respectively; and those with period 8 by $f(n) = 0, 1, -2, -1, 0, 1, 2, -1$ and $f(n) = 4 + t, 1, t, 1, -3t - 8, 1, t, 1$ for $n \equiv 0, 1, 2, 3, 4, 5, 6, 7 \bmod 8$, respectively, where $t \in \mathbb{Q}$ can be chosen arbitrarily. On the other hand, we now derive the following result.

Theorem 9. *Let $f\colon \mathbb{Z} \to \mathbb{Q}$ be multiplicative and have period q. If $|f(p^k)| < p - 1$ for every prime divisor p of q and every positive integer k, then $\sum_{n=1}^{\infty} f(n)/n \neq 0$.*

Proof. It follows from (5.7) that there is a prime divisor p of the period q such that $f(p^k) = -(p-1)$ for $k = 1, 2, \dots$. This yields a contradiction to (5.5), applied to this prime p. $\square$

It follows from Theorem 9 that Erdős' statement is true for multiplicative functions f.

If f is completely multiplicative and periodic mod q, then (5.7) implies that q is even and $f(p) = -p$ for some prime divisor p of q. Hence $f(p^k) = (-p)^k$ for every positive integer k, which is impossible for a periodic function. Thus we have:

Theorem 10. Let $f \colon \mathbb{Z} \to \mathbb{Q}$ be completely multiplicative and periodic. Then $\sum_{n=1}^{\infty} f(n)/n \neq 0$.

By combining this result with Theorem 5 we obtain Corollary 5.2.

Acknowledgements. I thank F. Beukers, Yu. F. Bilu, J.-H. Evertse, K. Győry, Yu. V. Nesterenko and T. N. Shorey for useful discussions and comments on early versions of the paper.

References

[1] S. D. Adhikari, N. Saradha, T. N. Shorey, and R. Tijdeman, *Transcendental infinite sums*, Indag. Math. **12** (2001), 1–14.

[2] A. Baker, *Transcendental number theory*, Cambridge University Press, 1975.

[3] ——, *The theory of linear forms in logarithms*, Transcendence Theory: Advances and Applications (A. Baker and D. W. Masser, eds.), Academic Press, 1977, pp. 1–27.

[4] A. Baker, B. J. Birch, and E. A. Wirsing, *On a problem of Chowla*, J. Number Theory **5** (1973), 224–236.

[5] A. Baker and J. Coates, *Integral points on curves of genus 1*, Proc. Cambridge Philos. Soc. **67** (1970), 595–602.

[6] A. Baker and G. Wüstholz, *Logarithmic forms and group varieties*, J. Reine Angew. Math. **442** (1993), 19–62.

[7] C. D. Bennett, J. Blass, A. M. W. Glass, D. B. Meronk, and R. P. Steiner, *Linear forms in the logarithms of three positive rational integers*, J. Théor. Nombres Bordeaux **9** (1997), 97–136.

[8] M. Bennett, *Rational approximation to algebraic numbers of small height: The Diophantine equation $|ax^n - by^n| = 1$*, J. Reine Angew. Math. **535** (2001), 1–49.

[9] F. Beukers, *The diophantine equation $Ax^p + By^q = Cz^r$*, Duke Math. J. **91** (1998), 61–88.

[10] F. Beukers and H. P. Schlickewei, *The equation $x + y = 1$ in finitely generated groups*, Acta Arith. **78** (1996), 189–199.

[11] Y. F. Bilu, G. Hanrot, and P. M. Voutier, *Existence of primitive divisors of Lucas and Lehmer numbers (with an appendix by M. Mignotte)*, J. Reine Angew. Math. **539** (2001), 75–122.

[12] C. Breuil, B. Conrad, F. Diamond, and R. Taylor, *On the modularity of elliptic curves over* $\mathbb{Q}$, J. Amer. Math. Soc. **14** (2001), 843–939.

[13] N. Bruin, *Chabauty methods and covering techniques applied to generalised Fermat equations*, Ph.D. thesis, Leiden University, 1999.

[14] ——, *The diophantine equations $x^2 \pm y^4 = \pm z^6$ and $x^2 + y^8 = z^3$*, Compositio Math. **118** (1999), 305–321.

[15] ——, *On powers as sums of two cubes*, Algorithmic Number Theory (W. Bosma, ed.), LNCS 1838, Springer, 2000, pp. 169–184.

[16] Y. Bugeaud and M. Laurent, *Minoration effective de la distance p-adique entre puissances de nombres algébriques*, J. Number Theory **61** (1996), 311–342.

[17] Y. Bugeaud and M. Mignotte, *On integers with identical digits*, Mathematika **46** (1999), 411–417.

[18] Y. Bugeaud, M. Mignotte, and Y. Roy, *On the diophantine equation $\frac{x^n-1}{x-1} = y^q$*, Pacific J. Math. **193** (2000), 257–268.

[19] Y. Bugeaud and T. N. Shorey, *On an equation of Goormaghtigh, II*, to appear.

[20] S. Chowla, *The Riemann zeta and allied functions*, Bull. Amer. Math. Soc. **58** (1952), 287–305.

[21] ——, *The nonexistence of nontrivial linear relations between the roots of a certain irreducible equation*, J. Number Theory **2** (1970), 120–123.

[22] H. Darmon, *The equations $x^n + y^n = z^2$ and $x^n + y^n = z^3$*, Int. Math. Res. Notices **10** (1993), 263–274.

[23] H. Darmon and A. Granville, *On the equations $z^m = F(x, y)$ and $Ax^p + By^q = Cz^r$*, Bull. London Math. Soc. **27** (1995), 513–543.

[24] H. Darmon and L. Merel, *Winding quotients and some variants of Fermat's last theorem*, J. Reine Angew. Math. **490** (1997), 81–100.

[25] P. Erdős, *On a diophantine equation*, J. London Math. Soc. **26** (1951), 176–178.

[26] P. Erdős, C. L. Stewart, and R. Tijdeman, *Some diophantine equations with many solutions*, Compositio Math. **66** (1988), 37–56.

[27] J.-H. Evertse, *On sums of S-units and linear recurrences*, Compositio Math. **53** (1984), 225–244.

[28] ———, *The number of solutions of decomposable form equations*, Invent. Math. **122** (1995), 559–601.

[29] J.-H. Evertse, H. P. Schlickewei, and W. M. Schmidt, *Linear equations in variables which lie in a multiplicative group*, to appear.

[30] J.-H. Evertse, C. L. Stewart, and R. Tijdeman, *Multivariate diophantine equations with many solutions*, to appear.

[31] K. Győry, *On the diophantine equation $\binom{n}{k} = x^l$*, Acta Arith. **80** (1997), 289–295.

[32] A. Kraus, *Sur l'équation $a^3 + b^3 = c^p$*, Experiment. Math. **7** (1998), 1–13.

[33] M. Laurent, M. Mignotte, and Y. V. Nesterenko, *Formes linéaires en deux logarithmes et déterminants d'interpolation*, J. Number Theory **55** (1995), 285–321.

[34] D. H. Lehmer, *Euler constants for arithmetical progressions*, Acta Arith. **27** (1975), 125–142.

[35] A. E. Livingston, *The series $\sum_1^\infty f(n)/n$ for periodic f*, Canad. Math. Bull. **8** (1965), 413–432.

[36] E.M. Matveev, *An explicit lower bound for a homogeneous rational linear form in logarithms of algebraic numbers. II*, Izv. Math. **64** (2000), 1217 – 1269.

[37] R. D. Mauldin, *A generalization of Fermat's Last Theorem: the Beal conjecture and prize problem*, Notices Amer. Math. Soc. **44** (1997), 1436–1437.

[38] M. Mignotte, *Lower bounds for Catalan's equation*, The Ramanujan J. **1** (1997), 351–356.

[39] _____ , *A corollary to the Laurent-Mignotte-Nesterenko theorem*, Acta Arith. **86** (1998), 101–111.

[40] P. Mihailescu, *A class number free criterion for Catalan's conjecture*, to appear.

[41] T. Okada, *On a certain infinite series for a periodic arithmetical function*, Acta Arith. **40** (1982), 143–153.

[42] H. P. Schlickewei, *The p-adic Thue-Siegel-Roth-Schmidt theorem*, Arch. Math. **29** (1977), 267–270.

[43] _____ , *The quantitative Subspace Theorem for number fields*, Compositio Math. **82** (1992), 245–274.

[44] _____ , *Multiplicities of recurrence sequences*, Acta Math. **176** (1996), 171–243.

[45] W. M. Schmidt, *Norm form equations*, Ann. Math. **96** (1972), 526–551.

[46] _____ , *The subspace theorem in diophantine approximation*, Compositio Math. **69** (1989), 121–173.

[47] _____ , *The zero multiplicity of linear recurrences*, Acta Math. **182** (1999), 243–282.

[48] T. N. Shorey, *Some conjectures in the theory of exponential diophantine equations*, Publ. Math. Debrecen **56** (2000), 631–641.

[49] T. N. Shorey and R. Tijdeman, *Exponential diophantine equations*, Cambridge University Press, 1986.

[50] C. L. Stewart, *On divisors of Fermat, Fibonacci, Lucas and Lehmer numbers*, Proc. Lond. Math. Soc. **35** (1977), 425–447.

[51] C. L. Stewart and R. Tijdeman, *On the Oesterlé-Masser conjecture*, Monatsh. Math. **102** (1986), 251–257.

[52] C. L. Stewart and K. R. Yu, *On the abc-conjecture*, Math. Ann. **291** (1991), 225–230.

[53] _____ , *On the abc-conjecture II*, Duke Math. J. **108** (2001), 169–181.

[54] R. L. Taylor and A. Wiles, *Ring-theoretic properties of certain Hecke algebras*, Ann. Math. **141** (1995), 553–572.

[55] R. Tijdeman, *On integers with many small prime factors*, Compositio Math. **26** (1973), 319–330.

[56] _____, *On the maximal distance between numbers composed of fixed primes*, Compositio Math. **28** (1974), 159–162.

[57] _____, *On the equation of Catalan*, Acta Arith. **29** (1976), 197–209.

[58] _____, *Diophantine equations and diophantine approximations*, Number Theory and Applications (R. Mollin, ed.), Kluwer, Dordrecht etc., 1989, pp. 215–243.

[59] A. J. van der Poorten and H. P. Schlickewei, *Additive relations in fields*, J. Austral. Math. Soc. Ser. A **51** (1991), 154–170.

[60] M. van Frankenhuysen, *A lower bound in the abc-conjecture*, J. Number Theory **82** (2000), 91–95.

[61] P. M. Voutier, *Primitive divisors of Lucas and Lehmer numbers III*, Math. Proc. Cambridge Philos. Soc. **123** (1998), 407–419.

[62] M. Waldschmidt, *Minorations de combinaisons linéaires de logarithmes de nombres algébriques*, Canadian J. Math. **45** (1993), 176–224.

[63] A. Wiles, *Modular elliptic curves and Fermat's Last Theorem*, Ann. Math. **141** (1995), 443–551.

[64] K. Yu, *p-adic logarithmic forms and group varieties I*, J. Reine Angew. Math. **502** (1998), 29–92.

[65] _____, *p-adic logarithmic forms and group varieties II*, Acta Arith. **89** (1999), 337–378.

On the Solutions of a Family of Sextic Thue Equations

Alain Togbé

1 Introduction

We consider the Thue equation

$$F_n(x, y) = x^6 - 2A_n x^5 y - 5(A_n + 3)x^4 y^2 - 20x^3 y^3$$
$$+ 5A_n x^2 y^4 + 2(A_n + 3)xy^5 + y^6 = 1 \quad (1.1)$$

with $A_n = 54n^3 + 81n^2 + 54n + 12$, $n \in \mathbb{Z}$. Our goal is to find all integer solutions to (1.1), for all values of the parameter n; for reasons we will detail, we are unable to achieve this. Indeed, we were unsuccessful in applying the classical approaches (see [9], [12], [13], [15], [16]), the method in [18] and the generalized method of Gáal-Pethő-Pohst (see [4]).

In this paper, we solve (1.1) for small values of n, via techniques of Bilu-Hanrot (see [2]).

In 1986, M. N. Gras (see [5] and [6]) studied the units of $\mathbb{Q}(\omega)$, where ω is a root of the polynomial $F_n(x, 1)$; such information is helpful in solving the equation (1.1). For the proof, we use estimates for linear forms in the logarithms of algebraic numbers, techniques from diophantine approximation and the computational method of Bilu-Hanrot.

This paper is based upon the results of our Ph.D. dissertation (see [17]), and is divided into 5 sections. In the second section, we give some elementary properties of the above polynomial and recall the results of Gras about an independent system of units of the number field associated with this Thue equation. We study the action of the Galois Group on this independent system of units. In the last two sections, we study the approximation properties of the solutions and we solve the equation for small values of n using the Bilu-Hanrot method (see [2]), with a program written in PARI/GP. The main result that we obtain is

Theorem 1.1. *If n is an integer such that $|n| \leq 2.03 \times 10^6$, then the equation (1.1) with $A_n = 54n^3 + 81n^2 + 54n + 12$, $n \in \mathbb{Z}$, has only the trivial solutions $(-1, 0)$, $(-1, 1)$, $(0, -1)$, $(0, 1)$, $(1, -1)$, $(1, 0)$.*

2 Elementary Properties of the Polynomial

We have the following properties.

(1) Set $A_n = A(n)$, a function of n. Then $A(-n-1) + 3 = -A(n)$, $n \in \mathbb{Z}$. So $-A(n)$ and $A(n)$ are symmetric with respect to the axis $n = -1/2$, and we may thus restrict our attention to $F_n(x, y)$ for $n \in \mathbb{N}$.

(2) We have

$$F_n(x, y) = F_n(-x, -y) = F_n(-y, x+y) = F_n(-x-y, x).$$

Hence, if (x, y) is an integral solution of (1.1), so are $(-x, -y)$, $(-y, x+y)$, $(-x-y, x)$.

(3) $(1, 0)$, $(-1, 0)$, $(0, 1)$, $(0, -1)$, $(1, -1)$ and $(-1, 1)$ are integral solutions of (1.1), called the trivial solutions.

(4) We consider only the integral solutions with y positive for small values of n.

Now we recall the work of Gras (see [5], [6]). Let θ be a root of the polynomial

$$P(x) = x^6 - \frac{t-6}{2}x^5 - 5\frac{t+6}{4}x^4 - 20x^3 + 5\frac{t-6}{4}x^2 + \frac{t+6}{2}x + 1, \quad (2.1)$$

where $t \in \mathbb{Z}$, and the other roots of $P(x)$ are

$$\sigma(\theta) = \frac{\theta - 1}{\theta + 2},$$

$$\sigma^2(\theta) = \frac{-1}{\theta + 1},$$

$$\sigma^3(\theta) = -\frac{\theta + 2}{2\theta + 1},$$

$$\sigma^4(\theta) = -\frac{\theta + 1}{\theta},$$

$$\sigma^5(\theta) = -\frac{2\theta + 1}{\theta - 1}.$$

Proposition 2.1. *If $t \in \mathbb{Z} \setminus \{0, \pm 6, \pm 26\}$, then $P(x)$ is irreducible in $\mathbb{Q}[x]$.*

Proof. See [5, p. 8, Prop. 3.3]. □

Remarks 2.2. Suppose that $t \in \mathbb{Z}$. Then $\mathbb{K}_t = \mathbb{Q}(\theta)$ is a cyclic number field with degree a divisor of 6.

(1) $[\mathbb{K}_t : \mathbb{Q}] = 3$ if and only if $t^2 + 108$ is square, i.e., $t = \pm 6$ or $t = \pm 26$.

(2) $[\mathbb{K}_t : \mathbb{Q}] = 2$ if and only if $x^3 - \frac{t-6}{4}x^2 - \frac{t+6}{4}x - 1$ has rational roots, i.e., $t = 0$.

(3) If t is changed to $-t$, then θ is changed to θ^{-1}; so we consider $t \geq 0$.

For $t \in \mathbb{N} \setminus \{0, 6, 26\}$, let $\omega \in \mathbb{K}_t$ be defined by

$$\omega = \theta^{1-\sigma^3} = -\frac{\theta(2\theta + 1)}{\theta + 2}; \tag{2.2}$$

then $\mathrm{Irr}(\omega, \mathbb{Q}) = (x - 1)^6 - (t^2 + 108)(x^2 + x)^2$.

Proposition 2.3. *There exists ν such that $\omega = \nu^{1+\sigma}$ if and only if $t = s(s^2 + 9)$, $s \in \mathbb{Z}$.*

Proof. See [5, p. 14, Prop. 4.3]. □

Put $t = s(s^2 + 9)$, and $L_s = \mathbb{K}_t = \mathbb{Q}(\omega)$, where ω is defined by (2.2).

Proposition 2.4. *We have $L_s = \mathbb{Q}(\nu)$, where*

$$\mathrm{Irr}(\nu, \mathbb{Q}) = (x - 1)^6 + (s^2 + 12)(x^5 - x^4 - s^2 x^3 - x^2 + x). \tag{2.3}$$

Proof. See [5, p. 23, Prop. 4.25]. □

We consider $t = 4(54n^3 + 81n^2 + 54n + 12) + 6$ and

$$t^2 + 108 = (s^2 + 3)^2(s^2 + 12) = (36n^2 + 36n + 12)^2(36n^2 + 36n + 21).$$

In this case, the number field

$$\mathbb{K}_t = \mathbb{Q}(\theta) = \mathbb{Q}(\omega) = \mathbb{Q}(\nu)$$

is defined by the following three polynomials:

$$f_n(x) = x^6 - 2A_n x^5 - 5(A_n + 3)x^4 - 20x^3 + 5A_n x^2 + 2(A_n + 3)x + 1,$$

where $A_n = \frac{t-6}{4} = 54n^3 + 81n^2 + 54n + 12$;

$$g_n(x) = x^6 - 6x^5 - B_n x^4 - 2(B_n + 25)x^3 - B_n x^2 - 6x + 1,$$

where $B_n = 46656n^6 + 139968n^5 + 198288n^4 + 163296n^3 + 81648n^2 + 23328n + 3009$;

$$h_n(x) = x^6 + C_n x^5 - (C_n - 9)x^4 - (C_n + 4)(C_n - 4)x^3 - (C_n - 9)x^2 + C_n x + 1,$$

where $C_n = 36n^2 + 36n + 15$, $n \in \mathbb{Z}$.

Its quadratic subfield is $k_2 = \mathbb{Q}\left(\sqrt{36n^2 + 36n + 21}\right)$ with a unit

$$\varepsilon_2 = \frac{(12n^2 + 12n + 5) + (2n + 1)\sqrt{36n^2 + 36n + 21}}{2}, \tag{2.4}$$

(see [14]); its cubic subfield is $k_3 = \mathbb{Q}(\phi)$ with

$$\mathrm{Irr}(\phi, \mathbb{Q}) = x^3 - (54n^3 + 81n^2 + 54n + 12)x^2 - (54n^3 + 81n^2 + 54n + 15)x - 1.$$

Thus $\{\varepsilon_2, \phi, \phi^\sigma, \nu, \nu^\sigma\}$ is an independent system of units of $\mathbb{K}_t$.

Remark 2.5. In fact, ε_2 is a fundamental unit of k_2 if $s = 6n + 3$, with $(6n+3)^2 + 3$ and $(6n+3)^2 + 12$ square-free, up to powers of 2. But by the method of Bilu-Hanrot, it is sufficient to work with an independent system of units. Here we would like to thank Gary Walsh and Guillaume Hanrot for helping to clarify this point.

3 Galois Group Action on This Independent System of Units

Let $S = \{\varepsilon_2, \varepsilon_3, \varepsilon_3', \nu, \nu'\}$ be the independent system of units of $\mathbb{K}_t$ obtained previously, with

$$\varepsilon_3 = |\phi| = -\phi = -\theta^{-1-\sigma^3} = \frac{2\theta + 1}{\theta(\theta + 2)},$$

$$\varepsilon_3' = |\phi^\sigma| = -\phi^\sigma = -\theta^{-\sigma-\sigma^4} = \frac{\theta(\theta + 2)}{(\theta - 1)(\theta + 1)},$$

$$\nu^{3\sigma} = \omega^{1+\sigma} = \theta^{(1+\sigma)(1-\sigma^3)} = \theta^{1+\sigma-\sigma^3-\sigma^4} = \frac{\theta^2(\theta-1)(2\theta+1)}{(\theta+1)(\theta+2)^2}, \tag{3.1}$$

$$\nu^3 = (\nu^{3\sigma})^{\sigma^5} = \theta^{1-\sigma^2-\sigma^3+\sigma^5} = -\frac{\theta(\theta+1)(2\theta+1)^2}{(\theta-1)(\theta+2)}.$$

Proposition 3.1. *The action of the Galois group $G = \mathrm{Gal}(\mathbb{K}_t/\mathbb{Q})$ on the independent system of units is as given in the following table.*

	ε_2	ε_3	ε_3'	ν	ν'
1	ε_2	ε_3	ε_3'	ν	ν'
σ	ε_2^{-1}	ε_3'	$1/(\varepsilon_3\varepsilon_3')$	ν'	$\nu^{-1}\nu'$
σ^2	ε_2	$1/(\varepsilon_3\varepsilon_3')$	ε_3	$\nu^{-1}\nu'$	ν^{-1}
σ^3	ε_2^{-1}	ε_3	ε_3'	ν^{-1}	$1/\nu'$
σ^4	ε_2	ε_3'	$1/(\varepsilon_3\varepsilon_3')$	$1/\nu'$	ν/ν'
σ^5	ε_2^{-1}	$1/(\varepsilon_3\varepsilon_3')$	ε_3	ν/ν'	ν

Proof. We have $\mathrm{Gal}(\mathbb{K}_t/k_2) = G_3 = \{1, \sigma^2, \sigma^4\}$ and $\mathrm{Gal}(k_2/\mathbb{Q}) \simeq G_2 = \{1, \sigma^3\}$, so

$$\varepsilon_2 = \sigma^2(\varepsilon_2) = \sigma^4(\varepsilon_2),$$

and
$$\varepsilon_2^{-1} = \sigma(\varepsilon_2) = \sigma^3(\varepsilon_2) = \sigma^5(\varepsilon_2)$$

because $N_{k_2/\mathbb{Q}}(\varepsilon_2) = 1$. Also, $\mathrm{Gal}(\mathbb{K}_4/k_3) = G_2 = \{1, \sigma^3\}$ and $\mathrm{Gal}(k_3/\mathbb{Q}) \simeq G_3 = \{1, \sigma^2, \sigma^4\}$, and therefore
$$\varepsilon_3 = \sigma^3(\varepsilon_3), \varepsilon_3' = \sigma(\varepsilon_3) = \sigma^4(\varepsilon_3)$$

and
$$\sigma^2(\varepsilon_3) = \sigma^5(\varepsilon_3) = 1/(\varepsilon_3 \varepsilon_3')$$

because $N_{k_3/\mathbb{Q}}(\phi) = -1$. Finally, we have

$$\nu^{\sigma^2} = \nu^{-1+\sigma} = \nu^{-1}\nu', \qquad\qquad \nu^{\sigma^3} = \nu^{-1},$$
$$\nu^{\sigma^4} = \nu^{-\sigma} = 1/\nu', \qquad\qquad \nu^{\sigma^5} = \nu^{-\sigma^2} = \nu/\nu'$$

because $\nu' = \nu^\sigma$ and $\nu^{1-\sigma+\sigma^2} = 1$. $\qquad\qquad\square$

Now we determine θ_i, ω_i and ν_i, $1 \le i \le 6$. Let θ be the greatest root of $f_n(x)$. One can check that $\frac{t-2}{2} < \theta < \frac{t}{2}$, so we obtain

$$\theta_1 = \theta, \qquad\qquad \frac{t-2}{2} < \theta_1 < \frac{t}{2}, \qquad\qquad \text{i.e., } \theta_1 \in \left(\frac{t-2}{2}, \frac{t}{2}\right);$$

$$\theta_2 = \frac{\theta-1}{\theta+2}, \qquad \frac{t-4}{t+4} < \theta_2 < \frac{t-2}{t+2}, \qquad \text{i.e., } \theta_2 \in (0,1);$$

$$\theta_3 = -\frac{1}{\theta+1}, \qquad -\frac{2}{t} < \theta_3 < -\frac{2}{t+2}, \qquad \text{i.e., } \theta_3 \in (-1,0);$$

$$\theta_4 = -\frac{\theta+2}{2\theta+1}, \qquad -\frac{t+4}{2(t-1)} < \theta_4 < -\frac{t+2}{2(t+1)}, \qquad \text{i.e., } \theta_4 \in (-1,0);$$

$$\theta_5 = -\frac{\theta+1}{\theta}, \qquad -\frac{t+2}{t-2} < \theta_5 < -1, \qquad \text{i.e., } \theta_5 \in (-2,-1);$$

$$\theta_6 = -\frac{2\theta+1}{\theta-1}, \qquad -\frac{2(t+1)}{t-4} < \theta_6 < -\frac{2(t-1)}{t-2}, \qquad \text{i.e., } \theta_6 \in (-3,-2).$$

We deduce
$$\theta_6 < \theta_5 < \theta_4 < \theta_3 < \theta_2 < \theta_1. \tag{3.2}$$

Similarly, we have:

$$\omega_1 = -\frac{\theta(2\theta+1)}{\theta+2}, \qquad -\frac{t(t+1)}{t+2} < \omega_1 < -\frac{(t-2)(t-1)}{t+4},$$
$$\text{i.e. } \omega_1 \in (-t+3, -t+4);$$

$$\omega_2 = -\frac{\theta(\theta-1)}{(\theta+1)(\theta+2)}, \qquad -\frac{t-2}{t+2} < \omega_2 < -\frac{(t-2)(t-4)}{(t+2)(t+4)},$$

i.e., $\omega_2 \in (-1,0)$;

$$\omega_3 = \frac{\theta-1}{(\theta+1)(2\theta+1)}, \qquad \frac{t-4}{(t+2)(t+1)} < \omega_3 < \frac{t-2}{t(t-1)},$$

i.e., $\omega_3 \in (0,1)$;

$$\omega_4 = -\frac{\theta+2}{\theta(2\theta+1)}, \qquad -\frac{t+4}{(t-2)(t-1)} < \omega_4 < -\frac{t+2}{t(t+1)},$$

i.e., $\omega_4 \in (-1,0)$;

$$\omega_5 = -\frac{(\theta+1)(\theta+2)}{\theta(\theta-1)}, \qquad -\frac{(t+2)(t+4)}{(t-2)(t-4)} < \omega_5 < -\frac{t+2}{t-2},$$

i.e., $\omega_5 \in (-2,-1)$;

$$\omega_6 = \frac{(\theta+1)(2\theta+1)}{\theta-1}, \qquad \frac{t(t-1)}{t-2} < \omega_6 < \frac{(t+2)(t+1)}{t-4},$$

i.e., $\omega_6 \in (t+4, t+5)$;

and

$$-\frac{t(t+1)^2}{t-4} < \nu_1^3 < -\frac{t(t-1)^2}{t+4},$$

$$\text{i.e., } \nu_1^3 \in (-(t+1/2)^2, -t^2);$$

$$\frac{(t-2)^2(t-4)(t-1)}{(t+2)(t+4)^2} < \nu_2^3 < \frac{t(t-2)(t+1)}{(t+2)^2},$$

$$\text{i.e., } \nu_2^3 \in (t-12, t-11);$$

$$-\frac{(t-2)^2}{t(t+2)(t-1)} < \nu_3^3 < -\frac{(t-2)(t-4)^2}{(t+4)(t+1)(t+2)^2},$$

$$\text{i.e., } \nu_3^3 \in (-1,0);$$

$$-\frac{t+4}{t(t-1)^2} < \nu_4^3 < -\frac{t-4}{t(t+1)^2},$$

$$\text{i.e., } \nu_4^3 \in (-1,0);$$

$$\frac{(t+2)^2}{t(t-2)(t+1)} < \nu_5^3 < \frac{(t+2)(t+4)^2}{(t-2)^2(t-4)(t-1)},$$

$$\text{i.e., } \nu_5^3 \in (0,1);$$

$$-\frac{(t+4)(t+1)(t+2)^2}{(t-2)(t-4)^2} < \nu_6^3 < -\frac{t(t+2)(t-1)}{(t-2)^2},$$

$$\text{i.e., } \nu_6^3 \in (-t-13, -t-12).$$

Also we have

$$\omega_1 < \omega_5 < \omega_2 < \omega_4 < \omega_3 < \omega_6, \quad \nu_1 < \nu_6 < \nu_3 < \nu_4 < \nu_5 < \nu_2. \qquad (3.3)$$

Finally, we have

$$\tfrac{1}{3}t^{\frac{2}{3}} < \varepsilon_2 < \tfrac{1}{3}\left(t+\tfrac{1}{2}\right)^{\frac{2}{3}}, \text{ for } n > 0. \qquad (3.4)$$

Now we use Cardano's formula to determine θ_i, ω_i and ν_i. Put

$$\Delta = 4p^3 + 27q^2,$$

$$j = -1/2 + i\sqrt{3}/2.$$

Then

$$h_n(x) = x^3\left[\left(x + \frac{1}{x}\right)^3 + C_n\left(x + \frac{1}{x}\right)^2\right.$$

$$\left. - (C_n - 6)\left(x + \frac{1}{x}\right) - (C_n^2 + 2C_n - 16)\right].$$

Setting $X = x + \frac{1}{x}$, we have

$$X^3 + C_n X^2 - (C_n - 6)X - (C_n^2 + 2C_n - 16) = 0,$$

and with $X = Y - C_n/3$, we obtain

$$Y^3 + (6 - C_n - C_n^2/3)Y + (16 - 4C_n - 2C_n^2/3 + 2C_n^3/27) = 0.$$

Put $Y = u + v$, $p = 6 - C_n - C_n^2/3$ and $q = 16 - 4C_n - 2C_n^2/3 + 2C_n^3/27$. Thus u^3 and v^3 are solutions of $Z^2 + qZ - p^3/27 = 0$, and so

$$u = \sqrt[3]{-27q/2 + 3\sqrt{3\Delta}/2},$$

$$a_1 = \frac{u + \bar{u} - C_n}{3}, \quad a_2 = \frac{ju + \overline{ju} - C_n}{3}, \quad a_3 = \frac{j^2 u + \overline{j^2 u} - C_n}{3},$$

where $\bar{u}$ is the complex conjugate of u. Therefore

$$\nu_1 = \frac{a_2 - \sqrt{a_2^2 - 4}}{2}, \quad \nu_2 = \frac{a_1 + \sqrt{a_1^2 - 4}}{2}, \quad \nu_3 = \frac{a_3 + \sqrt{a_3^2 - 4}}{2},$$

$$\nu_4 = \frac{a_2 + \sqrt{a_2^2 - 4}}{2}, \quad \nu_5 = \frac{a_1 - \sqrt{a_1^2 - 4}}{2}, \quad \nu_6 = \frac{a_3 - \sqrt{a_3^2 - 4}}{2},$$

whereby $\varepsilon = |\nu_1|, \varepsilon' = |\nu_2|$.

We use the same approach with g_n to obtain

$$a_1 = \frac{u + \bar{u}}{3} + 2, \quad a_2 = \frac{ju + \overline{ju}}{3} + 2, \quad a_3 = \frac{j^2 u + \overline{j^2 u}}{3} + 2,$$

and so

$$\omega_1 = \frac{a_2 - \sqrt{a_2^2 - 4}}{2}, \quad \omega_2 = \frac{a_3 + \sqrt{a_3^2 - 4}}{2}, \quad \omega_3 = \frac{a_1 - \sqrt{a_1^2 - 4}}{2},$$

$$\omega_4 = \frac{a_2 + \sqrt{a_2^2 - 4}}{2}, \quad \omega_5 = \frac{a_3 - \sqrt{a_3^2 - 4}}{2}, \quad \omega_6 = \frac{a_1 + \sqrt{a_1^2 - 4}}{2}.$$

For $\mathrm{Irr}(\phi, \mathbb{Q})$, we have

$$\phi_3 = \frac{u + \bar{u}}{3} + (18n^3 + 27n^2 + 18n + 4),$$

$$\phi_2 = \frac{ju + \overline{ju}}{3} + (18n^3 + 27n^2 + 18n + 4),$$

$$\phi_1 = \frac{j^2 u + \overline{j^2 u}}{3} + (18n^3 + 27n^2 + 18n + 4).$$

We know that $\phi_1 = \phi = \theta^{-1-\sigma^3}$, $\omega_1 = \omega = \theta^{1-\sigma^3}$. Hence $\theta^2 = \frac{\omega}{\phi}$, i.e., $\theta = \sqrt{\frac{\omega}{\phi}}$.

4   Approximation Properties of the Solutions of (1.1)

We begin this section with the following result.

Proposition 4.1. *Let (x, y) be an integral solution of* (1.1); *then we have* $|y| \geq 2$ *or* $(x, y) = (-1, 0), (-1, 1), (0, -1), (0, 1), (1, 0), (1, -1)$.

Proof. We know that $(1, 0)$, $(-1, 0)$, $(0, 1)$, $(0, -1)$, $(1, -1)$, $(-1, 1)$ are trivial solutions of (1.1). If $y = 1$, then $x = 0$, $x = -1$ or

$$P_n(x) = x^4 - (2A_n + 1)x^3 - (3A_n + 14)x^2 + 3(A_n - 2)x + 2(A_n + 3) = 0.$$

For $n \geq 0$, using Maple V, we have

$$P_n(-3) = 246 + 1080n + 1620n^2 + 1080n^3 > 0,$$

$$P_n(-2) = -14 < 0,$$

$$P_n(-1) = -24 - 108n - 162n^2 - 108n^3 < 0,$$

$$P_n(0) = 30 + 108n + 162n^2 + 108n^3 > 0,$$

$$P_n(1) = -14 < 0,$$

$$P_n(2A_n + 2) = -15414 - 163728n - 811296n^2 - 2490696n^3$$
$$- 5238594n^4 - 7838208n^5 - 8360172n^6$$
$$- 6141096n^7 - 2834352n^8 - 629856n^9 < 0,$$

$$P_n(2A_n + 3) = 3756 + 70416n + 484704n^2 + 1837512n^3$$
$$+ 4445442n^4 + 7278336n^5 + 8173548n^6$$
$$+ 6141096n^7 + 2834352n^8 + 629856n^9 > 0.$$

Therefore the roots of P_n are in $(-3, -2)$, $(-1, 0)$, $(0, 1)$, $(2A_n + 2, 2A_n + 3)$. $\qquad\square$

Now put

$$\gamma_i = x - \theta_i y. \tag{4.1}$$

Then

$$F_n(x, y) = \prod_{i=1}^{6} \gamma_i = 1. \tag{4.2}$$

This means that the γ_i are units of $\mathbb{K}_t$. So there exist b_0, b_1, b_2, b_3, b_4, $b_5 \in \mathbb{Z}$, such that

$$\gamma_1 = (-1)^{b_0} \varepsilon_2^{b_1} \varepsilon_3^{b_2} (\varepsilon_3')^{b_0} \nu^{b_4} (\nu')^{b_5}.$$

We deduce the following system:

$$|\gamma_1| = \varepsilon_2^{b_1} \varepsilon_3^{b_2} (\varepsilon_3')^{b_3} \varepsilon^{b_4} (\varepsilon')^{b_5},$$

$$|\gamma_2| = \varepsilon_2^{-b_1} \varepsilon_3^{-b_3} (\varepsilon_3')^{b_2-b_3} \varepsilon^{-b_5} (\varepsilon')^{b_4+b_5},$$

$$|\gamma_3| = \varepsilon_2^{b_1} \varepsilon_3^{-b_2+b_3} (\varepsilon_3')^{-b_2} \varepsilon^{-b_4-b_5} (\varepsilon')^{b_4},$$

$$|\gamma_4| = \varepsilon_2^{-b_1} \varepsilon_3^{b_2} (\varepsilon_3')^{b_3} \varepsilon^{-b_4} (\varepsilon')^{-b_5}, \tag{4.3}$$

$$|\gamma_5| = \varepsilon_2^{b_1} \varepsilon_3^{-b_3} (\varepsilon_3')^{b_2-b_3} \varepsilon^{b_5} (\varepsilon')^{-b_4-b_5},$$

$$|\gamma_6| = \varepsilon_2^{-b_1} \varepsilon_3^{-b_2+b_3} (\varepsilon_3')^{-b_2} \varepsilon^{b_4+b_5} (\varepsilon')^{-b_4}.$$

There are six possibilities for the minimum of $|\gamma_i|$, $1 \le i \le 6$. We say that (x,y) is of

- type I, if all $\gamma_i's$ are positive and $\gamma_1 = \min_{1 \le i \le 6}\{|\gamma_i|\}$;

- type II, if γ_1, γ_2 are negative, γ_3, γ_4, γ_5, γ_6 are positive and $|\gamma_2| = \min_{1 \le i \le 6}\{|\gamma_i|\}$;

- type III, if γ_1, γ_2 are negative, γ_3, γ_4, γ_5, γ_6 are positive and $\gamma_3 = \min_{1 \le i \le 6}\{|\gamma_i|\}$;

- type IV, if γ_1, γ_2, γ_3, γ_4 are negative, γ_5, γ_6 are positive and $|\gamma_4| = \min_{1 \le i \le 6}\{|\gamma_i|\}$;

- type V, if γ_1, γ_2, γ_3, γ_4 are negative, γ_5, γ_6 are positive and $\gamma_5 = \min_{1 \le i \le 6}\{|\gamma_i|\}$;

- type VI, if all $\gamma_i's$ are negative and $|\gamma_6| = \min_{1 \le i \le 6}\{|\gamma_i|\}$.

The above types exhaust all possibilities.

5   Proof of Theorem 1.1

Suppose that $(x,y) \in \mathbb{Z}^2$ is non-trivial solution of type I. We consider the Siegel identity

$$0 < \frac{\gamma_2(\theta_1 - \theta_3)}{\gamma_3(\theta_1 - \theta_2)} = 1 + \frac{\gamma_1(\theta_2 - \theta_3)}{\gamma_3(\theta_1 - \theta_2)}. \tag{5.1}$$

so

$$\varepsilon_2^{-2b_1} \varepsilon_3^{b_2-2b_3} (\varepsilon_3')^{2b_2-b_3} \varepsilon^{b_4} (\varepsilon')^{b_5} \left(\frac{\theta+2}{\theta+1}\right) = 1 + \left(\frac{1}{\theta+1}\right) \frac{\gamma_1}{\gamma_3},$$

Putting

$$\xi_1 = \frac{\theta+2}{\theta+1}, \qquad \tau_1 = \left(\frac{1}{\theta+1}\right) \frac{\gamma_1}{\gamma_3}, \tag{5.2}$$

we obtain

$$\Lambda_1 = b_1 \log(\varepsilon_2^{-2}) + b_2 \log(\varepsilon_3(\varepsilon_3')^2) + b_3 \log(\varepsilon_3^{-2}(\varepsilon_3')^{-1})$$
$$+ b_4 \log(\varepsilon) + b_5 \log(\varepsilon') + \log(\xi_1) = \log(1 + \tau_1). \tag{5.3}$$

Lemma 5.1. $\Lambda_1 \neq 0$.

Proof. If $\Lambda_1 = 0$, then $\log(1 + \tau_1) = 0$ and $\tau_1 = 0$. This is impossible because $\tau_1 > 0$ $\square$

We use the Baker-Wüstholz's theorem (see [1]) to obtain a lower bound for Λ_1.

Lemma 5.2. *We have*

$$\log(\Lambda_1) > - 7 \times 5 \times 3^{14} \times 2^{59} \log(72) \log(\varepsilon_2)$$
$$\times \log^2(\varepsilon_3^{-2}(\varepsilon_3')^{-1}) \log^2(\varepsilon) \log[9(\theta + 2)] \log(B),$$

where $B = \max\{|b_1|, |b_2|, |b_3|, |b_4|, |b_5|, 1\}$.

Proof. We have

$$h'(\varepsilon_2^{-2}) = h'(\varepsilon_2^2) = \log(\varepsilon_2), \qquad h'(\varepsilon_3(\varepsilon_3')^2) = h'(\varepsilon_3^2\varepsilon_3') = -\tfrac{1}{3}\log(\varepsilon_3^2\varepsilon_3'),$$
$$h'(\varepsilon) = h'(\varepsilon') = \tfrac{1}{3}\log(\varepsilon), \qquad h'(\xi_1) = \tfrac{1}{6}\log[9(\theta + 2)],$$

because the conjugates of ξ_1 are

$$\xi_1 = \frac{\theta + 2}{\theta + 1}, \qquad \xi_1^\sigma = \frac{3(\theta + 1)}{2\theta + 1}, \qquad \xi_1^{\sigma^2} = \frac{2\theta + 1}{\theta},$$
$$\xi_1^{\sigma^3} = \frac{3\theta}{\theta - 1}, \qquad \xi_1^{\sigma^4} = -(\theta - 1), \qquad \xi_1^{\sigma^5} = \frac{3}{\theta + 2}.$$

Moreover,

$$B = \max_{1 \le i \le 5}\{|b_i|, 1\}, \qquad C(6, 6) = 7 \times 5 \times 3^{19} \times 2^{60} \times \log(72).$$

Hence we obtain the desired result. $\square$

In the next step we use the method of Y. Bilu and G. Hanrot (see [2]). We define the constants $X_0, X_1, X_2, X_3, c_1, \ldots, c_{12}, \delta_i$ and λ_i as follows:

$$X_0 - 1,$$
$$c_1 = \frac{16\theta(\theta - 1)(\theta + 1)(\theta + 2)(2\theta + 1)^5}{9(\theta^2 + \theta + 1)^5}, \qquad c_2 = \frac{\theta^2 + \theta + 1}{(2\theta + 1)(\theta + 1)},$$

$$X_1 = \max\left(X_0, (7c_1/c_2)^{1/6}\right),$$

$$c_3 = \frac{7}{6(\theta^2 + \theta + 1)}c_1, \qquad c_4 = 1.39\, c_3,$$

$$X_2 = \max\left(X_1, (2c_3)^{1/6}\right),$$

$$c_5 = \begin{cases} \left|\frac{1}{U}\right| \left|\log(\varepsilon_3^2 \varepsilon_3')\right| & \text{if } n = 0 \\ \left|\frac{1}{V}\right| \left|\log(\varepsilon^2/\varepsilon')\right| & \text{otherwise,} \end{cases}$$

$$c_6 = \begin{cases} -\frac{1}{6U}\log(\varepsilon_3^2\varepsilon_3') + \left|\frac{1}{2U}\right|\left|-\log(\varepsilon_3)\log(T_2 T_5)\right. \\ \qquad\qquad \left. - \log(\varepsilon_3\varepsilon_3')\log(T_3 T_6)\right| & \text{if } n = 0 \\[4pt] \frac{1}{6V}\log(\varepsilon^2/\varepsilon') + \left|\frac{1}{6V}\right|\left|\log((\varepsilon')^2/\varepsilon)\log(T_2)\right. \\ \qquad\qquad + 3\log(\varepsilon'/\varepsilon)\log(T_3) \\ \qquad\qquad + 2\log(\varepsilon'/\varepsilon^2)\log(T_4) \\ \qquad\qquad - 3\log(\varepsilon)\log(T_5) \\ \qquad\qquad \left. - \log(\varepsilon\varepsilon')\log(T_6)\right| & \text{otherwise,} \end{cases}$$

where
$$T_j = \theta - \theta_j, j \neq 1,$$
$$U = \log^2(\varepsilon_3) + \log^2(\varepsilon_3') + \log(\varepsilon_3)\log(\varepsilon_3'),$$
$$V = \log^2(\varepsilon) + \log^2(\varepsilon') - \log(\varepsilon)\log(\varepsilon'),$$

$$c_7 = c_5, \qquad c_8 = c_6, \qquad c_9 = c_4 \exp(6c_6/c_5), \qquad c_{10} = 6/c_5$$

because $\mathbb{K}_t$ is real;

$$c_{11} = 7 \times 5 \times 3^{14} \times 2^{59} \log(72) \log(\varepsilon_2) \log^2(\varepsilon_3^{-2}(\varepsilon_3')^{-1}) \log^2(\varepsilon) \log[9(\theta + 2)],$$
$$B_0 = \max\left(e, 2c_{10}^{-1}c_{11} \log\left(c_{10}^{-1}c_{11}c_9^{1/c_{11}}\right)\right),$$

$$i_1 = \begin{cases} 2 & \text{if } n = 0, \\ 4 & \text{otherwise,} \end{cases}$$

$$i_2 = 1,$$

$$\delta = \begin{cases} \delta_1/\delta_2 & \text{if } n = 0, \\ \delta_1/\delta_4 & \text{otherwise,} \end{cases}$$

$$\lambda = \begin{cases} (\delta_1\lambda_2 - \delta_2\lambda_1)/\delta_2 & \text{if } n = 0, \\ (\delta_1\lambda_4 - \delta_4\lambda_1)/\delta_4 & \text{otherwise,} \end{cases}$$

$$c_{14} = c_{12} \exp(6c_6/c_5),$$
$$c_{15} = 6/c_5,$$
$$\delta_1 = -\frac{1}{\log(\varepsilon_2)},$$
$$\delta_2 = -\frac{1}{U} \log(\varepsilon_3^2 \varepsilon_3'),$$
$$\delta_3 = -\frac{1}{U} \log(\varepsilon_3 (\varepsilon_3')^2),$$
$$\delta_4 = -\frac{1}{V} \log(\varepsilon^2/\varepsilon'),$$
$$\delta_5 = \frac{1}{V} \log(\varepsilon/(\varepsilon')^2),$$

and

$$\lambda_1 = -\frac{\log(T_2 T_4 T_6)}{3 \log(\varepsilon_2)},$$
$$\lambda_2 = \frac{1}{2U}[-\log(\varepsilon_3) \log(T_2 T_5) - \log(\varepsilon_3 \varepsilon_3') \log(T_3 T_6)],$$
$$\lambda_3 = \frac{1}{2U}[-\log(\varepsilon_3 \varepsilon_3') \log(T_2 T_5) - \log(\varepsilon_3') \log(T_3 T_6)],$$
$$\lambda_4 = \frac{1}{6V}[\log((\varepsilon')^2/\varepsilon) \log(T_2) + 3\log(\varepsilon'/\varepsilon) \log(T_3)$$
$$+ 2\log(\varepsilon'/\varepsilon^2) \log(T_4) - 3\log(\varepsilon) \log(T_5) - \log(\varepsilon\varepsilon') \log(T_6)],$$
$$\lambda_5 = \frac{1}{6V}[-\log(\varepsilon\varepsilon') \log(T_2) - 3\log(\varepsilon') \log(T_3)$$
$$- 2\log((\varepsilon')^2/\varepsilon) \log(T_4) - 3\log(\varepsilon'/\varepsilon) \log(T_5) - \log(\varepsilon'/\varepsilon^2) \log(T_6)].$$

Using the system

$$\gamma_1 = x - \theta y = \varepsilon_2^{b_1} \varepsilon_3^{b_2} (\varepsilon_3')^{b_3} \varepsilon^{b_4} (\varepsilon')^{b_5},$$
$$\gamma_2 = x - \theta_2 y = \varepsilon_2^{-b_1} (\varepsilon_3')^{b_2} (\varepsilon_3 \varepsilon_3')^{-b_3} (\varepsilon')^{b_4} \left(-\varepsilon^{-1}\varepsilon'\right)^{b_5},$$

we obtain

$$x = \frac{-(\theta - 1)\gamma_1 + \theta(\theta + 2)\gamma_2}{\theta^2 + \theta + 1},$$
$$y = \frac{(\theta + 2)(-\gamma_1 + \gamma_2)}{\theta^2 + \theta + 1}.$$

Then we determine solutions (x, y) with $|x| \geq X_3 = \max(X_2, (2c_{12})^{1/6})$.

We use a similar approach to determine solutions of types II–VI; see [17, pp. 147–163].

Here are a few remarks about the computation. A program in PARI/GP is written to compute all the above constants. We know that

$$\theta < t/2, \qquad c_2 < \frac{1}{3}(t + 1/2)^{2/3}, \qquad \varepsilon_3 < 1,$$
$$\varepsilon_3' < 2, \qquad \varepsilon < (t + 1/2)^{2/3}, \qquad \varepsilon' < (t - 11)^{1/3},$$

so we put

$$n_2 = (n_1 + 2)(|b_1| + |b_2| + |b_3| + |b_4| + |b_5|) + 50$$

as the computation's precision, where n_1 is the number of digits of $t + 1$.

After two or three reductions, we obtain $B_0 \leq 4$ and $X_3 = 2$. The integral solutions (x, y) satisfying $|x| \leq X_3$, $|y| \leq X_3$ are:

$$(-1, 0), \quad (-1, 1), \quad (0, -1), \quad (0, 1), \quad (1, -1), \quad (1, 0).$$

In fact, we let b_1, b_2, b_3, b_4, b_5 run between $-B_0$ and B_0, and we observe the exponents associated with the integral solutions. Then we obtain

$$b_1 = b_2 = b_3 = b_4 = b_5 = 0,$$

so $(0, 1)$ is a solution. Property (2) of Section 2 gives us the other solutions. The computations were done with a SUN SPARC ULTRA1, and take roughly 9 seconds for each value of n.

Acknowledgements. We thank Professors Cornelius Greither, Günther Frei, Maurice Mignotte of our dissertation committee for their helpful advice, and particularly Professor Claude Levesque for, amongst many things, supervising us. We also thank the referee for the constructive suggestions on an earlier draft of this paper.

References

[1] A. Baker and G. Wüstholz, *Logarithmic forms and group varities*, J. Reine Angew. Math. **442** (1993), 19–62.

[2] Y. Bilu and G. Hanrot, *Solving Thue equations of high degree*, J. Number Theory **60** (1996), 373–392.

[3] V. Ennola, S. Mäki, and R. Turunen, *On real cyclic sextic fields*, Math. Comp. **45** (1985), 591–611.

[4] I. Gaál, A. Pethő, and M. Pohst, *On the resolution of index form equations in biquadratic number fields, III, the bicyclic biquadratic case*, J. Number Theory **53** (1995), 100–114.

[5] M. N. Gras, *Familles d'unités dans les extensions cycliques réelles de degré 6 de* $\mathbb{Q}$, Publ. Math. Fac. Sci. Besançon, 1984/85–1985/86.

[6] ———, *Special units in real cyclic sextic fields*, Math. Comp. **48** (1987), 179–182.

[7] F. Halter-Koch, G. Lettl, A. Pethő, and R. F. Tichy, *Thue equations associated with Ankeny-Brauer-Chowla number fields*, J. London Math. Soc. (2) **60** (1999), 1–20.

[8] G. Hanrot, *Résolution effective d'équations diophantiennes: algorithmes et applications*, 1997, Thèse, Université Bordeaux 1.

[9] G. Lettl and A. Pethő, *Complete solution of a family of quartic Thue equations*, Abh. Math. Sem. Univ. Hambourg, vol. 65, 1995, pp. 365–383.

[10] S. Mäki, *The determination of units in real cyclic sextic fields*, Lecture Notes in Maths., vol. 797, Springer-Verlag, Berlin and New York, 1980.

[11] M. Mignotte, A. Pethő, and F. Lemmermeyer, *On the family of Thue equations $x^3 - (n-1)x^2y - (n+2)xy^2 - y^3 = k$*, Acta Arith. **LXXVI.3** (1996), 245–269.

[12] M. Mignotte, A. Pethő, and R. Roth, *Complete solutions of a family of parametric quartic Thue and index form equations*, Math. Comp. **65** (1996), 341–354.

[13] A. Pethő, *Complete solutions to families of quartic Thue equations*, Math. Comp. **57** (1991), 777–798.

[14] H.-J. Stender, *Lösbare Gleichungen $ax^n - bx^n = c$ und Grundeinheiten für einige algebraische Zahlkörper von Grade $n = 3, 4, 6$*, J. Reine Angew. Math. **290** (1977), 24–62.

[15] E. Thomas, *Complete solutions to a family of cubic diophantine equations*, J. Number Theory **34** (1990), 235–250.

[16] ———, *Solutions to certain families of Thue equations*, J. Number Theory **43** (1993), 319–369.

[17] A. Togbé, *Sur la résolution de familles d'équations diophantiennes*, Ph.D. thesis, Université Laval, Québec, Canada, Décembre 1997.

[18] ———, *On the solutions of a family of Quartic Thue equations*, Math. Comp. **69** (2000), 839–849.

Waring's Problem: A Survey

R. C. Vaughan[1] and T. D. Wooley[2]

1 The Classical Waring Problem

"Omnis integer numerus vel est cubus, vel e duobus, tribus, 4, 5, 6, 7, 8, vel novem cubis compositus, est etiam quadrato-quadratus vel e duobus, tribus, &c. usque ad novemdecim compositus, & sic deinceps."

Waring [150, pp. 204-5].

"Every integer is a cube or the sum of two, three, ... nine cubes; every integer is also the square of a square, or the sum of up to nineteen such; and so forth."

Waring [152, p. 336].

It is presumed that by this, in modern notation, Waring meant that for every $k \geq 3$ there are numbers s such that every natural number is the sum of at most s k-th powers of natural numbers and that the smallest such number $g(k)$ satisfies $g(3) = 9$, $g(4) = 19$.

By the end of the nineteenth century, the existence of $g(k)$ was known for only a finite number of values of k. There is an account of this work in Dickson [48], and as far as we have been able to ascertain, by 1909 its existence was known for $k = 3, 4, 5, 6, 7, 8, 10$, but not for any larger k (of course, with the natural extension of the definition of $g(k)$, Lagrange proved in 1770 that $g(2) = 4$). However, starting with Hilbert [69], who showed that $g(k)$ does indeed exist for every k, the twentieth century has seen an almost complete solution of this problem. Let $[x]$ denote the greatest integer not exceeding x and write $\{x\}$ for $x - [x]$. As the result of the work of many mathematicians we now know that

$$g(k) = 2^k + [(3/2)^k] - 2,$$

provided that

$$2^k \{(3/2)^k\} + [(3/2)^k] \leq 2^k. \tag{1.1}$$

If this fails, then

$$g(k) = 2^k + [(3/2)^k] + [(4/3)^k] - \theta$$

[1]Research supported by NSF grant DMS-9970632.
[2]Packard Fellow, and supported in part by NSF grant DMS-9970440.

where θ is 2 or 3 according as

$$[(4/3)^k][(3/2)^k] + [(4/3)^k] + [(3/2)^k]$$

equals or exceeds 2^k.

The condition (1.1) is known to hold (Kubina & Wunderlich [85]) whenever $k \leq 471,600,000$, and Mahler [91] has shown that there are at most a finite number of exceptions. To complete the proof for all k it would suffice to know that $\{(3/2)^k\} \leq 1-(3/4)^{k-1}$. Beukers [3] has shown that whenever $k > 5,000$ one has $\{(3/2)^k\} \leq 1 - a^k$, where $a = 2^{-0.9} = 0.5358...$, and this has been improved slightly by Dubitskas [49] to $a = 0.5769...$, so long as k is sufficiently large (see also Bennett [1] for associated estimates). A problem related to the evaluation of $g(k)$ now has an almost definitive answer. Let $g_r(k)$ denote the smallest integer s with the property that every natural number is the sum of at most s elements from the set $\{1^k, r^k, (r+1)^k, \dots\}$. Then Bennett [2] has shown that for $4 \leq r \leq (k+1)^{1-1/k} - 1$, one has $g_r(k) = r^k + [(1+1/r)^k] - 2$.

By the way, before turning to the modern form of Waring's problem, it is perhaps worth observing that in the 1782 edition of *Meditationes Algebraicæ*, Waring makes an addition:

"confimilia etiam affirmari possunt (exceptis excipiendis) de eodem numero quantitatum earundem dimensionum."

Waring [151, p. 349].

"similar laws may be affirmed (exceptis excipiendis) for the correspondingly defined numbers of quantities of any like degree."

Waring [152, p. 336].

It would be interesting to know exactly what Waring had in mind. This, taken with some of the observations which immediately follow the remark, suggest that for more general polynomials than the k-th powers he was aware that some kind of local conditions can play a rôle in determining when representations occur.

2   The Modern Problem

The value of $g(k)$ is determined by the peculiar behaviour of the first three or four k-th powers. A much more challenging question is the value, for $k \geq 2$, of the function $G(k)$, the smallest number t such that every *sufficiently large* number is the sum of at most t k-th powers of positive integers. The function $G(k)$ has only been determined for two values of k, namely $G(2) = 4$, by Lagrange in 1770, and $G(4) = 16$, by Davenport [30]. The bulk of what is known about $G(k)$ has been obtained through the medium of the Hardy-Littlewood method. This has its genesis in a celebrated paper of Hardy and Ramanujan [64] devoted to the partition function. In this paper (section 7.2) there is also a brief discussion about the representation of a natural number as the sum of a fixed number of squares of integers, and there seems little doubt that it is the methods described therein which inspired the later work of Hardy and Littlewood.

Our knowledge concerning the function $G(k)$ currently leaves much to be desired. If, instead of insisting that all sufficiently large numbers be represented in a prescribed form, one rather asks that *almost all* numbers (in the sense of natural density) be thus represented, then the situation is somewhat improved. Let $G_1(k)$ denote the smallest number u such that almost every number n is the sum of at most u k-th powers of positive integers. The function $G_1(k)$ has been determined in five non-trivial instances as follows:

$$\begin{aligned}
\text{Davenport [29]}, \quad & G_1(3) = 4, \\
\text{Hardy and Littlewood [62]}, \quad & G_1(4) = 15, \\
\text{Vaughan [121]}, \quad & G_1(8) = 32, \\
\text{Wooley [155]}, \quad & G_1(16) = 64, \\
\text{Wooley [155]}, \quad & G_1(32) = 128
\end{aligned}$$

(of course, the conclusion $G_1(2) = 4$ is classical).

3   General Upper Bounds for G(k)

The first explicit general upper bound for $G(k)$, namely

$$G(k) \le (k-2)2^{k-1} + 5,$$

was obtained by Hardy and Littlewood [61] (in [58] and [59], only the existence of $G(k)$ is stated, although it is already clear that in principle their method gave an explicit upper bound). In Hardy and Littlewood [62] this is improved to

$$G(k) \le (k-2)2^{k-2} + k + 5 + [\zeta_k],$$

where
$$\zeta_k = \frac{(k-2)\log 2 - \log k + \log(k-2)}{\log k - \log(k-1)}.$$

There has been considerable activity reducing this upper bound over the years, and Table 1 below presents upper bounds for $G(k)$ that were probably the best that were known at the time they appeared, at least for

Vinogradov [136], $32(k \log k)^2$,

Vinogradov [137][139], $k^2 \log 4 + (2 - \log 16)k \ (k \geq 3)$,

Vinogradov [135] [138] [140] [143], $6k \log k + 3k \log 6 + 4k \ (k \geq 14)$,

Vinogradov [147], $k(3 \log k + 11)$,

Tong [114], $k(3 \log k + 9)$,

Jing-Run Chen [24], $k(3 \log k + 5.2)$,

Vinogradov [148], $2k(\log k + 2 \log \log k + O(\log \log \log k))$,

Vaughan [124], $2k(\log k + \log \log k + O(1))$,

Wooley [155], $k(\log k + \log \log k + O(1))$.

Table 1. General upper bounds for $G(k)$

k sufficiently large. This list is not exhaustive. In particular, there is a long sequence of papers by Vinogradov between 1934 and 1947, and for further details we refer the reader to the Royal Society obituary of I. M. Vinogradov (see Cassels and Vaughan [23]).

The last entry on this list has been refined further by Wooley [159], and this provides the estimate

$$G(k) \leq k(\log k + \log \log k + 2 + O(\log \log k / \log k))$$

that remains the sharpest available for larger exponents k.

4 Cubes

For small values of k there are many special variants of the Hardy-Littlewood method that have been developed. However, in the case of cubes, until recently the best upper bounds were obtained by rather different methods that related cubes to quadratic forms, especially sums of squares. Thus Landau [86] had shown that $G(3) \leq 8$, and this bound was reduced by Linnik [87][88] to $G(3) \leq 7$, with an alternative and simpler proof given by Watson [153]. Only with the advent of refinements to the circle method utilising efficient differencing did it become feasible (Vaughan [119]) to give a proof of the bound $G(3) \leq 7$ via the Hardy-Littlewood method. Subsequent developments involving the use of smooth numbers (see Vaughan [124][125] and Wooley [158]) have provided a more powerful approach to this problem that, from a practical point of view, is more direct than earlier treatments. Complicated nonetheless, these latter proofs yield much more information concerning Waring's problem for cubes. We can illustrate the latter observation with two examples which, in the absence of foreseeable progress on the upper bound for $G(3)$, provide the problems central to current activity surrounding Waring's problem for cubes.

When X is a large real number, denote by $E(X)$ the number of positive integers not exceeding X that *cannot* be written as a sum of four positive integral cubes. Then the conclusion $G_1(3) = 4$, attributed above to Davenport [29], is an immediate consequence of the estimate $E(X) \ll X^{29/30+\epsilon}$ established in the latter paper. Following subsequent work of Vaughan [119], Brüdern [10][12], and Wooley [158], the sharpest conclusion currently available (see Wooley [162]) shows that $E(X) \ll X^{1-\beta}$ for any positive number β smaller than

$$(422 - 6\sqrt{2833})/861 = 0.119215\ldots.$$

It is conjectured that $G(3) = 4$ (see §10 below), and this would imply that $E(X) \ll 1$.

Consider next the density of integers represented as a sum of three positive integral cubes. When X is a large real number, let $N(X)$ denote the number of positive integers of the latter type not exceeding X. It is conjectured that $N(X) \gg X$, and following work of Davenport [29][33], Vaughan [118][119], Ringrose [100], Vaughan [124] and Wooley [158], the sharpest currently available conclusion due to Wooley [162] establishes that $N(X) \gg X^{1-\alpha}$ for any real number α exceeding

$$(\sqrt{2833} - 43)/123 = 0.083137\ldots.$$

We remark that, subject to the truth of an unproved Riemann Hypothesis concerning certain Hasse-Weil L-functions, one has the conditional estimate $N(X) \gg X^{1-\epsilon}$ due to Hooley [73][74], and Heath-Brown [67]. Unfortunately, the underlying L-functions are not yet known even to have an analytic continuation inside the critical strip.

We finish our discussion of Waring's problem for cubes by noting that, while Dickson [47] was able to show that 23 and 239 are the only positive integers not represented as the sum of eight cubes of natural numbers, no such conclusion is yet available for sums of seven or fewer cubes (but see McCurley [92] for sums of seven cubes, and Deshouillers, Hennecart and Landreau [44] for sums of four cubes).

5 Biquadrates

Davenport's definitive statement that $G(4) = 16$ is not the end of the story for sums of fourth powers (otherwise known as biquadrates). Let

$G^{\#}(4)$ denote the least integer s_0 such that whenever $s \geq s_0$, and $n \equiv r$ (mod 16) for some integer r with $1 \leq r \leq s$, then n is the sum of at most s biquadrates. Then Davenport [30] showed that $G^{\#}(4) \leq 14$, and this has been successively reduced by Vaughan [121][124] to $G^{\#}(4) \leq 12$. In an ironic twist of fate, the polynomial identity

$$x^4 + y^4 + (x + y)^4 = 2(x^2 + xy + y^2)^2,$$

reminiscent of identities employed in the nineteenth century, has recently been utilised to make yet further progress. Thus, when $n \equiv r$ (mod 16) for some integer r with $1 \leq r \leq 10$, Kawada and Wooley [82] have shown that n is the sum of at most 11 biquadrates. This and allied identities have also permitted the proof of an effective version of Davenport's celebrated theorem. Thus, as a consequence of work of Deshouillers, Hennecart and Landreau [45] and Deshouillers, Kawada and Wooley [46], it is now known that all integers exceeding 13, 792 may be written as the sum of at most sixteen biquadrates. A detailed history of Waring's problem for biquadrates is provided in Deshouillers, Hennecart, Kawada, Landreau and Wooley [43].

6   Upper Bounds for G(k) when $5 \leq k \leq 20$

Although we have insufficient space to permit a comprehensive account of the historical evolution of available upper bounds for $G(k)$ for smaller values of k, in Table 2 we have recorded many of the key developments, concentrating on the past twenty-five years. Each row in this table presents the best upper bound known for $G(k)$, for the indicated values of k, at the time of publication of the cited work. We note that the claimed bound $G(7) \leq 52$ of Sambasiva Rao [99] is based on an arithmetical error, and hence we have attributed the bound $G(7) \leq 53$ parenthetically to Davenport's methods [28][32]. Also, it is worth remarking that the work of Vaughan and Wooley [130][131] and [132] appeared in print in an order reversed from its chronological development (indeed, this work was first announced in 1991). The bounds parenthetically attributed to Vaughan and Wooley [132] follow directly from the methods therein, and were announced on that occasion, though details (with additional refinements) appeared only in Vaughan and Wooley [134]. Meanwhile, the bounds recorded in Wooley [159] were an immediate consequence of the methods of Wooley [155] combined with the new estimates for smooth Weyl sums obtained in the former work (no attempt was made therein to exploit the methods of Vaughan and Wooley [132]).

k	5	6	7	8	9	10	11	12	13	14	15	16	17	18	19	20
Hardy & Littlewood [62]	41	87	192	425	949	2113										
James [81]	35		164		824											
Heilbronn [68]				164	190	217	244	272	300	329	359	388	418	449	480	511
Estermann [52]	29	42	59	78	101	125	153	184	217	253	292	333	377	424	474	
Hua [76]	28															
Davenport [28][32]	23	36	(53)													
Narasimhamurti [96]				73	99	122										
Chen [24]						121										
Cook [26]					96											
Vaughan [115]					91	107	122	137	153	168	184	200	216			
Thanigasalam [107]					90	106	121	136	152	167	183	199	215	231	248	264
Thanigasalam [108]					88	104	119	134	150	165	181	197	213	229	245	262
Thanigasalam [109]			50	68	87	103										
Vaughan [120][121]	21	31	45	62	82											
Vaughan [124][126]	19	29	41	57	75	93	109	125	141	156	171	187	202	217	232	248
Brüdern [11]	18															
Vaughan & Wooley [129]	18	28				92	108	124	139	153	168	183	198	213	228	243
Wooley [155]		27	36	47	55	63	70	79	87	95	103	112	120	129	138	146
Wooley [159]							62	78	86	94	102	110	118	127	135	144
Vaughan & Wooley [132]	17	25	33	43	51	(59)	(67)	(76)	(84)	(92)	(100)					
Vaughan & Wooley [130]				42												
Vaughan & Wooley [131]		24														
Meng [94]															143	
Vaughan & Wooley [134]												109	117	125	134	142
Conjectured	6	9	8	32	13	12	12	16	14	15	16	64	18	27	20	25

Table 2. Upper bounds for $G(k)$ when $5 \le k \le 20$

7   The Hardy-Littlewood Method

Practically all of the above conclusions have been obtained via the Hardy-Littlewood method. Here is a quick introduction. Let n be a large natural number, and write $P = n^{1/k}$ and

$$f(\alpha) = \sum_{x \le P} e(\alpha x^k)$$

(here we follow the standard convention of writing $e(z)$ for $e^{2\pi i z}$). Then on writing $R(n)$ for the number of representations of n as the sum of s kth powers of natural numbers, it follows from orthogonality that

$$R(n) = \int_0^1 f(\alpha)^s e(-\alpha n) d\alpha.$$

When α is "close" to a rational number a/q with $(a, q) = 1$ and q "small", we expect that

$$f(\alpha) \sim q^{-1} S(q, a) v(\alpha - a/q),$$

where

$$S(q, a) = \sum_{r=1}^q e(ar^k/q) \quad \text{and} \quad v(\beta) = \int_0^P e(\beta \gamma^k) d\gamma.$$

This relation is straightforward to establish in an interval about a/q, so long as "close" and "small" are interpreted suitably. Now put

$$R_A(n) = \int_{\mathcal{A}} f(\alpha)^s e(-\alpha n) d\alpha.$$

For a suitable union $\mathfrak{M}$ of such intervals centred on a/q (the *major arcs*), and for s sufficiently large in terms of k, one can establish that as $n \to \infty$, the asymptotic relation

$$R_{\mathfrak{M}}(n) \sim \frac{\Gamma(1 + 1/k)^s}{\Gamma(s/k)} n^{s/k-1} \mathfrak{S}(n)$$

holds, where $\mathfrak{S}(n)$ is the *singular series*

$$\mathfrak{S}(n) = \sum_{q=1}^{\infty} T(q; n)$$

and

$$T(q; n) = \sum_{\substack{a=1 \\ (a,q)=1}}^q q^{-s} S(q, a)^s e(-an/q).$$

8 The Necessary Congruence Condition

For each prime p, let

$$U(p;n) = \sum_{k=0}^{\infty} T(p^k;n).$$

The function $T(q;n)$ is multiplicative. Thus, when the singular series converges absolutely, one has

$$\mathfrak{S}(n) = \prod_p U(p;n).$$

It is helpful to the success of the circle method that the singular series should satisfy $\mathfrak{S}(n) \gg 1$. With this observation in mind, Hardy and Littlewood [61] defined $\Gamma(k)$ to be the least integer s with the property that, for every prime number p, there is a positive number $C(p)$ such that $U(p;n) \geq C(p)$ uniformly in n. Subsequently, Hardy and Littlewood [62] showed that indeed $\mathfrak{S}(n) \gg 1$ whenever $s \geq \max\{\Gamma(k), 4\}$. Next let $\Gamma_0(k)$ be the least number s with the property that the equation

$$x_1^k + \cdots + x_s^k = n \tag{8.1}$$

has a non-singular solution in $\mathbb{Q}_p$ (or rather, that the corresponding congruence modulo q always has a solution with $(x_1, q) = 1$). Hardy and Littlewood [63] were able to show that $\Gamma_0(k) = \Gamma(k)$ (see Theorem 1 of the aforementioned paper). Thus one sees that the singular series reflects the local properties of sums of k-th powers. In particular, the singular series is zero whenever the equation (8.1) fails to have a p-adic solution, for some prime p, and this reflects the trivial observation that the equation can be soluble over $\mathbb{Z}$ only if it is soluble everywhere locally.

Hardy and Littlewood [63] conjecture that $\Gamma(k) \to \infty$ as $k \to \infty$, but it is not even known whether or not one has

$$\liminf_{k\to\infty} \Gamma(k) \geq 4.$$

When $k > 2$, they showed that $\Gamma(k) = 4k$ when k is a power of 2 and that $\Gamma(k) \leq 2k$ otherwise. They also computed $\Gamma(k)$ exactly when $3 \leq k \leq 36$, and established that $\Gamma(k) \geq 4$ when $3 \leq k \leq 3000$. Here they showed that equality occurs only when $k = 3$, 7, 19, and possibly (but improbably) when $k = 1163$, 1637, 1861, 1997, 2053. These values of k can probably be settled by modern computing methods, and doubtless the calculations could be carried a good deal further. As far as we are aware, nothing has been done in this direction.

For a more detailed exposition of the Hardy-Littlewood method and the analysis of the major arcs and the singular series, see Vaughan [128] (especially Chapters 2 and 4).

9 The Minor Arcs

In order to establish an asymptotic formula for $R(n)$ it suffices to show that $R_{\mathfrak{m}}(n) = o(n^{s/k-1})$, where $\mathfrak{m} = [0,1) \setminus \mathfrak{M}$ (the *minor arcs*). One needs to show that the minor arc contribution $R_{\mathfrak{m}}(n)$ is smaller by a factor $o(n^{-1})$, or equivalently $o(P^{-k})$, than the trivial estimate P^s. Routinely this is established via an inequality of the kind

$$\int_{\mathfrak{m}} |f(\alpha)|^s d\alpha \leq \left(\sup_{\mathfrak{m}} |f(\alpha)| \right)^{s-2t} \int_0^1 |f(\alpha)|^{2t} d\alpha.$$

The integral on the right hand side of this inequality may be interpreted as the number of solutions of an underlying diophantine equation, and it is from here that most of the savings usually come. On the other hand, non-trivial estimates for $|f(\alpha)|$, when $\alpha \in \mathfrak{m}$, may be obtained from estimates stemming from work of Weyl [154] and Vinogradov [144] (see Vaughan [128] for more modern estimates). When successful, this leads to the relation

$$R(n) \sim \frac{\Gamma(1+1/k)^s}{\Gamma(s/k)} n^{s/k-1} \mathfrak{S}(n). \tag{9.1}$$

It is in finding ways of dealing with the minor arcs, or in modifying the method so as to make the minor arcs more amenable, that most of the research has concentrated in the eighty years that have elapsed since the pioneering investigations of Hardy and Littlewood.

10 The Asymptotic Formula

Define $\widetilde{G}(k)$ to be the smallest natural number s_0 such that whenever $s \geq s_0$, the asymptotic relation (9.1) holds. Work of Hardy and Littlewood [62], described already in §3, established the general upper bound

$$\widetilde{G}(k) \leq (k-2)2^{k-1} + 5.$$

Progress on upper bounds for $\widetilde{G}(k)$ has since been achieved on two fronts. In Table 3 we present upper bounds for $\widetilde{G}(k)$ relevant for small values of k.

When k is large, bounds stemming from Vinogradov's mean value theorem provide dramatic improvements over the estimates recorded in Table 3.

$$\text{Hua [75],} \quad 2^k + 1,$$
$$\text{Vaughan [119][122],} \quad 2^k \quad (k \geq 3),$$
$$\text{Heath-Brown [65][66],} \quad 7 \cdot 2^{k-3} + 1 \quad (k \geq 6),$$
$$\text{Boklan [5],} \quad 7 \cdot 2^{k-3} \quad (k \geq 6).$$

Table 3. Upper bounds for $\widetilde{G}(k)$: smaller k

$$\text{Vinogradov [142],} \quad 183k^9 (\log k + 1)^2 + 1,$$
$$\text{Vinogradov [141],} \quad 91k^8 (\log k + 1)^9 + 1 \quad (k \geq 20),$$
$$\text{Vinogradov [147],} \quad 10k^2 \log k,$$
$$\text{Hua [77],} \quad 4k^2 (\log k + \tfrac{1}{2} \log \log k + 8),$$
$$\text{Wooley [156],} \quad 2k^2 (\log k + O(\log \log k)),$$
$$\text{Ford [53],} \quad k^2 (\log k + \log \log k + O(1)).$$

Table 4. Upper bounds for $\widetilde{G}(k)$: larger k

In Table 4 we present upper bounds for $\widetilde{G}(k)$ of use primarily when k is large.

We note that the methods underlying the last two bounds can be adapted to give explicit bounds for $\widetilde{G}(k)$ when k is of moderate size. Thus the method of Ford yields a bound for $\widetilde{G}(k)$ that is superior to the best recorded in Table 3 as soon as $k \geq 9$, and indeed unpublished work of Boklan and Wooley pushes this transition further to $k \geq 8$.

The bounds recorded in Tables 3 and 4 are likely to be a long way from the truth. One might expect that $G(k) = \max\{k+1, \Gamma_0(k)\}$, and, with an appropriate interpretation of the asymptotic formula when $\mathfrak{S}(n) = 0$, that $\widetilde{G}(k) = k + 1$.

One curiosity is that when $k = 3$ and $s = 7$, it can be shown that $R(n) \gg n^{4/3}$ (see Vaughan [125]), yet we are unable to show that $R(n) \ll n^{4/3}$. Indeed, it is currently the case that, quite generally, when s lies in the range between the known upper bounds for $G(k)$ and $\widetilde{G}(k)$, we can show that $R(n) \gg n^{s/k-1}$, but not $R(n) \ll n^{s/k-1}$.

11   Diminishing Ranges

In the Hardy-Littlewood method as outlined above the main problem is that of obtaining a suitable estimate for the mean value

$$\int_0^1 |f(\alpha)|^{2t} d\alpha,$$

that is, the number of integral solutions of the equation

$$x_1^k + \cdots + x_t^k = y_1^k + \cdots + y_t^k, \qquad (11.1)$$

with $1 \le x_j, y_j \le P$ ($1 \le j \le t$). Available estimates can be improved significantly if the variables are restricted, an idea already present in Hardy and Littlewood [62]. Define

$$P_1 = \tfrac{1}{2}P, \quad P_j = \tfrac{1}{2}P_{j-1}^{1-1/k} \quad (2 \le j \le t),$$

and consider the equation (11.1) subject to the constraints $P_j < x_j, y_j \le 2P_j$ ($1 \le j \le t$). Inspecting the expression $|x_j^k - y_j^k|$ successively for $j = 1, 2, \ldots$ when $x_j \ne y_j$, we find that only the diagonal solutions in which $x_j = y_j$ ($1 \le j \le t$) occur. Thus we find that the number of solutions of this type is at most $O(P_1 \ldots P_t)$. This saves $P_1 \ldots P_t$ over the trivial bound, which is of order $(P_1 \ldots P_t)^2$. Now $P_1 \ldots P_t \approx P^\lambda$, where

$$\lambda = 1 + (1 - 1/k) + \cdots + (1 - 1/k)^{t-1} = k - k(1 - 1/k)^t.$$

Already when $t \sim Ck \log k$, for a suitable positive constant C, this exponent is close to k. Vinogradov and Davenport have exploited and developed this idea in a number of ways (see, for example, Davenport [27][34], and Vinogradov [147][149]; see also Davenport and Erdős [35]).

There is a "p-adic" analogue of this idea, first exploited by Davenport [31], in which one considers expressions of the kind

$$x_1^k + p_2^k(x_2^k + p_3^k(x_3^k + \cdots))$$

on each side of the equation. Here the p_i denote suitably chosen prime numbers. The analysis rests on congruences of the type $x_1 \equiv y_1 \pmod{p^k}$. When $p > P^{1/k}$ and $1 \le x_1, y_1 \le P$, this congruence implies that $x_1 = y_1$, and so on, just as in the diminishing ranges device. This idea has the merit of returning the various k-th powers $p_2^k \ldots p_j^k x_j^k$ to being in comparable size ranges. However in each of these methods the variables in (11.1) have varying natures and the homogeneity is essentially lost.

12 Smooth Numbers and Efficient Differences

In modern variants of the circle method as applied to Waring's problem, starting with the work of Vaughan [124], homogeneity is restored by considering the number of solutions $S_t(P, R)$ of the equation (11.1) with $x_j, y_j \in \mathcal{A}(P, R)$, where $\mathcal{A}(P, R)$ denotes the set of R-smooth numbers up to P, namely

$$\mathcal{A}(P, R) = \{n \in [1, P] \cap \mathbb{Z} : p|n \implies p \leq R\}.$$

In applications, one takes R to be a suitably small, but positive power of P. The set $\mathcal{A}(P, R)$ has the extremely convenient property that, given any positive number M with $M \leq P$, and an element $x \in \mathcal{A}(P, R)$ with $x > M$, there is always an integer m with $m \in [M, MR]$ for which $m|x$. Moreover, this integer m can be coaxed into playing the rôle of the prime p in the p-adic argument mentioned above. Finally, and of great importance in what follows, the set $\mathcal{A}(P, R)$ has positive density whenever R is no smaller than a positive power of P.

The objective now is to find good exponents λ_s with the property that whenever $\varepsilon > 0$, there exists a positive number $\eta_0 = \eta_0(s, k, \varepsilon)$ such that whenever $R = P^\eta$ with $0 < \eta \leq \eta_0$, one has

$$S_s(P, R) \ll P^{\lambda_s + \varepsilon}.$$

Such exponents are established via an iterative process in which a sequence of sets of exponents $\boldsymbol{\lambda}^{(n)} = (\lambda_1^{(n)}, \lambda_2^{(n)}, \ldots)$ is constructed by finding an expression for each $\lambda_s^{(n+1)}$ in terms of the elements of $\boldsymbol{\lambda}^{(n)}$. Boundedness is trivial, so there is always a convergent subsequence. In fact, our arguments produce monotonicity, and the convergence is fairly rapid. For a more detailed introduction and motivation for the underlying ideas in using smooth numbers in Waring's problem, see the survey article Vaughan [127].

Beginning with the work of Wooley [155], a key element in the iterations is the repeated use of *efficient differencing*, and this procedure is fully exploited in subsequent work of Vaughan and Wooley [130] [131] [132] [134]. For each $s \in \mathbb{N}$, we take $\phi_i = \phi_{i,s}$ $(i = 1, \ldots, k)$ to be real numbers with $0 \leq \phi_i \leq 1/k$. For $0 \leq j \leq k$, we then define

$$P_j = 2^j P, \quad M_j = P^{\phi_j}, \quad H_j = P_j M_j^{-k}, \quad Q_j = P_j(M_1 \ldots M_j)^{-1},$$

$$\widetilde{H}_j = \prod_{i=1}^j H_i \quad \text{and} \quad \widetilde{M}_j = \prod_{i=1}^j M_i R.$$

Define the modified forward difference operator, Δ_j^*, recursively by taking

$$\Delta_1^*(f(x); h; m) = m^{-k}(f(x + hm^k) - f(x)),$$

and when $j \geq 1$, by inductively defining

$$\Delta^*_{j+1}(f(x); h_1, \ldots, h_{j+1}; m_1, \ldots, m_{j+1})$$
$$= \Delta^*_1(\Delta^*_j(f(x); h_1, \ldots, h_j; m_1, \ldots, m_j); h_{j+1}; m_{j+1}).$$

For $0 \leq j \leq k$, let

$$f(z) = (z - h_1 m_1^k - \cdots - h_j m_j^k)^k,$$

and define the difference polynomial

$$\Psi_j = \Psi_j(z; h_1, \ldots, h_j; m_1, \ldots, m_j)$$

by taking

$$\Psi_j = \Delta^*_j(f(z); 2h_1, \ldots, 2h_j; m_1, \ldots, m_j).$$

Here we adopt the convention that $\Psi_0(z; \mathbf{h}; \mathbf{m}) = z^k$. We write

$$f_j(\alpha) = \sum_{x \in \mathcal{A}(Q_j, R)} e(\alpha x^k),$$

and

$$F_j(\alpha) = \sum_{z, \mathbf{h}, \mathbf{m}} e(\alpha \Psi_j(z; \mathbf{h}; \mathbf{m})),$$

where the summation is over z, $\mathbf{h}$, $\mathbf{m}$ with

$$1 \leq z \leq P_j, \quad 1 \leq h_i \leq 2^{j-i} H_i,$$

$$M_i < m_i \leq M_i R, \quad m_i \in \mathcal{A}(P, R),$$

for $1 \leq i \leq j$. Finally, we define

$$T(j, s) = \int_0^1 |F_j(\alpha) f_j(\alpha)^{2s}| d\alpha.$$

Now, on considering the underlying diophantine equation, we have

$$S_{s+1}(P, R) \leq \int_0^1 |F_0(\alpha)^2 f_0(\alpha)^{2s}| d\alpha.$$

The starting point in the iterative process is to bound the latter expression in terms of $S_s(Q_1, R)$ and $T(1, s)$. This corresponds to taking the first difference in the classical Weyl differencing argument, and extracting the contribution arising from those terms with $x_1 = y_1$. Thus one obtains

$$S_{s+1}(P, R) \ll P^\varepsilon M_1^{2s-1}(PM_1 S_s(Q_1, R) + T(1, s)),$$

and this inequality we write symbolically as

$$F_0^2 f_0^{2s} \longmapsto F_1 f_1^{2s}.$$

One way to proceed is by means of a repeated efficient differencing step. In principle this is based on the Cauchy-Schwarz inequality, applied in the form

$$\int_0^1 |F_j f_j^{2s}| d\alpha \le \left(\int_0^1 |f_j|^{2t} d\alpha \right)^{1/2} \left(\int_0^1 |F_j^2 f_j^{4s-2t}| d\alpha \right)^{1/2},$$

where for the sake of concision we have written f_j for $f_j(\alpha)$ and likewise F_j for $F_j(\alpha)$. Thus, for $j = 1, 2, \ldots$, the mean value $T(j, s)$ can be related to $S_t(Q_j, R)$ and $T(j+1, 2s-t)$, where $t < 2s$ is a parameter at our disposal, via inequalities of the shape

$$T(j, s) \ll P^\varepsilon (S_t(Q_j, R))^{1/2} (\widetilde{H}_j \widetilde{M}_j M_{j+1}^{4s-2t-1} \Xi_{j+1})^{1/2},$$

where we write

$$\Xi_{j+1} = P \widetilde{H}_j \widetilde{M}_{j+1} S_{2s-t}(Q_{j+1}, R) + T(j+1, 2s-t).$$

This is the $(j+1)$-th step in the differencing process and can be portrayed by

$$F_j f_j^{2s} \longrightarrow F_{j+1} f_{j+1}^{4s-2t}$$
$$\downarrow$$
$$f_j^{2t}$$

There are more sophisticated variants of this procedure wherein it may be useful to restrict some of the variables to a range $(\frac{1}{2} Q_j R^{-j}, Q_j]$, or to replace the set $\mathcal{A}(Q_j, R)$ by $\mathbb{N} \cap [1, Q_j]$ (see §2 of Vaughan and Wooley [134] for details, and a more complete discussion).

Another option is to use Hölder's inequality to bound $T(j, s)$. Thus we obtain an inequality of the type

$$T(j, s) = \int_0^1 |F_j(\alpha) f_j(\alpha)^{2s}| d\alpha \ll I_l^a I_{l+1}^b U_v^c U_w^d,$$

where

$$I_m = \int_0^1 |F_j(\alpha)|^{2^m} d\alpha \quad (m = l, l+1),$$

$$U_u = \int_0^1 |f_j(\alpha)|^{2u} d\alpha \quad (u = v, w),$$

and l, v, w, a, b, c, d are non-negative numbers satisfying the equations

$$a + b + c + d = 1, \quad 2^l a + 2^{l+1} b = 1, \quad vc + wd = s.$$

There is clearly great flexibility in the possible choices of the parameters here. We can summarise this process by

$$F_j f_j^{2s} \implies (F_j^{2^l})^a (F_j^{2^{l+1}})^b (f_j^{2v})^c (f_j^{2w})^d.$$

Yet another option is to apply the Hardy-Littlewood method to $T(j, s)$. In practice we expect that the minor arc contribution dominates, although this is not guaranteed. But if it does, then

$$T(j, s) \ll \left(\sup_{\alpha \in \mathfrak{m}} |F_j(\alpha)| \right) S_s(Q_j, R), \qquad (12.1)$$

and this we abbreviate to

$$F_j f_j^{2s} \implies (F_j)(f_j^{2s}).$$

By optimising choices for the parameters in order to obtain the sharpest estimates at each stage of the iteration process, one ultimately obtains relations describing $\lambda^{(n+1)}$ in terms of $\lambda^{(n)}$. The sharpest permissible exponents λ attainable by these methods are in general not easy to describe, and require substantial computations to establish (see, for example, Vaughan and Wooley [134]). However, one can describe in general terms the salient features of the permissible exponents λ_s. When s is rather small compared to k, it transpires that permissible exponents $\lambda_s = s + \delta_s$ can be derived with δ_s positive but small (see the next section for a consequence of this fact). Further, the simplest versions of the repeated efficient differencing method (see Wooley [155][157]) establish that the exponent $\lambda_s = 2s - k + ke^{1-2s/k}$ is permissible for every natural number s. Roughly speaking, therefore, one may compare the respective strengths of the diminishing ranges argument, and the repeated efficient differencing method, by comparing how rapidly the respective functions $k(1 - 1/k)^s \sim ke^{-s/k}$ and $ke^{1-2s/k}$ tend to zero as s increases.

The improvements in the most recent work (Vaughan and Wooley [134]) come about mostly through the following technical improvements:

- Better use of the Hardy-Littlewood method to estimate

$$T(j, s) = \int_0^1 \left| F_j(\alpha) f_j(\alpha)^{2s} \right| d\alpha.$$

In particular, tighter control is exercised in mean on the behaviour of the exponential sum $F_j(\alpha)$ on the major arcs, and this permits the assumption of (12.1) for a larger range of parameters ϕ than previously available.

- Better estimates for

$$I_m = \int_0^1 |F_j(\alpha)|^{2^m} \, d\alpha,$$

established largely via estimates for the number of integral points on certain affine plane curves.

- Better estimates (see Wooley [159]) for

$$\sup_{\alpha \in \mathfrak{m}} |f_0(\alpha)|.$$

Such estimates might be described as playing a significant rôle in the estimation of $G(k)$ in the final stages of the analysis.

13   Breaking Classical Convexity

All of the methods thus far described depend, in a fundamental manner, on the natural interpretation of even moments of exponential sums in terms of the number of solutions of certain underlying diophantine equations. In §12, for example, one is limited to permissible exponents λ_s corresponding to integral values of s, and in this setting the most effective method for bounding odd and fractional moments of smooth Weyl sums is to apply Hölder's inequality to interpolate between even moments. With the natural extension of the notion of a permissible exponent λ_s from integral values of s to arbitrary positive numbers s, the resulting exponents form the convex hull of the set of permissible exponents $\{\lambda_s : s \in \mathbb{N}\}$. A perusal of §12 reveals that extra flexibility in choice of parameters, and therefore the potential for further improvements, will be achieved by the removal of this "classical convexity" barrier, and such has recently become available.

In Wooley [158], a method is established which, loosely speaking, enables one to replace the inequality

$$S_{s+1}(P, R) \ll P^\varepsilon M_1^{2s-1}(P M_1 S_s(Q_1, R) + T(1, s))$$

that occurred in §12 with s restricted to be a natural number, by the new inequality

$$S_{s+t}(P, R) \ll P^\varepsilon M_1^{2s-t}(P^t M_1^t S_s(Q_1, R) + T^*),$$

where

$$T^* = \int_0^1 |F(\alpha)^t f_1(\alpha)^{2s}| d\alpha$$

and

$$F(\alpha) = \sum_{\substack{u \in \mathcal{A}(M_1 R, R) \\ u > M_1}} \sum_{\substack{z_1, z_2 \in \mathcal{A}(P, R) \\ z_1 \equiv z_2 \pmod{u^k} \\ z_1 \neq z_2}} e(\alpha u^{-k}(z_1^k - z_2^k)).$$

In this latter estimate, the parameter s is no longer restricted to be integral, and the parameter t may be chosen freely with $0 < t \le 1$. Moreover, the mean value T^* is very much reminiscent of $T(1, s)$, with $F(\alpha)$ substituted for $F_1(\alpha)$ and exhibiting similar properties. Thus, in addition to removing the integrality constraint on s, one may also iterate with a fractional number $2t$ of variables.

As might be expected, the additional flexibility gained in this way leads to improved permissible exponents λ_s even for integral s, since our methods are so highly iterative. The overall improvements are usually quite small and are largest for smaller values of k. Such progress has not yet delivered sharper bounds for $G(k)$, but this work provides the sharpest results available concerning sums of cubes (see §4, and also Brüdern and Wooley [18]), and has also permitted new conclusions to be derived in certain problems involving sums of mixed powers (see Brüdern and Wooley [16] [19] [17], and also §15 below). Also, this "breaking convexity" device provides the best available lower bounds for the number $\mathcal{N}_{k,s}(X)$ of natural numbers not exceeding X that are the sum of s kth powers of positive integers, at least when s is small compared to k. Thus, when $2 < s \le 2e^{-1}\sqrt{k}$, one has

$$\mathcal{N}_{k,s}(X) \gg X^{s/k - e^{-\gamma_s k}},$$

where $\gamma_s = 16/(es)^2$ (see Theorems 1.3 and 1.4 of Wooley [158], and the associated discussion). For comparison, one conjecturally has the lower bound $\mathcal{N}_{k,s}(X) \gg X^{s/k}$ whenever $s < k$.

Much remains to be investigated for fractional moments, in part owing to the substantial increase in complexity of the underlying computations (see Wooley [161] for more on this). However, such developments presently appear unlikely to have a large impact on the central problem of bounding $G(k)$ in the classical version of Waring's problem.

14   Variants of Waring's Problem: Primes

Much work has been devoted to various generalisations of the classical version of Waring's problem, and it seems appropriate to discuss some of the more mainstream variants.

We begin with the Waring-Goldbach problem, in which one seeks to represent integers as sums of kth powers of prime numbers. In order to describe the associated local conditions, suppose that k is a natural number and p is a prime number. We denote by $\theta = \theta(k, p)$ the integer with $p^\theta | k$ and $p^{\theta+1} \nmid k$, and then define $\gamma = \gamma(k, p)$ by

$$\gamma(k, p) = \begin{cases} \theta + 2, & \text{when } p = 2 \text{ and } \theta > 0, \\ \theta + 1, & \text{otherwise.} \end{cases}$$

Finally, we put

$$K(k) = \prod_{(p-1)|k} p^\gamma.$$

Denote by $H(k)$ the least integer s such that every sufficiently large positive integer congruent to s modulo $K(k)$ may be written as a sum of s kth powers of prime numbers. Note that when $(p - 1) | k$, one has $p^\theta(p - 1) | k$, whence $a^k \equiv 1 \pmod{p^\gamma}$ whenever $(p, a) = 1$. This explains our seemingly awkward definition of $H(k)$, since whenever n is the sum of s kth powers of primes exceeding $k + 1$, then necessarily $n \equiv s \pmod{K(k)}$. Naturally, further congruence conditions may arise from primes p with $(p - 1) \nmid k$.

Following the pioneering investigations of Vinogradov [145][146] (see also Vinogradov [147]), Hua comprehensively investigated additive problems involving prime numbers in his influential book (see Hua [79], but also Hua [78]). Thus, it is known that for every natural number k one has

$$H(k) \leq 2^k + 1,$$

and, when k is large, that

$$H(k) \leq 4k(\log k + \tfrac{1}{2} \log \log k + O(1)).$$

In the conventional plan of attack on the Waring-Goldbach problem, one applies the Hardy-Littlewood method in a manner similar to that outlined above, but in interpreting the number of solutions of an analogue of the equation (11.1) over prime numbers, one obtains an upper bound by discarding the primality condition. With sufficiently many variables employed to save a factor of n via such an approach, one additional variable suffices to save the extra power of $\log n$ required by primality considerations. Although this strategy evidently prohibits the use of smooth numbers, the diminishing ranges technology perfected by Davenport, and refined by Vaughan [121] and Thanigasalam [109]–[113] plays a prominent rôle in establishing the best available upper bounds for $H(k)$ when k is small. We should also mention that recent progress depends on good estimates of Weyl-type for exponential sums over primes, and allied sums, available from the use of Vaughan's identity (see Vaughan [117]), combined with the linear sieve equipped with a switching principle (see Kawada and Wooley [83]). Thus, for $4 \leq k \leq 10$, the best known upper bounds for $H(k)$ are as follows:

$$\text{Kawada and Wooley [83],} \quad H(4) \leq 14, \; H(5) \leq 21,$$
$$\text{Thanigasalam [111],} \quad H(6) \leq 33, \; H(7) \leq 47, \; H(8) \leq 63,$$
$$H(9) \leq 83, \; H(10) \leq 107.$$

Despite much effort on the Waring-Goldbach problem for exponents 1, 2 and 3, further progress remains elusive. Improvements are feasible, however, if one is prepared to accept almost-primes in place of prime numbers (see, in particular, Chen [25], Brüdern [13][14], and Brüdern and Fouvry [15]). Difficulties related to those associated with the Waring-Goldbach problem are encountered when other sequences are substituted for prime numbers. For Waring's problem with smooth variables, see Harcos [57] and Brüdern and Wooley [18]. Also, see Nechaev [97] for work on Waring's problem with polynomial summands (Wooley [160] and Ford [55] have restricted improvements employing smooth numbers).

15   Variants of Waring's Problem: Sums of Mixed Powers

Suppose that $k_1, k_2, \ldots, k_s$ are natural numbers with $2 \leq k_1 \leq k_2 \leq \cdots \leq k_t$. Then an optimistic counting argument suggests that whenever the equation

$$x_1^{k_1} + x_2^{k_2} + \cdots + x_s^{k_s} = n \tag{15.1}$$

has p-adic solutions for each prime p, and

$$k_1^{-1} + k_2^{-1} + \cdots + k_s^{-1} > 1, \tag{15.2}$$

then n should be represented as the sum of mixed powers of positive integers (15.1) whenever it is sufficiently large in terms of **k**. When $s = 3$ such an assertion may fail in certain circumstances (see Jagy and Kaplansky [80], or Exercise 5 of Chapter 8 of Vaughan [128]), but a heuristic application of the Hardy-Littlewood method suggests, at least, that the condition (15.2) should ensure that *almost all* integers in the expected congruence classes are thus represented. Moreover, subject instead to the condition

$$k_1^{-1} + k_2^{-1} + \cdots + k_s^{-1} > 2, \tag{15.3}$$

a formal application of the circle method suggests that *all* integers in the expected congruence classes should be represented in the form (15.1). Meanwhile, a simple counting argument shows that in circumstances in which the condition (15.2) does not hold, then arbitrarily large integers are not represented in the desired form.

The investigation of such analogues of Waring's problem for mixed powers has, since the early days of the Hardy-Littlewood method, stimulated progress in technology of use even in the classical version of Waring's problem. Additive problems in which the summands are restricted to be squares, cubes or biquadrates are perhaps of greater interest than those with higher powers, and here the current situation is remarkably satisfactory. We summarise below the current state of knowledge in the simpler problems of this nature. In Tables 5 and 6 we list constellations of powers whose sum represents, respectively, almost all, and all, integers subject to the expected congruence conditions. The tables are arranged, roughly speaking, starting with predominantly smaller exponents, and ending with predominantly larger exponents, and therefore not in chronological order of the results.

We have been unable to trace the origin in the literature of the conclusion on a square, two biquadrates and a kth power, but refer the reader to Exercise 6 of §2.8 of Vaughan [128] for related ideas (see also Roth

Davenport & Heilbronn [36],	two squares, one kth power,
Davenport & Heilbronn [37],	one square, two cubes,
Roth [101],	one square, one cube, one biquadrate,
Vaughan [116],	one square, one cube, one fifth power,
Folklore (?)	one square, two biquadrates, one kth power,
Hooley [70],	one square, one cube, one sixth power, one kth power,
Davenport [29],	four cubes,
Brüdern [9][8],	three cubes, one biquadrate,
Brüdern [8], Lu [89],	three cubes, one fifth power,
Brüdern & Wooley [19],	three cubes, one sixth power ,
Kawada & Wooley [82],	one cube, four biquadrates,
Vaughan [124],	six biquadrates,
Kawada & Wooley [82],	five biquadrates, one kth power (k odd).

Table 5. Representation of almost all integers

Gauss [56],	three squares,
Hooley [71],	two squares, three cubes,
Hooley [70],	two squares, assorted powers,
Vaughan [123],	one square, five cubes,
Brüdern & Wooley [16],	one square, four cubes, one biquadrate,
Ford [54],	one square, one cube, one biquadrate,...,
	one fifteenth power,
Linnik [87][88],	seven cubes,
Brüdern [9],	six cubes, two biquadrates,
Brüdern [9],	five cubes, three biquadrates,
Kawada & Wooley [82],	three cubes, six biquadrates,
Brüdern & Wooley [17],	two cubes, seven biquadrates,
Kawada & Wooley [82],	one cube, nine biquadrates
Vaughan [124],	twelve biquadrates,
Kawada & Wooley [82],	ten biquadrates, one kth power
	(k odd).

Table 6. Representation of all integers

[101]). We remark that all ternary problems of interest have been solved, since for non-trivial triples (k_1, k_2, k_3) not accounted for in Table 5, one has $k_1^{-1} + k_2^{-1} + k_3^{-1} \leq 1$. Also, energetic readers may be interested in tackling a problem which presently defies resolution only by the narrowest of margins, namely the problem of showing that almost all integers are represented as the sum of two cubes and two biquadrates of positive integers.

Note that although the three square theorem is commonly ascribed to Legendre, his "proof" depended on an unsubstantiated assumption only later established by Dirichlet, and the first complete proof is due to Gauss. We finish by noting that in problems involving sums of two squares, methods more effective than the circle method can be brought into play (see especially Hooley [70][71] and Brüdern [7]).

16 Variants of Waring's Problem: Beyond $\mathbb{Z}$

Given the considerable energy expended on the investigation of Waring's problem over the rational integers, it seems natural to extend this work to algebraic number fields. Here one encounters the immediate difficulty of deciding what precisely Waring's problem should mean in this broader context. It is possible, for example, that an algebraic integer in a number field K may not be a sum of any finite number of kth powers of algebraic integers of that field (consider, say, the parity of the imaginary part of $\mathfrak{a}^2$ when $\mathfrak{a} \in \mathbb{Z}[i]$). With this in mind, when K is a number field and $\mathfrak{O}_K$ is its ring of integers, we define J_k to be the subring of $\mathfrak{O}_K$ generated by the kth powers of integers of K. We must also provide an analogue of the positivity of the kth powers inherent in the classical version of Waring's problem. Thus we define $G_K(k)$ to be the smallest positive integer s with the property that, for some positive number $c = c(k, K)$, and for all totally positive integers $\nu \in J_k$ of sufficiently large norm, the equation

$$\nu = \lambda_1^k + \lambda_2^k + \cdots + \lambda_s^k \tag{16.1}$$

is always soluble in totally non-negative integers λ_j of K with $N(\lambda_j) \leq cN(\nu)^{1/k}$ $(1 \leq j \leq s)$.

Following early work of Meissner [93] and Mordell [95] for a restricted class of number fields, Siegel [102][103] was the first to obtain quite general conclusions for sums of squares, and hence, via the method of Hilbert [69], for sums of kth powers. Siegel later developed a proper generalisation of the Hardy-Littlewood method to number fields, and here the dissection into major and minor arcs is a particular source of difficulty. In this way, Siegel [104][105] obtained the upper bound

$$G_K(k) \leq dk(2^{k-1} + d) + 1,$$

where d denotes the degree of the field K. If one were to break the equation (16.1) into components with respect to an integral basis for $\mathfrak{O}_K$, then one would obtain d equations of degree k in ds variables, and so one might optimistically expect the analytic part of the circle method to apply with a number of variables roughly the same in both K and $\mathbb{Q}$. Perhaps motivated by such considerations, Siegel asked for reasonable bounds on $G_K(k)$ independent of d. This question was ultimately addressed through the work of Birch [4] and Ramanujam [98], who provided the upper bound

$$G_K(k) \leq \max\{8k^5, 2^k + 1\}. \tag{16.2}$$

It is evident that the uniform bound (16.2) is far from the above cited bound $G(k) \leq (1 + o(1))k \log k$ of Wooley [155], and the only slightly weaker precursors of Vinogradov. An important desideratum, therefore, is the reduction of the bound (16.2) to one of similar order to that presently available for $G(k)$, or at least a reduction to a bound polynomial in k.

Failing this, effort has been expended in pursuit of bounds of order $k \log k$, but with a modest dependence on the degree d of the field K. Progress towards this objective has mirrored developments in the classical version of Waring's problem. Thus, building on work generalising that of Vinogradov to the number field setting by Körner [84] and Eda [50], methods employing smooth numbers and repeated efficient differences have recently been applied by Davidson [39] to establish the bound

$$G_K(k) \le (3 + o(1))k \log k + c_d k,$$

where

$$c_d = 4d + 3 \log(\tfrac{1}{2}(d^2 + 1)) + 7,$$

and the term $o(1)$ is independent of d (apparently, the term $3 + o(1)$ here can be replaced by $2 + o(1)$ with only modest effort). Moreover, when K has class number one, Davidson [40][42] has obtained the bound

$$G_K(k) \le k(\log k + \log \log k + c_d),$$

where c_d is approximately $4d$. Finally, Davidson [41] has improved on earlier work of Tatuzawa [106] to establish the strikingly simple conclusion that, for every number field K, one has

$$G_K(k) \le 2d(\widehat{G}(k) + 2k),$$

where $\widehat{G}(k)$ denotes the least number s satisfying the property that the set of rational integers that can be expressed as the sum of s kth powers of natural numbers has positive density (in particular, of course, $\widehat{G}(k) \le G_1(k)$).

More exotic still than the variants of Waring's problem over number fields are those in which one works over the polynomial rings $\mathbb{F}_q[t]$. Here one must again impose restrictions on the size of polynomials employed in the representation (and in this situation, the degree of a polynomial provides a measure of its size). Analogues of the Hardy-Littlewood method have been devised in this polynomial ring setting (see, for example, Effinger and Hayes [51]). Unfortunately, however, Weyl differencing proves ineffective whenever kth powers are considered over $\mathbb{F}_q[t]$ with $k \ge \operatorname{char}(\mathbb{F}_q)$, for in such circumstances the factor $k!$, introduced into the argument of the exponential sum over kth powers via the differencing argument, is equal to zero in $\mathbb{F}_q[t]$. Consequently, one frequently restricts attention to kth powers with k smaller than the characteristic (but see Car and Cherly [22] for results on sums of 11 cubes in $\mathbb{F}_{2^h}[t]$). With this restriction, Car [20][21] has shown that every polynomial M, with $M \subset \mathbb{F}_q[t]$, of sufficiently large degree, can be written in the form

$$M = M_1^k + \cdots + M_s^k,$$

with $M_i \in \mathbb{F}_q[t]$ of degree smaller than $1 + (\deg(M))/k$ $(1 \le s \le k)$, provided that

$$s \ge \min\{2^k + 1, \ 2k(k-1)\log 2 + 2k + 3\}.$$

17 Open Problems and Conjectures

Returning temporarily to the methods of §12, there are a number of problems connected with the mean value

$$I_m = \int_0^1 |F_j(\alpha)|^{2^m} d\alpha$$

which suggest some interesting questions. We will concentrate on the situation with $m = 1$, and remark only that the available results become less satisfactory as m increases.

A simple combinatorial argument reveals that the difference polynomial $\Psi_j = \Psi_j(z; \mathbf{h}; \mathbf{m})$ is given explicitly by the formula

$$\Psi_j = k! 2^j h_1 \dots h_j \sum_{u \ge 0} \sum_{v_1 \ge 0} \cdots \sum_{v_j \ge 0} \frac{z^u (h_1 m_1^k)^{2v_1} \dots (h_j m_j^k)^{2v_j}}{u!(2v_1 + 1)! \dots (2v_j + 1)!},$$

where the summation is subject to the condition $u + 2v_1 + \cdots + 2v_j = k - j$. Consequently, one has

$$\Psi_j = h_1 \dots h_j z^d \sum_{r=0}^{\frac{1}{2}(k-j-d)} c_r(h_1 m_1^k, \dots, h_j m_j^k) z^{2r},$$

where

$$d = \begin{cases} 0, & \text{when } k - j \text{ is even,} \\ 1, & \text{when } k - j \text{ is odd,} \end{cases}$$

and for $0 \le r \le (k - j - d)/2$, the coefficients $c_r(\boldsymbol{\xi})$ are polynomials with positive integral coefficients that are symmetric in $\xi_1^2, \dots, \xi_j^2$, and have total degree $k - j - 2r - d$. Now the mean value

$$I_1 = \int_0^1 |F_j(\alpha)|^2 d\alpha$$

is equal to the number of solutions of the diophantine equation

$$\Psi_j(z; \mathbf{h}; \mathbf{m}) = \Psi_j(z'; \mathbf{h}'; \mathbf{m}'), \tag{17.1}$$

with the variables in ranges discernible from the definition of $F_j(\alpha)$ in §12, and one might hope that the total number of solutions is close to the number of diagonal solutions, which is to say that

$$I_1 \ll P^{1+\varepsilon} \widetilde{M}_j \widetilde{H}_j.$$

When $k-j$ is odd (so $d=1$), the presence of the term z^d makes it especially easy to deal with equation (17.1) by exploiting the inherent multiplicative structure, and indeed one can achieve the desired bound provided also that $j \leq (k-d)/3$. The cases $j=1$ and $k-j=2$ or 4 are also doable. However, when $k-j$ is even and $1 < j \leq k-6$, the situation is not so easy. By the way, this difficulty already occurs in Davenport's work [31]. To illustrate this situation, the simplest special case that we cannot handle directly corresponds to $k=8$ and $j=2$, and here one has

$$\begin{aligned}
\Psi_j =\; & h_1 h_2 \big(224z^6 + 1120z^4(h_1^2 m_1^{16} + h_2^2 m_2^{16}) \\
& + z^2(672h_1^4 m_1^{32} + 2240h_1^2 h_2^2 m_1^{16} m_2^{16} + 672h_2^4 m_2^{32}) \\
& + 32h_1^6 m_1^{48} + 224h_1^4 h_2^2 m_1^{32} m_2^{16} + 224h_1^2 h_2^4 m_1^{16} m_2^{32} + 32h_2^6 m_2^{48}\big).
\end{aligned}$$

This suggests various general questions.

• Suppose that $f, g \in \mathbb{Z}[x]$. Are there simple conditions on f, g such that the number N of integral points $(x, y) \in [-P, P]^2$ for which $f(x) = g(y)$ satisfies

$$N \ll \big(P\mathcal{H}(f)\mathcal{H}(g)\big)^\varepsilon?$$

Here $\mathcal{H}(h)$ denotes the height of h. A qualitative version of this has already been considered by Davenport, Lewis and Schinzel [38], and if $f(x) - g(y)$ is irreducible over $\mathbb{C}$, then a celebrated theorem of Siegel shows that the number of solutions is finite unless there is a rational parametric solution of special form.

By the way, in view of the above, it is perhaps not surprising that in our treatment of I_m the bound of Bombieri and Pila [6] plays a rôle.

• Suppose that $\mathcal{A} \subset \mathbb{Z}^k \cap [-X, X]^k$. Let $R(n; \mathcal{A})$ denote the number of solutions of the equation

$$a_1 x + a_2 x^2 + \cdots + a_k x^k = n$$

with $x \in \mathbb{Z} \cap [-P, P]$ and $\mathbf{a} \in \mathcal{A}$. Are there any simple conditions under which it is true that

$$\sum_n R(n; \mathcal{A})^2 \ll \operatorname{card}(\mathcal{A}) P(XP)^\varepsilon?$$

• A well-known conjecture in connection with Waring's problem is Hypothesis K (Hardy and Littlewood [62]). Let

$$R_{k,s}(n) = \text{card}\{\mathbf{x} \in \mathbb{N}^s : x_1^k + \cdots + x_s^k = n\}.$$

Then Hypothesis K asserts that for each natural number k, one has

$$R_{k,k}(n) \ll n^\varepsilon. \tag{17.2}$$

From this it would follow that

$$G(k) \le \max\{2k + 1, \Gamma_0(k)\} \tag{17.3}$$

and

$$G_1(k) = \max\{k + 1, \Gamma_0(k)\}. \tag{17.4}$$

The conjecture (17.2) was later shown by Mahler [90] to be false for $k = 3$, and indeed his counter-example shows that, infinitely often, one has $R_{3,3}(n) > 9^{-1/3}n^{1/12}$. However, the conjecture is still open when $k \ge 4$, and for (17.3) and (17.4), it suffices to know that

$$\sum_{n \le N} R_{k,k}(n)^2 \ll N^{1+\varepsilon}.$$

Hooley [74] established this when $k = 3$, under the assumption of the Riemann Hypothesis for a certain Hasse-Weil L-function. As far as we know, no simple conjecture of this kind is known from which it would follow that $G(k) = \max\{k + 1, \Gamma_0(k)\}$.

• It may well be true that, when $k \ge 3$, one has

$$\sum_{n \le N} R_{k,k}(n)^2 \sim CN.$$

However Hooley [72] has shown, at least when $k = 3$, that the constant C here is larger than what would arise simply from the major arcs. This leads to some interesting speculations. The number of solutions of the equation

$$x_1^k + \cdots + x_s^k = y_1^k + \cdots + y_s^k \le N,$$

in which the variables on the right hand side are a permutation of those on the left hand side, is asymptotic to $C_1 N^{s/k}$, for a certain positive number $C_1 = C_1(k, s)$, and when $s < k$ the contribution arising from the major arcs is smaller. Maybe one should think of these solutions as being "trivial", "parametric", or as arising from some "degenerate" property of the geometry of the surface. Anyway, their contribution is mostly concentrated on the minor arcs. It seems rather likely that this phenomenon persists for $s \geq k$ and explains Hooley's discovery. This leads to the philosophy that the major arcs correspond to non-trivial solutions, and the minor arcs to trivial solutions. There is an example of this phenomenon in Vaughan and Wooley [133].

• Recall the definitions of $f(\alpha)$, $S(q, a)$ and $v(\beta)$ from §7. One can conjecture that, whenever $(a, q) = 1$, one has

$$f(\alpha) - q^{-1} S(q, a) v(\alpha - a/q) \ll (q + P^k |q\alpha - a|)^{1/k}. \tag{17.5}$$

Possibly the exponent has to be weakened to $1/k + \varepsilon$, but any counter-examples would be interesting.

From (17.5) it would follow that

$$\sum_{n \leq N} R_{k,k}(n)^2 \ll N.$$

Also it is just conceivable that (17.5), in combination with a variant of the Hardy-Littlewood-Kloosterman method, would achieve the bound $G(k) \leq \max\{\mathfrak{G}(k), \Gamma_0(k)\}$, where $\mathfrak{G}(k) < 2k + 1$.

• The inequality (17.5) is a special case of conjectures that can be made about the exponential sum

$$f(\boldsymbol{\alpha}) = \sum_{x \leq P} e(\alpha_1 x + \cdots + \alpha_k x^k)$$

that would have many consequences in analytic number theory. For example one can ask if something like

$$f(\boldsymbol{\alpha}) - q^{-1} S(q, \mathbf{a}) v(\boldsymbol{\alpha} - \mathbf{a}/q) \ll \sum_{j=2}^{k} \left(\frac{q + P^j |\alpha_j q - a_j|}{(q, a_j)} \right)^{1/j} \tag{17.6}$$

is true. Here

$$S(q, \mathbf{a}) = \sum_{r=1}^{q} e((a_1 r + \cdots + a_k r^k)/q),$$

$$v(\boldsymbol{\beta}) = \int_0^P e(\beta_1 \gamma + \cdots + \beta_k \gamma^k) d\gamma,$$

and $a_1 = c(q, a_2, \ldots, a_k)$, where $c = c(q, \mathbf{a})$ is the unique integer with

$$-\frac{1}{2} < c - \alpha_1 q/(q, a_2, \ldots, a_k) \leq \frac{1}{2}.$$

The inequality (17.6) is known to hold for $k = 2$ (Vaughan unpublished) with the right hand side weakened slightly to

$$\left(\frac{q}{(q, a_2)} \right)^{1/2} \left(\log \frac{2q}{(q, a_2)} + P|\alpha_2 - a_2/q_2|^{1/2} \right).$$

• One way of viewing the Hardy-Littlewood method is that we begin by considering the Fourier transform with respect to Lebesgue measure on $[0, 1)$ for an appropriate generating function defined in terms of the additive characters, and then approximate to it by a product of discrete measures. Since part of the problem when one has relatively few variables is that the geometry genuinely intrudes, one can ask whether we are using the best measure for the problem at hand. In this situation something more closely related to the underlying geometry might be more useful.

In conclusion, it is clear that although we have come a long way in the twentieth century, there remains plenty still to be done!

References

[1]   M. A. Bennett, *Fractional parts of powers of rational numbers*, Math. Proc. Cambridge Philos. Soc. **114** (1993), 191–201.

[2]   ———, *An ideal Waring problem with restricted summands*, Acta Arith. **66** (1994), 125–132.

[3]   F. Beukers, *Fractional parts of powers of rationals*, Math. Proc. Cambridge Philos. Soc. **90** (1981), 13–20.

[4]   B. J. Birch, *Waring's problem in algebraic number fields*, Proc. Cambridge Philos. Soc. **57** (1961), 449–459.

[5]   K. D. Boklan, *The asymptotic formula in Waring's problem*, Mathematika **41** (1994), 147–161.

[6]   E. Bombieri and J. Pila, *The number of integral points on arcs and ovals*, Duke Math. J. **59** (1989), 337–357.

[7]   J. Brüdern, *Sums of squares and higher powers*, J. London Math. Soc. (2) **35** (1987), 233–243.

[8]   ———, *Iterationsmethoden in der additiven Zahlentheorie*, Dissertation, Göttingen, 1988.

[9]   _____, *On Waring's problem for cubes and biquadrates*, J. London Math. Soc. (2) **37** (1988), 25–42.

[10]   _____, *Sums of four cubes*, Monatsh. Math. **107** (1989), 179–188.

[11]   _____, *On Waring's problem for fifth powers and some related topics*, Proc. London Math. Soc. (3) **61** (1990), 457–479.

[12]   _____, *On Waring's problem for cubes*, Math. Proc. Cambridge Philos. Soc. **109** (1991), 229–256.

[13]   _____, *Sieves, the circle method, and Waring's problem for cubes*, Habilitationschrift, Göttingen, 1991, Mathematica Göttingensis **51**.

[14]   _____, *A sieve approach to the Waring-Goldbach problem, II. On the seven cube theorem*, Acta Arith. **72** (1995), 211–227.

[15]   J. Brüdern and É. Fouvry, *Lagrange's four squares theorem with almost prime variables*, J. Reine Angew. Math. **454** (1994), 59–96.

[16]   J. Brüdern and T. D. Wooley, *On Waring's problem: a square, four cubes and a biquadrate*, Math. Proc. Cambridge Philos. Soc. **127** (1999), 193–200.

[17]   _____, *On Waring's problem: two cubes and seven biquadrates*, Tsukuba J. Math. **24** (2000), 387–417.

[18]   _____, *On Waring's problem for cubes and smooth Weyl sums*, Proc. London Math. Soc. (3) **82** (2001), 89–109.

[19]   _____, *On Waring's problem: three cubes and a sixth power*, Nagoya Math. J. **163** (2001), 13–53.

[20]   M. Car, *Le problème de Waring pour l'anneau des polynômes sur un corps fini*, Séminaire de Théorie des Nombres 1972–1973 (Univ. Bordeaux I, Talence), Lab. Théorie des Nombres, CNRS, Talence, 1973, Exp. No. 6, 13pp.

[21]   _____, *Waring's problem in function fields*, Proc. London Math. Soc. (3) **68** (1994), 1–30.

[22]   M. Car and J. Cherly, *Sommes de cubes dans l'anneau $\mathbb{F}_{2^h}[X]$*, Acta Arith. **65** (1993), 227–241.

[23]   J. W. S. Cassels and R. C. Vaughan, *Ivan Matveevich Vinogradov*, Biographical Memoirs of Fellows of the Royal Society **31** (1985), 613–631, reprinted in: Bull. London Math. Soc. **17** (1985), 584–600.

[24] J.-R. Chen, *On Waring's problem for n-th powers*, Acta Math. Sinica **8** (1958), 253–257.

[25] ———, *On the representation of a large even integer as the sum of a prime and the product of at most two primes*, Sci. Sinica **16** (1973), 157–176.

[26] R. J. Cook, *A note on Waring's problem*, Bull. London Math. Soc. **5** (1973), 11–12.

[27] H. Davenport, *Sur les sommes de puissances entières*, C. R. Acad. Sci. Paris, Sér. A **207** (1938), 1366–1368.

[28] ———, *On sums of positive integral kth powers*, Proc. Roy. Soc. London, Ser. A **170** (1939), 293–299.

[29] ———, *On Waring's problem for cubes*, Acta Math. **71** (1939), 123–143.

[30] ———, *On Waring's problem for fourth powers*, Ann. of Math. **40** (1939), 731–747.

[31] ———, *On sums of positive integral kth powers*, Amer. J. Math. **64** (1942), 189–198.

[32] ———, *On Waring's problem for fifth and sixth powers*, Amer. J. Math. **64** (1942), 199–207.

[33] ———, *Sums of three positive cubes*, J. London Math. Soc. **25** (1950), 339–343.

[34] ———, *The collected works of Harold Davenport, Volume III (B. J. Birch, H. Halberstam and C. A. Rogers, eds.)*, Academic Press, London, 1977.

[35] H. Davenport and P. Erdős, *On sums of positive integral kth powers*, Ann. of Math. **40** (1939), 533–536.

[36] H. Davenport and H. Heilbronn, *Note on a result in the additive theory of numbers*, Proc. London Math. Soc. (2) **43** (1937), 142–151.

[37] ———, *On Waring's problem: two cubes and one square*, Proc. London Math. Soc. (2) **43** (1937), 73–104.

[38] H. Davenport, D. J. Lewis, and A. Schinzel, *Equations of the form $f(x) = g(y)$*, Quart. J. Math. Oxford (2) **12** (1961), 304–312.

[39] M. Davidson, *Sums of k-th powers in number fields*, Mathematika **45** (1998), 359–370.

[40] _____, *A new minor-arcs estimate for number fields*, Topics in number theory (University Park, PA, 1997), Kluwer Acad. Publ., Dordrecht, 1999, pp. 151–161.

[41] _____, *On Waring's problem in number fields*, J. London Math. Soc. (2) **59** (1999), 435–447.

[42] _____, *On Siegel's conjecture in Waring's problem*, to appear, 2001.

[43] J.-M. Deshouillers, F. Hennecart, K. Kawada, B. Landreau, and T. D. Wooley, *Sums of biquadrates: a survey*, in preparation, 2001.

[44] J.-M. Deshouillers, F. Hennecart, and B. Landreau, $7,373,170,279,850$, Math. Comp. **69** (2000), 421–439.

[45] _____, *Waring's problem for sixteen biquadrates – Numerical results*, J. Théorie Nombres Bordeaux **12** (2000), 411–422, Colloque International de Théorie des Nombres (Talence, 1999).

[46] J.-M. Deshouillers, K. Kawada, and T. D. Wooley, *On sums of sixteen biquadrates*, in preparation, 2001.

[47] L. E. Dickson, *All integers except* 23 *and* 239 *are sums of eight cubes*, Bull. Amer. Math. Soc. **45** (1939), 588–591.

[48] _____, *History of the theory of numbers, Vol. II: Diophantine analysis*, Chelsea, New York, 1966.

[49] A. K. Dubitskas, *A lower bound on the value* $\|(3/2)^k\|$, Uspekhi Mat. Nauk **45** (1990), 153–154, translation in: Russian Math. Surveys **45** (1990), 163–164.

[50] Y. Eda, *On Waring's problem in an algebraic number field*, Rev. Columbiana Math. **9** (1975), 29–73.

[51] G. Effinger and D. Hayes, *Additive number theory of polynomials over a finite field*, Oxford University Press, Oxford, 1991.

[52] T. Estermann, *On Waring's problem for fourth and higher powers*, Acta Arith. **2** (1937), 197–211.

[53] K. B. Ford, *New estimates for mean values of Weyl sums*, Internat. Math. Res. Notices (1995), 155–171.

[54] ——, *The representation of numbers as sums of unlike powers. II*, J. Amer. Math. Soc. **9** (1996), 919–940.

[55] ——, *Waring's problem with polynomial summands*, J. London Math. Soc. (2) **61** (2000), 671–680.

[56] C. F. Gauss, *Disquisitiones arithmeticæ*, Leipzig, 1801.

[57] G. Harcos, *Waring's problem with small prime factors*, Acta Arith. **80** (1997), 165–185.

[58] G. H. Hardy and J. E. Littlewood, *A new solution of Waring's problem*, Quart. J. Math. Oxford **48** (1920), 272–293.

[59] ——, *Some problems of "Partitio Numerorum": I. A new solution of Waring's problem*, Göttingen Nachrichten (1920), 33–54.

[60] ——, *Some problems of "Partitio Numerorum": II. Proof that every large number is the sum of at most 21 biquadrates*, Math. Z. **9** (1921), 14–27.

[61] ——, *Some problems of "Partitio Numerorum": IV. The singular series in Waring's Problem and the value of the number $G(k)$*, Math. Z. **12** (1922), 161–188.

[62] ——, *Some problems of "Partitio Numerorum" (VI): Further researches in Waring's problem*, Math. Z. **23** (1925), 1–37.

[63] ——, *Some problems of "Partitio Numerorum" (VIII): The number $\Gamma(k)$ in Waring's problem*, Proc. London Math. Soc. (2) **28** (1928), 518–542.

[64] G. H. Hardy and S. Ramanujan, *Asymptotic formulae in combinatory analysis*, Proc. London Math. Soc. (2) **17** (1918), 75–115.

[65] D. R. Heath-Brown, *Weyl's inequality, Hua's inequality, and Waring's problem*, J. London Math. Soc. (2) **38** (1988), 396–414.

[66] ——, *Weyl's inequality and Hua's inequality*, Number Theory (Ulm, 1987), Lecture Notes in Math., vol. 1380, Springer-Verlag, Berlin, 1989, pp. 87–92.

[67] ——, *The circle method and diagonal cubic forms*, Philos. Trans. Roy. Soc. London Ser. A **356** (1998), 673–699.

[68] H. Heilbronn, *Über das Waringsche Problem*, Acta Arith. **1** (1936), 212–221.

[69]   D. Hilbert, *Beweis für Darstellbarkeit der ganzen Zahlen durch eine feste Anzahl nter Potenzen Waringsche Problem*, Nach. Königl. Ges. Wiss. Göttingen math.-phys. Kl. (1909), 17–36, Math. Ann. **67** (1909), 281–300.

[70]   C. Hooley, *On a new approach to various problems of Waring's type*, Recent progress in analytic number theory, Vol. 1 (Durham, 1979), Academic Press, London, 1981, pp. 127–191.

[71]   _____, *On Waring's problem for two squares and three cubes*, J. Reine Angew. Math. **328** (1981), 161–207.

[72]   _____, *On some topics connected with Waring's problem*, J. Reine Angew. Math. **369** (1986), 110–153.

[73]   _____, *On Waring's problem*, Acta Math. **157** (1986), 49–97.

[74]   _____, *On Hypothesis K* in Waring's problem*, Sieve methods, exponential sums, and their applications in number theory (Cardiff, 1995), Cambridge Univ. Press, Cambridge, 1997, pp. 175–185.

[75]   L.-K. Hua, *On Waring's problem*, Quart. J. Math. Oxford **9** (1938), 199–202.

[76]   _____, *Some results on Waring's problem for smaller powers*, C. R. Acad. Sci. URSS (2) **18** (1938), 527–528.

[77]   _____, *An improvement of Vinogradov's mean-value theorem and several applications*, Quart. J. Math. Oxford **20** (1949), 48–61.

[78]   _____, *Die Abschätzungen von Exponentialsummen und ihre Anwendung in der Zahlentheorie*, B.G. Teubner, Leipzig, 1959.

[79]   _____, *Additive theory of prime numbers*, American Math. Soc., Providence, Rhode Island, 1965.

[80]   W. C. Jagy and I. Kaplansky, *Sums of squares, cubes, and higher powers*, Experiment. Math. **4** (1995), 169–173.

[81]   R. D. James, *On Waring's problem for odd powers*, Proc. London Math. Soc. **37** (1934), 257–291.

[82]   K. Kawada and T. D. Wooley, *Sums of fourth powers and related topics*, J. Reine Angew. Math. **512** (1999), 173–223.

[83]   _____, *On the Waring-Goldbach problem for fourth and fifth powers*, Proc. London Math. Soc. (3) **83** (2001), 1–50.

[84] O. Körner, *Über das Waringsche Problem in algebraischen Zahlkörpern*, Math. Ann. **144** (1961), 224–238.

[85] J. M. Kubina and M. C. Wunderlich, *Extending Waring's conjecture to* 471,600,000, Math. Comp. **55** (1990), 815–820.

[86] E. Landau, *Über eine Anwendung der Primzahlen auf das Waringsche Problem in der elementaren Zahlentheorie*, Math. Ann. **66** (1909), 102–105.

[87] Ju. V. Linnik, *On the representation of large numbers as sums of seven cubes*, Dokl. Akad. Nauk SSSR **35** (1942), 162.

[88] ――――, *On the representation of large numbers as sums of seven cubes*, Mat. Sb. **12** (1943), 218–224.

[89] M. G. Lu, *On Waring's problem for cubes and fifth power*, Sci. China Ser. A **36** (1993), 641–662.

[90] K. Mahler, *Note on hypothesis K of Hardy and Littlewood*, J. London Math. Soc. **11** (1936), 136–138.

[91] ――――, *On the fractional parts of the powers of a rational number (II)*, Mathematika **4** (1957), 122–124.

[92] K. S. McCurley, *An effective seven cube theorem*, J. Number Theory **19** (1984), 176–183.

[93] O. E. Meissner, *Über die Darstellung der Zahlen einiger algebraischen Zahlkörper als Summen von Quadratzahlen des Körpers*, Arch. Math. Phys. (3) **7** (1904), 266–268.

[94] Z. Z. Meng, *Some new results on Waring's problem*, J. China Univ. Sci. Tech. **27** (1997), 1–5.

[95] L. J. Mordell, *On the representation of algebraic numbers as a sum of four squares*, Proc. Cambridge Philos. Soc. **20** (1921), 250–256.

[96] V. Narasimhamurti, *On Waring's problem for 8th, 9th, and 10th powers*, J. Indian Math. Soc. (N.S.) **5** (1941), 122.

[97] V. I. Nechaev, *Waring's problem for polynomials*, Trudy Mat. Inst. Steklov, Izdat. Akad. Nauk SSSR, Moscow **38** (1951), 190–243.

[98] C. P. Ramanujam, *Sums of m-th powers in p-adic rings*, Mathematika **10** (1963), 137–146.

[99] K. Sambasiva Rao, *On Waring's problem for smaller powers*, J. Indian Math. Soc. **5** (1941), 117–121.

[100] C. J. Ringrose, *Sums of three cubes*, J. London Math. Soc. (2) **33** (1986), 407–413.

[101] K. F. Roth, *Proof that almost all positive integers are sums of a square, a positive cube and a fourth power*, J. London Math. Soc. **24** (1949), 4–13.

[102] C. L. Siegel, *Darstellung total positiver Zahlen durch Quadrate*, Math. Z. **11** (1921), 246–275.

[103] _____, *Additive Theorie der Zahlkörper II*, Math. Ann. **88** (1923), 184–210.

[104] _____, *Generalisation of Waring's problem to algebraic number fields*, Amer. J. Math. **66** (1944), 122–136.

[105] _____, *Sums of m-th powers of algebraic integers*, Ann. of Math. **46** (1945), 313–339.

[106] T. Tatuzawa, *On Waring's problem in algebraic number fields*, Acta Arith. **24** (1973), 37–60.

[107] K. Thanigasalam, *On Waring's problem*, Acta Arith. **38** (1980), 141–155.

[108] _____, *Some new estimates for $G(k)$ in Waring's problem*, Acta Arith. **42** (1982), 73–78.

[109] _____, *Improvement on Davenport's iterative method and new results in additive number theory, I*, Acta Arith. **46** (1985), 1–31.

[110] _____, *Improvement on Davenport's iterative method and new results in additive number theory, II. Proof that $G(5) \leq 22$*, Acta Arith. **46** (1986), 91–112.

[111] _____, *Improvement on Davenport's iterative method and new results in additive number theory, III*, Acta Arith. **48** (1987), 97–116.

[112] _____, *On sums of positive integral powers and simple proof of $G(6) \leq 31$*, Bull. Calcutta Math. Soc. **81** (1989), 279–294.

[113] _____, *On admissible exponents for kth powers*, Bull. Calcutta Math. Soc. **86** (1994), 175–178.

[114] K.-C. Tong, *On Waring's problem*, Advancement in Math. **3** (1957), 602–607.

[115] R. C. Vaughan, *Homogeneous additive equations and Waring's problem*, Acta Arith. **33** (1977), 231–253.

[116] ———, *A ternary additive problem*, Proc. London Math. Soc. (3) **41** (1980), 516–532.

[117] ———, *Recent work in additive prime number theory*, Proceedings of the International Congress of Mathematicians (Helsinki, 1978), Acad. Sci. Fennica, 1980, pp. 389–394.

[118] ———, *Sums of three cubes*, Bull. London Math. Soc. **17** (1985), 17–20.

[119] ———, *On Waring's problem for cubes*, J. Reine Angew. Math. **365** (1986), 122–170.

[120] ———, *On Waring's problem for sixth powers*, J. London Math. Soc. (2) **33** (1986), 227–236.

[121] ———, *On Waring's problem for smaller exponents*, Proc. London Math. Soc. (3) **52** (1986), 445–463.

[122] ———, *On Waring's problem for smaller exponents. II*, Mathematika **33** (1986), 6–22.

[123] ———, *On Waring's problem: one square and five cubes*, Quart. J. Math. Oxford (2) **37** (1986), 117–127.

[124] ———, *A new iterative method in Waring's problem*, Acta Math. **162** (1989), 1–71.

[125] ———, *On Waring's problem for cubes II*, J. London Math. Soc. (2) **39** (1989), 205–218.

[126] ———, *A new iterative method in Waring's problem II*, J. London Math. Soc. (2) **39** (1989), 219–230.

[127] ———, *The use in additive number theory of numbers without large prime factors*, Philos. Trans. Roy. Soc. London Ser. A **345** (1993), 363–376.

[128] ———, *The Hardy-Littlewood Method, 2nd edition*, Cambridge University Press, Cambridge, 1997.

[129] R. C. Vaughan and T. D. Wooley, *On Waring's problem: some re-finements*, Proc. London Math. Soc. (3) **63** (1991), 35–68.

[130] _____, *Further improvements in Waring's problem, III: Eighth powers*, Philos. Trans. Roy. Soc. London Ser. A **345** (1993), 385–396.

[131] _____, *Further improvements in Waring's problem, II: Sixth powers*, Duke Math. J. **76** (1994), 683–710.

[132] _____, *Further improvements in Waring's problem*, Acta Math. **174** (1995), 147–240.

[133] _____, *On a certain nonary cubic form and related equations*, Duke Math. J. **80** (1995), 669–735.

[134] _____, *Further improvements in Waring's problem, IV: Higher powers*, Acta Arith. **94** (2000), 203–285.

[135] I. M. Vinogradov, *A new estimate for $G(n)$ in Waring's problem*, Dokl. Akad. Nauk SSSR **5** (1934), 249–253.

[136] _____, *A new solution of Waring's problem*, Dokl. Akad. Nauk SSSR **2** (1934), 337–341.

[137] _____, *On some new results in the analytic theory of numbers*, C. R. Acad. Sci. Paris, Sér. A **199** (1934), 174–175.

[138] _____, *On the upper bound for $G(n)$ in Waring's problem*, Izv. Akad. Nauk SSSR Ser. Phiz.-Matem. **10** (1934), 1455–1469.

[139] _____, *A new variant of the proof of Waring's theorem*, Trud. Matem. instituta it. V. A. Steklov **9** (1935), 5–15.

[140] _____, *A new variant of the proof of Waring's theorem*, C. R. Acad. Sci. Paris, Sér. A **200** (1935), 182–184.

[141] _____, *An asymptotic formula for the number of representations in Waring's problem*, Mat. Sb. **42** (1935), 531–534.

[142] _____, *New estimates for Weyl sums*, Dokl. Akad. Nauk SSSR **8** (1935), 195–198.

[143] _____, *On Waring's problem*, Ann. of Math. **36** (1935), 395–405.

[144] _____, *On Weyl's sums*, Mat. Sb. **42** (1935), 521–530.

[145] _____, *Representation of an odd number as a sum of three primes*, Dokl. Akad. Nauk SSSR **15** (1937), 6–7.

[146] ———, *Some theorems concerning the theory of primes*, Mat. Sb. **44** (1937), 179–195.

[147] ———, *The method of trigonometrical sums in the theory of numbers*, Trav. Inst. Math. Stekloff **23** (1947), 109 pp.

[148] ———, *On an upper bound for $G(n)$*, Izv. Akad. Nauk SSSR Ser. Mat. **23** (1959), 637–642.

[149] ———, *Selected Works (translated from the Russian by N. Psv)*, Springer-Verlag, Berlin, 1985.

[150] E. Waring, *Meditationes Algebraicæ, second edition*, Archdeacon, Cambridge, 1770.

[151] ———, *Meditationes Algebraicæ, third edition*, Archdeacon, Cambridge, 1782.

[152] ———, *Meditationes Algebraicæ*, American Math. Soc., Providence, 1991, translation by Dennis Weeks of the 1782 edition.

[153] G. L. Watson, *A proof of the seven cube theorem*, J. London Math. Soc. **26** (1951), 153–156.

[154] H. Weyl, *Über die Gleichverteilung von Zahlen mod Eins*, Math. Ann. **77** (1916), 313–352.

[155] T. D. Wooley, *Large improvements in Waring's problem*, Ann. of Math. **135** (1992), 131–164.

[156] ———, *On Vinogradov's mean value theorem*, Mathematika **39** (1992), 379–399.

[157] ———, *The application of a new mean value theorem to the fractional parts of polynomials*, Acta Arith. **65** (1993), 163–179.

[158] ———, *Breaking classical convexity in Waring's problem: sums of cubes and quasi-diagonal behaviour*, Invent. Math. **122** (1995), 421–451.

[159] ———, *New estimates for smooth Weyl sums*, J. London Math. Soc. (2) **51** (1995), 1–13.

[160] ———, *On exponential sums over smooth numbers*, J. Reine Angew. Math. **488** (1997), 79–140.

[161] ———, *Quasi-diagonal behaviour and smooth Weyl sums*, Monatsh. Math. **130** (2000), 161–170.

[162] ———, *Sums of three cubes*, Mathematika, to appear, 2001.

Chernoff Type Bounds for Sums of Dependent Random Variables and Applications in Additive Number Theory

V. H. Vu

1 Introduction

A common problem in additive number theory is to prove that a sequence with certain properties exists. One of the successful ways to obtain an affirmative answer for such a problem is to use the probabilistic method, established by Erdős. To show that a sequence with a property $\mathcal{P}$ exists, it suffices to show that a properly defined random sequence satisfies $\mathcal{P}$ with positive probability. The value of the probabilistic method has been demonstrated by the fact that in most problems solved by it, it seems almost impossible to come up with a constructive proof.

Quite frequently, the property $\mathcal{P}$ requires that for all sufficiently large $n \in \mathbb{N}$, some relation $\mathcal{P}(n)$ holds. The general strategy to handle this situation is the following. For each n, one first shows that $\mathcal{P}(n)$ fails with a small probability, say $s(n)$. If $s(n)$ is sufficiently small so that $\sum_{n=1}^{\infty} s(n)$ converges, then by the Borel-Cantelli lemma, $\mathcal{P}(n)$ holds for all sufficiently large n with probability 1 (see, for instance, [8], Chapter 3).

The main issue in the above argument is to show that for each n, $\mathcal{P}(n)$ holds with high probability. One of the key tools one usually use to achieve this goal is the famous Chernoff bound [2], which provides a large deviation bound for the sum of independent random variables. Chernoff's bound has many variants and the following (Theorem A.1.14, [1]) seems to best suit the purpose of this paper.

Theorem 1.1. *Let* $t_1, \ldots, t_n$ *be independent binary random variables. Consider* $Y = \sum_{i=1}^{n} t_i$ *and let* μ *be the mean of* Y. *Then for any positive constant* ε, *there is a positive constant* $c(\varepsilon)$ *such that*

$$\Pr(|Y - \mu| > \varepsilon\mu) \leq 2e^{-c(\varepsilon)\mu}.$$

Let us illustrate the use of Theorem 1.1 by presenting Erdős' solution to one of his favorite problems, Sidon's problem on the existence of thin bases.

A subset X of $\mathbb{N}$ is a basis of order k if every sufficiently large number $n \in \mathbb{N}$ can be represented as a sum of k elements of X; we denote by $R_X^k(n)$ the number of such representations of n. A basis X is thin if for all sufficiently large n, $R_X^k(n)$ is small. The notion of thin bases was introduced by Sidon, who, in the 1930's, posed the following question to Erdős:

Is there a basis X of order 2 such that $R_X^2(n) = n^{o(1)}$ for all large n?

As Erdős later recalled, he was thrilled by the question, and thought that he would come up with an answer within few days. It took a little longer; about 20 years later, in 1956, Erdős answered the question in the affirmative by showing that much more is true [4]:

Theorem 1.2. *There is a subset $X \subset \mathbb{N}$ such that $R_X^2(n) = \Theta(\log n)$, for all sufficiently large n.*

Erdős' proof goes roughly as follows. Define a sequence X randomly by choosing each number x to be in X with probability $p_x = c(\log x/x)^{1/2}$, where c is a large positive constant. (If x is small so that $p_x = c(\log x/x)^{1/2} > 1$, set $p_x = 1$.) Let t_x be the indicator of the event that x is chosen; it is obvious that the t_x are independent random variables. The number of representations of n can now be expressed as

$$Y_n = \sum_{i+j=n} t_i t_j. \tag{1.1}$$

A routine calculation shows that $\mu_n = \mathbb{E}(Y_n) = \Theta(\log n)$. Set $\varepsilon = 1/2$; by increasing c, we can assume that $\mu_n \geq 2 \log n/c(\varepsilon)$ where $c(\varepsilon)$ is the constant in Theorem 1.1. Notice that if $i + j = i' + j' = n$ then either $\{i, j\} = \{i', j'\}$ or the two sets are disjoint. Therefore, the products $t_i t_j$ in (1.1) are independent. Theorem 1.1 thus applies and implies

$$\Pr(|Y_n - \mu_n| \geq \mu_n/2) \leq 2e^{-2\log n}. \tag{1.2}$$

Since $\sum_{n=1}^{\infty} e^{-2\log n}$ converges, the proof is complete by Borel-Cantelli's lemma.

It is quite natural to ask whether Theorem 1.2 can be generalized to arbitrary k. Using the above approach, in order to obtain a basis X such that $R_X^k(n) = \Theta(\log n)$, we should set $p_x = cx^{1/k-1}\log^{1/k} x$. Similarly to (1.1), consider

$$Y_n = \sum_{x_1 + \cdots + x_k = n} t_{x_1} \ldots t_{x_k}. \tag{1.3}$$

Although Y_n has the right expectation $\Theta(\log n)$, we face a major problem: the products $t_{x_1} \ldots t_{x_k}$ with $k > 2$ are no longer independent. In fact, a typical number x appears in rather many (namely, $\Omega(n^{k-2})$) solutions of $x_1 + \cdots + x_k = n$. This completely dashes the hope that one can use Theorem 1.1 to conclude the argument.

It took a long time to overcome this problem of dependency. In 1990, Erdős and Tetali [7] successfully generalized Theorem 1.2 to arbitrary k.

Theorem 1.3. *For any fixed k, there is a subset $X \subset \mathbb{N}$ such that $R_X^k(n) = \Theta(\log n)$, for all sufficiently large n.*

The heart of Erdős-Tetali's proof is to show that Y_n (as defined in (1.3)) is sufficiently concentrated, and to do this the authors used a very delicate moment computation. In particular, the tail probabilities $\Pr(Y_n \leq \mu_n/2)$ and $\Pr(Y_n \geq 3\mu_n/2)$ are treated separately with rather different techniques.

The purpose of this paper is to introduce several new large deviation bounds, which can replace Theorem 1.1 when the events in the sum are dependent. These bounds are interesting for several reasons. First, they can be seen as extensions of Chernoff's bound to a larger class of functions. Moreover, compared to other classical large deviation results such as Azuma's and Janson's inequalities, our bounds possess certain advantages. Finally, they seem very useful in applications concerning random sequences; we shall illustrate this by several applications. The approach based on the bounds presented there is very general and easy to implement in other problems. Proofs for most of the results given here can be found in the papers [10], [16], [17], and [19]. We hope that this account will give the reader a more systematic view of our results and a better understanding of the applications.

The rest of the paper is organized as follows. In the next section, we describe our large deviation results. We shall also make a brief comparison between our bounds and classical large deviation results such as Azuma's and Janson's inequalities. In Section 3, we shall use these bounds to give a short proof of a further extension of Theorem 1.3. This section also serves as a prelude to a more difficult application which follows in Section 4, involving thin Waring bases. The main result of this section settles a 20 year old question of Nathanson [12].

Notations.   Throughout the paper, $\mathbb{N}$ denotes the set of positive integers and $\mathbb{N}^r$ denotes the set of all r-powers. The asymptotic notations such as Θ, O, o are used under the assumption that $n \to \infty$. All logarithms have natural base. $\mathbb{E}(Y)$ denotes the expectation of a random variable Y.

2   Concentration of a Sum of Dependent Variables

In this section, we assume that $t_1, \ldots, t_n$ are independent binary random variables. In many problems concerning random sequences, the random variable we are interested in can be expressed in the form $\sum_{j=1}^{m} I_j$, where the I_j are indicator random variables which are usually products of several atom variables t_i. Typically, the same t_i may appear in many I_j's, and therefore the I_j's are not independent.

Our results focus on random variables depending on $t_1, \ldots, t_n$ which can be expressed as polynomials with positive coefficients (a random variable depending on $t_1, \ldots, t_n$ can be viewed as a function from the n-dimensional unit hypercube, equipped with the product measure, to $\mathbb{R}$). It is clear that this class covers random variables of the above type, since each I_j can be regarded as a monomial. It is also clear that all random variables considered in the previous section are of this type. The sum of all the t_i, considered in Theorem 1.1, is a polynomial of degree 1.

2.1   Preliminaries

Consider a polynomial Y of degree k in $t_1, \ldots, t_n$; we say that Y is *regular* if its coefficients are positive and at most 1. Moreover, Y is *homogeneous* if the degree of every monomial is the same.

Given a (multi-)set A, $\partial_A(Y)$ is the partial derivative of Y with respect to the variables with indices in A. For instance, if $Y = t_1 t_2^2$ and $A_1 = \{1, 2\}$, $A_2 = \{2, 2\}$ then $\partial_{A_1}(Y) = 2t_2$ and $\partial_{A_2}(Y) = 2t_1$, respectively. If A is empty then $\partial_A(Y) = Y$. $\mathbb{E}_A(Y)$ denotes the expectation of $\partial_A(Y)$. Furthermore, set $\mathbb{E}_j(Y) = \max_{|A| \geq j} \mathbb{E}_A(Y)$, for all $j = 0, 1, \ldots, k$.

The results we are about to present are of the following general type: For a positive polynomial Y, if the expectations of its partial derivatives are relatively small, then Y is strongly concentrated. For the intuition behind the results and the general method to prove them, we refer to [19].

2.2   Polynomials with Expectation $\Theta(\log n)$

In this subsection, we consider the special case when the polynomial in question has expectation of order $\log n$. This is typical of applications concerning random sequences, and the following theorem, proved by the author in [17], seems to be very useful.

Theorem 2.1. *For any positive constants k, $\alpha, \beta, \varepsilon$ there is a constant $Q = Q(k, \varepsilon, \alpha, \beta)$ such that the following holds. If Y is a regular positive homogeneous polynomial of degree k, if $n/Q \geq \mathbb{E}(Y) \geq Q \log n$ and*

$\mathbb{E}(\partial_A(Y)) \le n^{-\alpha}$ *for all non-empty sets* A *of cardinality at most* $k-1$, *then*

$$\Pr\left(|Y - \mathbb{E}(Y)| \ge \varepsilon\mathbb{E}(Y)\right) \le n^{-\beta}.$$

One can interpret this theorem as follows: If Y has expectation $\Theta(\log n)$ and the condition $\mathbb{E}(\partial_A(Y)) \le n^{-\alpha}$ is satisfied, then in a certain range the tail distribution of Y is similar to that of the Poisson distribution. The case $k = 1$ is well-known and appears in most textbooks in probability. It is instructive for the reader to compare this result with Theorem 1.1.

2.3 General Results

Theorem 2.1 is a special case of a more general result. To this end, we assume that our polynomial Y is regular and has degree k. To state the result, we need some new definitions.

First define a function f by $f(K) = \max\{1, \lfloor(K/k!)^{1/k}\rfloor\} - 2$. Furthermore, let

$$r(k, K, n, \delta) = \left(\log_2 \frac{1}{\delta}\right) \frac{n^k \delta^{\frac{f(K/2)}{2}}}{f(K/2)!} + (\delta/K^8)^{\lfloor\frac{1}{8k}\log\frac{1}{\delta}\rfloor}.$$

Given a regular polynomial Y, define $h(k, K, n, \delta)$ recursively in the following way:

$$h(1, K, n, \delta) = 0,$$
$$h(k, K, n, \delta) = h(k-1, K, n + \lceil\mathbb{E}(Y)\rceil, \delta) + nr(k, K, n, \delta).$$

Given a set A, we denote by $\partial_A^* Y$ the polynomial obtained from the partial derivative $\partial_A Y$ by subtracting its constant coefficient. $\mathbb{E}_A^*(Y)$ is the expectation of $\partial_A^* Y$ and we define $\mathbb{E}_1^*(Y) = \max_{|A| \ge 1} \mathbb{E}_A^*(Y)$.

Finally, notice that since our atom random variables t_i are binary, every high degree term can be reduced to degree 1 (for instance $t_1^3 = t_1^2 = t_1$). A polynomial is *simplified* if every monomial is completely reduced. The following theorem is the main theorem of [17].

Theorem 2.2. *Let* Y *be a simplified regular polynomial of degree* k. *For any positive numbers* δ, λ *and* K *satisfying* $K \ge 2k$, $\mathbb{E}_1^*(Y) \le \delta \le 1$ *and* $4kK\lambda \le \mathbb{E}(Y)$ *we have*

$$\Pr\left(|Y - \mathbb{E}(Y)| \ge 2\sqrt{\lambda kK\mathbb{E}(Y)}\right) \le 2ke^{-\lambda/4} + h(k, K, n, \delta).$$

The reader might want to try to deduce Theorem 2.1 from Theorem 2.2 as an exercise. One can also deduce from Theorem 2.2 the following corollary, which gives a concentration result for functions with expectation $O(\log n)$.

Corollary 2.3. *Assume that Y is a regular positive homogeneous polynomial of degree k and the expectation of Y is at most $\log n$. Assume furthermore that for all A, $1 \le |A| \le k-1$, and $\mathbb{E}(\partial_A(Y)) \le n^{-\alpha}$ for some positive constant α. Then there are positive constants $c = c(\alpha, k)$ and $d = d(\alpha, k)$ such that for any $0 \le \varepsilon \le 1$,*

$$\Pr\left(|Y - \mathbb{E}(Y)| \ge \varepsilon \mathbb{E}(Y)\right) \le de^{-c\varepsilon^2 \mathbb{E}(Y)}.$$

Again, the reader is invited to compare this corollary with Theorem 1.1.

To conclude this subsection, let us mention a result of a little different flavor. This result was needed in the proof of Theorem 2.2 and proved useful in several other situations (see [1] or Section 3, for instance). The proof (using moments calculation) is simple and can be found in Chapter 8 of [1].

Consider a polynomial Y which is a sum of different monomials with coefficients 1. Two monomials of Y are disjoint if they do not share any atom variable. At a point t, we say that a monomial is *alive* if it is not zero. We denote by $\mathrm{Disj}(Y(t))$ the maximum number of pairwise disjoint alive monomials of Y at t.

Proposition 2.4. *For Y as above and any positive integer we have K*

$$\Pr(Y \ge K) \le \mathbb{E}(Y)^K / K! .$$

2.4   More General Results

The conditions of Theorems 2.1 and 2.2 require the expectation of any partial derivative of Y to be much smaller than 1. This is usually satisfied when the polynomial has expectation $O(\mathrm{polylog}\, n)$. However, for polynomials with larger expectations, one cannot always expect this to happen. The following theorem, proved by Kim and the present author [10] (see also [1], Chapter 7) deals with such a situation.

Theorem 2.5. *For every positive integer k there are positive constants a_k and b_k depending only on k such that the following holds. For any positive polynomial $Y = Y(t_1, \ldots, t_n)$ of degree k, where the t_i's are independent binary random variables, we have*

$$\Pr\left(|Y - \mathbb{E}(Y)| \ge a_k \lambda^{k-1/2} \sqrt{\mathbb{E}_0(Y)\mathbb{E}_1(Y)}\right) \le b_k e^{-\lambda/4 + (k-1)\log n}.$$

Theorem 2.5 is powerful when the ratio $\mathbb{E}(Y)/\mathbb{E}_A(Y)$ is large (for any A). This theorem does not require $\mathbb{E}_A(Y)$ to be bounded by 1. For instance, if $\mathbb{E}(Y)/\mathbb{E}_A(Y) \ge n^\delta$ for some positive constant δ, then $\mathbb{E}_0(Y) = \mathbb{E}(Y) \ge$

$\mathbb{E}_1(Y)n^\delta$. Thus, one can choose $\lambda = n^\varepsilon$ for some positive constant ε such that the tail $a_k\lambda^k\sqrt{\mathbb{E}_0(Y)\mathbb{E}_1(Y)}$ is still negligible compared to $\mathbb{E}(Y)$. On the other hand, the bound is roughly $e^{-n^\varepsilon/4}$.

Theorem 2.5 and its variants have several deep applications in combinatorics [15], [18], [11]. The interested reader might want to check [19] for a survey of the subject.

Remark. Theorem 2.5 was stated in [10] and [1] with λ^k instead of $\lambda^{k-1/2}$. It was remarked in [10] that the result holds in the above form, with essentially the same proof. In a certain range of λ, $\lambda^{k-1/2}$ could be further reduced to $\lambda^{1/2}$; see [19].

2.5 Comparison

Two well-known results, which are frequently used in the Erdős probabilistic method, are Azuma's inequality and Janson's inequality. In the following, we state these results and make a brief comparison with our bounds.

Let us start with Azuma's inequality. We say that a function $Y = Y(t_1, \ldots, t_n)$ has Lipschitz coefficient r if changing any t_i changes the value of Y by at most r. The following version of Azuma's inequality follows from [1].

Theorem 2.6. *If Y has Lipschitz coefficient r, then for any $\lambda > 0$*

$$\Pr(|Y - \mathbb{E}(Y)| \geq r\sqrt{\lambda n}) \leq 2e^{-\lambda/2}.$$

The advantage of Azuma's inequality is that it is very general: Y can be any function. On the other hand, restricted to the class of polynomials, our results are frequently stronger since they guarantee strong concentration under a usually much weaker condition. There are several situations where Y has a large Lipschitz coefficient, but the expectations of its partial derivatives are still very small. A concrete example will appear in the next section when we examine the function Y_n defined in (1.3).

Now let us present Janson's inequality. This inequality deals with functions which can be expressed as $\sum_{j=1}^m I_j$, where each I_j is the product of some t_i's. We write $I_i \sim I_j$ if the two monomials share a common atom variable. In [9], Janson proved:

Theorem 2.7. *With Y as above and $\Delta = \sum_{I_i \sim I_j} \mathbb{E}(I_i I_j)$, the following inequality holds:*

$$\Pr(Y \leq (1-\varepsilon)\mathbb{E}(Y)) \leq \exp\left\{-\frac{(\varepsilon\mathbb{E}(Y))^2}{2(\mathbb{E}(Y) + \Delta)}\right\}.$$

The advantage of this theorem is that there is no restriction on the degree of Y. On the other hand, it provides only the lower tail bound. Our results provide bounds in both directions. Moreover, the strength of these bounds are often comparable to what one gets from Janson's inequality.

3   Thin Linear Bases

In this section, we extend Theorem 1.3 by allowing a representation to be a linear combination with fixed coefficients.

Fix k positive integers $a_1, \ldots, a_k$, where $\gcd(a_1, \ldots, a_k) = 1$. Let $Q_X^k(n)$ be the number of representations of n in the form $n = a_1 x_1 + \ldots a_k x_k$, where $x_i \in X$. We shall prove:

Theorem 3.1. *There is a subset $X \subset \mathbb{N}$ such that $Q_X^k(n) = \Theta(\log n)$, for all sufficiently large n.*

The proof of Theorem 3.1 is based essentially on the proof of the more difficult Theorem 4.2 in [16]. However, since the proof of Theorem 3.1 is much simpler, we present it first in order to give the reader a better understanding of our method.

The assumption $\gcd(a_1, \ldots, a_k) = 1$ is necessary, for a simple number-theoretic reason. Theorem 1.3 follows from Theorem 3.1 by setting all $a_i = 1$.

To start the proof, we use Erdős' idea and define a random set X as follows. For each $x \in \mathbb{N}$, choose x with probability $p_x = cx^{1/k-1} \log^{1/k} x$, where c is a positive constant to be determined. Let t_x be the indicator random variable of this choice; thus, t_x is a $\{0,1\}$ random variable with mean p_x.

Fix a number n, and let $\mathbf{Q}_n$ be the set of all k-tuples $(x_1, \ldots, x_k)$, where the x_i are positive integers and $\sum_i a_i x_i = n$. The number of representations of n using elements from the random sequence X can be expressed as a random variable by

$$Y_n = \sum_{(x_1, \ldots, x_k) \in \mathbf{Q}_n} t_{x_1} \ldots t_{x_k}. \tag{3.1}$$

It is obvious that Y_n is a polynomial of degree k in $t_1, \ldots, t_n$. We now show that with probability close to 1, Y_n is $\Theta(\log n)$ for any sufficiently large n. It is easy to show that $\mathbb{E}(Y_n)$ is of the right order, namely $\log n$. Next, we want to make use of Theorem 2.1. The main obstruction here is that Y_n, as a polynomial, does have partial derivatives with large expectations which violate the condition of Theorem 2.1. For instance, consider the representation $a_1 K + a_2 x_2 + \cdots + a_k x_k$ where K is a constant. The partial derivative with respect to $t_{x_2}, \ldots, t_{x_k}$ has expectation $p_K = \Theta(1)$. However, we could easily overcome this obstruction by splitting Y_n into two parts, as follows. Set $a = 0.4$ (0.4 can be any small constant) and let $\mathbf{Q}_n^{[1]}$ be the subset of $\mathbf{Q}_n$ consisting of all tuples whose smallest element is at least n^a and $\mathbf{Q}_n^{[2]} = \mathbf{Q}_n \backslash \mathbf{Q}_n^{[1]}$. We break Y_n into the sum of two terms corresponding to $\mathbf{Q}_n^{[1]}$ and $\mathbf{Q}_n^{[2]}$, respectively:

$$Y_n = Y_n^{[1]} + Y_n^{[2]},$$

where

$$Y_n^{[j]} = \sum_{(x_1, \ldots, x_k) \in \mathbf{Q}_n^{[j]}} t_{x_1} \ldots t_{x_k}.$$

Intuitively, $Y_n^{[1]}$ should be the main part of Y_n, since in most solutions of $\sum_{i=1}^k a_i x_i = n$, all x_i are $\Theta(n)$. To finish the proof it suffices to show

(A) $\mathbb{E}(Y_n^{[1]}) = \Theta(\log n)$ and $\Pr\left(|Y_n^{[1]} - \mathbb{E}(Y_n^{[1]})| \geq \mathbb{E}(Y_n^{[1]})/2\right) \leq n^{-2}$.

(B) For almost every sequence X, there is a finite number $M(X)$ such that $Y_n^{[2]} \leq M(X)$ for all sufficiently large n.

(A) and (B) confirm our intuition. The main part of Y_n, which is of order $\log n$, indeed comes from $Y_n^{[1]}$; $Y_n^{[2]}$'s contribution is bounded by a constant.

We apply Theorem 2.1 to verify (A). In order to complete this, we first need the following lemma, which gives a bound for $\mathbb{E}(\partial_A Y_n^{[1]})$.

Lemma 3.2. *For all non-empty multi-sets A of size at most $k - 1$,*

$$\mathbb{E}(\partial_A Y_n^{[1]}) = O(n^{-a/2k}).$$

Proof. Consider a (multi-)set A of $k - l$ elements $y_1, \ldots, y_{k-l}$. For a permutation $\pi \in S_k$ (where S_k denotes the symmetric group on $\{1, 2, \ldots, k\}$) let $\mathbf{Q}_{n,l,\pi}^{[1]}$ be the set of l-tuples $(x_1, \ldots, x_l)$ of positive integers satisfying $x_i \geq n^a$ for all i and

$$\sum_{i=1}^l a_{\pi(i)} x_i = n - \sum_{j=1}^{k-l} a_{\pi(l+j)} y_j.$$

A simple consideration shows that

$$\partial_A(Y_n^{[1]}) \leq b(k) \sum_{\pi \in S_k} \sum_{(x_1, \ldots, x_l) \in \mathbf{Q}_{n,l,\pi}^{[1]}} t_{x_1} \ldots t_{x_l},$$

where $b(k)$ is a constant depending on k. By symmetry, it now suffices to verify

$$\mathbb{E}\left(\sum_{(x_1,\ldots,x_l)\in\mathbf{Q}^{[1]}_{n,l,\pi_0}} t_{x_1}\ldots t_{x_l}\right) = O(n^{-a/2k}),$$

where π_0 is the identity permutation. Without loss of generality, we can assume that $x_l = \max(x_1,\ldots,x_l)$. Set $m = n - \sum_{j=1}^{k-l} a_{l+j}y_j$; since $\sum_{i=1}^{l} a_i x_i = m$ and the a_i's are fixed numbers, it follows that $x_l = \Omega(m/l)$. Using the fact that $\sum_{x=1}^{m} x^{1/k-1} \approx \int_1^m z^{1/k-1}\partial z \approx m^{1/k}$, we have

$$\mathbb{E}\left(\sum_{(x_1,\ldots,x_l)\in\mathbf{Q}^{[1]}_{n,l,\pi_0}} t_{x_1}\ldots t_{x_l}\right) = O\left(\sum_{(x_1,\ldots,x_l)\in\mathbf{Q}^{[1]}_{n,l,\pi_0}} p_{x_1}\ldots p_{x_l}\right)$$

$$= O(\log n)\sum_{\substack{n^a\leq\min(x_1,\ldots,x_l)\\ a_1x_1+\cdots+a_lx_l=m}} x_1^{1/k-1}\ldots x_l^{1/k-1}$$

$$= O(\log n)O\left(\left(\sum_{x=1}^{m} x^{1/k-1}\right)^{l-1}(m/l)^{1/k-1}\right)$$

$$= O(\log n)O\left(m^{(l-1)/k}(m/l)^{1/k-1}\right)$$

$$= O(\log n)O(m^{(l-k)/k}) = O(n^{-a/2k}),$$

since $k-l \geq 1$ and $m \geq n^a$. This concludes the proof of the lemma. □

The last step in the previous calculation explains the restriction $\min(x_i)_{i=1}^{k} \geq n^a$. This assumption guarantees that every partial derivatives of $Y_n^{[1]}$ has small expectation.

From the above calculation, it follows immediately (by setting $l = k$ and $m = n$) that $\mathbb{E}(Y_n^{[1]}) = O(\log n)$. Moreover, a straightforward argument shows that if $c \to \infty$, then $\mathbb{E}(Y_n^{[1]})/\log n \to \infty$. Indeed, there are at least $\Theta(n^{k-1})$) k-tuples $(x_1, x_2\ldots,x_k)$ in $\mathbf{Q}_n^{[1]}$ with $x_i = \Theta(n)$ for all $i \leq k$, where the constants in the Θ's depend only on k and the a_i's. On the other hand, each such tuple contributes at least $c^k n^{1-k}\log n$ to $\mathbb{E}(Y_n^{[1]})$. Therefore, by increasing c, we can assume that $\mathbb{E}(Y_n^{[1]})$ satisfies the condition of Theorem 2.1. Theorem 2.1 then applies and implies (A).

Before proving (B), let us pause for a moment and show why we could not apply Azuma's inequality to prove (A). The reason is that the Lipschitz coefficient of $Y_n^{[1]}$ is way too large. It is clear that there is a number x which appears in $\Omega(n^{k-2})$ tuples in $\mathbf{Q}_n^{[1]}$ (as a matter of fact, many numbers do so). For such an x, changing t_x, in the worst case, might change $Y_n^{[1]}$ by $\Omega(n^{k-2})$. Thus, the Lipschitz coefficient of $Y_n^{[1]}$ is $\Omega(n^{k-2})$. This coefficient is clearly too large for Azuma's inequality to deliver a non-trivial bound.

Now we turn to the proof of (B), which is purely combinatorial. We say that an l-tuple $(x_1, \ldots, x_l)$ $(l \le k)$ is an l-representation of n if there is a permutation $\pi \in S_k$ such that $\sum_{i=1}^{l} a_{\pi(i)} x_i = n$. For all $l < k$, let $Q_X^l(n)$ be the number of l-representations of n. With essentially the same computation as in the previous lemma, one can show that $\mathbb{E}(Q_X^l(n)) = O(n^{-1/k} \log n) = O(n^{-1/2k})$. Proposition 2.4 then implies that for a sufficiently large constant M_1, with probability $1 - O(n^{-2})$, the maximum number of disjoint representations of n in $Q_X^l(n)$ is at most M_1. By Borel-Cantelli's lemma, we conclude that for almost every random sequence X there is a finite number $M_1(X)$ such that for any $l < k$ and all n, the number of disjoint l-representations of n from X is at most $M_1(X)$.

Using a computation similar to the one in the proof of Lemma 3.2, one can deduce that $\mathbb{E}(Y_n^{[2]}) = O(n^{(a-1)/k} \log n) = O(n^{-1/2k})$. Indeed, since $x_1 \le n^a$, instead of $(\sum_{x=1}^{n} x^{1/k-1})^{k-1}$, one can write $\sum_{x=1}^{n^a} x^{1/k-1} (\sum_{x=1}^{n} x^{1/k-1})^{k-1}$ and the bound follows. So, again by Proposition 2.4 and Borel-Cantelli's lemma, there is a constant M_2 such that almost surely, the maximum number of disjoint k-representations of n in $Y_n^{[2]}$ is at most M_2 for all large n. From now on, it would be useful to think of $Y_n^{[2]}$ as a family of sets of size k, each corresponding to a representation of n.

We say that a sequence X is *good* if it satisfies the properties described in the last two paragraphs.

To finish the proof, it suffices to show that if X is good, then $Y_n^{[2]}$ is bounded by a constant. This follows directly from a well-known combinatorial result of Erdős and Rado [6], stated below. A collection of sets $A_1, \ldots, A_r$ forms a sun flower if the sets have pairwise the same intersection (the A_i's are called petals of the flower). Erdős and Rado have shown:

Lemma 3.3. *If H is a collection of sets of size at most k and $|H| > (r-1)^k k!$, then H contains r sets in H forming a sun flower.*

Set $M(X) = (\max(M_1(X) k!, M_2))^k k!$ and assume that n is sufficiently large. It is clear that if $Y_n^{[2]} > M(X)$, then by the Erdős-Rado sunflower lemma, $Y_n^{[2]}$ contain a sunflower with $M_3 = \max(M_1(X) k!, M_2) + 1$ petals. If the intersection of this sunflower is empty, then the petals form a family of M_3 disjoint k-representations of n. Otherwise, assume that the intersection consists of $y_1, \ldots, y_j$ where $1 \le j \le k - 1$. By the pigeon hole principle, there is a permutation $\pi \in S_k$ such that one can find $M_1(X) + 1$ $(k - j)$-representations of $m = n - \sum_{i=1}^{j} a_{\pi(i)} y_i$ among the sets obtained by the petals minus their common intersection. These $M_1(X) + 1$ sets are disjoint due to the definition of the sun flower. Therefore, in both cases we obtain a contradiction.

Remark. Using the more general Theorems 2.2 or 2.5 one can prove a statement similar to Theorem 3.1 when $Q_X^k(n)$ is required to be $\Theta(g(n))$, for any "reasonable" function $g(n) \gg \log n$ (reasonable here means that one should be able to set p_x so that the expectation of $Q_X^k(n)$ is $g(n)$). In fact, in such a case, one can easily show that $Q_X^k(n)$ is asymptotically $g(n)$, since in the proof of (A) one can now obtain a deviation tail of size $o(g(n))$. The much harder direction is to go below $\log n$, and this seems to be beyond the reach of the probabilistic method. In order to prove that a function Y with probability $1 - n^{-2}$ does not deviate too much from its expectation, it does seem that one needs to assume that the expectation of Y is of order at least $\log n$.

4   Thin Waring Bases

4.1   The Problem

The most interesting bases in additive number theory are, perhaps, the set of all r^{th} powers, for arbitrary $r \geq 2$. The famous Waring's conjecture (proved by Hilbert, Hardy-Littlewood, Vinogradov, and many others at the beginning of the last century) asserts that for any fixed r and sufficiently large k, $\mathbb{N}^r$, the set of all r^{th} powers, is a k basis.

These bases are very far from being thin, due to the following deep theorem of Vinogradov (see Vaughan [14]).

Theorem 4.1. *For any fixed $r \geq 2$, there is a constant $k(r)$ such that if $k \geq k(r)$ then $R_{\mathbb{N}^k}^k(n)$, the number of representations of n as a sum of k r^{th} powers, satisfies*

$$R_{\mathbb{N}^r}^k(n) = \Theta(n^{k/r - 1})$$

for every positive integer n.

It is natural to ask whether $\mathbb{N}^r$ contains a thin subbasis and one may hope to obtain a result similar to Theorem 1.3. On the other hand, the problem appears much more difficult. For many years, researchers focused on a simpler question whether $\mathbb{N}^r$ contains a subbasis of small density. This question has been investigated intensively for $r = 2$; see [5],[3],[21],[22],[20], and [13]. Choi, Erdős and Nathanson [3] proved that $\mathbb{N}^2$ contains a subbasis X of order 4, with $X(m) \leq m^{1/3+\epsilon}$, where $X(m)$ denotes the number of elements of X not exceeding m. Improving this result, Zöllner [21][22] showed that for any $k \geq 4$ there is a subbasis $X \subset \mathbb{N}^2$ of order k satisfying $X(m) \leq m^{1/k+\epsilon}$ for arbitrary positive constant ϵ. Wirsing [20], sharpening Zöllner's theorem, proved that for any $k \geq 4$ there is a subbasis $X \subset \mathbb{N}^2$ of order k satisfying $X(m) = O(m^{1/k} \log^{1/k} m)$. It is easy to see, via the pigeon-hole principle, that Wirsing's result is best possible, up to the log term. A short proof of Wirsing's result for the case $k = 4$ was given by Spencer in [13]. For $r \geq 3$, much less was known. In 1980, Nathanson [12] proved that $\mathbb{N}^r$ contains a subbasis with density $o(m^{1/r})$. In the same paper, he raised the following question.

Question. *Let $r \geq 2$ and k be fixed, positive integers, where k is sufficiently large compared to r. What is the smallest density of a subbasis of order k of $\mathbb{N}^r$? Can it be $m^{1/k+o(1)}$?*

It is clear that the conjectured density $m^{1/k+o(1)}$ is best possible up to the $o(1)$ term.

Very recently, the author [16] succeeded in extending Theorem 1.3 to $\mathbb{N}^r$ for arbitrary r:

Theorem 4.2. *For any fixed $r \geq 2$, there is a constant $k(r)$ such that if $k \geq k(r)$ then $\mathbb{N}^r$ contains a subset X such that*

$$R_X^k(n) = \Theta(\log n),$$

for every positive integer $n \geq 2$, and $R_X^k(1) = 1$.

This theorem implies, via the pigeon hole principle, that X has density $O(m^{1/k} \log^{1/k} m)$, settling Nathanson's question.

The proof of Theorem 4.2 uses the framework presented in the previous subsection and again relies heavily on Theorem 2.1.

4.2   Sketch of the Proof

The proof of Theorem 4.2 follows closely the framework provided in the previous subsection, but it will require, in addition, a purely number theoretic lemma (Lemma 4.3), which is an extended version of Vinogradov's theorem (Theorem 4.1) and is interesting in its own right. To start, we define a random subset of $\mathbb{N}^r$ as follows. Choose, for each $x \in \mathbb{N}$, x^r with probability $p_x = cx^{-1+r/k} \log^{1/k} x$, where c is a positive constant to be determined. Again let t_x denote the characteristic random variable representing the choice of x^r: $t_x = 1$ if x^k is chosen and 0 otherwise. Similar to (3.1), the number of representations of n (not counting permutations), restricted to X, can be expressed by

$$Y_n = R_X^k(n) = \sum_{x_1^r + \cdots + x_k^r = n} \prod_{j=1}^{k} t_{x_j} = Y(t_1, \ldots, t_{\lfloor n^{1/r} \rfloor}). \qquad (4.1)$$

Given the framework presented in the last section, the main difficulty remaining is to estimate the expectations of Y_n and its partial derivatives. (To be precise, we should consider $Y_n^{[1]}$, where $Y_n^{[1]}$ is defined similarly with a properly chosen parameter a; however, let us put this technicality aside.) In the following, we shall focus on the expectation of Y_n. Notice that

$$\mathbb{E}(Y_n) = \sum_{x_1^r + \cdots + x_k^r = n} c^k \prod_{j=1}^{k} x_j^{-1+r/k} \log^{1/k} x_j.$$

To see that the right hand side has order $\log n$, one may argue as follows. A typical solution $(x_1, \ldots, x_k)$ of the equation $\sum_{j=1}^{k} x_j^r = n$ should satisfy $x_j = \Theta(n^{1/r})$, for all j. Thus, a typical term in the sum is $\Theta(n^{-k/r+1} \log n)$. On the other hand, by Theorem 4.1, the number of terms is $\Theta(n^{k/r-1})$ and we would be done by taking the product. However, there could be many non-typical solutions with a larger contribution. For instance, assume that $\mathbf{x} = (x_1, \ldots, x_k)$ is a solution where $1 \le x_j \le P_j$ and some of the P_j's are considerably smaller than $n^{1/r}$ (for example, $P_1 = n^{\varepsilon}$ with $\varepsilon \ll 1/r$). The contribution of the term corresponding to $\mathbf{x}$ is at least $\Omega(\prod_{j=1}^{s} P_j^{-1+r/k})$, which is significantly larger than the contribution of a typical term.

To overcome this problem, we need an upper bound on the number of solutions of the equation $\sum_{j=1}^{k} x_j^r = n$, with respect to additional constrains $x_j \le P_j$, for arbitrary positive integers we have $P_1, \ldots, P_k$. Denote this number by $\mathrm{Root}(P_1, \ldots, P_k)$. We have proved the following lemma, which generalizes the lower bound in Theorem 4.1.

Lemma 4.3. *For a fixed positive integer $r \ge 2$, there exists a constant k_r such that the following holds. For any constant $k \ge k_r$, there is a positive constant $\delta = \delta(r, k)$ such that for every sequence $P_1, \ldots, P_k$ of positive integers we have*

$$\mathrm{Root}(P_1, \ldots, P_k) = O\left(n^{-1} \prod_{j=1}^{k} P_j + \prod_{j=1}^{k} P_j^{1-r/k-\delta} \right),$$

for all n.

The proof of this lemma requires a sophisticated application of the Hardy-Littlewood's circle method and is beyond the scope of this paper. The reader may consult [16] for the full proof.

References

[1] N. Alon and J. Spencer, *The probabilistic methods*, second ed., Wiley, New York, 2000.

[2] H. Chernoff, *A measure of the asymptotic efficiency for tests of a hypothesis based on the sum of observation*, Ann. Math. Stat. **23** (1952), 493–509.

[3] S. L. G. Choi, P. Erdős, and M. Nathanson, *Lagrange's theorem with $N^{1/3}$ squares*, Proc. Am. Math. Soc. **79** (1980), 203–205.

[4] P. Erdős, *Problems and results in additive number theory*, Colloque sur la Théorie des Nombres, Bruxelles, 1955, Masson and Cie., Paris, 1956, pp. 127–137.

[5] P. Erdős and M. Nathanson, *Lagrange's theorem and thin subsequences of squares*, Contribution to Probability (J. Gani and V. K. Rohatgi, eds.), Academic Press, New York, 1981, pp. 3–9.

[6] P. Erdős and R. Rado, *Intersection theorems for systems of sets*, J. London Math. Soc. **35** (1960), 85–90.

[7] P. Erdős and P. Tetali, *Representations of integers as sum of k terms*, Random Structures and Algorithms **1** (1990), 245–261.

[8] H. Halberstam and K. F. Roth, *Sequences*, Springer-Verlag, New York, 1983.

[9] S. Janson, *Poisson approximation for large deviations*, Random Structures and Algorithms **1** (1990), 221–230.

[10] J. H. Kim and V. H. Vu, *Concentration of multivariate polynomials and its applications*, Combinatorica **20** (2000), 417–434.

[11] ——, *Small complete arcs in projective planes*, submitted.

[12] M. Nathanson, *Waring's problem for sets of density zero*, Analytic Number Theory (M. Knopp, ed.), Lecture Notes in Mathematics 899, Springer-Verlag, 1980, pp. 302–310.

[13] J. Spencer, *Four squares with few squares*, Number Theory, New York Seminar (D. V. Chudnovsky et al., ed.), Springer, 1991–1995, pp. 295–297.

[14] R. C. Vaughan, *The Hardy-Littlewood method*, Cambridge Univ. Press, 1981.

[15] V. H. Vu, *New bounds on nearly perfect matchings in hypergraphs: higher codegrees do help*, Random Structures and Algorithms **17** (2000), 29–63.

[16] ———, *On a refinement of Waring's problem*, Duke Math. J. **105** (2000), 107–134.

[17] ———, *On the concentration of multivariate polynomials with small expectation*, Random Structures and Algorithms **16** (2000), 344–363.

[18] ———, *An upper bound on the list chromatic number of locally sparse graphs*, Combinatorics, Probability, and Computing, to appear.

[19] ———, *Concentration of non-Lipschitz functions and applications*, Random Structures and Algorithms, to appear .

[20] E. Wirsing, *Thin subbases*, Analysis **6** (1986), 285–308.

[21] J. Zöllner, *Der Vier-Quadrate-Satz und ein Problem von Erdős and Nathanson*, Ph.D. thesis, Johannes Gutenberg-Universität, Mainz, 1984.

[22] ———, *Über eine Vermutung von Choi, Erdős and Nathanson*, Acta Arith. **45** (1985), 211–213.

Prime Divisors of the Bernoulli and Euler Numbers

Samuel S. Wagstaff, Jr.[1]

1 Factoring the Bernoulli and Euler Numbers

The Bernoulli numbers B_n may be defined by the generating function

$$\frac{t}{e^t - 1} = \sum_{n=0}^{\infty} B_n \frac{t^n}{n!}.$$

The B_n are all rational numbers, $B_{2k+1} = 0$ for all $k \geq 1$, and the non-zero B_n alternate in sign. The first few non-zero ones are: $B_0 = 1$, $B_1 = -1/2$, $B_2 = 1/6$, $B_4 = -1/30$, $B_6 = 1/42$, $B_8 = -1/30$, $B_{10} = 5/66$, $B_{12} = -691/2730$, $B_{14} = 7/6$, $B_{16} = -3617/510$, $B_{18} = 43867/798$. B_{20} is the first one with a composite numerator: $174611 = 283 \cdot 617$.

Write B_n as N_n/D_n with $D_n > 0$ and $\gcd(N_n, D_n) = 1$. It is easy to describe the denominators:

Theorem 1 (von Staudt-Clausen [34], [9] 1840). *If $n > 0$, then*

$$D_n = \prod_{\substack{p \text{ prime} \\ p-1 \mid n}} p, \quad and \quad B_n + \sum_{\substack{p \text{ prime} \\ p-1 \mid n}} \frac{1}{p} \quad is \ an \ integer.$$

If a prime p divides some numerator N_n, then it divides every $(p-1)$-st numerator after that:

Theorem 2 (Kummer [19] 1851). *If $n \geq 1$, p is a prime ≥ 5 and $p - 1 \nmid 2n$, then*

$$\frac{B_{2n+(p-1)}}{2n + (p-1)} \equiv \frac{B_{2n}}{2n} \mod p.$$

Another useful fact about the prime factors of N_n is this:

Theorem 3 (J. C. Adams [1] 1878). *If p is prime, $n \geq 1$, $p - 1 \nmid 2n$ and $p^e \mid 2n$ for some $e \geq 1$, then $p^e \mid N_{2n}$.*

[1]This work was supported in part by the Purdue University CERIAS and by the Lilly Endowment, Inc.

Slavutskii [27] attributes both Kummer's congruence and Adams' theorem to two obscure pamphlets [35] of von Staudt. See also [28]. The Bernoulli numbers and the prime factors of their numerators have been of fundamental importance in the study of cyclotomic fields since the time of Kummer. For example, see Iwasawa [16] and Ribenboim [25]. Before Wiles proved Fermat's Last Theorem, these numbers provided an important avenue of attack on that problem.

M. Ohm [22] made the first attempt to factor Bernoulli numerators in 1840. In unpublished work, J. Bertrand, J. L. Selfridge, M. C. Wunderlich, and others, factored more Bernoulli numerators. In 1978, we [36] published the factorizations through N_{60}, but there was a typo in the very last factor. Now we have factored N_{2k} for all $2k \leq 152$ and for many larger $2k \leq 300$. See Adams [1] for the unfactored N_{2k} and the D_{2k}. See Knuth and Buckholtz [18] for a simple method of computing these numbers. We used their method to compute the numbers. We publish the factors here to aid the study of cyclotomic fields.

Some other works which consider prime factors of Bernoulli numbers, mostly with large subscripts (far beyond the range of this paper), and which extend the work of [36], include [6], [4], [5] and pages 116ff of [10].

Ten tables, placed at the end of this paper to preserve continuity, summarize our efforts over many years to factor the Bernoulli numerators and the Euler numbers. The complete results are available at the web address: http://www.cerias.purdue.edu/homes/ssw/bernoulli/index.html.

In Tables 1 and 2, we give the complete factorization of N_{2k} for $60 \leq 2k \leq 132$. In the tables, Pxx and Cxx denote prime and composite numbers with xx digits, respectively. To keep the paper short, Tables 3–6 show only the large (> 11 digits) prime factors. We assume that anyone using the tables can compute the numerators and discover the small factors easily. Several modern computer algebra systems, such as Maple and Mathematica, have Bernoulli and Euler numbers and polynomials as built-in functions. If a numerator is omitted, then we know no large prime factor of it. But the numerator is not omitted if the final known factor is prime. Thus the line "144 P135" in Table 3 means that N_{144} is the product of one or more small primes (in fact, 6500309593) times a 135-digit prime—not that N_{144} is prime.

The Euler numbers E_n may be defined by the generating function

$$\frac{2e^{t/2}}{e^t + 1} = \sum_{n=0}^{\infty} \frac{E_n \cdot t^n}{2^n \cdot n!} = \sum_{n=0}^{\infty} \frac{E_n}{n!} \left(\frac{t}{2}\right)^n$$

or by the formula

$$\sec x = \sum_{n=0}^{\infty} (-1)^n E_{2n} \frac{x^{2n}}{(2n)!}.$$

The Euler numbers with odd subscripts vanish: $E_{2k+1} = 0$ for all $k \geq 0$. The non-zero Euler numbers are odd integers which alternate in sign. The first few non-zero Euler numbers are: $E_0 = 1$, $E_2 = -1$, $E_4 = 5$, $E_6 = -61$, $E_8 = 1385$, $E_{10} = -50521$, $E_{12} = 2702765$.

Since the Euler numbers are all integers, there is no analogue for them of the von Staudt-Clausen Theorem. Kummer's Theorem has an analogue for E_{2n}, also proved by Kummer. We state it as Theorem 4 below. Our search for an analogue to J. C. Adams' Theorem led to the work in the next section.

The prime factors of the Euler numbers determine the structure of certain cyclotomic fields. See Ernvall and Metsänkylä [13], for example.

Most of the above remarks about factoring Bernoulli numbers apply equally to Euler numbers. We published the factorizations through E_{42} in 1978 [36]. Now we have factored E_{2k} for all $2k \leq 88$ and for some larger $2k \leq 200$.

In Tables 7 and 8, we give the complete factorization (if known) of E_{2k} for $40 \leq 2k \leq 112$. To save space, Tables 9 and 10 show only the large ($> 10^{11}$) prime factors. We assume that anyone using the tables can compute the Euler numbers and discover the small factors easily. If an Euler number is omitted, then we know no large prime factor of it.

We found most of the factors in the five tables by trial division and the elliptic curve method [20]. The largest two of these factors found by the elliptic curve method were the P42 of E_{150} and the P40 of N_{206}. A few large composite cofactors were finished by the quadratic sieve factoring algorithm [23], including the C114 = P37 · P77 of N_{206} and the C112 = P44 · P69 of E_{116}. Large primes in these tables were proved prime by the methods of the Cunningham Project [3], including the elliptic curve prime proving method [2] for the large primes. The two largest prime divisors of Bernoulli numerators known to us are the P359 factor of N_{292} and the P332 divisor of N_{298}. The largest known prime divisor of an Euler number is the P278 of E_{194}. No doubt one could easily find larger prime divisors of the Bernoulli and Euler numbers by extending the tables a little. The first incomplete factorizations in the tables are the C123 of N_{154} and the C119 of E_{90}. The elliptic curve method, using several hundred curves with a first phase limit $2 \cdot 10^6$, has been tried on these numbers and on all the other composites in the tables.

2   Congruences for the Euler Numbers

In this section we state Kummer's theorem for Euler numbers and two little-known congruences for Euler numbers which we rediscovered by examining (the full version of) Tables 7–10 in search of an analogue for J. C. Adams' Theorem. We also give some historical remarks about these theorems.

The Euler polynomials may be defined by the generating function

$$\frac{2e^{xt}}{e^t + 1} = \sum_{n=0}^{\infty} E_n(x) \frac{t^n}{n!}.$$

It is easy to see that $E_n = 2^n E_n(1/2)$, for $n \geq 0$, and that $E'_n(x) = n E_{n-1}(x)$, for $n > 0$. These two facts lead easily to the Taylor expansion of $E_n(x)$ about $x = 1/2$:

$$E_n(x) = \sum_{k=0}^{n} \binom{n}{k} \frac{E_k}{2^k} \left(x - \frac{1}{2}\right)^{n-k}, \qquad (2.1)$$

which holds for all nonnegative integers n and all real x, and which was proved by Raabe [24] in 1851.

Euler, on page 499 of [14], introduced Euler polynomials to evaluate the alternating sum

$$A_n(m) = \sum_{k=1}^{m} (-1)^{m-k} k^n = m^n - (m-1)^n + \cdots + (-1)^{m-1} 1^n,$$

where m and n are nonnegative integers. The identity $E_n(x+1) + E_n(x) = 2x^n$ follows easily from the definition of Euler polynomials. Alternately adding and subtracting this identity with $x = m-1, x = m-2, \ldots, x = 1$, gives the formula

$$A_n(m) = \frac{1}{2}(E_n(m+1) - (-1)^m E_n(1)) \qquad (2.2)$$

for integers $m, n \geq 0$. In the same way, one can prove that

$$C_n(b, m) \stackrel{\text{def}}{=} \sum_{k=1}^{m} (-1)^{m-k} (k+b-1)^n = \frac{1}{2}(E_n(b+m) - (-1)^m E_n(b)) \quad (2.3)$$

for any real b and integers $m, n \geq 0$.

The next Proposition follows easily from Equations (2.1) and (2.2).

Proposition 1. If $p > 0$ is odd and $n > 0$ is even, then

$$A_n\left(\frac{p-1}{2}\right) = 2^{-n-1} \sum_{k=0}^{n} \binom{n}{k} E_k p^{n-k}.$$

Wells Johnson [17] began with a formula analogous to the one in Proposition 1 and gave p-adic proofs of many facts about Bernoulli numbers, including Theorems 1, 2 and 3. We have used similar methods to prove the results about Euler numbers in this section. We omit the proofs to keep the paper brief and because Carlitz has given proofs similar to ours. Our proofs appear in a paper at the web address mentioned earlier.

The analogue of Kummer's Theorem mentioned above is:

Theorem 4 (Kummer [19] 1851). *If $n \geq 1$ and $p \geq 3$ is prime, then $E_{2n+(p-1)} \equiv E_{2n} \bmod p$.*

One can derive this theorem from Proposition 1 and Fermat's Little Theorem. Carlitz and Levine [8] have also investigated Kummer's congruence for Euler numbers.

Here is the analogue of J. C. Adams' Theorem, which follows from Proposition 1 and Euler's generalization of Fermat's Little Theorem:

Theorem 5. *Let p be an odd prime, n a positive integer and e a nonnegative integer. Suppose $(p-1)p^e$ divides n. Then $E_n \equiv 0$ or $2 \bmod p^{e+1}$ according as $p \equiv 1$ or $3 \bmod 4$.*

Theorem 5 shows, for example, that $E_{2k} \equiv 2 \bmod 3$, $E_{4k} \equiv 0 \bmod 5$, $E_{6k} \equiv 2 \bmod 7$, $E_{6k} \equiv 2 \bmod 9$ and $E_{10k} \equiv 2 \bmod 11$ for all $k > 0$.

Carlitz [7] gave a proof very similar to the one we found.

Now define

$$D_n(m) = \sum_{k=1}^{m} (-1)^{m-k}(2k-1)^n = (2m-1)^n - (2m-3)^n + \cdots + (-1)^{m-1}1^n$$

for integers $m \geq 1$, $n \geq 0$.

It is easy to show:

Proposition 2. *If $m \geq 1$ and $n \geq 0$, then $D_n(m) = 2^n C_n\left(\frac{1}{2}, m\right)$.*

From Proposition 2 and Equations (2.1) and (2.3), one can prove:

Proposition 3. *If $m \geq 1$ and $n \geq 0$, then*

$$D_n(m) = \sum_{k=0}^{n-1} \binom{n}{k} 2^{n-k-1} E_k m^{n-k} + \frac{1-(-1)^m}{2} E_n.$$

The next theorem, whose proof uses Proposition 3, gives an interesting congruence for the Euler numbers modulo a power of 2.

Theorem 6. *For all integers $n \geq 0$ and $k \geq 0$ we have $E_{2n} \equiv E_{2n+2^k} + 2^k \bmod 2^{k+1}$.*

Corollary 1. *The set* $\{E_0, E_2, \ldots, E_{2^k-2}\}$ *forms a reduced set of residues modulo* 2^k *for* $k \geq 1$.

Theorem 5 and Corollary 1 were stated without proof by Sylvester [31], [33], [30], [32] in 1861. A few years later, Stern [29] gave brief sketches of proofs of these two results and of Theorem 6. In 1910, Frobenius [15] amplified Stern's sketches of these proofs. Ernvall [12] in 1979 said he couldn't understand Frobenius' outline of the proofs and gave his own proofs using the umbral calculus. The case $e = 0$ of Theorem 5 was proved by Ely [11] and mentioned by Nielsen [21]. These works of Sylvester, Stern and Ely are noted by Saalschütz [26]. Proposition 2 is in Nielsen [21]. The proofs of Theorems 4, 5 and 6 which we found have the p-adic flavor of proofs of similar statements for the Bernoulli numbers in Johnson [17].

n	Prime factorization of $|N_n|$
60	$2003 \cdot 5549927 \cdot 109317926249509865753025015237911$
62	$31 \cdot 157 \cdot 266689 \cdot 329447317 \cdot 28765594733083851481$
64	$1226592271 \cdot 8705731535452217918484989699791727$
66	$11 \cdot 839 \cdot 159562251828620181390358590156239282938769$
68	$17 \cdot 37 \cdot 101 \cdot 123143 \cdot 1822329343 \cdot$ $\cdot 552547336651093002822748 1$
70	$5 \cdot 7 \cdot 688531 \cdot 20210499584198062453 \cdot$ $\cdot 3090850068576441179447$
72	$3112655297839 \cdot$ $\cdot 18723419087606889767942264996363043575678 11$
74	$37 \cdot 923038305114085622008920911661422572613197507651$
76	$19 \cdot 58231 \cdot$ $\cdot 2228428593011623643012285556037270788516992470 9$
78	$13 \cdot 787388008575397 \cdot 33364652939596337 \cdot$ $\cdot 1214698595111676682009391$
80	$631 \cdot 10589 \cdot 5009593 \cdot 141795949 \cdot \text{P}39$
82	$41 \cdot 4003 \cdot 38189 \cdot \text{P}51$
84	$233 \cdot 271 \cdot 68767 \cdot 167304204004064919523 \cdot \text{P}37$
86	$43 \cdot 541 \cdot 21563 \cdot \text{P}55$

Table 1. Bernoulli numerators $|N_n|$ factored

n	Prime factorization of $|N_n|$
88	$11 \cdot 307 \cdot 2682679 \cdot \text{P}60$
90	$5 \cdot 587 \cdot 1758317910439 \cdot \text{P}57$
92	$23 \cdot 587 \cdot 108023 \cdot \text{P}63$
94	$47 \cdot 467 \cdot 1499 \cdot 2459153 \cdot 4217126617741589575995641 \cdot \text{P}34$
96	$7823741903 \cdot 4155593423131 \cdot 10017952436526113 \cdot$ $\cdot\, 96454277809515481 \cdot \text{P}25$
98	$7 \cdot 7 \cdot 2857 \cdot 3221 \cdot 1671211 \cdot 9215789693276607167 \cdot \text{P}43$
100	$263 \cdot 379 \cdot 28717943 \cdot 656771716927555564821811 33 \cdot \text{P}45$
102	$17 \cdot 59 \cdot 827 \cdot 17833331 \cdot 86023144558386407 \cdot$ $\cdot\, 29911635890983027644744 3337 \cdot \text{P}28$
104	$13 \cdot 37 \cdot 776253902057299 \cdot 6644689804135385589700423 \cdot \text{P}45$
106	$53 \cdot 3967 \cdot 37217 \cdot 77272435237709 \cdot \text{P}65$
108	$656884664663 \cdot 23657486502844933 \cdot \text{P}69$
110	$5 \cdot 157 \cdot 76493 \cdot 150235116317549231 \cdot$ $\cdot\, 369448188741168234283576 91 \cdot \text{P}44$
112	$7 \cdot 887569 \cdot 8065483 \cdot \text{P}86$
114	$19 \cdot \text{P}97$
116	$29 \cdot 7559 \cdot 7438099 \cdot 6795944986967 \cdot \text{P}77$
118	$59 \cdot \text{P}100$
120	$6495690221 \cdot 8070196213 \cdot \text{P}93$
122	$61 \cdot 1545314586433142560447 \cdot$ $\cdot\, 1545923474257037240728199709913 \cdot \text{P}54$
124	$31 \cdot 67 \cdot 74747 \cdot 162263 \cdot 14066893 \cdot 8262971607841 \cdot$ $\cdot\, 3498285428145163 \cdot 16743250272239551 \cdot \text{P}45$
126	$103 \cdot 409 \cdot 216363744721 \cdot \text{P}102$
128	$35089 \cdot 5953097 \cdot 12349588663 \cdot 13349390911530343 \cdot$ $\cdot\, 6996505560116602097773394576621473 \cdot \text{P}46$
130	$5 \cdot 13 \cdot 149 \cdot 463 \cdot 2264267 \cdot 3581984682522167 \cdot \text{P}92$
132	$11 \cdot 804889 \cdot 10462099 \cdot \text{P}112$

Table 2. Bernoulli numerators $|N_n|$ factored (continued)

n	Large prime factors of $|N_n|$
134	$338420464438865099 \cdot 6005440277888093849051345046242759 \cdot$ $\cdot \text{P65}$
136	$2983509658548393462 1 \cdot \text{P98}$
138	$554744941981 \cdot 756906736720877 \cdot$ $\cdot 995959666194215326642640313557460384737 9 \cdot \text{P48}$
140	$44124706530665069 \cdot 499190989552139944322431620 77 \cdot \text{P68}$
142	$11178195490847948438398 1 \cdot \text{P105}$
144	P135
146	$22639970526343 \cdot 6726159702783854797 \cdot$ $\cdot 37996324998547740539691528067877 \cdot$ $\cdot 1754821172656266926966923716442469 \cdot \text{P34}$
148	$4975417507662031677157 \cdot$ $\cdot 1248863436460860523032749 \cdot \text{P84}$
150	$5810708205829 \cdot 2166479673949953104094 7 \cdot$ $\cdot 2409795082015672566733218756037 \cdot \text{P72}$
152	$372034103782702933865518136371704496 1 \cdot$ $\cdot 3783571607405842689072559655030411819649815 9 \cdot \text{P52}$
154	$384785986561 \cdot \text{C123}$
156	$16760414993553486506490 7 \cdot$ $\cdot 9488426748329562220014361617994 7 \cdot \text{P101}$
160	$40094692599177383 \cdot 1283008671289189098343005994856 3 \cdot$ $\cdot 1744826505423362390046833266050403703791289 \cdot \text{P62}$
164	$104386532651 \cdot 2903061743891 \cdot 9898920431428993 \cdot \text{C117}$

Table 3. Large prime factors of Bernoulli numerators $|N_n|$

n	Large prime factors of $|N_n|$
166	311318618909 · 37074748512889 · · 605190683329889640846518910327717 · · 11709228761805923962023525960553218961 9 · P52
168	19254163575306510187 · 10094494587919631151637 · C128
170	751612064207 · P154
172	P174
174	6659961564676431900928667503 · C137
176	333026571343 · 110783038328477 · 124813394943812621 · C138
178	129180506448277 · 1823634234826012967 · · 39326836920802601519 · P129
180	249829228470043 · 207625243678748953583 3 · · 4241477436592626145879 · P127
182	73107144475261423 · 311089841618633327 · · 3627027615648746666477 · 2122174114227419648093461601 · · 83276165458323300429587071706402939815 92673849 · P59
184	21983088204089362967 · P169
186	922966808867 · 9161904079472101 · C156
190	60860762760882373 · 1742620927079710201 04538709609 · C152
196	58273617156601282072242637946609 · C173
198	723357738211 · P201
200	5370056528687 · C204

Table 4. Large prime factors of Bernoulli numerators $|N_n|$ (continued)

n	Large prime factors of $|N_n|$
202	$85704723183916799 \cdot C173$
204	$9131578873975602379 \cdot P207$
206	$4134128959054219 \cdot 28391723373218209 \cdot$ $\cdot 408428439912252710783201 \cdot$ $\cdot 4794779427824009051318510739603796493 \cdot$ $\cdot 3705636735000917624663544925511551624891 \cdot P77$
216	$P239$
218	$498630504627832848561390486831 \cdot C175$
220	$792913356669011 \cdot C224$
222	$270574469649607096339 \cdot C229$
224	$676535170838231166688886121 \cdot$ $\cdot 135687310586064972665288502 07 \cdot P187$
226	$226941007255811687 \cdot C229$
230	$948756114525995558524 9403 \cdot C234$
232	$2483032145171 \cdot 25905116405567190927047382 0520219 \cdot P225$
234	$48237362885215689907 \cdot C222$
236	$504680422913 \cdot 146568915231099952945767205094299 87 \cdot C219$
238	$30079831621249 \cdot C258$
240	$26230095767160160157 \cdot C260$
242	$4967552208919410364191 7241 \cdot C236$
246	$1015348391695196501 \cdot C267$
248	$11513470342710425729471126 5272763 \cdot C240$

Table 5. Large prime factors of Bernoulli numerators $|N_n|$ (continued)

n	Large prime factors of $|N_n|$
252	2028290804799829 · 650932177698080567099 · · 130625066385309173899099708579 · · 7542235420324865718852164333401349 · · 1000993741524774643539942570884595839 · C171
258	1236523928730271 · C292
262	6337971290361982570984 7 · P278
266	167825382335090242001 · 1727816222222026922465407 · · 1571264305785183471309381325703 · C216
270	2539833907837164114167 · P306
274	21804608848811 · 201500345265433 · 31628480989746829 · · 3277838401217446489 · 2572908479911 7836987901 · · 892008372912807309877541 · C222
276	116773511307223 · 928048176111241418044710 2368597 · C293
280	136100780239 · C338
282	4525048629470223385658435650031 · C311
284	792213846555737 · C331
286	8812943587829 · 16865476527940273 · · 34000751682694166738635652417 · C285
288	1259554461969878619108227 · C328
292	P359
296	146409753143342542769 · C351
298	371472263795653589766634977803 · P332
300	7985787872578979 · C352

Table 6. Large prime factors of Bernoulli numerators $|N_n|$ (continued)

n	Prime factorization of $|E_n|$
40	$5 \cdot 5 \cdot 41 \cdot 763601 \cdot 52778129 \cdot 359513962188687126618793$
42	$137 \cdot 5563 \cdot 1359952912756417481954933903061 9651971$
44	$5 \cdot 587 \cdot 32027 \cdot 9728167327 \cdot 36408069989737 \cdot 238716161191111$
46	$19 \cdot 285528427091 \cdot 1229030085617829967076190070873124909$
48	$5 \cdot 13 \cdot 17 \cdot$ $\cdot 551699424938329607121419524242248249 2286460673697$
50	$5639 \cdot 1508047 \cdot 10546435076057211497 \cdot$ $\cdot 6749451555259847962 2918721$
52	$5 \cdot 31 \cdot 53 \cdot 1601 \cdot 2144617 \cdot 537569557577904730817 \cdot P24$
54	$43 \cdot 2749 \cdot 3886651 \cdot 78383747632327 \cdot P36$
56	$5 \cdot 29 \cdot 5303 \cdot 7256152441 \cdot 52327916441 \cdot 2551319957161 \cdot P26$
58	$14598794767712473479610314450 01033 \cdot P34$
60	$5 \cdot 5 \cdot 13 \cdot 47 \cdot 61 \cdot 6821509 \cdot 14922423647156041 \cdot P42$
62	$101 \cdot 6863 \cdot 418739 \cdot 1042901 \cdot P56$
64	$5 \cdot 17 \cdot 19 \cdot 25349 \cdot 85297 \cdot P65$
66	$61 \cdot 105075119 \cdot 508679461 \cdot 155312172341 \cdot P51$
68	$5 \cdot 2039 \cdot 66041 \cdot 29487071944189 \cdot 15138431327918641 \cdot P45$
70	$353 \cdot 2586437056036336027701234101159 \cdot P54$
72	$5 \cdot 13 \cdot 37 \cdot 73 \cdot 2341 \cdot 4014623 \cdot 24259423 \cdot$ $\cdot 30601587075439337 \cdot P51$
74	$193 \cdot 34629826749613 \cdot 4207222848740394629 \cdot$ $\cdot 22060457167870794468746201 \cdot P34$
76	$5 \cdot 145007 \cdot 3460859370585503071 \cdot$ $\cdot 5816628272808637232395643 86159 \cdot P43$

Table 7. Euler numbers $|E_n|$ factored

n	Prime factorization of $|E_n|$
78	$274001956110391029122841712305499482531697987 \cdot \text{P55}$
80	$5 \cdot 5 \cdot 17 \cdot 41 \cdot 7701306020743 \cdot 357236360318890217539621 3 \cdot \text{P62}$
82	$19 \cdot 31 \cdot 4395659 \cdot \text{P98}$
84	$5 \cdot 13 \cdot 29 \cdot 4397 \cdot$ $\cdot 7397623352390151867065277351927955207 26707 \cdot \text{P62}$
86	$311 \cdot 390751 \cdot 46053168570671 \cdot \text{P92}$
88	$5 \cdot 89 \cdot 1019 \cdot 588528876550967927 \cdot$ $\cdot 16292380848703930709213 \cdot \text{P72}$
90	$307 \cdot \text{C119}$
92	$5 \cdot 67 \cdot 7096363493 \cdot 7308346963823 \cdot 1204768 13565517 \cdot \text{P85}$
94	$53089 \cdot 20609829625906839913745698187 \cdot$ $\cdot 180986288780569828566819992453 \cdot \text{P66}$
96	$5 \cdot 13 \cdot 17 \cdot 43 \cdot 79 \cdot 97 \cdot 835823 \cdot 2233081 \cdot$ $\cdot 1951860271597317997069749059 \cdot$ $\cdot 9416370608392625586845089085196635167 \cdot \text{P47}$
98	$71 \cdot 376003429 \cdot 5160267661 \cdot 436390726250 6552373343 \cdot \text{P94}$
100	$5 \cdot 5 \cdot 5 \cdot 19 \cdot 101 \cdot \text{C134}$
102	$8647 \cdot \text{C139}$
104	$5 \cdot 53 \cdot 761 \cdot 2477 \cdot \text{P138}$
106	$47 \cdot 4858416191 \cdot 98985829942673 \cdot$ $\cdot 1150887066548393492521971151372616707 \cdot \text{P88}$
108	$5 \cdot 13 \cdot 37 \cdot 109 \cdot 1462621 \cdot 8445961 \cdot 4675063 901 \cdot \text{C125}$
110	$509053 \cdot 116904299 \cdot 134912677 \cdot 748079839 770433 \cdot \text{P120}$
112	$5 \cdot 17 \cdot 29 \cdot 31 \cdot 113 \cdot 8185757 \cdot 617575481323 \cdot$ $\cdot 1522046069820268709 \cdot 26505314603042887 6430329 \cdot \text{P94}$

Table 8. Euler numbers $|E_n|$ factored (continued)

n	Large prime factors of $|E_n|$
114	$5290253211544727 \cdot 22557103319451713 \cdot$ $\cdot 2565948669867461313318215567 \cdot$ $\cdot 11897268445383513539263419255627345471818759570534 3 \cdot$ $\cdot \text{P}52$
116	$1098948437923935829829 \cdot$ $\cdot 17698520871521406115634951924463689 \cdot$ $\cdot 116619065933163530588469118477095110617775 23 \cdot \text{P}69$
118	$8661193890969663597268314978 1 \cdot \text{C}142$
124	$5458931108933632733743 39 \cdot \text{C}137$
126	$44305294819613 \cdot 16723717485156209220 1 \cdot \text{P}128$
128	$91486803609919 \cdot 33397018471037747 \cdot$ $\cdot 38280927951817207 \cdot 18236941888532279049499046 27 \cdot$ $\cdot 25218189671884291383279399150744135824 9 \cdot \text{P}64$
134	$321639994822891 \cdot 21407431771728232601749801895 3 \cdot \text{P}148$
136	$\text{P}200$
138	$12254459673349 \cdot 34356165690119899 \cdot \text{P}157$
140	$\text{P}199$
142	$2978734769 \cdot 8557612247 \cdot \text{P}197$
144	$9785760855589235011 79 \cdot 170513218370189155958048891371 \cdot$ $\cdot \text{P}149$
146	$\text{P}213$
148	$238661068231279 \cdot \text{C}202$
150	$13621373428254587 \cdot$ $\cdot 11138197399926022828223816743133558543305 9 \cdot \text{C}168$

Table 9. Large prime factors of Euler numbers $|E_n|$

n	Large prime factors of $|E_n|$
152	1805155617412973535918 1 · 3957666449530267510589053 · · 4383213340951838246582947093 67 · C149
154	139668927262709710013 · C210
156	227071134239 · P198
158	5519160811451003 · C220
162	174175655449 · C242
164	634888487743565027305430606 33 · C209
166	50150236900098278077 · C214
168	86771436435012390277 · C230
170	70727223023077 · 1034326231547973051559 · P239
172	743155422133 · 2840083403239 · C243
180	6923483330327017 · C269
184	2804389579706797633 · C284
186	22658461432253 · 54342802734882461 · · 1086110887390889008410968159777 · C229
190	559570609330768709 · 63860147345993694105869027 68943 · · C265
192	1469840300183 · 6895766514961118059 · · 12696721063842186926157906929 11 · C249
194	28024555486506389 · 2436437750204310804841 · P278
198	2507798651531 · 49639305210453901009432031 · C277
200	16640782677056849 · C306

Table 10. Large prime factors of Euler numbers $|E_n|$ (continued)

References

[1] J. C. Adams, *Table of the first sixty-two numbers of Bernoulli*, J. Reine Angew. Math. **85** (1878), 269–272.

[2] A. O. L. Atkin and F. Morain, *Elliptic curves and primality proving*, Math. Comp. **61** (1993), 29–68.

[3] J. Brillhart, D. H. Lehmer, J. L. Selfridge, B. Tuckerman, and S. S. Wagstaff, Jr., *Factorizations of $b^n \pm 1$, b = 2, 3, 5, 6, 7, 10, 11, 12 up to high powers*, Amer. Math. Soc., Providence, 1988.

[4] J. P. Buhler, R. E. Crandall, R. Ernvall, and T. Metsänkylä, *Irregular primes and cyclotomic invariants to four million*, Math. Comp. **61** (1993), 151–153.

[5] J. P. Buhler, R. E. Crandall, R. Ernvall, T. Metsänkylä, and A. Shokrollahi, *Irregular primes and cyclotomic invariants to twelve million*, J. Symbolic Comput. **31** (2001), 89–96.

[6] J. P. Buhler, R. E. Crandall, and R. W. Sompolski, *Irregular primes to one million*, Math. Comp. **59** (1992), 717–722.

[7] L. Carlitz, *A note on Euler numbers and polynomials*, Nagoya Math. J. **7** (1954), 35–43.

[8] L. Carlitz and J. Levine, *Some problems concerning Kummer's congruences for the Euler numbers and polynomials*, Trans. Amer. Math. Soc. **96** (1960), 23–37.

[9] T. Clausen, *Lehrsatz aus einer Abhandlung über die Bernoullischen Zahlen*, Astronomische Nachrichten **17** (1840), 351–352.

[10] R. E. Crandall, *Topics in Advanced Scientific Computation*, TELOS/-Springer-Verlag, Santa Clara, CA, 1996.

[11] G. S. Ely, *Some notes on the numbers of Bernoulli and Euler*, Amer. J. Math. **5** (1882), 337–341.

[12] R. Ernvall, *Generalized Bernoulli numbers, generalized irregular primes, and class number*, Ann. Univ. Turku. Ser. A **I** (1979), 1–72.

[13] R. Ernvall and T. Metsänkylä, *Cyclotomic invariants and E-irregular primes*, Math. Comp. **32** (1978), 617–629.

[14] L. Euler, *Institutiones Calculi Differentialis*, Petersburg, 1755.

[15] F. G. Frobenius, *Über die Bernoullischen Zahlen und die Eulerschen Polynome*, Sitzungsberichte der Königlich Preussischen Akademie der Wissenschaften zu Berlin, pages 809–847, 1910, also in Gesammelte Abhandlungen III, 1968, Springer-Verlag, pp. 440–478.

[16] K. Iwasawa, *A class number formula for cyclotomic fields*, Ann. Math. **76** (1962), 171–179.

[17] W. Johnson, *p-adic proofs of congruences for the Bernoulli numbers*, J. Number Theory **7** (1975), 251–265.

[18] D. E. Knuth and T. J. Buckholtz, *Computation of tangent, Euler, and Bernoulli numbers*, Math. Comp. **21** (1967), 663–688.

[19] E. E. Kummer, *Über eine allgemeine Eigenschaft der rationalen Entwickelungscoëfficienten einer bestimmten Gattung analytischer Functionen*, J. Reine Angew. Math. **41** (1851), 368–372.

[20] H. W. Lenstra, Jr., *Factoring integers with elliptic curves*, Ann. Math. **126** (1987), 649–673.

[21] N. Nielsen, *Traité élémentaire des Nombres de Bernoulli*, Gauthier-Villars, Paris, 1923.

[22] M. Ohm, *Etwas über die Bernoulli'schen Zahlen*, J. Reine Angew. Math. **20** (1840), 11–12.

[23] C. Pomerance, *The quadratic sieve factoring algorithm*, Advances in Cryptology, Proceedings of EUROCRYPT 84 (T. Betha, N. Cot, and I. Ingemarsson, eds.), Lecture Notes in Computer Science, vol. 209, Springer-Verlag, Berlin, Heidelberg, New York, 1985, pp. 169–182.

[24] J. L. Raabe, *Zurückführung einiger Summen und bestimmten Integrale auf die Jacob-Bernoullische Function*, J. Reine Angew. Math. **42** (1851), 348–367.

[25] P. Ribenboim, *13 Lectures on Fermat's Last Theorem*, Springer-Verlag, New York, 1979.

[26] L. Saalschütz, *Vorlesungen über die Bernoullischen Zahlen*, Springer-Verlag, Berlin, 1893.

[27] I. Sh. Slavutskii, *Staudt and arithmetical properties of Bernoulli numbers*, Historica Sci. **5** (1995), 69–74.

[28] ———, *About von Staudt congruences for Bernoulli numbers*, Comment. Math. Univ. St. Paul **48** (1999), 137–144.

[29] M. A. Stern, *Zur Theorie der Eulerschen Zahlen*, J. Reine Angew. Math. **79** (1875), 67–98.

[30] J. J. Sylvester, *Addition à la précédente note*, C. R. Acad. Sci., Paris **52** (1861), 212–214.

[31] ———, *Note on the numbers of Bernoulli and Euler, and a new theorem concerning prime numbers*, Phil. Magaz. **21** (1861), 127–136.

[32] ———, *Note relative aux communications faites dans les séances de 28 Janvier et 4 Fèvrier 1861*, C. R. Acad. Sci., Paris **52** (1861), 307–308.

[33] ———, *Sur une propriété de nombres premiers qui se rattache au dernier théorème de Fermat*, C. R. Acad. Sci., Paris **52** (1861), 161–163.

[34] K. G. C. von Staudt, *Beweis eines Lehrsatzes, die Bernoullischen Zahlen betreffend*, J. Reine Angew. Math. **21** (1840), 372–374.

[35] ———, *De numeris Bernoullianis, I, II*, Junge, Erlangen, 1845.

[36] S. S. Wagstaff, Jr., *The irregular primes to 125000*, Math. Comp. **32** (1978), 583–591.

An Improved Method for Solving the Family of Thue Equations $X^4 - 2rX^2Y^2 - sY^4 = 1$

P. G. Walsh

1 Introduction

Since the groundbreaking work of Thue [13], the subject of solving those Diophantine equations of the form

$$F(X,Y) = m,$$

with $F \in Z[X,Y]$ a binary form of degree at least 3 and m a nonzero integer, has received enormous interest, with many significant results. For a detailed history on this topic, the reader may wish to consult Halter-Koch et. al. [4], or Heuberger [5]. Under suitable hypotheses, such Diophantine equations have only finitely many solutions in integers X and Y, and methods for determining all such solutions have been developed by Tzanakis and de Weger [14], Bilu-Hanrot [2], and others. Much of this research has centered around the problem of solving parametric families of Thue equations, as described in the work of E. Thomas [12]. We remark that in almost all examples solved in the literature, the coefficients are defined in terms of a single parameter.

At the 1988 NATO-ASI meeting in Banff, Stroeker [11] described a method to solve a class of quartic Thue equations given in the form

$$X^4 - 2rX^2Y^2 - sY^4 = 1. \tag{1.1}$$

If r, s in (1.1) are such that $r^2 + s = t^2 d$, with $d > 0$, then Stroeker's method will work provided that $d \equiv 3 \pmod 4$, and also provided that certain other technical conditions hold during the course of his algorithm.

The purpose of the present paper is to provide a method to solve such equations using a combination of classical results of Ljunggren together with some recent refinements thereof. We will show how one can efficiently solve any equation of the type (1.1), thereby removing all of the special hypotheses required for Stroeker's method. Moreover, the method we describe can be used to deal not only with the more general equation

$$X^2 - 2rXY^2 - sY^4 = 1, \tag{1.2}$$

but also with

$$X^2 - rXY^2 - sY^4 = c \tag{1.3}$$

for $|c| \in \{1, 2, 4\}$. We will however restrict our attention to the former of these two equations. In the special case that $s = 1$ in (1.3), the complete solution in positive integers was determined in [17]. Thus the results of this paper also serve to improve on the results of [17].

We remark that in order to determine all solutions to (1.2) it is sufficient to determine all positive integer solutions Y, since the determination of corresponding values for X then becomes a trivial task. By defining $Z = X - rY^2$ and $D = r^2 + s$, integer solutions (X, Y) to (1.2) are in correspondence with integer solutions (Z, Y) to

$$Z^2 - DY^4 = 1. \tag{1.4}$$

We will therefore focus our attention on the problem of determining all integer solutions to (1.4).

In what follows, let r, s be the integers given in equation (1.2), and $D = r^2 + s$. We will assume that D is a nonsquare positive integer; otherwise solving equations (1.2) and (1.4) is trivial. Let $T_1 + U_1\sqrt{D}$ denote the minimal solution in positive integers to the Pell equation $X^2 - DY^2 = 1$. For $k \geq 1$ define

$$T_k + U_k\sqrt{D} = (T_1 + U_1\sqrt{D})^k.$$

Theorem 1. 1. *There are at most two positive integer solutions (Z, Y) to equation (1.4). If two solutions $Y_1 < Y_2$ exist, then $Y_1^2 = U_1$ and $Y_2^2 = U_2$, except only if $D = 1785$ or $D = 16 \cdot 1785$, in which case $Y_1^2 = U_1$ and $Y_2^2 = U_4$.*

2. If only one positive integer solution (Z, Y) to equation (1.4) exists, then $Y^2 = U_l$ where $U_1 = lv^2$ for some squarefree integer l, and either $l = 1$, $l = 2$, or $l = p$ for some prime $p \equiv 3 \pmod 4$.

Theorem 1 evidently provides an efficient algorithm for solving equation (1.4). Using the continued fraction expansion of $\sqrt{D} = \sqrt{r^2 + s}$, one can compute $T_1 + U_1\sqrt{D}$. Upon factoring U_1 in the form $U_1 = lv^2$, with l squarefree, one only needs to determine if any of U_1, U_2, U_4, U_l are squares. The shortcoming of this result is that evidence, such as an effective form of the abc conjecture, indicates that U_k cannot be a square if $k > 4$. The reader is referred to [15] for more details concerning this connection.

Conjecture 1. *If D is a nonsquare positive integer, $\epsilon_D = T + U\sqrt{D}$ the minimal solution to $X^2 - DY^2 = 1$, and $\epsilon_D^k = T_k + U_k\sqrt{D}$ for $k \geq 1$, then the equation*

$$U_k = y^2$$

implies that $k \leq 4$.

To be more precise, there are infinitely many values D for which U_1 (respectively U_2) is a square, namely $D = m^2 - 1$ (respectively $D = 4m^4 - 1$). Similarly, there are infinitely many values of D for which U_3 is a square. For $k \geq 1$, put $t_k + u_k\sqrt{3} = (2 + \sqrt{3})^k$. Then for each $D_i = ((t_{4i+1}/2)^2 - 1)/9$, $i \geq 1$, it is easy to verify that the corresponding value U_3 is a square. Lastly, by the first part of Theorem 1, it is easy to prove that U_4 is a square only for $D = 1785$ and $D = 16 \cdot 1785$.

If Conjecture 1 were known to be true, it would facilitate the complete solution of equations (1.1), (1.2) and (1.4) considerably, since one would only need to verify if any of the numbers U_i, $1 \leq i \leq 4$, is a square. We will provide an example at the end of the paper illustrating this.

2 Preliminary Results

In this section we will collect those results which will be needed in the course of proving Theorem 1. It is worth noting that Ljunggren [6] proved a result which is close to Theorem 1, but does not deal with the case that only one solution to (1.4) exists. We therefore provide a somewhat different proof that is also based on Ljunggren's work, but in addition uses recent results of our own.

Lemma 1. *Let $d > 1$ be a squarefree integer, and let $\epsilon_d = T + U\sqrt{d}$ denote the minimal unit (> 1) in $Q(\sqrt{d})$. Then*

$$\epsilon_d = \tau^2,$$

where

$$\tau = \frac{a\sqrt{m} + b\sqrt{n}}{\sqrt{c}},$$

$c \in \{1, 2\}$, $d = mn$, m *is not a square, and* $a^2m - b^2n = c$.

Proof. This is well known; see, for example, Nagell [10].   □

Lemma 2. *Let $d > 1$ denote a nonsquare positive integer, $\epsilon_d = T + U\sqrt{d}$ be the minimal solution to $X^2 - dY^2 = 1$, and $T_k + U_k\sqrt{d} = \epsilon_d^k$ for $k \geq 1$. If $T_k = x^2$ for some integer x, then either $k = 1$ or $k = 2$.*

Proof. This is a recent result of Cohn [3], which improves on previous work of Ljunggren. For a detailed account of the entire proof, the reader is referred to the first section of [18].   □

Let $\{T_k\}$ and $\{U_k\}$ be as above. For a positive integer b we define the rank of apparition of b in $\{T_k\}$ (resp. $\{U_k\}$) to be the minimal index k such that b divides some term T_k (resp. $\{U_k\}$), provided such an index exists, and denote it by $\beta(b)$ (resp. $\alpha(b)$). It is well known that $\alpha(b)$ exists for all positive integers b, but $\beta(b)$ may or may not exist.

Lemma 3. *Let $d > 1$ denote a nonsquare positive integer, and $b > 1$ a squarefree integer.*
   1. *If $T_k = bx^2$ for some integer x, then $k = \alpha(b)$. In particular, if $T_k = 2x^2$ for some integer x, then $k = 1$.*
   2. *If $U_k = bx^2$ and $\alpha(b)$ is even, then $k = \alpha(b)$, except only in the case that $2T_1^2 - 1 = v^2$ for some integer v and $T_1 U_1 = bu^2$ for some integer u, in which case $U_4 = U_{2\alpha(b)} = b(2uv)^2$.*

Proof. For part 1, see [1], Theorem 1 and Corollary 1, while for part 2 see [16]. $\qquad\square$

Lemma 4. *Let m and n be positive integers, m not a square, such that*

$$mX^2 - nY^2 = 1 \qquad\qquad (2.1)$$

is solvable in integers, and let $\tau = \tau_{m,n} = A\sqrt{m} + B\sqrt{n}$ denote its minimal solution with A and B positive integers. Then all positive integer solutions to (2.1) are given by

$$\tau^{2k+1} = A_{2k+1}\sqrt{m} + B_{2k+1}\sqrt{n}. \ (k \geq 0)$$

Assume that $B = lv^2$ with l squarefree, and suppose that (x, y) is a solution to

$$mX^2 - nY^4 = 1. \qquad\qquad (2.2)$$

Then l is odd and $y^2 = B_l$.

Proof. This is a theorem of Ljunggren [8]. The reader may wish to consult [18] for a detailed account of the proof and related problems. $\qquad\square$

Lemma 5. *Let m and n be positive integers, m not a square, such that*

$$mX^2 - nY^2 = 2 \qquad\qquad (2.3)$$

is solvable in odd integers X and Y. Let $\tau_{m,n} = \dfrac{A\sqrt{m} + B\sqrt{n}}{\sqrt{2}}$ denote its minimal solution with A and B odd positive integers. (There is no ambiguity in using the notation $\tau_{m,n}$ once again, since at most one of equations (2.1) and (2.3) are solvable in integers.) Then all positive integer solutions to (2.3) are given by

$$\tau^{2k+1} = \dfrac{A_{2k+1}\sqrt{m} + B_{2k+1}\sqrt{n}}{\sqrt{2}} \ (k \geq 0).$$

If (x, y) is a solution to

$$mX^2 - nY^4 = 2, \tag{2.4}$$

then either $y^2 = B_1$ or $y^2 = B_3$.

Proof. This has recently been proved [9], improving upon previous work of Ljunggren.   □

Lemma 6. *Let m and n be positive integers, m not a square, such that*

$$mX^2 - nY^2 = 1 \tag{5}$$

is solvable in integers, and let $\tau_{m,n} = A\sqrt{m} + B\sqrt{n}$ denote its minimal solution. If

$$mX^4 - nY^2 = 1 \tag{2.5}$$

has a solution (x, y), then $x^2 = A_{2k+1}$ for some k, where A_{2k+1} is defined in Lemma 4, and either $2k + 1 = 1$ or $2k + 1$ is a prime $p \equiv 3 \pmod 4$. Furthermore, if A_{2k+1} is a square for some index $2k + 1$, then A_1 is also a square.

Proof. The reader is referred to [18] for the details of this.   □

3 Proof of Theorem 1

Let k denote an index with the property that $U_k = y^2$ for some integer y. If k is even, $k = 2r$, then $y^2 = U_k = U_{2r} = 2T_rU_r$, from which it follows that T_r is either a square or twice a square. By Lemma 2 and Lemma 3, we see that $r = 1$ or $r = 2$, and hence $k = 2$ or $k = 4$.

Let $D = w^2d$ with d squarefree. Let $\epsilon_d = t + u\sqrt{d}$ denote the minimal solution to $X^2 - dY^2 = 1$, and define r_D to be the positive integer for which $\epsilon_d^{r_D} = \epsilon_D$, i.e., $r_D = \log(\epsilon_D)/\log(\epsilon_d)$.

Assume first that r_D is even. Let $t_k + u_k\sqrt{d} = (\epsilon_d)^k$. Then $U_k = y^2$ for some integer y if and only if $u_{2k_1} = wz^2$ for some integer z, where $2k_1 = kr_D$. If w divides u_{k_1}, then t_{k_1} is a square, and Lemma 2 implies that $k_1 = 1$ or $k_1 = 2$. It follows that $k \in \{1, 2, 4\}$. If w does not divides u_{k_1}, then the rank of apparition $\alpha^*(w)$ of w in the sequence $\{u_k\}$ is even, and so by Lemma 3, either $2k_1 = 4$ or $2k_1 = \alpha^*(w)$. Since $2k_1 = kr_D$ and since $\alpha^*(w)$ clearly divides r_D, we see that $k \in \{1, 2, 4\}$.

We henceforth assume that both k and r_D are odd positive integers. In this case we can assume that U_1 is properly divisible by 2 to an even power, otherwise the binomial theorem shows that U_k is properly divisible by 2 to an odd power for all odd positive integers k. Appealing to Lemma 1, we see that $\epsilon_D = \tau^2$ with

$$\tau = \frac{A_1\sqrt{m} + B_1\sqrt{n}}{\sqrt{c}}, \quad c \in \{1,2\},$$

m a nonsquare positive integer, $A_1^2 m - B_1^2 n = c$, $D = 4^{2-c} mn$, $U_k = A_k B_k$, and $\gcd(A_k, B_k) = 1$ for all odd $k \geq 1$, where

$$\tau^k = \frac{A_k\sqrt{m} + B_k\sqrt{n}}{\sqrt{c}}.$$

Therefore, U_k is a square precisely when both A_k and B_k are squares.

We will consider the cases $c = 1$ and $c = 2$ separately. Suppose first that $c = 1$. By Lemma 4, B_k can only be a square when $k = l$, where $B_1 = lv^2$ for some squarefree positive integer l and some integer v. By Lemma 6, if A_k is a square, then so is A_1, and $k = 1$ or k is a prime $p \equiv 3 \pmod 4$. Combining these two results shows that either $U_1 = W^2$ for some integer W and $k = 1$, or $U_1 = A_1 B_1 = pW^2$ and $k = p$.

Now assume that $c = 2$. If U_k is a square, then by the same reasoning as in the previous case, A_k and B_k must both be squares, and so Lemma 5 shows that $k = 1$ or $k = 3$. Since $A_3 = A_1(2A_1^2 - 3)$ is not a square for any value A_1, it follows that $k = 1$. This completes the proof of Theorem 1.

4 Remarks and Examples

Example 1. We consider the particular equation

$$X^4 + 2X^2Y^2 - 2Y^4 = 1. \tag{4.1}$$

This example was given in Stroeker's paper [11], and it was shown how the method described therein failed to produce all integer solutions.

Theorem 1 enables one to solve the more general equation

$$X^2 + 2XY^2 - 2Y^4 = 1. \tag{4.2}$$

In this case, $r = -1$, $s = 2$, $D = r^2 + s = 3$, $\epsilon_D = 2 + \sqrt{3}$, $U_1 = 1$, and so by Theorem 1, the only squares in the sequence $\{U_k\}$ are $U_1 = 1$ and $U_2 = 4$. Therefore, the only positive integer solutions (X, Y) to (4.2) are $\{(1,1), (3,2)\}$, and hence the only positive integer solution (X, Y) to (4.1) is $(1,1)$.

Example 2. In this example, consider the quartic equation

$$X^2 - 2rXY^2 - sY^4 = 1, \qquad (4.3)$$

where r and s are given as follows:

     $r = 1234567891234567891234567891234567891234567891234567891234 5$
     $6789123456789,$

     $s = 22636443969777031534601700421544940787344513741443275045163$
     $96123560459045172097933402233305651793692304469768012697740$
     $01264654883170241316205201587738460374694515537866338751697$
     $47721930570469642954844280606351400575451276412617414086433$
     $74963898542088957743482661219542646152262038545898685717649$
     $68720252089370408670503077118521291270834918124326660 5767.$

In this case, the discriminant $D = r^2 + s$ is given by

     $D = 22636443969777031534601700421544940787344513741443 27504516$
     $39612356045904517209793340223330565179369230446976 80126977$
     $40012646548831702413162052015877384603746945155378 66338751$
     $69747721930570469642954844280606351552991239083149 40287514$
     $42168650177661905893558588473882737682352552831565 41139050$
     $60432806779871913516005625077258005493967428864803 86401679$
     $6288.$

The minimal solution $\epsilon_D = T + U\sqrt{D}$ of $X^2 - DY^2 = 1$ has

     $T = 22163240162908610895966847308564651014744778299084 52902263$
     $21324518008264061127175002067476254221180493032160 56909611$
     $71591495157929619160276073492480133491993317513169 85850160$
     $60899261575836456385898348840126353293379238885549 74335871$
     $16053666058789858235207579819910506665517418329031 91640251$
     $6501150983810394065085355962838817943 3778367$

and

     $U = 46583181997946016356103241149178527659808457814534 025660634$
     $64031689606394749016576136000056394396733973832943 289987057$
     $8769409981217335597765471997152395407099.$

In order to completely solve (4.3) using Theorem 1, we need to determine first if any of U_1, U_2 or U_4 are squares, and a simple calculation on Maple shows that they are not. Next, we need to write U in the form $U = lv^2$ for some squarefree integer l, and check whether U_l is a square. Unfortunately, the problem of writing U in the prescribed form is polynomially equivalent to the problem of factoring U, which is (currently) a difficult computational problem. We leave it as a challenge for the reader to determine if (4.3) has any integer solutions with $Y \neq 0$.

References

[1] M. A. Bennett and P. G. Walsh, *The Diophantine equation $b^2 X^4 - dY^2 = 1$*, Proc. Amer. Math. Soc. **127** (1999), 3481–3491.

[2] Y. Bilu and G. Hanrot, *Solving Thue equations of high degree*, J. Number Theory **60** (1996), 373–392.

[3] J. H. E. Cohn, *The Diophantine equation $x^4 - Dy^2 = 1$ II*, Acta Arith. **78** (1997), 401–403.

[4] F. Halter-Koch, G. Lettl, A. Pethö, and R. F. Tichy, *Thue equations associated with Ankeny-Brauer-Chowla number fields*, J. London Math. Soc. **60** (1999), 1–20.

[5] C. Heuberger, *On a conjecture of E. Thomas concerning parametrized Thue equations*, Acta Arith. **98** (2001), 375–394.

[6] W. Ljunggren, *Einige Eigenschaften der Einheiten reeller quadratischer und rein biquadratischer Zahl-Körper usw.*, Oslo Vid.-Akad. Skrifter (1936).

[7] _____, *Über die Gleichung $x^4 - Dy^2 = 1$*, Arch. Math. Naturv. **45** (1942), No. 5.

[8] _____, *Ein Satz über die Diophantische Gleichung $Ax^2 - By^4 = C$ $(C = 1, 2, 4)$*, Tolfte Skand. Matemheikerkongressen, Lund, 1954, pp. 188–194.

[9] F. Luca and P. G. Walsh, *Squares in Lucas sequences with Diophantine applications*, Acta Arith. **100** (2001), 47–62.

[10] T. Nagell, *On a special class of Diophantine equations of the second degree*, Ark. Math. **3** (1954), 51–65.

[11] R. J. Strocker, *On the Diophantine equations of type $x^4 - 2ax^2y^2 - by^4 = 1$*, Number Theory and Applications (R. A. Mollin, ed.), NATO-ASI Series C, vol. 265, Kluwer Publishers, Boston, 1989, pp. 547–554.

[12] E. Thomas, *Complete solutions to a family of cubic Diophantine equations*, J. Number Theory **34** (1990), 235–250.

[13] A. Thue, *Ein Fundamentaltheorem zur Bestimmung von Annäherungswerten aller Wurzeln gewisser ganzer Funktionen*, J. Reine Angew. Math. **138** (1910), 96–108.

[14] N. Tzanakis and B. M. de Weger, *On the practical solution of the Thue equation*, J. Number Theory **31** (1989), 99–132.

[15] P. G. Walsh, *A note on Ljunggren's theorem about the Diophantine equation* $aX^2 - bY^4 = 1$, C. R. Math. Acad. Sci. Soc. R. Canada **20** (1998), 113–119.

[16] ———, *The Diophantine equation* $X^2 - db^2Y^4 = 1$, Acta Arith. **57** (1999), 179–188.

[17] ———, *A note on a theorem of Ljunggren and the Diophantine equations* $x^2 - kxy^2 + y^4 = 1, 4$, Arch. Math. **72** (1999), 1–7.

[18] ———, *Diophantine equations of the form* $aX^4 - bY^2 = \pm 1$, Proceedings of the International Conference held in Graz, August 30 to September 5, 1998 (F. Halter-Koch and R. F. Tichy, eds.), Walter de Gruyter, 2000, pp. 531–554.

Constructing Hyperelliptic Curves Using Complex Multiplication

Annegret Weng[1]

1 Introduction

To ensure the security of a cryptosystem based on the discrete logarithm problem in a finite abelian group G it is necessary that the group order $|G|$ contains a large prime factor. If G is the group of $\mathbb{F}_q$-rational points on an abelian variety, the efficient determination of the group order is a non-trivial problem.

One possible solution to this problem is the construction of curves for which we know the group order in advance. Atkin [1] suggested the construction of elliptic curves over finite fields with prescribed order using the theory of complex multiplication on elliptic curves. The complex multiplication method (CM–method) for elliptic curves can be efficiently implemented and is very attractive for practical applications (see [4]).

Koblitz [6] suggested the use of hyperelliptic curves for cryptography to provide a larger class of curves. We can consider the discrete logarithm problem in the group of $\mathbb{F}_q$-rational elements of the Jacobian. At present there exists no algorithm for counting points on a Jacobian defined over a finite prime field that has a high enough group order (say 2^{160}) to be of interest in cryptography. Gaudry and Harley were able to count the number of points on the Jacobian of a curve of genus two defined over $\mathbb{F}_p$ with $p = 10^{19} + 51$ and of group order approximately 10^{38}. The current record in the genus three case is due to Stein and Teske and involves a group order of size 10^{29} [12].

In this article we discuss a generalization of the CM–method. We construct hyperelliptic curves with given group order.

Spallek [11] proposed an algorithm in the case of genus two. Her work was later improved by Wang [14] and Weber [15] who replaced the computation of Groebner bases with an efficient algorithm by Mestre [8]. Based on Spallek's work, Van Wamelen [13] constructed all curves of genus two

---

[1]Supported by the NRW Forschungsverbund Datensicherheit and the DFG (Graduiertenkolleg).

over the rationals having complex multiplication by the maximal order of a CM-field.

Until now there had not been an implementation of the CM–method for genus two curves. All curves that had been constructed had Igusa invariants defined over the rationals.

In this article we give a complete description of an algorithm for low genus curves (see Section 2). We describe the results of our implementation for $g = 2$ (Section 3) and give an example which is relevant for cryptography. The Igusa invariants of our curve lie in a number field of degree four over the rationals.

In the last section (Section 4) we discuss the generalization to curves of higher genus and present an example for $g = 3$. We also suggest directions for future work.

2   The CM–Method for g $\geq$ 2

For background on abelian varieties with complex multiplication see [10] and [7].

We restrict ourselves to hyperelliptic curves defined over a finite prime field $\mathbb{F}_p$, although it might be possible to take a field $\mathbb{F}_q = \mathbb{F}_{p^n}$, where n is small.

We denote the CM-field by K and its maximal real subfield by K_0. We restrict ourselves to CM-fields K whose maximal real subfield has class number one and a monogenic ring of integers, i.e., $\mathcal{O}_{K_0} = \mathbb{Z}[w]$ for some w.

Our algorithm consists of two parts.

The first part is a precomputational step. We suppose that we have already a curve C defined over $\mathbb{F}_p$ and that we know that the Jacobian J_C of C has complex multiplication by the maximal order $\mathcal{O}_K$ of a CM-field K. The isogeny

$$(x, y) \mapsto (x^p, y^p)$$

on the curve induces an endomorphism π on the Jacobian J_C, called the Frobenius endomorphism. It corresponds to some integer w in the field K, i.e., the characteristic polynomial of the Frobenius is the minimal polynomial of an algebraic integer w of degree $2g$ over $\mathbb{Q}$.

By the Hasse-Weil theorem for hyperelliptic curves we have

$$p = w\overline{w}, \quad \#J_C(\mathbb{F}_p) = \prod_{i=1}^{2g}(1 - \xi w_i) \text{ for some root of unity } \xi \in K,$$

where the w_i are the roots of the characteristic polynomial of w.

We assume that K does not contain a cyclotomic field. In this case there are only two (resp. four) possibilities for the group order $\#J_C(\mathbb{F}_p)$, depending on the splitting behavior of p in the absolute extension $K/\mathbb{Q}$.

Therefore, given a CM-field K and a Jacobian with complex multiplication by the maximal order $\mathcal{O}_K$, we only need to solve a relative norm equation with respect to K/K_0. This will give us two (resp. four) possibilities for the group order. By checking with a random $\mathbb{F}_p$-rational divisor D on the Jacobian we decide which group order is the correct one. Since we are only interested in a group order which is almost prime, this test will suffice.

The second part is the main part of the algorithm. It deals with the construction of a hyperelliptic curve over $\mathbb{F}_p$ whose Jacobian has complex multiplication by $\mathcal{O}_K$ for a given CM-field K. This is the interesting, but time consuming step. Note that we perform the second part of the algorithm only if we know that one of the two group orders is prime or almost prime.

Assume that we are given a CM-field K where $[K : \mathbb{Q}] = 2g$ and a prime p which satisfies a relative norm equation with respect to K/K_0, and that there are two possible group orders n_1 and n_2.

The output of the algorithm is either a hyperelliptic curve C over $\mathbb{F}_p$ having group order n_1 or n_2, or the information that the algorithm fails on the given input.

1. Determine a complete set (up to isomorphisms) of period matrices Ω_i of all simple principally polarized abelian varieties having complex multiplication by $\mathcal{O}_K$.

   Recall that the Jacobian of a curve is a principally polarized abelian variety in a natural way. A principally polarized abelian variety A over $\mathbb{C}$ is isomorphic to a lattice

   $$\mathbb{Z}^g + \Omega_i \mathbb{Z}^g,$$

   where Ω_i lies in the Siegel upper half plane $\mathbb{H}_g = \{z \in \mathbb{C}^{g^2}, z^t = z, \operatorname{Im} z > 0\}$.

2. Compute a complete set (up to isomorphism) of period matrices Ω_i of all simple principally polarized abelian varieties over $\mathbb{C}$ having complex multiplication by $\mathcal{O}_K$.

3. Compute (up to a given precision) all even theta characteristics

   $$\theta \begin{bmatrix} \delta \\ \epsilon \end{bmatrix} (\Omega, 0), \quad \delta\epsilon^t \equiv 0 \mod 2 \text{ with } \delta, \epsilon \in \{0,1\}^g,$$

where

$$\theta \begin{bmatrix} \delta \\ \epsilon \end{bmatrix} (\Omega, 0) = \sum_{n \in \mathbb{Z}^g} \exp(\pi i (n + \tfrac{1}{2}\delta)^t \Omega (n + \tfrac{1}{2}\delta) + 2(n + \tfrac{1}{2}\delta)(\tfrac{1}{2}\epsilon)^t).$$

4. Decide whether Ω is the period matrix of the Jacobian of a hyperelliptic curve or not. If not, the algorithm failed and is terminated. Otherwise continue.

5. Compute a curve

$$y^2 = x(x - 1) \prod_{i=1}^{2g-1} (x - \lambda_i)$$

over $\mathbb{C}$, where the Jacobian has period matrix corresponding to Ω. This can be done by use of the values of the theta constants ([15]).

6. At this point there are several possible methods to continue, described in the following table. We can choose the method that is most appropriate for the given case. For genus two curves it turns out that the combination of Methods 1 and 2 is the most efficient way of computing the curve.

Method 1	**Method 2**	**Combining Methods 1 and 2**
Compute the generating absolute invariants	Compute the absolute invariants needed for Mestre's algorithm [8].	Compute $j_1, \ldots, j_{2g-1}$
Reduce $j_1, \ldots, j_{2g-1}$ mod p.	Reduce them mod p.	Reduce them mod p. Express Mestre's invariants through the j-invariants.
Find a curve having these j-invariants by making use of Groebner bases.	Apply Mestre's algorithm.	Apply Mestre's algorithm.

The algorithm fails if there does not exist a hyperelliptic curve whose Jacobian has complex multiplication by $\mathcal{O}_K$. As we will see in Section 4, this is a serious problem.

3 Implementation of the Case g = 2

3.1 Details for $g = 2$

We first give the details in the case $g = 2$. The steps correspond to those of the general algorithm described in the previous section.

1. Over $\mathbb{C}$, up to isomorphism, the Jacobian can be given as a quotient $\mathbb{C}^2/L$ for some lattice
$$L = \mathbb{Z}^2 + \Omega\mathbb{Z}^2,$$

   where $\Omega \in \mathbb{C}^{2\times 2}$ is symmetric and $\mathrm{Im}(\Omega)$ is positive definite. Spallek [11] gives a complete representation system of all principally polarized abelian varieties of dimension two having complex multiplication by the maximal order of a CM-field K whose real subfield has class number one.

2. For the computation of the theta constants $\theta \begin{bmatrix} \delta \\ \epsilon \end{bmatrix}$, one needs an algorithm for listing all vectors $n \in \mathbb{Z}^2$ with

$$\left| \exp\left(-\pi \left(n + \frac{1}{2}\delta\right) \mathrm{Im}(\Omega) \left(n + \frac{1}{2}\delta\right)^t\right) \right| \le C.$$

   The constant $C \in \mathbb{R}$ depends on the precision needed to compute the invariants.

3. Every principally polarized abelian varieties *of dimension two* is the Jacobian of a hyperelliptic curve. Therefore we do not have to check whether a period matrix Ω corresponds to the Jacobian of a hyperelliptic curve.

   We can shorten our calculation in this case by computing Igusa's j-invariants j_1, j_2 and j_3 (see [5]) directly from the theta constants. The formulae are given in Spallek ([14]).

   From the j-invariants we compute the class polynomials

$$H_k(X) = \prod_{i=1}^{s} (X - j_k^{(i)}), k = 1, \dots, 3.$$

   In contrast to elliptic curves with complex multiplication, where the j-invariants are always integers in a number field $\mathcal{O}_M$, the j-invariants of genus two curves are in general not integers. However, they lie in $\mathcal{O}_M(D^{-1})$ for some rational integer D.

The denominator D is not too large if the discriminant of the CM-field is small. In most cases it is enough to apply the continued fraction algorithm to the second highest coefficient of the polynomial. Once we have found the denominator, we get a polynomial.

5. We factor the class polynomials modulo p to get j-invariants j_1, j_2 and j_3 in $\mathbb{F}_p$. It is possible to express Mestre's invariants by the j-invariants (see [8], p. 317). We apply Mestre's algorithm to get a possible hyperelliptic curve. Finally we test whether this curve has the right group order.

3.2   Example

Consider the CM-field

$$K = \mathbb{Q}(\alpha) = \mathbb{Q}\left(i\sqrt{7 + 2\frac{-1 + \sqrt{33}}{2}} \right).$$

It has class number two and two polarizations.
   The class polynomials are given by

$$\begin{aligned}
\mathbf{H_1}(X) = {}& w^4 + 125426939904w^3 + 206483140868310761472w^2 \\
& - 3777735852531193527889035264w \\
& + 4880287864430944225048694259449856,
\end{aligned}$$

$$\begin{aligned}
\mathbf{H_2}(X) = {}& w^4 + 660000960w^3 + 106952268616185600w^2 \\
& + 27255466149375338496000w \\
& + 837300145473346170101760000,
\end{aligned}$$

$$\begin{aligned}
\mathbf{H_3}(X) = {}& w^4 + 189766368w^3 + 7505309625975360w^2 \\
& + 434631556065843035136w - 45329807190376508829696.
\end{aligned}$$

This is one of the rare cases in which the Igusa-invariants lie in $\mathcal{O}_K$. Take the prime

$$p = 590001860371546761181989109202421.$$

It satisfies a relative norm equation $p = w\overline{w}$, where

$$\begin{aligned}
w = {}& (-76811578202943299 + 107438053(\frac{-1 + \sqrt{33}}{2})) \\
& + (521777257258 + 990510120225(\frac{-1 + \sqrt{33}}{2}))\alpha.
\end{aligned}$$

We find the curve

$$C : y^2 = t^5 + 2251831303237605767657618195346350t^4$$
$$+ 139598757092657807798091055038175 5t^3$$
$$+ 3449986084090239803090552184527208t^2$$
$$+ 1071704234696277993751073168935 95t$$
$$+ 2770857204236068378720416405312357.$$

The group order $\#J_C(\mathbb{F}_p)$ is given by

$$3481021952418861785390626980854276431541396337102367100426$$
$$3947730632 = 8 \cdot q_{prime},$$

where q_{prime} is a prime with 67 decimal digits.

The complete computation of this example (including the class polynomials) took 82 seconds on a Pentium III with 450 Mhz.

3.3 Complexity of the Computation of the Class Polynomial

The table below will give an idea of the complexity of the computation of the class polynomials $H_k(X)$. It shows a small selection of CM-fields whose class polynomial we were able to compute.

The first three columns of the table describe the CM-field K defined as

$$\mathbb{Q}\left(i\sqrt{a + b\sqrt{d}}\right), \qquad\qquad d = \frac{D}{4}, \text{ if } D \equiv 0 \mod 4.$$

$$\mathbb{Q}\left(i\sqrt{a + b\frac{-1 + \sqrt{d}}{2}}\right), \qquad d = D, \text{ if } D \equiv 1 \mod 4.$$

The fourth column gives the class number h_K of K and the fifth column the number of possible polarizations. The degree of the class polynomial is equal to $h_k \cdot \#\text{Pol}$.

Recall that the monic class polynomial has rational coefficients. The sixth column gives the number of decimal digits of the denominator of the second highest coefficient of the class polynomial. The seventh column gives the precision which was necessary for our computations to get an exact result. The eighth column gives the time (in seconds) needed for the computation.

D	a	b	h_K	# Pol.	Denom. dec. digits	Prec.	Time (in sec.)
5	3	1	1	1	1	20	1
5	35	8	5	2	29	400	315
8	2	1	1	1	1	20	1
8	6	1	4	2	17	300	111
8	13	6	3	2	4	100	7
8	15	4	5	2	4	300	193
12	9	4	2	2	1	20	1
12	11	4	4	2	4	100	15
13	6	2	2	2	16	150	6
24	17	6	6	2	16	600	1973
29	12	3	2	2	16	200	25
33	95	28	4	2	40	400	38
41	10	2	4	2	8	200	238
53	5	1	4	2	26	400	1497
76	279	64	4	2	8	200	880
97	8	1	3	2	11	150	30
137	10	1	4	2	33	400	2498
269	9	1	1	2	17	100	2

The time depends on the precision needed to find the right denominator and on the degree of the class polynomials. Another factor influencing the time is the size of the first successive minimum of the imaginary part of the period matrix. The smaller the successive minimum the longer the computation of the theta constants to a given precision C will take.

We had to choose CM-fields whose class numbers were small enough; the tables in [9] were very helpful for that purpose. For more information on the case $g = 2$ see [16].

Remark 1. Our method works also for extension fields $\mathbb{F}_q$ of small degree. Let C be a hyperelliptic curve whose j-invariants lie in $\mathbb{F}_q$, where $q = p^n$ but not in a smaller subfield. Suppose that the Jacobian of C has complex multiplication by $\mathcal{O}_K$ and that we are in the case where $\mathcal{O}_K$ has two polarizations. It follows that $n \leq 2h_K$.

4   The Cases g = 3 and g = 4

4.1   Details for $g = 3, 4$

We are mainly concerned with the case $g = 3$. The cryptosystems based on curves of genus $g \geq 4$ appear to be slightly weaker than those for $g \leq 3$ [3]. We discuss the steps of the algorithm in detail.

1. In the first step we generalize Spallek's work and determine a complete set of period matrices of simple principal polarized abelian varieties of dimension three.

   This can also easily be done for dimension four. However, it will be more difficult to check whether the abelian varities are simple, since we have to consider the Galois closure and in dimension four there are 38 possibilities for the Galois closure of K [2].

2. The computation of the even theta constants is analogous to $g = 2$.

3. For $g = 3$ a period matrix corresponds to a hyperelliptic curve if and only if exactly one of the 36 even theta constants $\begin{pmatrix} \delta \\ \epsilon \end{pmatrix}$, $\delta \epsilon^t \equiv 0$ mod 2, vanishes at Ω.

   The case $g = 4$ is more complicated. A certain subset of 10 even theta constants has to vanish in order to have a hyperelliptic Jacobian.

   It is a non-trivial task to find a hyperelliptic Jacobian. Principal polarized abelian varieties of dimension three are Jacobians of curves, but not necessary Jacobians of hyperelliptic curves. In fact, the moduli space of hyperelliptic curves of genus three has codimension one in the moduli space of curves of genus three. This means that the proportion of hyperelliptic curves is negligible.

4. From the theta constants we are able to compute the hyperelliptic curve $y^2 = f(x)$ over $\mathbb{C}$ [15].

5. The algorithm of Mestre has one limitation. It only works if the curve does not have a non-trivial automorphism, i.e., an automorphism other than the hyperelliptic involution.

4.2 Example

Consider the CM-field $K = \mathbb{Q}(i)K_0$ where K_0 is generated by $w^3 - w^2 - 2w + 1$. The prime

$$p = 123456776543211236173$$

satisfies a relative norm equation $p = w\overline{w}$ with respect to K/K_0. One of the four possible group orders is given by

$$n = 18816758018643798911143395355645388052746925947685906882211848 = 8l,$$

where l is a prime with 60 decimal digits. By our construction we find the curve

$$C : y^2 = x^7 + 7x^5 + 14x^3 + 7x.$$

The group order $\#J_C(\mathbb{F}_p)$ is n.

Here we used Buchberger's algorithm in the last step. Note that we could not apply Mestre's algorithm since the curve has non-trivial automorphisms.

4.3 Problems and Extensions

Non-trivial automorphisms on the curve imply that the CM-field is either Galois or has a Galois closure of degree 12 over $\mathbb{Q}$. We could not find an example of a CM-field K whose Galois closure has degree 24 or 48. It could be the case that there does not exist such a CM-field whose set of simple principally polarized abelian varieties contains the Jacobian variety of a hyperelliptic curve and whose class number is small enough to make computations possible. This is a possibility for future research.

As we have seen in the case $g = 2$, the absolute invariants are not integers. The denominator grows with the discriminant of the CM-field. Since the CM-fields of degree six have a larger discriminant, it is more difficult to find the denominator.

A natural question is whether this method can be generalized to other classes of abelian varieties, or at least to Jacobians of arbitrary curves.

For a general curve we have no formulae to get a model of the curve from its theta constants. Since it is no longer hyperelliptic, we can no longer use the theory of binary forms. Furthermore, we need invariants of a nice form, i.e., almost integers. Thus the CM–method does not seem to work for a more general class of curves.

Acknowledgements. I thank my supervisor Professor G. Frey for his support, help and encouragement. This work is part of my PhD-thesis.

I do not want to forget all the people who helped me by pointing out interesting questions and reading the preliminary versions: S. Galbraith, M. Müller, K. Nguyen, H.-G. Rück, A. Stein, M. Wessler.

I am grateful to the NRW Verbundsprojekt Datensicherheit and the DFG (Graduiertenkolleg) who gave me the financial support making this work possible. I thank the Center for Applied Cryptographic Research in Waterloo for their hospitality.

For my implementations I used the C library PARI (`ftp://megrez.math.u-bordeaux.fr/pub/pari/`) and the C++ library NTL (`http://www.shoup.net/ntl/`).

References

[1]   A. O. L. Atkin, *The number of points on an elliptic curve modulo a prime*, Preprint, 1991.

[2]   B. Dodson, *The structure of Galois groups of CM-fields*, Trans. American Math. Soc. **283** (1984), 1–32.

[3]   P. Gaudry, *An algorithm for solving the discrete log problem on hyperelliptic curves*, Advances in cryptology—EUROCRYPT 2000 (Bruges), Springer, Berlin, 2000, pp. 19–34.

[4]   IEEE P1363, *Standard specifications for public key cryptography*, 2000, http://grouper.ieee.org/groups/1363/.

[5]   J. I. Igusa, *Arithmetic variety of moduli of genus two*, Ann. Math. **72** (1960), 612–649.

[6]   N. Koblitz, *Hyperelliptic cryptosystems*, J. Cryptology **1** (1989), 139–150.

[7]   S. Lang, *Complex multiplication*, Springer-Verlag, New York, 1983.

[8]   J.-F. Mestre, *Construction de courbes de genre 2 à partir de leurs modules*, Effective methods in algebraic geometry (Castiglioncello, 1990), Birkhäuser Boston, Boston, MA, 1991, pp. 313–334.

[9]   R. Okazaki, *On evaluation of L-functions over real quadratic fields*, J. Math. Kyoto Univ. **31** (1991), 1125–1153.

[10]   G. Shimura, *Abelian varieties with complex multiplication and modular functions*, Princeton Univ. Press, Princeton, 1998, revised edition.

[11] A.-M. Spallek, *Kurven vom Geschlecht 2 und ihre Anwendung in Public-Key-Kryptosystemen*, Ph.D. thesis, Institut für Experimentelle Mathematik, Universität GH Essen, 1994.

[12] A. Stein and E. Teske, *The parallelized Pollard's kangaroo method in real quadratic function fields*, Math. Comp. **71** (2002), 793–814.

[13] P. Van Wamelen, *Examples of genus two CM curves defined over the rationals*, Math. Comp. **68** (1999), 307–320.

[14] X. Wang, *2-dimensional simple factors of $J_0(N)$*, Manuscr. Math. **87** (1995), 179–197.

[15] H.-J. Weber, *Hyperelliptic simple factors of $J_0(N)$ with dimension at least 3*, Experiment. Math. **6** (1997), 273–287.

[16] A. Weng, *Constructing hyperelliptic curves of genus 2 suitable for cryptography*, to appear in Math. Comp.

Solving the Pell Equation

H. C. Williams

1 Introduction

Let D be a positive nonsquare integer. The misnamed Pell equation is an expression of the form

$$T^2 - DU^2 = 1, \tag{1.1}$$

where we constrain the values of T and U to be integers. For example, the Pell equation

$$T^2 - 7U^2 = 1 \tag{1.2}$$

has the solutions $(\pm 1, 0), (8, 3), (-8, 3), (8, -3), (-8, -3), (127, 48)$, etc. In fact, Lagrange, completing earlier work of Euler, had established by 1768 (see Weil [95, pp. 314–315]) that (1.1) will always have a non-trivial solution (a solution with $U \neq 0$). Furthermore, there exists an infinitude of such solutions (T, U) given by

$$T + \sqrt{D}U = \pm \left(t + \sqrt{D}u \right)^n, \tag{1.3}$$

where n is any integer and (t, u) is the *fundamental solution* (with $t + u\sqrt{D} > 1$) of (1.1), and no other solutions of (1.1) exist except those given by (1.3). The fundamental solution of (1.1) is the least positive integral solution; for example, the fundamental solution of (1.2) is $(8, 3)$. In view of these observations, we will say that the problem of solving the Pell equation (1.1) is that of determining t and u.

This very simple Diophantine equation has been the object of study by mathematicians for over 2000 years. It is named after John Pell because of an error in attribution made by Euler [20] to a method of solving it in "Wallis's works." This was most likely the result of a cursory reading by him of Wallis's *Algebra*. As noted by many authorities, most recently Weil [95, p. 174], Pell's name occurs frequently in *Algebra*, but never in connection with the Pell equation. In fact, it seems most likely that the method referred to by Euler for solving (1.1) is a technique that Wallis credits to Brouncker. In spite of ample evidence attesting to Euler's carelessness (see also footnote 3 on p. 59 of [97]), there has been an effort made to connect Pell with (1.1) (see, for example, Scott [73, p. 208]). This seems to have begun with a misunderstanding of a remark of Hankel [27, p. 203], who actually stated that "Pell has done it no other service than to set it forth again in a much read work." The "much read work" was the English translation [70] of the *Teutschen Algebra* of Rahn. However, a careful examination of this work by Konen [42, pp. 33–34, footnote 1], Wertheim [96] and Eneström [19] did not result in the discovery of any mention of this equation. There can be little doubt that much of this book, particularly pages 100–192, were due to Pell (see Scriba [74]), yet the only mention of anything resembling the Pell equation in it is the equation.

$$x = 12y^2 - z^2 \qquad (1.4)$$

on p. 143. This caused Whitford [97, p. 2] to believe that Pell had some acquaintance with (1.1) and seems also to have served as the reason that Pell's biography [3, pp. 1973–1975] suggests that this might have been the case. A thorough inspection of the context in which (1.4) arises in [70] reveals that it is to be used to find x after values for y and z have been selected. This can scarcely be regarded as the Pell equation. Thus, there is no evidence whatsoever linking Pell with (1.1). Nevertheless, as Weil [95, p. 174] asserts, the "traditional designation [of (1.1)] as 'Pell's equation' is unambiguous and convenient." Consequently, it is the term used here for (1.1), even though it is both historically wrong and unjust to those early individuals who did make important contributions to its study.

Much of the very early work on (1.1) was episodic, but since the rediscovery of the equation by Fermat in 1657, research involving it has been continual, and a considerable literature has accumulated, including two books: *Geschichte der Gleichung $t^2 - Du^2 = 1$* by H. Konen [42] and *The Pell Equation* by E. E. Whitford [97]. Indeed, research on the Pell equation continues to be very active today; at least one hundred articles dealing with it in various contexts have appeared within the last decade. This is simply because it is fundamental to the problem of solving the general second degree Diophantine equation in two unknowns:

$$g(x,y) = ax^2 + bxy + cy^2 + dx + ey + f = 0, \qquad (1.5)$$

which means that it finds many applications in number theory and elsewhere. Furthermore, the problem of solving (1.1) is very much connected to the problem of solving the discrete logarithm problem in a real quadratic field (see Jacobson [33], [34]), a problem of much interest to cryptographers [36].

The purpose of this paper is to provide a survey of results concerning the problem of solving (1.1) as we have characterized that problem above. As such, it will be largely influenced by the history of this problem.

Before beginning this, however, it will be convenient to make one further simplifying assumption. We will assume that D has no square integer factor, i.e., D is square-free. We can do this because Lehmer [48], [49] has shown that if p is any prime and we define

$$\psi_D\left(p^k\right) = \begin{cases} 2^k & \text{when } p = 2 \\ p^k & \text{when } p \text{ is odd and } p|D \\ p^{k-1}\left(p - (D/p)\right)/2 & \text{when } p \text{ is odd and } p \nmid D, \end{cases}$$

then $p^k|U_s$, where $s = \psi_D\left(p^k\right)$,

$$T_n + \sqrt{D}U_n = \left(t + u\sqrt{D}\right)^n \qquad (n \in \mathbb{Z})$$

and (D/p) is the Legendre symbol. If

$$m = \prod_{i=1}^{k} p_i^{\alpha_i},$$

where p_i $(i = 1, 2, \ldots, k)$ are distinct primes and we define

$$\psi_D(m) = LCM\left[\psi_D\left(p_i^{\alpha_i}\right), i = 1, 2, \ldots, k\right],$$

then $m|U_s$, where $s = \psi_D(m)$. Furthermore, if $\omega_D(m)$ is the least positive value of s such that $m|U_s$, then

$$\omega_D(m)|\psi_D(m). \tag{1.6}$$

Now let $D = EF^2$, where E has no square factor. If (t_1, u_1) is the fundamental solution of $T^2 - EU^2 = 1$ and

$$T_n + \sqrt{E}U_n = \left(t_1 + \sqrt{E}u_1\right)^n,$$

then the fundamental solution (t, u) of (1.1) is given by

$$t = T_n, u = U_n/F,$$

where $n = \omega_E(F)$. Since by (1.5) all the possibilities for $\omega_E(F)$ must be divisors of an easily computed integer, it is not difficult to find (t, u) once (t_1, u_1) is known, provided we have the prime factorization of F.

2 Contributions of the Greeks and Indians

There is no direct evidence extant that the ancient Greeks knew how to solve the Pell equation. In fact, there is little mention of the equation at all in their surviving work. Theon of Smyrna (c. 130 AD) knew how to find successive solutions to

$$T^2 - 2U^2 = \pm 1$$

by putting $T_1 = U_1 = 1$ and finding solution (T_{n+1}, U_{n+1}) from (T_n, U_n) by

$$T_{n+1} = 2T_n + U_n, \qquad U_{n+1} = T_n + U_n.$$

Tannery [87] showed that this idea could be extended to any equation of the form

$$T^2 - DU^2 = r, \tag{2.1}$$

but it is unclear whether the Greeks knew this, and even if they did, it would provide little evidence that they knew how to solve (1.1) for t and u.
Diophantus (c. 250 AD) in his *Arithmetica* [28] did solve

$$T^2 - 26U^2 = 1$$

and

$$T^2 - 30U^2 = 1$$

for integral values of T and U (see §§9 and 11 in Book V of [28]), but the method given would, in general, only find rational values of T and U for other values of D. Also, if r is a square, he showed how to find a second rational solution of (2.1) from a given rational solution (see the lemma in §14 of Book VI). This concentration on techniques that produce rational solutions suggests the possibility that the Greeks had not developed an interest in computing integral solutions of (1.1), and in spite of Tannery's [86, p. 101ff] suggestion that Diophantus might have considered such problems in the lost seven books of the *Arithmetica*, we might feel justified in thinking that the Greeks made little contribution to the study of (1.1). However, there is an important piece of evidence that we have not yet considered: the Cattle Problem of Archimedes.

This problem (see Vardi [91] for an English translation of the problem, further discussion and references) consists of two parts. The first of these asks us to determine 8 unknowns (numbers of bulls and cows in each of four herds of cattle), given seven relatively simple linear relationships among these numbers. From this it is not difficult to deduce that the solution of this part of the problem can be given by

$$(10366482, 7460514, 4149387, 7358060, 7206360, 4893246, 543913, 3515820)n,$$

where n is any positive integer. The second, far more difficult part constrains two of the numbers to add to a perfect square and two others to add to a triangular number. Symbolically, this means that $n = 4456749U^2$ where

$$T^2 - 41028642327842U^2 = 1 \qquad (2.2)$$

and T, U are integers.[1]

This problem was attributed to Archimedes (c. 250 BC) in ancient times, and Krumbiegel [44] has argued that, while the text of the problem which we currently possess did not likely originate with Archimedes, the problem itself is likely due to him. However, Fraser [23, p. 402, pp. 407–409] has provided good reasons for accepting Archimedes, not only as the source of the problem, but also as the likely composer of the poetic form in which we now have it. Certainly, Weil [95, p. 17] is correct when he asserts that "there is indeed every reason to accept the attribution to Archimedes, and none for putting it in doubt."

There has been some dispute about the exact wording of the problem (see, for example, Schreiber [72] and Waterhouse [93]) and some concerning the first condition of the second part (see, however, Dijksterhuis [17, pp. 399–400, footnote 3]), but by and large the bulk of expert opinion on the problem is that it reduces to the form stated above. It was first solved by Amthor in 1880, but he was unable to write down the numbers because the value of u for the fundamental solution of (2.2) is approximately 1.86×10^{103265}. The total number of cattle was first computed (using a computer) in 1965 [101] and published later by Nelson [63] in 12 pages of fine print. Vardi has given a very elegant representation of the total number of cattle as

$$\left\lceil \tfrac{25194541}{184119142} \left(10993198673282973497986623282143354390108804 9 \right. \right.$$
$$\left. \left. + 50549485234315033074477819735540408986340\sqrt{4729494} \right)^{4658} \right\rceil.$$

This number by almost any measure is very large $\left(\approx 7.76 \times 10^{206544} \right)$, but compared to some of the values for t and u which we will find later, it will appear to be rather small.

While it is extremely unlikely that Archimedes was able to solve the cattle problem, we are left with the puzzle of why it was posed and what Archimedes knew when he posed it. The lightly satirical tone of the text of the problem prompted Hultsch to suggest that the problem was Archimedes' response to Apollonius' improvement to his measurement of

---

[1]This is a case of (1.1) in which D is not squarefree. Amthor [4] noted that

$$41028642327842 = 4729494(9314)^2,$$

a fact he was able to exploit in deriving his solution of (2.2).

the circle and his work on naming large numbers. On the other hand, Knorr has speculated that Eratosthenes composed the first part of the problem and that the second part is Archimedes' response. Whatever the reason, the truly interesting problem is determining the state of Archimedes' knowledge of how to solve the Pell equation.

In his work on the measurement of the circle, he gives without any explanation the inequality

$$265/153 < \sqrt{3} < 1351/780.$$

This suggests that he knew of some method of finding good rational approximations to the square roots of integers, but as there are many possible methods he might have used (see Knorr [41, pp. 136–137]) to do this, it is unclear as to how much he might have known concerning discovering integer solutions of (1.1). Nevertheless, as knowledge of how to solve problems of this kind is important for solving the Pell equation, it is certainly reasonable to infer that Archimedes might have been interested in problems like solving (1.1) and likely had devised some technique for accomplishing this for small values of D, at least. Possibly these investigations prompted him to believe that the Pell equation is always solvable, but that when D is large it is a difficult problem.

Knorr [41] is certainly correct when he states that "Greek mathematics was founded on a strong tradition of practical arithmetic competence." Further, Fowler [22] has developed a convincing reconstruction of some of the arithmetic techniques that might have been known to the Greeks. In particular, he has provided strong evidence that continued fractions (see §3) were fundamental to the way that ratios were understood to them. Thus, it is certainly possible that the Greeks possessed much more information about the Pell equation than that suggested by the pitifully few fragments that have survived. Unfortunately, we simply cannot be sure of what it was they did know.

The situation is much different when we consider the achievements of the Indian mathematicians of the Middle Ages. As early as the fifth century, Aryabhata I had developed a method for solving the linear Diophantine equation

$$ax - by = c \tag{2.3}$$

for integers x, y, given integers a, b, c. This technique was subsequently refined and extended by later workers until the Indian mathematicians had essentially produced what is known today as the extended Euclidean algorithm for solving (2.3). It is often assumed by number theorists that the Greeks had found a method of solving (2.3); this may well be true, but very little direct evidence has come down to us to justify this assumption.

In 628 AD Brahmagupta discovered that if

$$A^2 - DB^2 = Q \qquad (2.4)$$

and

$$P^2 - DR^2 = S, \qquad (2.5)$$

then

$$(AP + DBR)^2 - D(AR + BP)^2 = QS. \qquad (2.6)$$

He made use of this observation to develop an *ad hoc* method of solving (1.1) for integers. However, the crowning achievement of Indian mathematics with respect to the Pell equation was the development of the cyclic method for solving it. The technique, described by Bhaskara II in 1150 AD, and its history are well described by Selenius [75], [76], and the interested reader should consult this work for further details and references. We will only sketch, with additional information, one variant (there were several) of the algorithm here.

We will assume that $Q, A, B \in \mathbb{Z}$ and that $(A, B) = 1$ in (2.4); this means that $(B, Q) = 1$. As the technique for solving (2.3) was known, the step of finding an integer P such that $Q | BP + A$ could be easily achieved. It follows that since $(B, Q) = 1$, we must have $Q | P^2 - D$ and $Q | AP + DB$. By putting $R = 1$ in (2.5) we see from (2.6) that

$$\left(\frac{AP + DB}{Q} \right)^2 - D \left(\frac{A + BP}{Q} \right)^2 = \frac{P^2 - D}{Q}. \qquad (2.7)$$

From this simple observation we can develop the cyclic method for solving the Pell equation.

Given integers n, A_{n-1}, B_{n-1}, Q_n where $(A_{n-1}, B_{n-1}) = 1$ such that

$$\left| A_{n-1}^2 - DB_{n-1}^2 \right| = Q_n,$$

find an integer P_{n+1} such that $Q_n | P_{n+1} B_{n-1} + A_{n-1}$ and $\left| P_{n+1}^2 - D \right|$ is minimal. Put $Q_{n+1} = \left| P_{n+1}^2 - D \right| / Q_n$,

$$A_n = (A_{n-1} P_{n+1} + DB_{n-1}) / Q_n, \, B_n = (B_{n-1} P_{n+1} + A_{n-1}) / Q_n. \qquad (2.8)$$

By (2.7) we get

$$\left| A_n^2 - DB_n^2 \right| = Q_{n+1}, \qquad (2.9)$$

and $(A_n, B_n) = 1$. The latter result follows easily by observing that $|A_n B_{n-1} - B_n A_{n-1}| = 1$. The method terminates when, for some n, $Q_{n+1} = 1, 2, 4$ because Brahmagupta had previously developed methods of solving (1.1) for integers from an integer solution of

$$T^2 - DU^2 = -1, \pm 2, \pm 4.$$

Concerning this technique Hankel [27, p. 202], stated, "It is beyond all praise; it is certainly the finest thing that was achieved in the theory of numbers before Lagrange." Unfortunately, the Indians did not provide a proof that the cyclic method would always work. They were content, it seems, in the empirical knowledge that it always seemed to do so. They used it to solve (1.1) for $D = 61, 67, 97, 103$.

A number of misconceptions continue to circulate concerning the cyclic method. One of these is that it was rediscovered by Lagrange. This, as Selenius has vigorously pointed out, is not strictly speaking the case, but Lagrange did make use of a method similar to it. Often the algorithm is attributed to Bhaskara II, but as mentioned by Shankar Shukla [84, p. 1, p. 20], Bhaskara made no claim to being the originator of the method, and as Jayadeva, who worked in the 10th century or earlier, had discovered a variant of the technique, it seems that it must have been developed much earlier than the time of Bhaskara. Finally, there is the belief, apparently due to Tannery [87], that the cyclic method derives from Greek influences. There seems, apart from possible wishful thinking on the part of Tannery, to be little solid evidence in support of this. The simple fact is that, as mentioned earlier, we don't really know what the Greeks knew about the Pell equation. What we do know, however, is that the Indian methods display a history of steady development and refinement up to and including the discovery of the cyclic method, and this very strongly suggests that Hankel's [27, pp. 203–204] position that the Indians evolved the technique by themselves is the correct one.

3 Continued Fractions

The story of the Pell equation resumes with the challenge [21, pp. 333–335], issued in 1657 to Frénicle in particular and mathematicians in general by Fermat. Fermat had most likely, through his research, come to recognize the fundamental nature of the Pell equation. An English translation of his general challenge can be found in [28, pp. 285–286]. It asks for a proof of the following statement:

> Given any [positive] number [D] whatever that is not a square, there are also given an infinite number of squares such that, if the square is multiplied into the given number and unity is added to the product, the result is a square.

It next requests a general rule by which solutions of this problem could be determined, and as examples asks for solutions when $D = 109, 149, 433$.

The story of how the second part of this challenge was answered by Brouncker and Wallis has been very well told by Weil [95] and Mahoney [57] and needs no elaboration here. Suffice it to say that Brouncker developed a technique for solving (1.1) in integers, but neither he nor Wallis nor Frénicle was able to provide a proof that the Pell equation could always be solved (non-trivially) for any positive non-square value of D. Fermat [21, p. 433], took notice of this and stated that he had such a proof "by means of *descente* duly and appropriately applied." Unfortunately, Fermat provided no further information concerning his proof than this. Hofmann [29] and, with greater success, Weil [94], [95, §XIII] have attempted to reconstruct what Fermat's method might have been. While we may never really know what this was it is nevertheless very likely that Fermat did have a proof. The fact that he selected 109, 149, 433 for values of D as challenge examples is particularly suggestive because the corresponding Pell equations have large values of t and u.

The method of Brouncker was modified and extended by Euler, who realized that continued fractions could be used to provide an efficient algorithm for solving the Pell equation. However, even though he had devised all of the important tools, he just fell short of proving that his method would work for any non-square D. As mentioned earlier, the development of such a technique was first done by Lagrange in a rather clumsy work, which he later improved. For further information on this particularly interesting part of mathematical history, the reader is referred to [95].

While the technique of the Indians and those of Brouncker, Euler and Lagrange for solving the Pell equation are different to some degree, they can all be unified by considering the theory of semiregular continued fractions. For a given real number ϕ, put $\phi_0 = \phi$ and define

$$\phi_{i+1} = \frac{a_{i+1}}{\phi_i - b_i},$$

where $a_{i+1} = \pm 1$ and $b_i \in \mathbb{Z}$ such that $\phi_{i+1} > 1$ $(i = 0, 1, 2, \ldots)$. Then

$$\phi = b_0 + \cfrac{a_1}{b_1 + \cfrac{a_2}{b_2 + \cfrac{a_3}{b_3 + \cdots + \cfrac{a_n}{b_n + \cfrac{a_{n+1}}{\phi_{n+1}}}}}} \qquad (3.1)$$

If $b_i > 1$, $b_i + a_{i+1} \geq 1$ $(i \geq 1)$, the above expression is called a *semiregular continued fraction expansion of* ϕ. (See, for example, Perron [67, ch. 5].) For example, in the case of $b_i = \lfloor \phi_i \rfloor$ and $a_{i+1} = 1$ $(i \geq 0)$, we say that (3.1) is the *regular* continued fraction (RCF) expansion of ϕ; in the case of $b_i =$

$\lfloor \phi_i + 1/2 \rfloor$, the nearest integer to ϕ_i, and $sign\,(a_{i+1}) = sign\,(\phi_i - b_i)$ $(i \geq 0)$, we have what is called the *nearest integer* continued fraction (NICF) expansion of ϕ.

If

$$C_n = b_0 + \cfrac{a_1}{b_1 + \cfrac{a_2}{b_2 + \cfrac{a_3}{b_3 + \ddots + \cfrac{a_n}{b_n}}}},$$

we call C_n a *convergent* of (3.1), and it can be shown that $C_n = A_n/B_n$, where $A_{-2} = 0, A_{-1} = 1, B_{-2} = 1, B_{-1} = 0, a_0 = 1$, and

$$A_{i+1} = b_{i+1}A_i + a_{i+1}A_{i-1}, B_{i+1} = b_{i+1}B_i + a_{i+1}B_{i-1} \quad (i \geq 1).$$

Also,

$$A_n B_{n-1} - B_n A_{n-1} = (-1)^{n-1} a_1 a_2 \dots a_n \quad (n \geq 1).$$

If $\phi = \left(P_0 + \sqrt{D}\right)/Q_0$, where $P_0, Q_0 \in \mathbb{Z}$ and $Q_0 | D - P_0^2$, then

$$\phi_{n+1} = \left(P_{n+1} + \sqrt{D}\right)/Q_{n+1},$$

where,

$$P_{n+1} = b_n Q_n - P_n, Q_{n+1} = a_{n+1}\left(D - P_{n+1}^2\right)/Q_n > 0.$$

Let $\alpha \in \mathbb{Q}\left(\sqrt{D}\right)$. We denote by $\bar{\alpha}$, the conjugate of α in $\mathbb{Q}\left(\sqrt{D}\right)$ and by $N(\alpha)$ (the norm of α) the value of $\alpha\bar{\alpha} \in \mathbb{Q}$. It is easy to deduce that

$$N\,(A_n - \phi B_n) = (-1)^{n+1}\,(a_1 a_2 \dots a_{n+1})\,Q_{n+1}/Q_0$$

and

$$A_n - \phi B_n = \prod_{i=1}^{n+1} \frac{P_i - \sqrt{D}}{Q_{i-1}}.$$

If we define $G_n = Q_0 A_n - P_0 B_n$ $(= A_n$ when $\phi = \sqrt{D})$, then

$$Q_n G_n = P_{n+1} G_{n-1} + D B_{n-1}, Q_n B_n = G_{n-1} + P_{n+1} B_{n-1}, \quad (3.2)$$

$$G_n^2 - D B_n^2 = (-1)^{n+1}\,(a_1 a_2 \dots a_{n+1})\,Q_{n+1} Q_0, \quad (3.3)$$

and

$$\left(G_n + \sqrt{D} B_n\right)/Q_0 = A_n - \bar{\phi} B_n = \prod_{i=1}^{n+1} \frac{P_i + \sqrt{D}}{Q_{i-1}}. \quad (3.4)$$

Notice that the formulas (3.2) are generalizations of (2.7) and (3.3) is a generalization of (2.8).

Now put $c_n = \lfloor \phi_n \rfloor$, $P'_{n+1} = c_n Q_n - P_n$, $Q'_{n+1} = \left(D - P'^2_{n+1} \right) / Q_n$, $P''_{n+1} = P'_{n+1} + Q_n$, $Q''_{n+1} = \left(P''^2_{n+1} - D \right) / Q_n$. We next define $b_n = c_n + 1$, $a_{n+1} = -1$ when $Q''_{n+1} \leq Q'_{n+1}$ and $b_n = c_n$, $a_{n+1} = 1$ otherwise. Note that in the former case we get $P_{n+1} = P''_{n+1}$, $Q_{n+1} = Q''_{n+1}$ and in the latter case we get $P_{n+1} = P'_{n+1}$, $Q_{n+1} = Q'_{n+1}$. If b_i and a_{i+1} are defined in this way for $i = 0, 1, 2, \ldots$, we get the semiregular continued fraction that is implicitly produced by the cyclic method when $\phi = \phi_0 = \sqrt{D}$.

Indeed, when $\phi = \sqrt{D}$ it can be shown in the case of the cyclic method, the RCF (used by Euler and Lagrange) and the NICF that

$$t + u\sqrt{D} = A_n + B_n\sqrt{D},$$

where n is the least non-negative integer such that

$$(-1)^{n+1} (a_1 a_2 \ldots a_{n+1}) Q_{n+1} = 1. \tag{3.5}$$

Furthermore, for any nonsquare positive value of D, (3.5) must always occur for some $n \geq 0$.

Subsequent to the development of the RCF algorithm for solving the Pell equation, several authors, starting with Legendre in 1798, produced tables of values of t, u for certain ranges of values of D. These tables are described in Lehmer [50, pp. 54–59]. This work culminated in the unpublished table of Lehmer [47] in 1926 which dealt with all nonsquare D in the range $1700 < D \leq 2000$. At this point (1.1) had been solved for all positive $D \leq 2000$. Considering that these tables were all produced by hand calculation, it is easy to see why this was about as much as could be done for several years. In 1941 and more extensively in 1955 Patz [65], [66] produced tables of the RCF of $\sqrt{D}$ for all nonsquare values of D such that $1 < D < 10000$. Finally, in 1961 Kortum and McNeil [43], [77] used a computer to produce a very large table, giving, among other things, the RCF of $\sqrt{D}$ and the least solution of $T^2 - DU^2 = \pm 1$ for all nonsquare D such that $1 < D < 10000$. For reasons that will be made clear in the next section, no further tables have ever been published. It should, however, be mentioned that before the advent of computers, several authors attempted to solve the Pell equation for values of D in excess of 2000. One of the more impressive of these is the solution for $D = 9817$ by Martin [58] in 1877. Not only is his solution correct, but the value of t is a number of 97 digits.

4   The Regulator

One of the main reasons why it is not practicable to produce large tables of solutions of the Pell equation is that the values of t and u tend to become very large. For example, the values of t and u when $D = 95419$ have 376 and 374 decimal digits, respectively. Of course, another important reason why tables are no longer produced is the easy availability of special purpose computer software which will very quickly and accurately produce a solution of the Pell equation on input of the value of D. Nevertheless, large values of D still present a problem, as we shall see.

Let $\mathcal{K} = \mathbb{Q}\left(\sqrt{D}\right)$, where D is a square-free positive integer and put

$$r = \begin{cases} 2 & \text{when } D \equiv 1 \pmod 4, \\ 1 & \text{otherwise.} \end{cases}$$

We define $\omega = \left(r - 1 + \sqrt{D}\right)/r$ and d, the *discriminant* of $\mathcal{K}$, as $(\omega - \overline{\omega})^2 = 4D/r^2$. The ring of algebraic integers $\mathcal{O}$ of $\mathcal{K}$ is given by $\mathbb{Z} + \omega\mathbb{Z}$. If $\eta \in \mathcal{O}$ and $N(\eta) = \pm 1$, then η is a *unit* of $\mathcal{K}$ (or $\mathcal{O}$). It is well known that the set $\mathcal{O}^*$ of all units of $\mathcal{O}$ is an abelian group under multiplication with generators -1 and $\varepsilon_d = \left(x + y\sqrt{d}\right)/2$, where $x, y \in \mathbb{Z}$ and $\varepsilon_d(> 1)$ is called the *fundamental unit* of $\mathcal{K}$ (or $\mathcal{O}$). If we define $\varepsilon(D) = t + u\sqrt{D}$, then

$$\varepsilon(D) = \varepsilon_d^\nu,$$

where $\nu \in \{1, 2, 3, 6\}$.

It is a relatively straight-forward exercise to show that the value of ν can be determined from the table below. Thus, we can always easily determine ν when we know the value of $d \pmod{16}$ and the values of $x, y \pmod 8$. It follows that, given ε_d, we can find $\varepsilon(D)$, and as $t = \lceil \varepsilon(D)/2 \rceil$ and $u = \lfloor \varepsilon(D)/\left(2\sqrt{D}\right) \rfloor$, we can compute t and u. However, the problem with computing ε_d is that the values of x and y also tend to become very large as D increases. So instead we work with the *regulator*[2] R_d of $\mathcal{K}$, which is defined by $R_d = \log \varepsilon_d$. This is much easier to deal with from the point of view of the storage of information; for example, the regulator[3] when $D = 95419$ is 865.5675.

---

[2] The term "regulator" was first used (in a much wider context than this) by Dedekind [51, p. 597], in 1893. However, as pointed out in a footnote, he borrowed the term from an 1844 paper of Eisenstein [18, p. 313], who used it in a somewhat different context.

[3] The regulator is an irrational number; therefore, we cannot provide its exact value. The values given in this paper are approximations rounded to the number of figures written.

$d \pmod{16}$	y	x	ν
12	—	—	1
8	$2\mid y$	—	1
8	$2\nmid y$	—	2
$1, 5, 9, 13$	$4\mid y$	—	1
$1, 5, 9, 13$	$2\parallel y$	—	2
5	$2\nmid y$	$x \equiv \pm 3y \pmod 8$	3
		$x \equiv \pm y \pmod 8$	6
13	$2\nmid y$	$x \equiv \pm y \pmod 8$	3
		$x \equiv \pm 3y \pmod 8$	6

One of the interesting aspects of the RCF expansion of $\phi = \omega$ or $\sqrt{D}$ is its symmetry property. We find that if $n+1$ is the least positive integer such that $Q_{n+1} = Q_0 \ (= 1 \text{ or } 2)$, then

$$Q_i = Q_{n+1-i}, \ P_{i+1} = P_{n+1-i} \quad (i = 0, 1, 2, \ldots, n),$$

$$b_i = b_{n+1-i} \quad (i = 1, 2, \ldots, n).$$

These results mean that we can rewrite (3.4) as

$$\frac{\left(G_n + \sqrt{D}B_n\right)}{Q_0} = \begin{cases} \frac{P_{k+1}+\sqrt{D}}{Q_0} \prod_{i=1}^{k} \left(\frac{P_i+\sqrt{D}}{Q_i}\right)^2 & \text{when } n+1 = 2k+1, \\ \frac{Q_k}{Q_0} \prod_{i=1}^{k} \left(\frac{P_i+\sqrt{D}}{Q_i}\right)^2 & \text{when } n+1 = 2k. \end{cases}$$

Furthermore, we know that $n+1 = 2k+1$ whenever we find some minimal $k (\geq 0)$ such that $Q_k = Q_{k+1}$; on the other hand we get $n+1 = 2k$ whenever we find some minimal $k (> 0)$ such that $P_k = P_{k+1}$. In the case where $\phi = \omega$, we get

$$\varepsilon_d = \left(G_n + \sqrt{D}B_n\right)/Q_0;$$

hence,

$$R_d = \begin{cases} \log\left(\left(P_{k+1} + \sqrt{D}\right)/Q_0\right) + 2S & \text{when } Q_k = Q_{k+1}, \\ \log\left(Q_n/Q_0\right) + 2S & \text{when } P_k = P_{k+1}, \end{cases} \tag{4.1}$$

where

$$S = \sum_{i=1}^{k} \log\left(\left(P_i + \sqrt{D}\right)/Q_i\right).$$

Formulas like (4.1) were used in 1976 by Williams and Broere [99] to determine all the values of R_d for $D < 1.5 \times 10^5$.

The NICF expansion of ω also has symmetry properties which make it possible to produce a formula like (4.1) for R_d. However, instead of only 2 mid-period criteria, there are 6. Williams and Buhr [100] made use of the NICF to compute R_d for all square-free values of $D < 10^6$. The use of the NICF is recommended over that of the RCF because, as noted by Adams [2], the value for n for the NICF tends to be about 70% of that for the RCF expansion of ϕ. However, the extra work needed to determine the proper mid-period criterion tends to cause the run-time of the NICF algorithm to increase to 75% of the RCF algorithm. For larger values of D of a certain type (the type that tends to produce large values of n), the results of Williams [98] could be used to improve this speed, but as we shall see below, there are much better methods for evaluating R_d for large values of D. Before beginning to deal with these, it should be noted that all the continued fraction methods for computing R_d described to this point execute in time which is $O(R_d) = O\left(d^{1/2+\varepsilon}\right)$ for any $\varepsilon > 0$.

If we let $h(d)$ denote the class number of $\mathcal{K}$, we know from the analytic class number formula that

$$2h(d)R_d = \sqrt{d}L(1, \chi_d),$$

where $\chi_d(j)$ is the Kronecker symbol (d/j) and

$$L(1, \chi_d) = \sum_{j=1}^{\infty} \frac{\chi_d(j)}{j} = \prod_{q}\left(\frac{q}{q - (d/q)}\right). \tag{4.2}$$

Here the (Euler) product is taken over all the primes q. Hua [30] has shown that $L(1, \chi_d) < (\log d)/2 + 1$, a result that has recently been improved by Louboutin [56], but under a generalized Riemann Hypothesis (GRH), we know from a result of Littlewood [55] and a small improvement by Shanks [81] that

$$\{1 + o(1)\}(c_1 \log \log d)^{-1} < L(1, \chi_d) < \{1 + o(1)\} c_2 \log \log d, \tag{4.3}$$

where

$$c_2 = re^{\gamma}, c_1 = 4(r+1)e^{\gamma}/\pi^2, e^{\gamma} = 1.781072418.$$

This result has been tested numerically by Shanks [81], Jacobson, Lukes and Williams [35] and Jacobson [31], [32], and it has always been found to hold.

We now note that the Cohen-Lenstra heuristics [15], [16] on the distribution of the values of $h^*(d)$ (the odd part of the class number; $h(d) = 2^s h^*(d)$, $2 \nmid h^*(d)$) strongly suggest that $h^*(d)$ tends to be small. Indeed, about 75% of the values of $h^*(d)$ are 1. Also, under these heuristics we find that

$$P_r(h^*(d) > x) = \frac{1}{2x} + O\left(\frac{\log x}{x^2}\right)$$

(see [31]), where $P_r(h^*(d) > x)$ is the probability that $h^*(d) > x$. Since there are many values of d for which $h(d) = h^*(d)$ (for example, prime values of d), we would expect that the value of R_d can be large very often, and this was certainly found to be the case in the numerical study [35].

To gain some understanding of just how large R_d can get, we define the *regulator index RI* for $\mathcal{K}$ by

$$RI = 2R_d/(c_2\sqrt{d}\log\log d) = L(1,\chi_d)/(c_2 h(d)\log\log d).$$

From our previous observations we would expect that RI should only infrequently be very small and that $RI < 1 + o(1)$. For example, when $D = 95419$, the value of RI is .61612. In fact, the largest value of RI for all $d < 10^9$ is $RI = .68698$ [32] for $d = 513215704$ and the largest value of RI currently known is $RI = .78354$ for the 18 digit $D = 574911115184562766$ (see [31, p. 93]). Although it has been shown that as d runs through prime values ($\equiv 1 \pmod 4$)

$$\limsup L(1,\chi_d)/\log\log d \geq e^\gamma$$

(see, for example, Joshi [40]), it's usually rather difficult to produce a very large value of D such that $RI > 1/2$. Indeed, very recent investigations by Granville and Soundararajan [24] have led them to believe that there exists a constant C such that if

$$f(d) = 2c_2^{-1}L(1,\chi_d) - \log\log d - \log\log\log d,$$

then

$$\limsup f(d) = C.$$

This, of course, suggests that, as d becomes very large, we would not expect the value of RI to be much larger than $1/2$.

What is particularly important to observe here is that if we are getting values of R_d that are roughly of the same order of magnitude as $\sqrt{d}$, then for such values of d, particularly as d becomes large, the problem of simply writing t and u down is of exponential complexity. In other words we cannot record the values of t and u in the conventional way in polynomial time;

indeed, we may very often need exponential time in order to do this. Thus, we seem to be forced in such cases to record the value of R_d only. Now R_d is certainly a fundamental invariant of $\mathcal{K}$, but it's a transcendental number (and therefore can only be written to a certain precision) and, on the face of it, does not seem to reveal much about t and u. However, as we will point out in §7 we can compute an approximation to R_d which will allow us to develop an expression for $\varepsilon(D)$ that can be written in polynomial time. Before doing that, we must examine the problem of computing R_d more efficiently than we have done so far.

5   Ideals and Ideal Classes

In this section we will review several results that will be important for the discussion of the infrastructure of the class group of $\mathcal{K}$, which will follow. Most of this material can be found in any algebraic number theory textbook. There is also a computationally oriented discussion in Mollin and Williams [61]. Thus, our treatment of this information will be brief. We first note that $\mathcal{O}$ (the maximal order of $\mathcal{K}$) is the set of algebraic integers in $\mathcal{K}$ and serves in $\mathcal{K}$ in much the same way as $\mathbb{Z}$ does in the rationals $\mathbb{Q}$. Since $\mathcal{O}$ is a ring, it has ideals. If $\mathfrak{a}$ is any (integral) ideal of $\mathcal{O}$, then $\mathfrak{a}$ can be written

$$\mathfrak{a} = a\mathbb{Z} + \beta\mathbb{Z},$$

where $\beta = b + c\omega$; $a, b, c \in \mathbb{Z}$; $c|a$, $c|b$; $a, c > 0$; $|b| < a$. We call $\{a, \beta\}$ a *normal* $\mathbb{Z}$-*basis* of $\mathfrak{a}$. The quantity a is unique for $\mathfrak{a}$; it is the least positive rational integer in $\mathfrak{a}$. Thus, if m is any other rational integer in $\mathfrak{a}$, then $a|m$; this means that if $\gamma \in \mathfrak{a}$, then $a|N(\gamma)$. The norm of $\mathfrak{a}$, written $N(\mathfrak{a})$, is equal to ac. If $c = 1$, we say that $\mathfrak{a}$ is a *primitive* ideal; that is, $\mathfrak{a}$ has no rational integer divisors. It is well known that if $\mathfrak{a}$ is primitive, then

$$\left(N\left(\mathfrak{a}\right), \beta + \overline{\beta}, N\left(\beta\right)/N\left(\mathfrak{a}\right)\right) = 1. \tag{5.1}$$

Let $\theta_1, \theta_2, \ldots, \theta_k \in \mathcal{O}$ and define $\mathfrak{a} = (\theta_1, \theta_2, \theta_3, \ldots, \theta_k)$ to be $\{\theta_1\gamma_1 + \theta_2\gamma_2 + \ldots + \theta_k\gamma_k : \gamma_1, \gamma_2, \ldots, \gamma_k \in \mathcal{O}\}$. This is clearly an ideal of $\mathcal{O}$, and we call it the ideal *generated* by $\theta_1, \theta_2, \ldots, \theta_k$. It can be shown that if $\mathfrak{a}$ is any ideal of $\mathcal{O}$, then there exists $\theta_1, \theta_2 \in \mathcal{O}$ such that $\mathfrak{a} = (\theta_1, \theta_2)$. If $\mathfrak{a} = (\theta)$, we say that $\mathfrak{a}$ is the *principal ideal generated* by θ. For such an ideal we get $N(\mathfrak{a}) = |N(\theta)|$. If $\mathfrak{a} = (\theta_1, \theta_2, \ldots, \theta_k)$ and $\mathfrak{b} = (\phi_1, \phi_2, \ldots, \phi_n)$, then we define the product ideal $\mathfrak{a}\mathfrak{b}$ to be the ideal generated by the kn generators $\theta_i\phi_j$ ($i = 1, 2, \ldots, k; j = 1, 2, \ldots, n$). Given normal $\mathbb{Z}$-bases for the ideals $\mathfrak{a}$ and $\mathfrak{b}$, we can compute a normal $\mathbb{Z}$-basis for the ideal $\mathfrak{a}\mathfrak{b}$ by a deterministic algorithm that executes in $O(\log N(\mathfrak{a}) + \log N(\mathfrak{b}) + \log d)$

multiplication/division operations.[4] This algorithm is essentially the algorithm that Gauss discovered for the composition of quadratic forms and can be found, for example, on pp. 284–285 of [61]. Also, if $\mathfrak{a} = a\mathbb{Z} + \beta\mathbb{Z}$, then we define the ideal $\bar{\mathfrak{a}}$ conjugate to $\mathfrak{a}$ by $\bar{\mathfrak{a}} = a\mathbb{Z} + \bar{\beta}\mathbb{Z}$. It is easy to show that $\mathfrak{a}\bar{\mathfrak{a}} = (N(\mathfrak{a}))$, a principal ideal. If $\mathfrak{a} = \mathfrak{b}\mathfrak{c}$, we say that the ideal $\mathfrak{b}$ divides $\mathfrak{a}$; a necessary and sufficient condition that $\mathfrak{b}|\mathfrak{a}$ is that $\mathfrak{b} \supseteq \mathfrak{a}$. A property of the norm is that $N(\mathfrak{a}\mathfrak{b}) = N(\mathfrak{a})N(\mathfrak{b})$.

If $\mathfrak{a}$ and $\mathfrak{b}$ are non-zero ideals of $\mathcal{O}$ and there exist $\alpha, \beta \in \mathcal{O}$ such that $\alpha, \beta \neq 0$ and

$$(\alpha)\mathfrak{a} = (\beta)\mathfrak{b},$$

we say that $\mathfrak{a}$ and $\mathfrak{b}$ are *equivalent* and denote this by $\mathfrak{a} \sim \mathfrak{b}$. This equivalence relationship partitions all the ideals in $\mathcal{O}$ into $h(d)$ disjoint equivalence classes. If we denote by $[\mathfrak{a}]$ the class of all ideals of $\mathcal{O}$ equivalent to $\mathfrak{a}$, we can define an operation of multiplication of these classes by $[\mathfrak{a}][\mathfrak{b}] = [\mathfrak{a}\mathfrak{b}]$. Under this operation, it is easy to see that the set of all these ideal classes forms a group Cl with identity $[(1)]$, the class of all principal ideals.

We say that a primitive ideal $\mathfrak{a}$ of $\mathcal{O}$ is *reduced* if there does not exist any non-zero $\alpha \in \mathfrak{a}$ such that

$$|\alpha| < N(\mathfrak{a}), |\bar{\alpha}| < N(\mathfrak{a}).$$

An interesting and important consequence of this definition is that if $\mathfrak{a}$ is a reduced ideal, then $N(\mathfrak{a}) < \sqrt{d}$. It follows that there can only be a finite number of reduced ideals in $\mathcal{O}$. Also, if $\mathfrak{a}$ is any primitive ideal and $N(\mathfrak{a}) < \sqrt{d}/2$, then $\mathfrak{a}$ must be reduced.

If $\{a, \beta\}$ is a normal basis of $\mathfrak{a}$ and we produce the RCF expansion of $\beta/a = \left(P_0 + \sqrt{D}\right)/Q_0$, then we can produce a sequence of primitive ideals

$$\mathfrak{a}_1, \mathfrak{a}_2, \mathfrak{a}_3 \ldots \tag{5.2}$$

such that

$$\mathfrak{a}_i = (Q_{i-1}/r)\,\mathbb{Z} + \left(\left(P_{i-1} + \sqrt{D}\right)/r\right)\mathbb{Z} \tag{5.3}$$

and $\mathfrak{a}_i \sim \mathfrak{a}\,(=\mathfrak{a}_1)$ $(i = 1, 2, \ldots)$. Here, the values of the Q_{i-1} and P_{i-1} are those defined in §3. It can be shown that for a value $i = O(\log N(\mathfrak{a}))$, we must have $N(\mathfrak{a}_i) < \sqrt{d}/2$. Thus, the RCF algorithm provides us with a technique that will produce a reduced ideal $\mathfrak{b}$ such that $\mathfrak{b} \sim \mathfrak{a}$ and this algorithm will execute in $O(\log N(\mathfrak{a}))$ operations.

Let $\mathfrak{a}_1 = (1) = \mathcal{O}$. We can use the RCF algorithm on ω to produce a sequence of reduced ideals (5.2) ((1) is evidently a reduced ideal) such that

---

[4]We will use the term "operation" in the sequel to refer to an operation of multiplication or division (or addition or subtraction).

$\mathfrak{a}_i = (\theta_i)$, where $\theta_i = \left(G_{i-2} + \sqrt{D}B_{i-2}\right)/r$ (see §3). We can also show that

$$1 + 1/\sqrt{d} < \theta_{i+1}/\theta_i < \sqrt{d}$$
$$\theta_{i+2}/\theta_i > 2 \quad (i = 1, 2, \ldots). \tag{5.4}$$

Hence $\theta_1 = 1$ and

$$\theta_1 < \theta_2 < \theta_3 < \ldots < \theta_n < \theta_{n+1} < \ldots.$$

Indeed, we must have $\mathfrak{a}_{n+1} = \mathfrak{a}_1$ for some minimal $n > 0$. It follows, as mentioned above, that $\varepsilon_d = \theta_{n+1}$. If we put $\mathcal{R} = \{\mathfrak{a}_1, \mathfrak{a}_2, \ldots, \mathfrak{a}_n\}$, then it can be shown that $\mathcal{R}$ is the set of all reduced principal ideals of $\mathcal{O}$. That is, the RCF algorithm produces all the reduced principal ideals of $\mathcal{O}$, something that other semiregular continued fractions, like the NICF, do not in general do.

6   The Infrastructure of Cl

As mentioned earlier, in the early 1970s Shanks was testing the truth of (4.3) by doing some numerical work. He had defined what he called the upper Littlewood index (ULI) as

$$\text{ULI} = L\left(1, \chi_d\right) / \left(c_2 \log \log d\right)$$

and was attempting to find values of it that were large. An examination of (4.2) suggests that $L\left(1, \chi_d\right)$ will be large when $(d/q) = 1$ for as many of the small (in particular) primes as possible. In October of 1971, Lehmer had found the 14 digit $d = D = 26437680473689$ on his delay line sieve DLS-157; this number is a prime such that $(d/q) = 1$ for all primes $q \leq 149$. Shanks knew through the empirical work of Kloss et al. (see Shanks [78]) that it was likely that $h(d) = 1$ when d is a prime congruent to 1 modulo 4; thus, by estimating $L\left(1, \chi_d\right)$ and using the analytic class number formula, assuming $h(d) = 1$, he concluded that $R_d \approx 2.17 \times 10^7$. With the algorithms for evaluating R_d at his disposal, however, he realized that he would not be able to compute an accurate value of R_d, which he needed to improve his estimate of $L\left(1, \chi_d\right)$. It was this problem which caused him to examine the inner structure of the various ideal classes, structures that he collectively referred to as the infrastructure of Cl (or $\mathcal{K}$).

Although the ideas described below can be applied to any ideal class, we will confine (as did Shanks in [80]) our attention here to the principal class.

Let $c \in \mathcal{R}$; then $c = a_i$ for some i such that $1 \le i \le n$. Define $\delta(c) = \log \theta_i$. By (5.4) we see that $\delta(a_{i+2}) > \delta(a_i) + \log 2$; hence $j = O(\delta(a_j))$. Now let $a, b \in \mathcal{R}$ and consider the ideal ab. This ideal is the product of two principal ideals and must therefore be principal, but it is not necessarily reduced. Suppose we apply our reduction algorithm to ab until we get c, the first ideal such that $N(c) < \sqrt{d}/2$. We will denote this c by $\mathrm{red}(ab)$ and observe that c is reduced.[5] Thus, $c \in \mathcal{R}$. Furthermore

$$\delta(c) \equiv \delta(a) + \delta(b) + \kappa \pmod{R_d} \tag{6.1}$$

and

$$-\log d < \kappa < \log 2.$$

What is important to note here is that the values of $\delta(a), \delta(b), \delta(c)$ can be quite large (near R_d, say), but κ is quite small. Also, there is a simple formula for κ which allows it to be computed quickly and accurately. Thus, if we start with some ideal $b \in \mathcal{R}$ such that $\delta(b)$ is large by comparison to $\log d$, but smaller than R_d, and define

$$b_{i+1} = \mathrm{red}(bb_i), \tag{6.2}$$

where $b_1 = b$, then by (6.1) we get

$$\delta(b_j) \ge j\delta(b) - (j-1)\log d.$$

Having arrived at this deduction, Shanks was next able to apply his baby-step giant-step [79] technique[6] to find R_d.

First, use the RCF to compute the set of ideals $\mathcal{L} = \{a_1 = (1), a_2, a_3, \ldots, a_t, a_{t+1}, a_{t+2}\} \subseteq \mathcal{R}$. These are the $t + 2$ baby-steps. Here we need t to be large enough that $\delta(a_t) > \sqrt[4]{d}$. Next, put $b_1 = a_t$ and compute $b_2, b_3, \ldots$ as in (6.2). These are the giant steps. Now notice that

$$\delta(b_{i+1}) - \delta(b_i) < \delta(b_1) + \log 2 = \delta(a_t) + \log 2 < \delta(a_{t+2})$$

---

[5]It should be noted that Shanks [83] also developed a method of computing a reduced ideal $\sim ab$ which did not first compute ab and then perform a reduction technique on ab. He called this algorithm NUCOMP. At first it was thought that this algorithm was not much more efficient than the technique described above, but a new analysis by van der Poorten [89] and unpublished computational results of Jacobson suggest that it is at least 20% faster when $d > 0$.

[6]Curiously, it was also numbers produced from Lehmer's sieve (the DLS-127, an earlier version of the DLS-157) which inspired Shanks' discovery of the baby-step giant-step technique. In 1968 Lehmer had produced 3 values of d for which he wanted the value of $h(-d)$, the class number of $\mathbb{Q}(\sqrt{-d})$. These values were selected in order to minimize the value of $h(-d)/\sqrt{d}$ (see Teske and Williams [88]), but they were too large for computing $h(-d)$ by the means currently available. The first time the baby-step giant-step idea was used was in the determination of $h(-229565917267) = 29351$ in August of 1968.

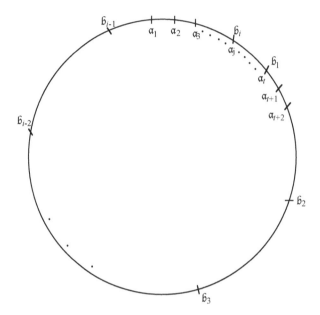

by (6.1) and (5.4). Consider the diagram above where each ideal is placed on the circumference of the circle according to the value of $\delta(\mathfrak{c})$. Since $\mathfrak{a}_{n+1} = \mathfrak{a}_1$, the length of the circumference of the circle is R_d. If we examine this diagram we see that we must get some $\mathfrak{b}_i \in \mathcal{L}$. This means that $\mathfrak{b}_i = \mathfrak{a}_j$ and we can compute

$$R_d = \delta(\mathfrak{b}_i) - \delta(\mathfrak{a}_j).$$

The complexity of computing R_d by this procedure can be easily estimated. We know that

$$d^{1/2+\varepsilon} > R_d > \delta(\mathfrak{b}_{i-1})$$
$$> (i-1)\delta(\mathfrak{a}_t) - (i-2)\log d > (i-1)\sqrt[4]{d} - (i-2)\log d.$$

Hence $i = O\left(d^{1/4+\varepsilon}\right)$. Furthermore, if $\delta(\mathfrak{a}_{t-1}) < \sqrt[4]{d}$ and $\delta(\mathfrak{a}_t) > \sqrt[4]{d}$, then $t = O\left(d^{1/4+\varepsilon}\right)$. It follows that the total complexity of determining R_d is $O\left(d^{1/4+\varepsilon}\right)$, a considerable saving over the previous methods of using continued fractions, particularly when d is large.

By using this technique, Shanks was able to compute R_d for Lehmer's value of d as $R_d = 21737796.43$ and $L(1, \chi_d) = 8.45539$. Also $RI = .69180$. In fact, as computed later [82], $n = 18334815$. This is much too large to compute R_d by the continued fraction method with the computer power available to Shanks at the time. It should be stressed here that although we have used an ideal theoretic setting in which to describe Shanks' ideas and methods, he used the language and techniques of the theory of quadratic forms.

Shanks' discovery led to a number of results of both theoretic and computational interest. For example, in 1979 Lagarias [45] used the infrastructure in his sketch of a proof of the following interesting result.

Theorem 6.1. *Let $L(m)$ denote the length of the binary encoding of the integer m and let $L(g)$ be the length of the binary encoding (the total number of bits in $a, b, \ldots, f$) of the quadratic Diophantine equation (1.5). Suppose (1.5) has a non-negative integer solution (x, y) satisfying*

$$x \equiv x_1 \ (mod \ m)$$
$$y \equiv y_1 \ (mod \ m).$$

Then there exists a certificate (proof) showing that $f(x, y) = 0$ has such a solution which requires only $O\left(L^5 \log L \log \log L\right)$ elementary (bit) operations to verify, where $L = L(g) + 3L(m)$.

This result, in view of what we have said about the solution of the Pell equation in §4, is surprising because we know that these are equations of the form (1.5) whose smallest non-negative integer solution is so large that it takes time which is exponential in $L(g)$ just to write the solution down in the usual binary or base 10 representation. Unfortunately, the importance of this result was not recognized at the time and a full proof of it never appeared in any refereed journal. It was published as a technical report [46], but did not circulate sufficiently for the result to receive the attention it merited.

Later, in 1982, Lenstra [54] showed how to use the infrastructure, together with estimates on the value of $L(1, \chi_d)$, to improve the speed of computing R_d to $O\left(d^{1/5+\varepsilon}\right)$. The value of R_d that Lenstra's algorithm produces is correct (to the number of figures written); but the complexity of his algorithm is conditional on the truth of an appropriate generalization of the Riemann hypothesis (GRH). A year later Schoof [71] implemented Lenstra's algorithm on a computer in order to determine the class groups of certain real quadratic fields. He also showed that under the GRH any method that can compute R_d can be converted to a method which will obtain the integer factorization of D and that this only requires a polynomial (in $\log D$) multiple of the time needed to find R_d to execute. Thus, computing R_d is at least (from a computational complexity point of view) as difficult as factoring D under the GRH.

In 1988 Buchmann and Williams [9] showed that R_d could be computed unconditionally with complexity $O\left(R_d^{1/2} d^\varepsilon\right)$. This is still the best unconditional complexity measure known for the computation of R_d. More recently, Lenstra's technique, incorporating a method of Bach [6] for obtaining more accurate estimates for $L(1, \chi_d)$, was implemented on a computer by Jacobson, Lukes and Williams [35] and Jacobson [32] to compute all R_d for $d < 10^9$. Indeed, to illustrate the power of this technique, we computed R_d for the 30 digit

$$D = 990676090995853870156271607886.$$

Here $R_d = 4770372955851343.43$ and $RI = .63242$. This is the largest value of D with a regulator index $> .5$ for which a non-conditional regulator has ever been computed.

Finally, we point out that Srinivasan [85] has recently produced a version of Lenstra's algorithm in which she obtains an estimate of $L(1, \chi_d)$ without assuming any Riemann hypotheses, but by using a new method called the random summation technique. As a result she is able to compute the regulator deterministically in expected time $O(d^{1/5+\varepsilon})$. The regulator that the algorithm produces is correct — only the running time is probabilistic.

7   Compact Representation

In view of the result of Lagarias mentioned above, it should be possible to write $\varepsilon_d = (x + y\sqrt{d})/2$ in a more compact way than by using the standard binary or base 10 representation of x and y. Indeed, Buchmann, Thiel and Williams [8] have shown that there always exists a *compact representation* of ε_d given by

$$\varepsilon_d = \prod_{j=1}^{k} (\alpha_j / d_j)^{2^{k-j}}, \tag{7.1}$$

where

1) $k < \log_2 R_d + 3$,

2) $d_j \in \mathbb{Z}; \alpha_j = \left(a_j + b_j \sqrt{d} \right)/2 \in \mathcal{O}; a_j, b_j \in \mathbb{Z}$ $(j = 1, 2, \ldots k)$,

3) $0 < d_j < d^{1/2}, |a_j| \le 2d^{5/2}, |b_j| < 2d^2$ $(j = 1, 2, \ldots, k)$.

Thus, to represent ε_d we require only $O((\log d)^2)$ bits instead of the possibly exponential number of bits needed by the standard representation. Also, a refinement of the methods of [8] will yield better bounds of $|a_j| < 30d$ and $|b_j| < 30\sqrt{d}$ $(j = 1, 2, \ldots, k)$. We also have the following important theorem.

Theorem 7.1. *There is a deterministic polynomial (in* log *d) time algorithm which, given* $\hat{R}_d$ *such that* $\left| \hat{R}_d - R_d \right| < 1$, *will compute a compact representation of* ε_d.

In fact, this algorithm, which is easily implemented on a computer, executes in $O\left((\log d)^{3+\varepsilon}\right)$ operations and requires only $O\left((\log d)^{2+\varepsilon}\right)$ space.

Thus, we have solved the two problems mentioned in §4 concerning the use of R_d instead of $\varepsilon(D)$. We need only compute R_d to within 1 of its actual value in order to compute ε_d. On the face of it, this seems an utterly amazing result because we can compute the enormous ε_d from very little information, but the constraint that $N\left(\varepsilon_d\right) = \pm 1$ is so restrictive that even the relatively tiny amount of data represented by $\hat{R}_d$ is sufficient to allow for the complete determination of ε_d.

To give some idea of how effective the notion of a compact representation is, we point out that for the 30 digit value of D mentioned in the last section, we get $\varepsilon(D) = \varepsilon_d$ and, from the value of R_d, we can easily deduce that $t, u > 10^{2 \cdot 10^{15}}$. This means that it would require over 6,000,000 books, each of 1000 pages, of the same format used by Nelson to record the solution of the Cattle Problem, to record t and u. However, it requires less than one page to write $\varepsilon(D)$ as a compact representation.

We also have a companion theorem to Theorem 7.1

Theorem 7.2. *Given any positive integer* q *and a compact representation of* ε_d, *there exists a polynomial (in* log *d and* q) *time algorithm which will compute* R_d *such that*

$$\left| \hat{R}_d - R_d \right| < 2^{-q}.$$

This means that we can easily obtain a very good approximation to R_d, once we have a compact representation of ε_d.

It should be emphasized that there is nothing unique about an expression of the form (7.1). We can easily find others, although we may not necessarily have the inequalities given in (3). In the process of determining a compact representation by the algorithm given in [8], we produce a set of reduced ideals of $\mathcal{O}$:

$$\mathfrak{g}_0, \mathfrak{g}_1, \mathfrak{g}_2, \cdots \mathfrak{g}_k$$

such that $\mathfrak{g}_0 = (1)$, $\mathfrak{g}_j = (\gamma_j)$ has a normal basis $\{d_j, \mu_j\}$ ($\mathfrak{g}_i = d_i \mathfrak{A}_i$ in [8]) ($j = 1, 2, \ldots, k$), where

$$\gamma_j = \alpha_j \prod_{i=1}^{j-1} (\alpha_i/d_i)^{2^{j-i}} \in \mathcal{O}.$$

We also have

$$(d_{j-1})^2 \, \mathfrak{g}_j = (\alpha_j) \, \mathfrak{g}_{j-1}^2 \quad (j = 1, 2, \ldots, k). \tag{7.2}$$

Now for any i $(0 \le i \le k)$, let $\lambda_i \in \mathfrak{g}_i$; there must exist some ideal $\mathfrak{h}_i$ of $\mathcal{O}$ such that

$$(N(\mathfrak{h}_i)) \, \mathfrak{g}_i = (\lambda_i) \, \mathfrak{h}_i; \tag{7.3}$$

hence

$$|N(\lambda_i)| = N(\mathfrak{h}_i) \, N(\mathfrak{g}_i)$$

by the multiplicative property of the norm. Also,

$$\mathfrak{h}_i = (\kappa_i),$$

where

$$\kappa_i = \overline{\lambda}_i \gamma_i / d_i \in \mathcal{O} \tag{7.4}$$

Put $\lambda_0 = 1$. Then $\mathfrak{h}_0 = \mathfrak{g}_0$, $\kappa_0 = 1$ and since

$$\gamma_i = \alpha_i \left(\gamma_{i-1} / d_{i-1} \right)^2 \quad (i = 1, 2, \ldots k),$$

we get

$$\kappa_i = \nu_i \left(\kappa_{i-1} / N(\mathfrak{h}_{i-1}) \right)^2 \quad (i = 1, 2, \ldots, k), \tag{7.5}$$

where

$$\nu_i = \frac{\overline{\lambda}_i \alpha_i N(\mathfrak{h}_{i-1})^2}{d_i \overline{\lambda}_{i-1}^2} = \frac{\overline{\lambda}_i \alpha_i \lambda_{i-1}^2}{d_{i-1}^2} \pm \frac{N(\mathfrak{h}_i) \, \alpha_i \lambda_{i-1}^2}{d_{i-1}^2 \lambda_i} \, .$$

By (7.2) and (7.3) we get

$$\left(N(\mathfrak{h}_{i-1})^2 \, d_{i-1}^2 \lambda_i \right) \mathfrak{h}_i = \left(\alpha_i \lambda_{i-1}^2 N(\mathfrak{h}_i) \right) \mathfrak{h}_{i-1}^2;$$

hence $\nu_i \in \mathfrak{h}_i \subseteq \mathcal{O}$. If we put $\lambda_k = N(\mathfrak{g}_k) = 1$, then $\mathfrak{h}_k = \mathfrak{g}_k = (1)$ and from (7.4) we get

$$\varepsilon_d = \kappa_k = \prod_{i=1}^{k} \left(\nu_i / N(\mathfrak{h}_i) \right)^{2^{k-i}} \, . \tag{7.6}$$

Since α_k and ν_k must have the same sign, we may assume that

$$\nu_i = \frac{N(\mathfrak{h}_i) \, \alpha_i \lambda_{i-1}^2}{d_{i-1}^2 \lambda_i} \quad (i = 1, 2, \ldots, k) \tag{7.7}$$

in (7.6).

It may be thought that a representation of the form (7.1) for ε_d might be somewhat unwieldy for computing, but as shown in [8] there are many operations that can be performed quite expeditiously on such representations. For example, we have the following theorem.

Theorem 7.3. *Let $\varepsilon_d = \left(x + y\sqrt{d}\right)/2$. If we are given a compact representation of ε_d and some positive integer m, the values of x and y modulo m can be determined in $O(\log d)$ operations on numbers of $O\left(\log m + (\log d)^2\right)$ bits.*

It is also possible to develop an algorithm to perform this same function which requires $O(\log m \log d)$ operations on numbers of only $O(\max(\log m, \log d))$ bits. The only real problem in doing this is the possibility that $(d_j, m) > 1$ for certain values of j. In order to circumvent this, we put $\lambda_j = x d_j + y \mu_j$ and select $x, y \pmod{m}$ such that

$$N(\mathfrak{h}_j) = |N(\lambda_j)| / d_j = \left| x^2 d_j + xy \left(\mu_j + \overline{\mu}_j\right) + y^2 N(\mu_j) / d_j \right|$$

is relatively prime to m. Since by (5.1) we know that $\left(d_j, \mu_j + \overline{\mu}_j, N(\mu_j)/d_j\right) = 1$, this can be done in no more than $O(\log m)$ operations. In the case that $(d_j, m) = 1$, we use $x = 1$, $y = 0$ and get $N(\mathfrak{h}_j) = d_j$. This process allows us to find a sequence of values of $\nu_i \in \mathcal{O}$ by (7.7) such that (7.6) can be used to find $\varepsilon_d \pmod{m}$. (See Jacobson and Williams [39].)

We conclude this section by mentioning that the idea behind expressing ε_d in compact representation was used recently by van der Poorten, te Riele and Williams [90] to verify the Ankeny-Artin-Chowla conjecture for all primes $p\,(p \equiv 1 \pmod 4)$ up to 10^{11}. This conjecture states that if $D = p$, then $p \nmid u$.

8   The Subexponential Method

Undoubtedly, the most exciting recent development for accelerating the computation of R_d for large d has been the discovery of the subexponential method. Up to now all of the methods for doing this that we have discussed have been of exponential complexity, but as we shall see here, there is a subexponential algorithm for computing R_d if we are willing to accept the truth of certain generalized Riemann hypotheses. It is important to emphasize here that while the values of R_d that we find by this technique are very likely to be correct, they are only provably so if the GRH holds. The GRH is not just used to measure the complexity of computing R_d, but must also be assumed to establish rigorously the correctness of the value for R_d that the method produces.

The basic idea behind the subexponential method first occurred as Remark 2.15 in a technical report [52] (later a book chapter) by A. K. and H. W. Lenstra in 1987. Their idea was then elaborated by McCurley [60] and Hafner and McCurley [25]. It was originally intended as a technique to determine the class number of a quadratic field with negative discriminant; however, it can also be applied to real quadratic fields. As the details of this technique have been very well described elsewhere, we will provide only a brief sketch here.

We let $\mathcal{P} = \{p_1, p_2, \ldots, p_k\}$ be a set of rational primes such that $(d/p_i) = 1$ $(i = 1, 2, \ldots, k)$. We know that the ideal (p_i) must split into the product of two prime ideals $\mathfrak{p}_i$ and $\bar{\mathfrak{p}}_i$ in $\mathcal{O}$. We put $FB = \{\mathfrak{p}_1, \mathfrak{p}_2, \ldots, \mathfrak{p}_k\}$ and we suppose that the equivalence classes of the ideals in FB generate the class group Cl of $\mathbb{Q}\left(\sqrt{d}\right)$. For $\vec{v} = (v_1, v_2, v_3, \ldots, v_k) \in \mathbb{Z}^k$, we define $\mathfrak{p}_i^{-v_i}$ to be $\bar{\mathfrak{p}}_i^{v_i}$ $(v_i > 0)$ and

$$FB^{\vec{v}} = \prod_{i=1} \mathfrak{p}_i^{v_i}.$$

Let $\Lambda = \{\vec{v} \in \mathbb{Z}^k = FB^{\vec{v}} \sim (1)\}$. This is clearly a sublattice of $\mathbb{Z}^k$. For an introduction to the study of lattices, the reader is referred to any text on the geometry of numbers, such as Cassels [11, ch. 1]. Now let θ be the homomorphism

$$\theta \colon \mathbb{Z}^k \to \mathrm{Cl}$$

defined by

$$\vec{v} \mapsto \left[FB^{\vec{v}}\right].$$

θ is surjective and Λ is its kernel; thus,

$$\mathbb{Z}^k/\Lambda \cong \mathrm{Cl}.$$

The recognition of this important fact began in the work of Pohst and Zassenhaus in 1979 [68] and appears in a more fully developed form in [69]. The important information that follows is that

$$|\det(\Lambda)| = |\mathrm{Cl}| = h(d).$$

Let $\{\vec{v}_1, \vec{v}_2, \ldots, \vec{v}_n\}$ be a generating system for Λ and put

$$A_{k \times n} = \left(\vec{v}_1^T, \vec{v}_2^T, \ldots, \vec{v}_n^T\right).$$

Then the matrix A has rank k and there exists an integer matrix T such that multiplying A on the right by T will put A into Hermite normal form (see, for example, Newman [64]). That is,

$$AT = [0|H] , \tag{8.1}$$

where H is an upper triangular $k \times k$ matrix. The product of the elements on the diagonal of H is $h(d)$.

One of the problems in implementing this idea is guaranteeing that the classes of the ideals in FB will generate Cl. At the time that [25] was written it was known that under the GRH there exists an effectively computable constant c such that if $\mathcal{P}$ contains all the primes $\leq c(\log d)^2$, then FB will contain ideals whose classes will generate Cl. Later, in 1990, Bach [5] provided a value for c. In fact, he gives a table (Table 3, based on the size of d) of values of c_1 and c_2 such that if $\mathcal{P}$ contains all the primes $< (c_1 \log d + c_2)^2$, then FB can be used to generate Cl. This, unfortunately, is also only proved under the assumption of the GRH, but calculations done by Jacobson [32] suggest that Bach's constants are likely a good deal larger than they need be. This is important because keeping k small is essential for the effective execution of the algorithm. Thus, any substantial lowering of Bach's constants would be of enormous benefit to those utilizing the subexponential method.

One consequence of the development of the subexponential method is that under the GRH it is possible to show the existence of a short proof (verifiable in polynomial time) for the value of R_d (see Buchmann and Williams [10]). This, however, is a far cry from having a polynomial algorithm for computing R_d, even under the GRH. However, Buchmann [7] pointed out how the subexponential technique could be applied to the problem of computing R_d. His ideas were later modified and implemented by Cohen, Diaz y Diaz and Olivier [13], [14]. The important ingredient in all of this is that we must now compute a generator γ for any $FB^{\vec{v}} \sim (1)$ such that we have $FB^{\vec{v}} = (\gamma)$; we call this a *relation*. Suppose we have a set of n such relations

$$FB^{\vec{v}_i} = (\gamma_i), \tag{8.2}$$

with generators $\gamma_1, \gamma_2, \ldots, \gamma_n$. We put

$$\vec{r} = \frac{1}{2} \left(\log |\gamma_1/\overline{\gamma}_1|, \log |\gamma_2/\overline{\gamma}_2|, \ldots, \log |\gamma_n/\overline{\gamma}_n| \right)$$

and compute

$$\vec{r}T = (r_1, r_2, \ldots, r_n) \quad,$$

where T is the matrix in (8.1). The values of $r_1, r_2, \ldots, r_n$ are all real numbers and, in particular, r_i ($i = 1, 2, \ldots, n - k$) are integer multiples of R_d. Indeed, if we compute the real $\gcd(r_1, r_2, \ldots, r_{n-k})$ (see Cohen [12, p. 289]), this is R_d.

Of course, it is possible that the set of relations that we produce may only generate a sublattice K of Λ. However, in this case we know that

$$|\det(K)/\det(\Lambda)| \in \mathbb{Z}.$$

Also, we can (again under the GRH) get a sufficiently good approximation of $h(d)R_d$ by truncating the Euler product in (4.2) to produce a value h^* such that

$$h^*/2 < h(d)R_d < h^*. \tag{8.3}$$

Thus, if our computed value of $h(d)R_d$ lies in the interval between $h^*/2$ and h^*, we know that $K = \Lambda$ and our values for $h(d)$ and R_d must be correct (under the GRH).

The complexity of computing R_d by this technique was determined by Abel [1] in her doctoral thesis. Her result is that under the GRH the probabilistic running time to compute R_d is

$$\exp\left\{ (1.44 + o(1)) \left(\log d \log \log d \right)^{1/2} \right\}.$$

Very recently, Vollmer [92] noted that this result can be improved somewhat: Under the GRH there exists a probabilistic algorithm that with error probability ϵ, given in advance and independent of d, will compute R_d in expected time $\exp\left\{ (\sqrt{2} + o(1)) \left(\log d \log \log d \right)^{1/2} \right\}$.

Cohen et al. implemented their method and were able to compute values for R_d and $h(d)$ for values of d as large as 10^{45}. For example, at the author's request, they computed the regulator for the 38 digit (see [35])

$$D = 13208708795807603033522026252612243246.$$

The value of RI in this case is .66971.

9   Computational Results

Recently, Jacobson [33] has made a number of improvements to the subexponential algorithm which greatly improve its efficiency. As these are fully described in [33], we will concentrate here on his main improvement — the use of sieving to compute relations.

The main bottleneck in running the subexponential method is the determination of enough relations (8.2) such that we finally get (8.3). Because of the Bach bounds, when d gets large we may have to compute many relations. Jacobson's approach was to apply the techniques of polynomial sieving, which have proved to be so effective in the problem of integer factorization (see, for example, Montgomery [62]). We first compute a normal basis for a primitive ideal $\mathfrak{a}$, where

$$\mathfrak{a} = \prod_{i=1}^{k_0} p_i^{v_i} = a\mathbb{Z} + \beta\mathbb{Z},$$

k_0 is small and v_i $(i = 1, 2, \ldots, k_0)$ is selected at random from $\{0, 1, -1\}$. This is done in such a way that $a = N(\mathfrak{a}) \approx \sqrt{d}/M$, where M is a given sieving radius. If $\gamma \in \mathfrak{a}$, then $\gamma = ax + \beta y$ $(x, y \in \mathbb{Z})$, and since $\mathfrak{a}$ contains γ, there must exist an ideal $\mathfrak{b}$ such that

$$(\gamma) = \mathfrak{a}\mathfrak{b}$$

Thus, if $\mathfrak{b}$ factors over FB, we have a relation. As before we get

$$N(\mathfrak{b}) = ax^2 + \left(\beta + \overline{\beta}\right) xy + \left(N(\beta)/a\right) y^2 = f(x, y) \ .$$

If we put $y = 1$, $f(x, 1)$ is a quadratic polynomial in x with integer coefficients. We attempt to factor $\mathfrak{b}$ over FB by sieving $f(x, 1)$ over the range $-M$ to M using the primes p_i $(i = 1, 2, \ldots, k)$. Of course, we make use of the observation that if $p|f(x, 1)$, then $p|f(x + mp, 1)$. This simple adjustment to the algorithm has resulted in a very considerable increase in its speed of computing relations. For example, the method of Cohen et al. could take as long as 4.46 hours to generate enough relations for a 39 digit d, but Jacobson's method could do this in only 8.07 seconds [33, Table 5.26]. Furthermore, as we can use many different ideals as our initial $\mathfrak{a}$, Jacobson's technique can be readily parallelized to run on several computers at once.

When the values of d become very large $(> 10^{70})$, we need well over 5000 prime ideals in FB. This means that the amount of time needed to determine R_d becomes very large. The reason for this, besides the computation of a large number of relations, is the difficulty of the linear algebra which prevents us from computing the class number and regulator simultaneously. The linear algebra required to compute only the class number, even with a large factor base, is much more manageable than that required to simultaneously compute R_d. Thus, it is best to use a much smaller factor base to determine a value S which is likely to be R_d, but at the very worst is an integral multiple of R_d, and then use the large factor base (as determined by the Bach bounds) to find a value for $h(d)$. If $Sh(d)$ falls in the range (8.3), then $S = R_d$ and we are done.

We should also mention that if $(d/p) = 1$ for as many of the small primes p as possible, then not only is $L(1, \chi_d)$ likely to be large and, as a consequence, RI is probably big, but also more prime ideals with small norm will appear in FB. As it is more likely that a small prime will divide a given number than a large one, this means that we would expect to find the required number of relations needed to compute R_d and $h(d)$ much more quickly than in the case where $(d/p) = -1$ for many of the small primes. To illustrate this we consider the 72 digit [38]

$$D_1 = 13300724392227875124126003410285180354292513910059927613\backslash$$
$$9935498154029253.$$

We get $R_d = 66252913306616520534293587275456066.557249$, $h(d) = 4$, $L(1, \chi_d) = .145331$. These results required over 18 days of computing on a 296MHz SUN Ultra SPARC-II computer. However for the 80 digit

$$D_2 = 12779403100260586715025492824657916044067403863724697039\backslash$$
$$7197773038860596550553681$$

we get $R_d = 182871089219957536671992302657711142676945.486447$, $h(d) = 1$ and $RI = .551826$. These results required only 3.45 days [37] of computing on the same computer. The difference between D_1 and D_2 is that we have $(d/p) = -1$ for all primes $p \leq 337$ for D_1, where as for D_2 we have $(d/p) = 1$ for all primes $p \leq 239$.

Extrapolating from these and other results to larger values of d for which we expect to have large values of RI (the "easy" values of d) of 90 and 100 digits, we estimated that it would take about 112 days to compute R_d for a 90 digit d and well over five years to compute R_d for a 100 digit d on the same computer that we used for the smaller values of d. In fact, the technique was implemented by Jacobson in parallel on a cluster of 16 350Mhz Pentium II computers, and we found that for the 90 digit

$$d = 21522469810372840041048377124060167166863420091501850604 6263\backslash$$
$$9189777165915901265583086 31804$$

we get

$$R_d = 131411783793381336054345076740506011516668 6144.033218,$$

$h(d) = 1$ and $RI = .59718$. This computation was completed in 8 days. For the 101 digit value of

$$d = 13022194102190350410319085329793205127319464132884776163 3\backslash$$
$$615836657137909258356026308739718466909983 6,$$

we computed

$$R_d = 31780254623174755539291764915494863617276316347826 0.945231,$$

$h(d) = 1$ and $RI = .57483$, a result that required 87 days of computing [36].

10   Conclusion

We have seen that the hard part of solving the Pell equation (1.1) is the problem of determining the regulator of $Q(\sqrt{D})$. We have also seen that we are on the verge of being able to do this (conditionally, at least) for values of d of up to 100 digits. This remarkable achievement, however, is accompanied by a regrettable loss of rigor. Thus, the problem of solving the Pell equation is still far from being solved in any computationally satisfactory way. Nevertheless, it should be mentioned that Hallgren [26] has recently shown that there exists a polynomial (in $\log d$) time *quantum algorithm* for finding the regulator of a real quadratic field of discriminant d. This is a very exciting result which essentially solves the main problem of this paper, but until quantum computers become available, we are left with many questions such as:

(1) Is there a deterministic polynomial (in $\log d$) time algorithm for finding R_d? If there is, then there is (under the GRH) a polynomial time algorithm for factoring d, something that is not widely believed.

(2) Is there an unconditional, deterministic algorithm for evaluating R_d that runs in expected subexponential time? There very likely is such an algorithm, but no one has any idea currently as to how to tackle this problem.

(3) Is there an efficient method for determining R_d, given an integral multiple S of R_d? The answer to this is yes if S/R_d is small. For suppose $S/R_d < B$. If we compute the reduced principal ideal $\mathfrak{a}$ such that $\delta(\mathfrak{a}) = S/q$ and find that $\mathfrak{a} = (1)$, then S/q must be an integral multiple of R_d; otherwise, it is not. As Shanks well knew, the complexity of finding a reduced principal ideal $\mathfrak{a}$, given a value for $\delta(\mathfrak{a})$ requires only $O\left(\log \delta(\mathfrak{a})\right)$ operations. Thus, this technique works well enough if B is small, but we have no idea of how to deal with this problem when B is large. Jacobson has observed that there is an unconditional method for finding R_d from S that will execute in $O\left(S^{1/3+\varepsilon}\right)$ operations.

(4) Is there an unconditional, deterministic algorithm for computing R_d that executes with a smaller time complexity function than $O\left(R_d^{1/2} d^\varepsilon\right)$? The information that we have presented previously suggests that this is most likely, but we are still, it seems, a long way from answering this question.

(5) Can Jacobson's method be made to run significantly faster? Almost certainly. There is much room for improvement, but the problems

involved such as reducing the Bach bounds,[7] performing the linear algebra, and computing a value for S, are not easy. See Jacobson [33] and Maurer [59] for further details.

(6) Can the number field sieve methods (see Lenstra and Lenstra [53]) be applied to the problem of computing R_d with the success that they have enjoyed with respect to the integer factoring problem or the discrete logarithm problem? So far no one knows how to do this, and it is believed by some experts that it may not be possible.

Thus, the current state of the art in solving the Pell equation is far from satisfactory. In spite of the enormous progress that has been made on this problem in the last few decades, we are still without answers to many fundamental questions. However, we are, it seems, beginning to understand what the questions should be.

References

[1]   C. S. Abel, *Ein Algorithmus zur Berechnung der Klassenzahl und des Regulators reellquadratischer Ordnungen*, Ph.D. thesis, Universität des Saarlandes, Saarbrücken, Germany, 1994.

[2]   W. Adams, *On the relationship between the convergents of the nearest integer and regular continued fractions*, Math. Comp. **33** (1979), 1321–1331.

[3]   American Council of Learned Societies, *Biographical dictionary of mathematics*, vol. 4, Scribner's Sons, New York, 1991.

[4]   A. Amthor, *Das Problema bovinum des Archimedes*, Zeitschrift für Math. u. Physik (Hist. Litt. Abtheilung) **25** (1880), 153–171.

[5]   E. Bach, *Explicit bounds for primality testing and related problems*, Math. Comp. **55** (1990), 335–380.

[6]   ———, *Improved approximations for Euler products*, Number Theory: CMS Proc., vol. 15, Amer. Math. Soc., Providence, RI, 1995, pp. 13–28.

[7]   J. Buchmann, *A subexponential algorithm for the determination of class groups and regulators of algebraic number fields*, Séminaire de Théorie des Nombres (Paris), 1988–1989, pp. 27–41.

---

[7]Of course, even if the Bach bounds were considerably reduced, we still have to find a sufficient number of relations in order to guarantee (8.3). Thus, there is an interesting trade-off here that needs investigation.

[8]   J. Buchmann, C. Thiel, and H. C. Williams, *Short representation of quadratic integers*, Computational Algebra and Number Theory, Mathematics and its Applications, vol. 325, Kluwer, Dordrecht, 1995, pp. 159–185.

[9]   J. Buchmann and H. C. Williams, *On the infrastructure of the principal ideal class of an algebraic number field of unit rank one*, Math. Comp. **50** (1988), 569–579.

[10]   ――――, *On the existence of a short proof for the value of the class number and regulator of a real quadratic field*, Number Theory and Applications, NATO ASI Series C, vol. 265, Kluwer, Dordrecht, 1989, pp. 327–345.

[11]   J. W. S. Cassels, *An Introduction to the Geometry of Numbers*, Springer, Berlin, 1971.

[12]   H. Cohen, *A Course in Computational Algebraic Number Theory*, Graduate Texts in Mathematics, vol. 138, Springer, 1993.

[13]   H. Cohen, F. Diaz y Diaz, and M. Olivier, *Calculs de nombres de classes et de régulateurs de corps quadratiques en temps sous-exponentiel*, Séminaire de Théorie des Nombres (Paris), 1993, pp. 35–46.

[14]   ――――, *Subexponential algorithms for class and unit group computations*, J. Symb. Comp. **24** (1997), 443–441.

[15]   H. Cohen and H. W. Lenstra, Jr., *Heuristics on class groups*, Number Theory, Lecture Notes in Mathematics 1052, Springer, Berlin, 1984, pp. 26–36.

[16]   ――――, *Heuristics on class groups of number fields*, Number Theory, Lecture Notes in Mathematics 1068, Springer, Berlin, 1984, pp. 33–62.

[17]   E. J. Dijksterhuis, *Archimedes*, Princeton University Press, Princeton, 1987.

[18]   G. Eisenstein, *Allgemeine Untersuchengen über die Formen dritten Grades mit drei variabeln, welche der Kreistheilung ihre Enstehung verdanken*, J. Reine Angew. Math. **28** (1844), 289–374.

[19]   G. Eneström, *Über der Ursprung der Benennung "Pellsche Gleichung"*, Bibliotheca Math. **3** (1902), 204–207.

[20] L. Euler, *Correspondance, Mathématique et Physique*, St. Petersbourg, 1843, P. H. Fuss (ed.), T1, pp. 35–39.

[21] P. Fermat, *Oeuvres de Fermat*, vol. II, Gauthier-Villars, Paris, 1891–1912.

[22] D. H. Fowler, *The Mathematics of Plato's Academy: A New Reconstruction*, Clarendon Press, Oxford, 1999.

[23] P. M. Fraser, *Ptolemaic Alexandria*, vol. 1, The Clarendon Press, Oxford, 1972.

[24] A. Granville and K. Soundararajan, *The distribution of values of* $L(1, \chi)$, to appear.

[25] J. L. Hafner and K. S. McCurley, *A rigorous subexponential algorithm for computation of class groups*, J. Amer. Math. Soc. **2** (1989), 837–850.

[26] S. Hallgren, *Polynomial-time quantum algorithms for Pell's equation and the Principal Ideal Problem*, unpublished manuscript, 2001.

[27] H. Hankel, *Zur Geschichte der Mathematik in Altertum und Mittelalter*, 2nd ed., Hildesheim, 1965.

[28] T. L. Heath, *Diophantus of Alexandria*, 2nd ed., Dover, New York, 1964.

[29] J. E. Hofmann, *Studien zur Zahlentheorie Fermats*, Abh. Preuss. Akad. Wiss. (1994), no. 7.

[30] L.-K. Hua, *Introduction to number theory*, Springer-Verlag, Berlin, 1982, pp. 326–329.

[31] M. J. Jacobson, Jr., *Computational techniques in quadratic fields*, Master's thesis, University of Manitoba, Winnipeg, Manitoba, 1995.

[32] _____, *Experimental results on class groups of real quadratic fields (extended abstract)*, Algorithmic Number Theory - ANTS-III (Portland, Oregon), Lecture Notes in Computer Science, vol. 1423, Springer-Verlag, Berlin, 1998, pp. 463–474.

[33] _____, *Subexponential Class Group Computation in Quadratic Orders*, Ph.D. thesis, Technische Universität Darmstadt, Darmstadt, Germany, 1999.

[34] ———, *Computing discrete logarithms in quadratic orders*, J. Cryptology **13** (2000), 473–492.

[35] M. J. Jacobson, Jr., R. F. Lukes, and H. C. Williams, *An investigation of bounds for the regulator of quadratic fields*, Experimental Mathematics **4** (1995), 211–225.

[36] M. J. Jacobson, Jr., R. Scheidler, and H. C. Williams, *The efficiency and security of a real quadratic field based-key exchange protocol*, Public-Key Cryptography and Computational Number Theory, Walter de Gruyter, Berlin, to appear.

[37] M. J. Jacobson, Jr. and H. C. Williams, *The size of the fundamental solutions of consecutive Pell equations*, Experimental Math. **9** (2000), 631–640.

[38] ———, *New Quadratic Polynomials with High Densities of Prime Values*, to appear.

[39] ———, *Modular Arithmetic on Elements of Small Norm in Quadratic Fields*, unpublished MS.

[40] P. T. Joshi, *The size of $L(1, \chi)$ for real nonprincipal residue characters χ with prime modulus*, J. Number Theory **2** (1970), 58–73.

[41] W. Knorr, *Archimedes and the measurement of the circle: a new interpretation*, Arch. Hist. Exact Sci. **15** (1975), 115–140.

[42] H. Konen, *Geschichte der Gleichung $t^2 - Du^2 = 1$*, Leipzig, 1901.

[43] R. Kortum and G. McNeil, *A Table of Periodic Continued Fractions*, Lockheed Aircraft Corp., Sunnyvale, CA, 1961.

[44] B. Krumbiegel, *Das problema bovinum des Archimedes*, Zeitschrift für Math. u. Physik (Hist. Litt. Abtheilung) **25** (1880), 121–136.

[45] J. C. Lagarias, *Succinct Certificates for the Solvability of Binary quadratic Diophantine Equations (Extended Abstract)*, Proc. 20th IEEE Symp. on Foundations of Computer Science, 1979, pp. 47–54.

[46] ———, *Succinct Certificates for the Solvability of Binary quadratic Diophantine Equations*, Technical Memorandum 81-11216-54, Bell Labs, Sept. 28, 1981.

[47] D. H. Lehmer, *A list of errors in tables of the Pell equation*, Bull. Amer. Math. Soc. **32** (1926), 545–550.

[48]   _____, *On the indeterminate equation* $t^2 - p^2 D u^2 = 1$, Annals of Math. **27** (1926), 471–476.

[49]   _____, *On the multiple solutions of the Pell equation*, Annals of Math. **30** (1928), 66–72.

[50]   _____, *Guide to Tables in the Theory of Numbers*, National Research Council, Washington, 1941.

[51]   P. G. Lejeune-Dirichlet, *Vorlesungen über Zahlentheorie*, 4th ed., Braunschweig, 1893.

[52]   A. K. Lenstra and H. W. Lenstra, Jr., *Algorithms in Number Theory*, Technical Report 87-008, University of Chicago, 1987.

[53]   A. K. Lenstra and H. W. Lenstra, Jr. (eds.), *The Development of the Number Field Sieve*, Lecture Notes in Mathematics, vol. 1554, Springer, Berlin, 1993.

[54]   H. W. Lenstra, Jr., *On the calculation of regulators and class numbers of quadratic fields*, London Math. Soc. Lecture Note Series **56** (1982), 123–150.

[55]   J. E. Littlewood, *On the class number of the corpus* $P(\sqrt{-k})$, Proc. London Math. Soc. **27** (1928), 358–372.

[56]   S. Louboutin, *Explicit upper bounds for* $|L(1, \chi)|$ *for primitive even Dirichlet characters*, Acta Arith. **101** (2002), 1–18.

[57]   M. S. Mahoney, *The Mathematical Career of Pierre de Fermat*, 2nd ed., Princeton, New Jersey, 1994.

[58]   A. Martin, *Solution*, The Analyst **4** (1877), 154–155.

[59]   M. Maurer, *Regulator approximation and fundamental unit computation for real quadratic orders*, Ph.D. thesis, Technische Universität Darmstadt, Darmstadt, Germany, 2000.

[60]   K. S. McCurley, *Cryptographic key distribution and computation in class groups*, Proc. NATO ASI on Number Theory and Applications, NATO ASI series C, vol. 265, Kluwer, Dordrecht, 1989, pp. 459–479.

[61]   R. A. Mollin and H. C. Williams, *Computation of the class number of a real quadratic field*, Utilitas Mathematica **41** (1992), 259–308.

[62]   P. L. Montgomery, *A survey of modern integer factoring algorithms*, CWI Quarterly **7** (1994), 337–366.

[63] H. L. Nelson, *A solution to Archimedes' Cattle Problem*, J. Recreational Math. **13** (1981), 162–176.

[64] M. Newman, *Integral Matrices*, Pure and Applied Mathematics, vol. 45, Academic Press, New York, 1972.

[65] W. Patz, *Tafel der Regelmässigen Kettenbrüche für die Quadratwurzeln aus den Natürlichen Zahlen von 1 − 10000*, Becker and Erler, Leipzig, 1941.

[66] _____, *Tafel der Regelmässigen, Kettenbrüche und ihres vollständigen Quotienten für die Quadratwuzeln aus den Natürlichen Zahlen von 1 − 10000*, Akademie-Verlag, Berlin, 1955.

[67] O. Perron, *Die Lehre von den Kettenbrüchen*, 2nd ed., Chelsea, New York, undated.

[68] M. Pohst and H. Zassenhaus, *On unit computation in real quadratic fields*, Symbolic and Algebraic Computation, Lecture Notes in Computer Science, vol. 72, Springer, Berlin, 1979.

[69] _____, *Über die Berechnung von Klassenzahlen und Klassengruppen algebraischer Zahlkörper*, J. reine angew. Math. **361** (1985), 50–72.

[70] J. H. Rahn, *An Introduction to Alegbra*, T. Brancker (Tr.), London, 1668.

[71] R. Schoof, *Quadratic fields and factorization*, Computational Methods in Number Theory (H. W. Lenstra, Jr. and R. Tijdeman, eds.), no. 155, Part II, Math. Centre Tracts, Amsterdam, 1983, pp. 235–286.

[72] P. Schreiber, *A note on the Cattle Problem of Archimedes*, Historia Math. **20** (1993), 304–306.

[73] J. F. Scott, *The Mathematical Work of John Wallis*, London, 1938.

[74] Ch. J. Scriba, *John Pell's English edition of J. H. Rahn's Teutsche Algebra*, For Dirk Struick, R. S. Cohen et al., (eds.), pp. 261–274, Reidel, Dordrecht, 1974.

[75] C.-O. Selenius, *Kettenbruchtheoretische Erklärung der zyklischen Methode zur Lösung der Bhaskara-Pell-Gleichung*, Acta acad. Aboensis, math. phys. **23** (1963), no. 10.

[76] _____, *Rationale of the chakravala process of Jayadeva and Bhaskara II*, Historia Math. **2** (1975), 167–184.

[77] D. Shanks, *Review RMT30*, Math. Comp. **16** (1962), 377–379, also **23** (1969), 217, 219.

[78] ———, *Review RMT10*, Math. Comp. **23** (1969), 213–214.

[79] ———, *Class number, a theory of factorization and genera, 1969 Number Theory Institute*, Proc. Symp. Pure Math. 20, AMS, Providence, R.I., 1971, pp. 415–440.

[80] ———, *The infrastructure of real quadratic fields and its applications*, Proc. 1972 Number Theory Conf. (Boulder, Colorado), 1972, pp. 217–224.

[81] ———, *Systematic examination of Littlewood's bounds on $L(1,\chi)$*, Analytic Number Theory, Proc. Symp. Pure Math. (Providence, R.I.), vol. 24, AMS, 1973, pp. 267–283.

[82] ———, *Review RMT11*, Math. Comp. **28** (1974), 333–334.

[83] ———, *On Gauss and composition I, II*, Number Theory and Applications, NATO ASI series C, vol. 265, Kluwer Academic Press, Dordrecht, 1989, pp. 163–179.

[84] K. Shankar Shukla, *Acarya Jayadeva, the mathematician*, Ganita **5** (1954), 1–20.

[85] A. Srinivasan, *Computations of class numbers of real quadratic fields*, Math. Comp. **67** (1998), 1285–1308.

[86] P. Tannery, *L'Arithmétique des Grecs dans Pappus*, Mémoires Scientifiques, T.I., Toulouse, 1912–37, pp. 80–105.

[87] ———, *Sur la mesure de circle d'Archimède*, Mémoires Scientifiques, T.I., Toulouse, 1912–37, pp. 226–253.

[88] E. Teske and H. C. Williams, *A problem concerning a character sum*, Experimental Math. **8** (1999), 63–72.

[89] A. J. van der Poorten, *A Note on NUCOMP*, unpublished MS.

[90] A. J. van der Poorten, H. te Riele, and H. C. Williams, *Computer verification of the Ankeny-Artin-Chowla conjecture for all primes less than 100000000000*, Math. Comp. **70** (2001), 1311–1328.

[91] I. Vardi, *Archimedes' Cattle Problem*, Amer. Math. Monthly **105** (1998), 305–319.

[92] U. Vollmer, *Asymptotically fast discrete logarithms in quadratic number fields*, Algorithmic Number Theory - ANTS-IV (Leiden, The Netherlands), Lecture Notes in Computer Science, vol. 1838, Springer, 2000, pp. 581–594.

[93] W. Waterhouse, *On the Cattle Problem of Archimedes*, Historia Math. **22** (1995), 186–187.

[94] A. Weil, *Fermat et l'équation de Pell*, Collected Papers, vol. III, Springer, Berlin, 1979, pp. 413–419.

[95] ———, *Number Theory. An Approach Through History*, Birkhäuser, Boston, 1984.

[96] G. Wertheim, *Die Algebra des Johann Heinrich Rahn (1659) und die englische übersetzung derselben*, Bibliotheca Math. **3** (1902), no. 3, 113–126.

[97] E. E. Whitford, *The Pell Equation*, College of the City of New York, New York, 1912.

[98] H. C. Williams, *On mid period criteria for the nearest integer continued fraction expansion of $\sqrt{D}$*, Utilitas Math. **27** (1985), 169–185.

[99] H. C. Williams and J. Broere, *A computational technique for evaluating $L(1, \chi)$ and the class number of a real quadratic field*, Math. Comp. **30** (1976), 887–893.

[100] H. C. Williams and P. A. Buhr, *Calculation of the regulator of $\mathbb{Q}\left(\sqrt{D}\right)$ by use of the nearest integer continued fraction algorithm*, Math. Comp. **33** (1979), 369–381.

[101] H. C. Williams, R. A. German, and C. R. Zarnke, *Solution of the Cattle Problem of Archimedes*, Math. Comp. **19** (1965), 671–674.

A Central Limit Theorem for the Number of Distinct Degrees of Prime Factors in Additive Arithmetical Semigroups

Wen-Bin Zhang

In this paper, we will prove a central limit theorem for the number of distinct degrees of prime factors of elements of additive arithmetical semigroups. Besides information on the distribution of distinct degrees of prime factors, this provides an interesting natural example of a central limit theorem for a non-additive function in an additive arithmetical semigroup.

Let $\mathbb{F}_q[x]$ be the ring of polynomials in one indeterminate x over a finite field $\mathbb{F}_q$. Investigation of the distribution of distinct degrees of prime factors in additive arithmetic semigroups is motivated by the study of the distribution of distinct degrees of irreducible factors in the factorization in $\mathbb{F}_q[x]$. Let $f(x) \in \mathbb{F}_q[x]$ be a monic polynomial and let $\rho(f)$ denote the number of distinct degrees of monic irreducible factors of $f(x)$. Many deterministic as well as probabilistic factorization algorithms for polynomials in $\mathbb{F}_q[x]$ require, as an initial step, that a distinct degree factorization of a given polynomial be performed. This method factors the monic polynomial f into factors f_d such that each f_d is the product of all the irreducible factors of f of degree d. If f has k irreducible factors of $\rho(f)$ distinct degrees, then at most $\rho(f)$ polynomials f_d require factorization (if they are not all already irreducible), and a further $k - \rho(f)$ irreducible factors need to be extracted in order to obtain the complete factorization. It is therefore of interest to determine the average value of $\rho(f)$, as well as its distribution law.

A. Knopfmacher [3] proved that

$$q^{-n} \sum_{\partial(f)=n} \rho(f) = S(n) + c_2(q) + O\left(\frac{1}{n}\right),$$

where $\partial(f)$ denotes the degree of $f(x)$,

$$S(n) := \sum_{1 \le k \le n} \frac{1}{k},$$

and $c_2(q)$ is a constant (see (3) below). Further, in [5], R. Walimont discussed this problem in the context of additive arithmetical semigroups.

Let $(\mathcal{G}, \partial)$ be an additive arithmetical semigroup. By definition, this means that $\mathcal{G}$ is a free commutative semigroup with identity 1 such that $\mathcal{G}$ has a countable generating set $\mathcal{P}$ of "primes" p and $\partial : \mathcal{G} \to \mathbb{N} \cup \{0\}$ is a "degree" mapping satisfying

(i) $\partial(ab) = \partial(a) + \partial(b)$ for all $a, b \in \mathcal{G}$;

(ii) the total number $G(n)$ of elements of degree n in $\mathcal{G}$ is finite for each $n \geq 0$.

Then, from (i) and (ii), $\partial(a) = 0$ if and only if $a = 1$ and $G(0) = 1$.

The ring $\mathbb{F}_q[x]$ as well as many familiar structures from algebra provide examples of additive arithmetical semigroups [4, pp. 7–15].

For $a \in \mathcal{G}$, $a \neq 1$, let

$$\omega(a) := \#\{p \in \mathcal{P} : p \mid a\},$$
$$\omega_m(a) := \#\{p \in \mathcal{P} : p \mid a \quad \text{and} \quad \partial(p) = m\},$$

and

$$\rho(a) := \sum_{\substack{m \\ \omega_m(a)>0}} 1.$$

Thus $\omega(a)$ is the number of distinct prime divisors of a and $\rho(a)$ is the number of distinct degrees of prime divisors of a. We note that $\omega(a)$ is additive, whereas $\rho(a)$ is not. For convenience, let $\omega(1) = 0$ and $\rho(1) = 0$.

Let $P(n)$ denote the number of prime elements p in $\mathcal{P}$ of degree n. Assume that there exist constants $A > 0$ and $q > 1$ such that

$$\sum_{n=0}^{\infty} \sup_{m \geq n} |G(m)q^{-m} - A| < \infty. \tag{1}$$

Then

$$P(n) \ll q^n n^{-1}$$

and

$$\sum_{m \leq n} P(m)q^{-m} = S(n) + c_1 + O\left(\frac{1}{n}\right), \tag{2}$$

where

$$c_1 = \sum_{m=1}^{\infty} \left(P(m)q^{-m} - m^{-1}\right).$$

Warlimont [5] showed that

$$\bar{\rho}(n) := \frac{1}{G(n)} \sum_{\partial(a)=n} \rho(a) \tag{3}$$

$$= S(n) + c_2 + O\left(t(n)\log n + \frac{1}{n}\right)$$

and

$$\frac{1}{G(n)} \sum_{\partial(a)=n} (\rho(a) - \bar{\rho}(n))^2 \tag{4}$$

$$= S(n) + c_2 - c_3 - \zeta(2) + O\left(t(n)\log^2 n + \frac{\log n}{n}\right),$$

where

$$c_2 = \sum_{m=1}^{\infty} \left(1 - m^{-1} - (1 - q^{-m})^{P(m)}\right),$$

$$c_3 = \sum_{m=1}^{\infty} \left(1 - (1 - q^{-m})^{P(m)}\right)^2,$$

and

$$t(n) := \sup_{m \geq \frac{n}{2}} |G(m)q^{-m} - A|.$$

Formula (3) is similar to the well-known formula (see, e.g., [5])

$$\bar{\omega}(n) := \frac{1}{G(n)} \sum_{\partial(a)=n} \omega(a) \tag{5}$$

$$= S(n) + c_1 + O\left(t(n)\log n + \frac{1}{n}\right)$$

and a similar formula of $\Omega(a)$. In view of the central limit theorem with remainder estimate for $\omega(a)$ on $\mathcal{G}$ proved recently in [6], [7] (see also [2]), which is an analogue of the celebrated theorem of Erdös and Kac [1] in some sense, this similarity and the intrinsic connection between the two functions $\rho(a)$ and $\omega(a)$ suggest that there should also exist a central limit theorem for $\rho(a)$. This suggestion is confirmed by the following theorem.

For convenience, let $P(a, x, m)$ denote a proposition, with parameters x and m, on elements a in an additive arithmetical semigroup $\mathcal{G}$, and let

$$\nu_m(P(a, x, m)) := \sum_{\substack{\partial(a)=m \\ P(a,x,m)}} 1$$

be the number of true values of $P(a, x, m)$ among those a with $\partial(a) = m$. Also, let

$$\Phi(x) := \frac{1}{\sqrt{2\pi}} \int_{-\infty}^{x} e^{-\frac{t^2}{2}} dt.$$

Theorem 1. *Suppose there exist constant $A > 0$, $q > 1$, and $\gamma > 2$ such that*

$$G(m) = Aq^m + O(q^m m^{-\gamma}), \quad m \geq 1.$$

Then

$$\frac{1}{G(m)} \nu_m \left(\frac{\rho(a) - \log m}{\sqrt{\log m}} \leq x \right) = \Phi(x) + O\left((\log m)^{-\frac{1}{3}} \right) \qquad (6)$$

uniformly for all real numbers x and integers $m \geq 2$.

Remark. The error term in (6) is not the best possible. We conjecture that (6) holds with error term $O((\log m)^{-1/2})$.

Theorem 1 with an error estimate $O((\log m)^{-1/4})$ can be proved by appealing to [5] and [7] directly. However, the better remainder $O((\log m)^{-1/3})$ requires an asymptotic estimate for $(G(n))^{-1} \sum_{\partial(a)=n} (\omega(a) - \rho(a))^2$, which is proved in Lemma 2.

Let

$$Z(x) := \sum_{a \in \mathcal{G}} x^{\partial(a)} = \prod_{p \in \mathcal{P}} (1 - x^{\partial(p)})^{-1}$$

be the zeta function of $(\mathcal{G}, \partial)$. We first seek a generating function of $\sum_{\partial(a)=n} \omega(a)\rho(a)$ via a combinatorial argument.

Lemma 1. *We have*

$$\sum_{n=0}^{\infty} \left(\sum_{\partial(a)=n} \omega(a)\rho(a) \right) x^n \qquad (7)$$

$$= Z(x) \left[\left(\sum_{m=1}^{\infty} (1 - (1 - x^m)^{P(m)}) \right) \left(\sum_{m=1}^{\infty} P(m)x^m \right) \right.$$

$$\left. + \sum_{m=1}^{\infty} P(m)x^m (1 - x^m)^{P(m)} \right].$$

Proof. Let

$$K(x, y, z) := \sum_{a \in \mathcal{G}} x^{\partial(a)} y^{\rho(a)} z^{\omega(a)}.$$

Then

$$K(x, y, z) = 1 + \sum_{n,k,\ell=1}^{\infty} w(n, k, \ell) x^n y^k z^\ell, \qquad (8)$$

where

$$w(n,k,\ell) := \sum_{\substack{a \in \mathcal{G} \\ \partial(a)=n, \rho(a)=k \\ \omega(a)=\ell}} 1 = \#\{a \in \mathcal{G} : \partial(a) = n, \rho(a) = k, \omega(a) = \ell\}.$$

The function $K(x,y,z)$ can be written as an infinite product. To this end, let

$$\mathcal{P}_m = \{p \in \mathcal{P} : \partial(p) = m\}$$

and

$$\mathcal{G}_m = \{a \in \mathcal{G} : a \neq 1, p \,|\, a \quad \text{implies} \quad p \in \mathcal{P}_m\}.$$

Then

$$K(x,y,z) = 1 + \sum_{k=1}^{\infty} y^k \sum_{\substack{a \in \mathcal{G} \\ \rho(a)=k}} z^{\omega(a)} x^{\partial(a)}$$

$$= 1 + \sum_{k=1}^{\infty} y^k \sum_{0<m_1<\cdots<m_k} \sum_{\substack{a=b_1\cdots b_k \\ b_i \in \mathcal{G}_{m_i} \\ i=1,\ldots,k}} z^{\omega(a)} x^{\partial(a)}.$$

Note that, given $0 < m_1 < \cdots < m_k$, the innermost sum on the right-hand side equals

$$\sum_{\substack{b_i \in \mathcal{G}_{m_i} \\ i=1,\ldots,k}} \prod_{i=1}^{k} z^{\omega(b_i)} x^{\partial(b_i)} = \prod_{i=1}^{k} \sum_{b_i \in \mathcal{G}_{m_i}} z^{\omega(b_i)} x^{\partial(b_i)}.$$

Hence

$$K(x,y,z) = \prod_{m=1}^{\infty} \left(1 + y \sum_{b \in \mathcal{G}_m} z^{\omega(b)} x^{\partial(b)} \right).$$

The sums appearing in the factors on the right-hand side can be rewritten as

$$\sum_{\ell=1}^{\infty} z^{\ell} \sum_{\substack{b \in \mathcal{G}_m \\ \omega(b)=\ell}} x^{\partial(b)}.$$

By the unique factorization, the inner sum equals

$$\sum_{\substack{p_i \in \mathcal{P}_m \\ p_i \neq p_j \quad \text{for} \quad i \neq j \\ i,j=1,\ldots,\ell}} \sum_{n_1,\ldots,n_\ell=1}^{\infty} x^{n_1 \partial(p_1)+\cdots+n_\ell \partial(p_\ell)}$$

$$= \sum_{\substack{p_i \in \mathcal{P}_m \\ p_i \neq p_j \quad \text{for} \quad i \neq j \\ i,j=1,\ldots,\ell}} \prod_{i=1}^{\ell} \frac{x^{\partial(p_i)}}{1 - x^{\partial(p_i)}}.$$

Hence

$$\sum_{b \in \mathcal{G}_m} z^{\omega(b)} x^{\partial(b)} = \prod_{p \in \mathcal{P}_m} \left(1 + \frac{z x^{\partial(p)}}{1 - x^{\partial(p)}}\right) - 1$$

$$= \left(1 + \frac{z x^m}{1 - x^m}\right)^{P(m)} - 1.$$

Then substitution yields

$$K(x, y, z) = \prod_{m=1}^{\infty} (1 + y M(m, x, z)), \tag{9}$$

where

$$M(m, x, z) = \left(1 + \frac{z x^m}{1 - x^m}\right)^{P(m)} - 1.$$

To deduce (7), we note that, on the one hand, (8) gives, by partial differentiation,

$$\frac{\partial^2 K}{\partial y \partial z} = \sum_{n,k,\ell=1}^{\infty} k \ell w(n, k, \ell) x^n y^{k-1} z^{\ell-1}$$

and hence

$$K_{yz}(x, 1, 1) = \sum_{n=1}^{\infty} \left(\sum_{k,\ell=1}^{\infty} k \ell w(n, k, \ell)\right) x^n$$

$$= \sum_{n=1}^{\infty} \left(\sum_{\partial(a)=n} \omega(a)\rho(a)\right) x^n.$$

On the other hand, from (9), we obtain by logarithmic differentiation,

$$\frac{\partial K}{\partial y} = K(x, y, z) S(x, y, z),$$

where

$$S(x, y, z) = \sum_{m=1}^{\infty} \frac{M(m, x, z)}{1 + yM(m, x, z)}.$$

Then

$$\frac{\partial^2 K}{\partial y \partial z} = \frac{\partial K}{\partial z} S + K \frac{\partial S}{\partial z}$$

$$= K \left[S \sum_{m=1}^{\infty} \frac{y \frac{\partial}{\partial z} M(m, x, z)}{1 + yM(m, x, z)} + \sum_{m=1}^{\infty} \frac{\frac{\partial}{\partial z} M(m, x, z)}{(1 + yM(m, x, z))^2} \right]$$

by a further application of logarithmic differentiation. Thus

$$K_{yz}(x, 1, 1) = K(x, 1, 1) \times$$

$$\times \left[S(x, 1, 1) \sum_{m=1}^{\infty} \frac{M_z(m, x, 1)}{1 + M(m, x, 1)} + \sum_{m=1}^{\infty} \frac{M_z(m, x, 1)}{(1 + M(m, x, 1))^2} \right]. \quad (10)$$

It is easily seen that

$$M(m, x, 1) = (1 - x^m)^{-P(m)} - 1,$$

$$M_z(m, x, 1) = P(m)x^m(1 - x^m)^{-P(m)},$$

$$S(x, 1, 1) = \sum_{m=1}^{\infty} (1 - (1 - x^m)^{P(m)}),$$

$$\sum_{m=1}^{\infty} \frac{M_z(m, x, 1)}{1 + M(m, x, 1)} = \sum_{m=1}^{\infty} P(m)x^m,$$

$$\sum_{m=1}^{\infty} \frac{M_z(m, x, 1)}{(1 + M(m, x, 1))^2} = \sum_{m=1}^{\infty} P(m)x^m(1 - x^m)^{P(m)},$$

and

$$K(x, 1, 1) = \sum_{a \in \mathcal{G}} x^{\partial(a)} = Z(x).$$

Therefore, substitution of the last four formulas into (10) yields (7). $\quad \square$

Lemma 2. *Assume (1). Then*

$$\frac{1}{G(n)} \sum_{\partial(a)=n} (\omega(a) - \rho(a))^2 = c + O\left(t(n)\log^2 n + \frac{\log n}{n}\right), \quad (11)$$

where c is an explicit constant.

Remark. The constant c is given by

$$c = 2 \sum_{m=1}^{\infty} q^{-m} P(m) \left(1 - (1 - q^{-m})^{P(m)}\right) \tag{12}$$

$$+ \sum_{m=1}^{\infty} \left(1 - q^{-m} P(m) - (1 - q^{-m})^{P(m)}\right)$$

$$- \sum_{m=1}^{\infty} P(m) q^{-2m} - \sum_{m=1}^{\infty} \left(1 - (1 - q^{-m})^{P(m)}\right)^2$$

$$+ \left(\sum_{m=1}^{\infty} \left(1 - q^{-m} P(m) - (1 - q^{-m})^{P(m)}\right) \right)^2.$$

Proof. We first show that

$$\frac{1}{G(n)} \sum_{\partial(a)=n} \omega(a) \rho(a)$$

$$= S^2(n) + (c_1 + c_2 + 1) S(n) + c_1 (c_2 + 1) - \zeta(2) - c_4 \tag{13}$$

$$+ O\left(t(n) \log^2 n + \frac{\log n}{n}\right),$$

where

$$c_4 = \sum_{m=1}^{\infty} q^{-m} P(m) \left(1 - (1 - q^{-m})^{P(m)}\right).$$

Using the generating function (7) with $x = q^{-1} z$, we have

$$\sum_{n=0}^{\infty} \left(q^{-n} \sum_{\partial(a)=n} \omega(a) \rho(a) \right) z^n$$

$$= Z(q^{-1} z) \left[\left(\sum_{m=1}^{\infty} (1 - (1 - q^{-m} z^m)^{P(m)}) \right) \left(\sum_{m=1}^{\infty} q^{-m} P(m) z^m \right) \right.$$

$$\left. + \sum_{m=1}^{\infty} q^{-m} P(m) z^m (1 - q^{-m} z^m)^{P(m)} \right]. \tag{14}$$

Now,

$$Z(q^{-1} z) = \frac{A}{1 - z} + R(z),$$

where

$$R(z) = \sum_{m=0}^{\infty} r_m z^m$$

with $r_m = G(m)q^{-m} - A$. Setting $p_m = q^{-m}P(m)$, we have

$$\sum_{m=1}^{\infty} \left(1 - (1 - q^{-m}z^m)^{P(m)}\right) = \hat{P}(z) + B(z),$$

say. Here

$$\hat{P}(z) = \sum_{m=1}^{\infty} p_m z^m$$

and

$$B(z) = \sum_{m=1}^{\infty} b_m z^m$$

with

$$b_m = -q^{-m} \sum_{\substack{k|m \\ k<m}} \binom{P(k)}{m/k}(-1)^{\frac{m}{k}}.$$

Also,

$$\sum_{m=1}^{\infty} q^{-m}P(m)z^m(1 - q^{-m}z^m)^{P(m)} = \hat{P}(z) - Q(z),$$

where

$$Q(z) = \sum_{m=1}^{\infty} q^{-m}P(m)z^m \left[1 - (1 - q^{-m}z^m)^{P(m)}\right]$$

$$= \sum_{n=2}^{\infty} q_n z^n$$

with

$$q_n = q^{-n} \sum_{\substack{1 \le k \le P(m) \\ (k+1)m=n}} (-1)^{k+1}\binom{P(m)}{k}P(m).$$

Then the right-hand side of (14) can be written as

$$\left(\frac{A}{1-z} + R(z)\right)\left(\hat{P}^2(z) + (B(z)+1)\hat{P}(z) - Q(z)\right). \qquad (15)$$

Note that (see [5])

$$b_m \ll m^{-2}.$$

Also,

$$q_n \leq q^{-n} \left[\left(\frac{Kq^{\frac{n}{2}}}{n/2} \right)^2 + \sum_{\substack{2 \leq k \leq P(m) \\ (k+1)m=n}} \binom{P(m)}{k} P(m) \right]$$

$$\ll \frac{1}{n^2} + q^{-n} \sum_{\substack{k \geq 2 \\ (k+1)m=n}} \frac{(P(m))^{k+1}}{k!}$$

$$\leq \frac{1}{n^2} + q^{-n} \sum_{k \geq 2} \frac{1}{k!} \left(\frac{Kq^{n/(k+1)}}{n/(k+1)} \right)^{k+1}$$

$$\ll \frac{1}{n^2} + \sum_{k \geq 2} (k+1) \left(\frac{Ke}{n} \right)^{k+1}$$

$$\ll \frac{1}{n^2}$$

for $n \geq 2Ke$, by Stirling's formula, since $P(m) \leq Kq^m m^{-1}$ for all $m \in \mathbb{N}$. Let

$$\hat{P}^2(z) + (B(z) + 1)\hat{P}(z) - Q(z) = \sum_{n=1}^{\infty} s_n z^n.$$

Then

$$s_n = \sum_{k+\ell=n} p_k p_\ell + \sum_{k+\ell=n} b_k p_\ell + p_n - q_n$$

$$\ll \frac{\log n}{n}$$

since

$$\sum_{k+\ell=n} b_k p_\ell \ll \frac{1}{n} \sum_{1 \leq k \leq \frac{n}{2}} |b_k| + \frac{1}{n^2} \sum_{\frac{n}{2} < k < n} |p_{n-k}| \ll \frac{1}{n}.$$

Now, from (14) and (15), we have

$$q^{-n} \sum_{\partial(a)=n} \omega(a)\rho(a)$$

$$= A \left(\sum_{k+\ell \leq n} p_k p_\ell + \sum_{k+\ell \leq n} b_k p_\ell + \sum_{k \leq n} p_k - \sum_{k \leq n} q_k \right) + \sum_{k+\ell=n} r_k s_\ell.$$

Note that (see [5])

$$\sum_{k+\ell \leq n} p_k p_\ell = S^2(n) + 2c_1 S(n) + c_1^2 - \zeta(2) + O\left(\frac{\log n}{n}\right),$$

$$\sum_{k+\ell \leq n} b_k p_\ell = B(1)(S(n) + c_1) + O\left(\frac{\log n}{n}\right),$$

and

$$\sum_{q \leq n} q_k = Q(1) + O\left(\frac{1}{n}\right).$$

Moreover,

$$\sum_{k+\ell=n} |r_k||s_\ell| = \left(\sum_{1 \leq k \leq \frac{n}{2}} + \sum_{\frac{n}{2} < k < n}\right) |r_k||S_{n-k}|$$

$$\ll \frac{\log n}{n} + t(n) \log^2 n.$$

Therefore, we obtain

$$q^{-n} \sum_{\partial(a)=n} \omega(a)\rho(a)$$

$$= A\Big(S^2(n) + (2c_1 + B(1) + 1)S(n) + c_1^2 - \zeta(2)$$

$$+ (B(1) + 1)c_1 - Q(1)\Big) + O\left(t(n)\log^2 n + \frac{\log n}{n}\right).$$

Then (13) follows from

$$\frac{1}{G(n)} \sum_{\partial(a)=n} \omega(a)\rho(a) = \frac{q^{-n}}{A + r_n} \sum_{\partial(a)=n} \omega(a)\rho(a)$$

and

$$B(1) = c_2 - c_1, \quad Q(1) = c_4.$$

To prove (11), we note that the left-hand side of (11) equals

$$\frac{1}{G(n)} \left(\sum_{\partial(a)=n} (\omega(a) - \bar{\omega}(n))^2 + \sum_{\partial(a)=n} (\rho(a) - \bar{\rho}(n))^2 \right. \tag{16}$$

$$\left. -2 \sum_{\partial(a)=n} \omega(a)\rho(a)\right) + 2\bar{\omega}(n)\bar{\rho}(n) + (\bar{\omega}(n) - \bar{\rho}(n))^2.$$

It is known [5] that

$$\frac{1}{G(n)} \sum_{\partial(a)=n} (\omega(a) - \bar{\omega}(n))^2 = S(n) + c_1 - c_5 - \zeta(2) \tag{17}$$

$$+ O\left(t(n) \log^2 n + \frac{\log n}{n}\right),$$

where

$$c_5 = \sum_{m=1}^{\infty} P(m) q^{-2m}.$$

Thus (11) follows from (16), (17), (4), (13), (3), and (5). □

Proof of Theorem 1. We have

$$\rho(a) = \sum_{\substack{m \\ \omega_m(a)>0}} 1 \leq \sum_{m=1}^{\infty} \omega_m(a) = \omega(a).$$

Hence

$$\nu_m \left(\rho(a) - \log m \leq x\sqrt{\log m}\right)$$

$$= \nu_m \left(\omega(a) - \log m \leq x\sqrt{\log m}\right) + U(m), \tag{18}$$

where

$$U(m) := \sum_{\substack{\partial(a)=m \\ \omega(a)-\log m > x\sqrt{\log m} \\ \rho(a)-\log m \leq x\sqrt{\log m}}} 1.$$

By (5.2) of [7],

$$\frac{1}{G(m)} \nu_m \left(\frac{\omega(a) - \log m}{\sqrt{\log m}} \leq x\right) = \Phi(x) + O\left((\log m)^{-\frac{1}{2}}\right) \tag{19}$$

uniformly for all real numbers x and integers $m \geq 2$. It suffices to show that

$$\frac{1}{G(m)} U(m) = O\left((\log m)^{-\frac{1}{3}}\right). \tag{20}$$

Let

$$U_1(m) := \sum_{\substack{\partial(a)=m \\ \omega(a)-\rho(a)>T}} 1$$

and

$$U_2(m) := \sum_{\substack{\partial(a)=m \\ \omega(a)-\log m > x\sqrt{\log m} \\ \rho(a)-\log m \leq x\sqrt{\log m} \\ \omega(a)-\rho(a) \leq T}} 1$$

with $T > 0$ to be chosen. Then

$$U(m) \leq U_1(m) + U_2(m). \tag{21}$$

We have

$$\frac{1}{G(m)} U_1(m) \leq \frac{1}{T^2 G(m)} \sum_{\partial(a)=m} (\omega(a) - \rho(a))^2 = O(T^{-2}) \tag{22}$$

by Lemma 2. Also, by (19) and an application of the mean-value theorem of differential calculus,

$$\frac{1}{G(m)} U_2(m)$$

$$\leq \frac{1}{G(m)} \left[\nu_m \left(\frac{\omega(a) - \log m}{\sqrt{\log m}} \leq x + \frac{T}{\sqrt{\log m}} \right) - \nu_m \left(\frac{\omega(a) - \log m}{\sqrt{\log m}} \leq x \right) \right]$$

$$= \int_x^{x + \frac{T}{\sqrt{\log m}}} e^{-\frac{t^2}{2}} \, dt + O\left((\log m)^{-\frac{1}{2}} \right) \tag{23}$$

$$= O\left(\frac{T}{\sqrt{\log m}} + (\log m)^{-\frac{1}{2}} \right).$$

Thus (20) follows from (21), (22), and (23) when $T = (\log m)^{\frac{1}{6}}$ is chosen.

$\square$

References

[1] P. D. T. A. Elliott, *Probabilistic number theory*, Springer-Verlag, New York, 1979.

[2] K.-H. Indlekofer and E. Manstavicius, *Additive and multiplicative functions on arithmetical semigroups*, Publ. Math. Debecen **45** (1994), 1–17.

[3] A. Knopfmacher, *On the degrees of irreducible factors of polynomials over a finite field*, Discrete Mathematics **196** (1999), 197–206.

[4] J. Knopfmacher and W.-B. Zhang, *Number theorey arising from finite fields*, Monographs and Textbooks in Pure and Applied Math., vol. 241, Marcel Dekker, New York, 2001.

[5] R. Warlimont, *On the distribution of the degrees of prime element factors in additive arithmetical semigroups*, Monatshefte Math. **129** (2000), 63–81.

[6] W.-B. Zhang, *The probabilistic number theory in additive arithmetic semigroups I*, Analytic Number Theory (B. C. Berndt et al., ed.), Prog. Math. 139, Birkhäuser, 1996, pp. 839–885.

[7] _____ , *The probabilistic number theory in additive arithmetic semigroups II*, Math. Z. **235** (2000), 747–816.